Molecular Biology Biochemistry and Biophysics

35

Light Reaction Path of Photosynthesis

Edited by Francis K. Fong

With Contributions by
G.S. Beddard · R.H. Clarke · R.J. Cogdell
F.K. Fong · A. Hoff · H. Levanon
J.R. Norris · H. Scheer · M.R. Wasielewski

With 118 Figures

Springer-Verlag
Berlin Heidelberg New York 1982

Professor FRANCIS K. FONG
Purdue University
Department of Chemistry
Chemistry Building
West Lafayette, IN 47907, USA

ISBN 3-540-11379-7 Springer-Verlag Berlin Heidelberg New York
ISBN 0-387-11379-7 Springer-Verlag New York Heidelberg Berlin

Library of Congress Cataloging in Publication Data. Main entry under title: Light reaction path of photosynthesis. (Molecular biology, biochemistry, and biophysics; v. 35). Bibliography: p. Includes index. Contents: Free energy change for quantum storage in photosynthesis / F.K. Fong - Phycobiliproteins, molecular aspects of photosynthetic antenna system/Hugo Scheer - The antenna chlorophylls and the light harvesting process/Godfrey S. Beddard, Richard J. Cogdell - [etc.] 1. Photosynthesis. I. Fong, Francis K., 1938-. II. Beddard, G.S. (Godfrey S.). III. Series. QK882.L48. 581.1'3342. 82-5526. AACR2.

Typesetting and printing: Beltz, Offsetdruck, Hemsbach/Bergstr.
Bookbinding: Brühlsche Universitätsdruckerei, Giessen

2131/3130-543210

Preface

This monograph deals with the light reaction pathway in photosynthesis. The photophysico-chemical events are presented in the order of their occurrence, beginning with the collection of sunlight by antenna systems, ending with the reduction of CO_2 to carbohydrates. Relationships between the structural properties and kinetic effects of primary and secondary events spanning time domains in the range 10^{-12}-1s are explored. Photosynthesis is examined in terms of a light-induced redistribution of reaction intermediates common to the biosynthesis and metabolic degradation of carbohydrates.

The experimental procedures and results reviewed in the book are representative of developments in instrumental methods and conceptual formulations in this area during the past decade. In particular, picosecond spectroscopy, time-resolved and magnetic resonance techniques, along with structural and photoelectrochemical models of photosynthesis, have provided clues for the molecular mechanisms of energy migration from the antenna systems to the reaction centers, and of succeeding stages of photochemical events leading to the carbon-reduction cycle.

The preparation of this monograph resulted from the efforts of workers in distantly separated institutions. The writer gratefully acknowledges the responsive collaboration of the contributing authors and members of the Springer editorial staff that made possible completion of the manuscript.

West Lafayette, IN, USA
August, 1982

F.K. Fong

Contents

Chapter 4: Photooxidation of the Reaction Center Chlorophylls and Structural Properties of Photosynthetic Reaction Centers. A.J. Hoff (With 27 Figures)

Chapter 5: Triplet State and Chlorophylls. H. Levanon and J.R. Norris (With 10 Figures)

Chapter 6: The Chlorophyll Triplet State and the Structure of Chlorophyll Aggregates. R.H. Clarke (With 9 Figures)

Chapter 1

Free Energy Change for Quantum Storage in Photosynthesis

Francis K. Fong[1]

1 Introduction

In green plant photosynthesis sunlight is captured by pigment molecules and, through a sequence of primary and secondary reactions initiating from excited chlorophyll reaction centers, is transformed into stored chemical energy. Stages of photochemical events are discernible in widely separated time domains. Photoexcitation of the light-collecting pigments followed by excitation migration spans a range $10^{-14} - 10^{-12}$ s. The photooxidation of the chlorophyll, a subnanosecond reaction, results in the reduction of carbon dioxide by water within $\lesssim 1$s, whereby the chlorophyll is regenerated in its ground state [1]. The Chl *a* water-splitting reaction is given by the reaction cycle [2]:

$$\text{Light:}\quad 2(\text{Chl}\ a\cdot 2\text{H}_2\text{O})_n + 2\text{H}_2\text{O} \longrightarrow 2(\text{Chl}\ a\cdot 2\text{H}_2\text{O})_n{}^{+\cdot} + 2\text{OH}^- + \text{H}_2 \qquad (1)$$

$$\text{Dark:}\quad 4(\text{Chl}\ a\cdot 2\text{H}_2\text{O})_n{}^{+\cdot} + 2\text{H}_2\text{O} \longrightarrow 4(\text{CHl}\ a\cdot 2\text{H}_2\text{O})_n + 4\text{H}^+ + \text{O}_2 \qquad (2)$$

where $(\text{Chl}\ a\cdot 2\text{H}_2\text{O})_n$ is an aggregate of Chl *a* dihydrate, whose structural and photochemical properties provide a model description for the P680 water-splitting reaction center chlorophyll in plant photosynthesis. The electrochemical potential that drives the water oxidation reaction in (2) is the standard-state redox potential, ϵ, of the Chl *a* radical cation

$$(\text{Chl}\ a\cdot 2\text{H}_2\text{O})_n{}^{+\cdot} + \text{e} \longrightarrow (\text{Chl}\ a\cdot 2\text{H}_2\text{O})_n\,. \qquad (3)$$

From experimental effects we observed [3] that the magnitude of ϵ is directly related to the $(\text{Chl}\ a\cdot 2\text{H}_2\text{O})_n$ aggregate size, so that, according to relaxation theory [3], the rate of reaction (2) is an exponential function of the physical dimensions of $(\text{Chl}\ a\cdot 2\text{H}_2\text{O})_n$. Reactions (1) and (2) as the primary mechanism for solar energy storage in photosynthesis are given quantitative definition in this chapter.

1 Department of Chemistry, Purdue University, West Lafayette, IN 47907, USA

2 Definition of the Primary Photochemical Reaction

Photochemical reactions may be discussed in terms of a reactive complex formed as a result of photoexcitation [4]. Consider the photoactivation of molecules AB to electronically excited states:

$$AB + h\nu \longrightarrow (A. \ldots .B)^* \tag{4}$$

where the asterisk denotes the ractive complex. In a condensed fluid medium, $(A\cdot\cdot\cdot B)$ is conceivably held together by a solvent cage at appropriately small values of R_{AB}, the separation of the centers of mass of A and B. The state $(A\cdot\cdot\cdot)^*$ may be deactivated via photophysical relaxation with rate constant k_{nr}

$$(A\cdot\cdot\cdot B)^* \xrightarrow{k_{nr}} (AB). \tag{5}$$

Reaction (5) completes with the dissociation process

$$(A\cdot\cdot\cdot B)^* \xrightarrow{k_r} A^* + B^* \tag{6}$$

having a rate constant k_r. Process (6), formally defined [4] as the *primary* photochemical reaction, results in the diffusion of A* and B* away from each other, which eventually are consumed in secondary reactions. The above description is suitable for simple photodissociation processes in condensed fluids, such as the photodissociation of I_2 in organic solvents [5]. On recombination of the A and B fragments, the reaction cycle is completed as the ground state molecule AB is regenerated.

The photodissociation of AB, followed by geminate recombination of the A and B fragments, produces no net chemical reaction, resulting in no photon energy storage. The primary and secondary mechanisms represent a sequence of relaxation processes through which the excitation energy is dissipated as heat in transitions between the various bound and/or quasi-bound states of the A-B system. In analyzing the kinetic behavior of the primary and secondary processes attending reactions (1) and (2), the catalytic cycle of water photolysis, it is necessary to generalize the above model. In place of the atomic or molecular fragments in (4–6) we envisage the coupling of a photoactivated, reversible redox cycle $D \rightleftarrows D^+ + e$ with decomposition of the photolyte, A:

$$D \cdot mA \underset{k_{nr}}{\overset{h\nu}{\rightleftarrows}} [D^+ \cdot (m - p) A\cdot\cdot\cdot pA^-] \tag{7}$$

$$[D^+ \cdot (m-p) A\cdot\cdot\cdot pA^-] \xrightarrow{k_r} D^+ \cdot (m - p)A + [pA]^- (\equiv P_1) \tag{8}$$

$$(p + \ell)A + D^+ \cdot (m - p)A \xrightarrow{k_r'} D \cdot mA + [\ell A]^+ \qquad (\equiv P_2) \tag{9}$$

where D is the agent for catalyzing the net reaction

$$(p + \ell) A \xrightarrow[D]{h\nu} P_1 + P_2 \tag{10}$$

and the subcripts nr and r respectively denote nonreactive and reactive pathways. The definition of $[D^+ \cdot (m\text{-}p)A \cdot \cdot \cdot pA^-]$ as the reactive complex is tantamount to identifying (8) as the primary photochemical reaction. A photochemically inactive system is given by $k_r = 0$. The photoxidation of D and subsequent reduction of D^+ in reactions (8) and (9), respectively, provide the mechanism for the photolysis of A. On completion of the reaction cycle, $(p + \ell)$ molecules of A are decomposed to yield products P_1 and P_2.

In plant photosynthesis water splitting and carbon reduction are mediated by pheophytin and Chl *a* as electron acceptors [1, 6-8]. One thus cannot rule out the possibility that Chl *a* and/or Ph, present in minute quantities as impurity, also act as intermediate electron acceptors in water photolysis in vitro [2]. However, in considering energy storage here and in Sect. 7 of Chap. 8, D and A are given by $(\text{Chl } a \cdot 2H_2O)_n$ and H_2O, respectively. The radical cation $(\text{Chl } a \cdot 2H_2O)_n{}^+ \cdot$ replaces D^+ in (8) and (9). The products P_1 and P_2 are respectively written:

$$H_2O^- \equiv [H] + OH^- \tag{11}$$

and

$$H_2O^+ \equiv [OH] + H^+ \tag{12}$$

where [H] and [OH] are respectively precursors to hydrogen and oxygen evolution in the photolytic process [1]. The above simplifications are justified if [H] and [OH] denote the primary storage products in photosynthesis.

For a single cycle, reaction (10) becomes

$$2H_2O \xrightarrow[(\text{Chl } a \cdot 2H_2O)_n]{h\nu} \frac{1}{2} H_2 + \frac{1}{4} O_2 + \frac{3}{2} H_2O. \tag{13}$$

The one-electron reaction scheme is consistent with the g = 2.0023 esr signals observed [3, 9] for $(\text{Chl } a \cdot 2H_2O)_n{}^+\cdot$ in (1) and (2). The stoichiometry of water coordination in $(\text{Chl } a \cdot 2H_2O)_n$ is maintained at successive stages of the water-splitting cycle. As the complexed H_2O molecules are depleted in photolytic decomposition, they are replenished by a continuous supply of water molecules from the bulk solvent medium.

3 Photopotential, Photooverpotential and Free Energy Change for Quantum Storage in Photosynthesis

In plant photosynthesis light is converted into an electromotive force (EMF) that drives energy uphill chemical processes. The formation of reducing and oxidizing equivalents, [H] and Chl $a^+\cdot$, sets in motion enzymatic reaction sequences for carbon reduction and oxygen evolution [1]. In order to illustrate the energy storage concept, we consider the oxygen evolution redox couple:

$$2H_2O \underset{k_c}{\overset{k_a}{\rightleftharpoons}} 4H^+ + O_2 + 4e \tag{14}$$

where k_a and k_c are the anodic and cathodic rate constants, respectively. Under equilibrimium conditions we have $k_a = k_c = k^o$, where k^o is defined [10] as the value of k_a or k_c at the standard potential of the redox couple in (14). With minor rearrangement the following development is also applicable to the carbon reduction complement of the photosynthetic process.

The application of an electric potential $E_{applied}$ results in a displacement $\Delta G = -4FE_{applied}$, where F is the Faraday, in the free energy contents of the reactant and product systems from the equilibrium values. Under the influence of an applied field that favors the forward direction of (14) the dependence of k_a and k_c on the electrode potential may, according to standard treatment [10] be given

$$k_a = k^o{}_a \exp\left[-\frac{aF}{RT}\left(E_{applied} - E_a\right)\right] \tag{15}$$

$$k_c = k^o{}_c \exp\left[\frac{(1\text{-}a)F}{RT}\left(E_{applied} - E_a\right)\right] \tag{16}$$

where E_a is the mid-point potential of reaction (14) (equal to –0.81V at pH7) and a, known as the transfer coefficient, is the fraction of - $FE_{applied}$ that acts to descrease the height of the anodic energy barrier for the forward reaction in (14).

In calculating the rate of water splitting we replace $E_{applied}$ by the photopotential, $E_{h\nu}$, due to the production of $(\text{Chl } a \cdot 2H_2O)_n{}^{+\cdot}$ under illumination. Given standard conditions the rate of reaction (2) is

$$\begin{aligned} R_2 &= R_a - R_c \\ &= k^o \left\{C^+(\nu)\exp\left[\frac{-aF}{RT}\left(E_{h\nu} - E_a\right)\right] - C^o(\nu)\exp\left[\frac{(1-a)F}{RT}\left(E_{h\nu} - E_a\right)\right]\right\} \end{aligned} \tag{17}$$

where R_a and R_c are the forward and backward rates of reaction (2); $C^+(\nu)$ and $C^o(\nu)$ are the concentrations in mol l^{-1} of $(\text{Chl } a \cdot 2H_2O)_n{}^{+\cdot}$ and $(\text{Chl } a \cdot 2H_2O)_n$ under steady illumination by light of frequency ν, respectively. Assuming equilibrium conditions, under which $R_2 = O$, we obtain from equating the first and second terms in Eq. (17)

$$E'_{h\nu} = E_{eq} = E_a - \frac{RT}{F} \ln \frac{C^o(\nu)}{C^+(\nu)}. \tag{18}$$

The exchange rate R^o, defined to be either R_a or R_c at equilibrium, is written on substitution of $E_{h\nu}$ by E_{eq} in Eq. (17):

$$R^o = k^o\, C^+(\nu)^{(1-a)}\, C^o(\nu)^a. \tag{19}$$

The photoactivation *overpotential* η for observing a measurable rate of oxygen evolution is defined by

$$\eta(\nu) = E_{h\nu} - E_{eq} = E_{h\nu} - E_a - \frac{RT}{F} \ln \frac{C^o(\nu)}{C^+(\nu)} . \quad (20)$$

Combining Eqs. (17), (19), and (20) we get

$$R_2 = R^o \exp\left[\frac{-aF\eta(\nu)}{RT}\right] - \exp\left[\frac{(1-a)F\eta(\nu)}{RT}\right] . \quad (21)$$

In Eq. (20) $E_{h\nu}$ is the electrochemical potential resulting from photon energy storage by the conversion of (Chl $a \cdot 2H_2O)_n$ to (Chl $a \cdot 2H_2O)_n{}^{+}\cdot$. An energy supply is released upon reduction of (Chl $a \cdot 2H_2O)_n{}^{+}\cdot$ according to (3). The chemical free energy thus availed is equal to that stored from the photoexcitation:

$$\Delta G_{h\nu} = -nFE_{h\nu} = nF\epsilon = nF(E^+ - E^o) \quad (22)$$

where E^+ and E^O are the energies of (Chl $a \cdot 2H_2O)_n{}^{+}\cdot$ and (Chl $a \cdot 2H_2O)_n$, respectively. By convention, the electrochemical potentials $E_{h\nu}$ and ϵ are related by the equation

$$E_{h\nu} = -\epsilon = E^O - E^+ = -E^+ < 0 \quad (23)$$

so the preservation of light as chemical free energy is given in Eq. (22) by positive values of $\Delta G_{h\nu}$, consistent with the traditional notion that the spontaneous expenditure of energy, a process opposite to energy storage, is given by negative free energy changes. The study of photosynthesis is concerned with the molecular mechanics leading to such energy storage.

4 Outline of this Book

In photosynthetic organisms sunlight is collected by antenna systems. The energy, thus harvested, is then directed to reaction centers where the chlorophyll initiates a series of reactions resulting in conversion of the excitation into stored chemical energy in the form of [H] and [OH], precursors to CO_2 reduction and O_2 evolution, respectively. The present volume reviews current topics on the structural and kinetic properties governing the path of photoexcited constituents in photosynthesis.

Chapters 2 and 3 are concerned with antenna systems such as the bacterial chlorophyll antenna complex and the biliproteins of cyanobacteria and red algae. Energy transport phenomena in photosynthesis, examined by picosecond techniques, are described in terms of exciton migration. The experimental behavior of Chl a triplet states in vitro and in vivo is treated in Chapters 4 and 5. Optical-magnetic resonance spectroscopic methods applicable to the study of the triplet state are reviewed. Chapter 6 outlines methods for synthesizing artificial porphyrin and chlorophyll aggregates related to proposed models for Chl a complexes in vivo. Chapter 7 is a comprehensive review of experimental observations of the photochemical transformations in plant and bacterial photosynthesis. The range of instrumental methods discussed circular and linear dichroism, nano- and picosecond spectroscopy, absorption difference spectroscopy, and magnetic resonance techniques. The observations of the photooxidation of reaction centers, P700 and P680, are described, as are those on the intermediary acceptors in plant photosynthesis. Chapter 8 provides an account of the light path of carbon reduction in plant photosynthesis. The physiological observations are analyzed in view of corresponding phenomena in vitro.

Photosynthesis in vivo is a multistep reaction in which light is *efficiently* transformed into stored energy according to the photoelectrochemical considerations leading to Eq. (22). The following chapters bring into view various elements of this reaction which provide a plausible, though incomplete, interpretation of the complex process. Improved experimental techniques and conceptions of structure-photoreactivity relationships, as well as a proper sequencing of the primary and secondary photoevents are being developed. We thus expect, with measured confidence, an accelerated pace toward filling in the details still missing at the present writing.

References

1. Fong, F.K.: This volume, Chap. 8
2. Fong, F.K., Galloway, L.: J. Am. Chem. Soc. **100**, 3594 (1978)
3. Fong, F.K., Galloway, L., Matthews, T.G., Lytle, F.E., Hoff, A.J., Brinkman, F.A.: J. Am. Chem. Soc. **104**, 2759 (1982)
4. Diestler, D.J., Fong, F.K.: J. Am. Chem. Soc. **100**, 1992 (1978)
5. Chaung, T.J., Hoffmann, G.W., Eisenthal, K.B.: Chem. Phys. Lett. **25**, 201 (1974)
6. Klimov, V.V., Klevanik, A.V., Shuvalov, V.A., Krasnovsky, A.A.: FEBS Lett. **82**, 183 (1977)
7. Klimov, V.V., Allakhverdiev, S.I., Pashchenco, V.Z.: Doll. Akad. Nauk **242**, 1204 (1978)
8. Shuvalov, V.A., Dolan, Ed, Ke, B.: Proc. Natl. Acad. Sci. USA, Biophysics **76**,770 (1979)
9. Fong, F.K., Hoff, A.J., Brinkman, F.A.: J. Am. Chem. Soc. **100**, 619 (1978)
10. Laitinen, H.A., Harris, W.E.: Chemical Analysis, 2 nd ed., p. 262. New York: McGraw Hill 1975

Chapter 2
Phycobiliproteins: Molecular Aspects of Photosynthetic Antenna System

Hugo Scheer[1]

1 Introduction

Harvesting the sun requires both the absorption of the dilute energy, light, and its transformation into chemical energy. With one exception (e. g., the halobacteria), the organisms capable of photosynthesis have these two functions also physically divided. Antenna systems collect the light and guide the excitation energy to the reaction centers, where it is transformed into electrochemical energy. The reaction centers are the conservative part of the photosynthetic apparatus, whereas the size, organization and composition of the antenna varies widely as a developmental and often also individual response to the environmental light quality (Table 1). The reaction centers are always integral parts of the photosynthetic membranes. The antenna may be part of the membrane, too, but it may also be attached on either its inner or outer surface, or even in separate particles or vesicles.

Irrespective of their location, the function of all antenna systems is to store excitation as a temporary buffer, and at the same time guide it to the reaction centers. Depending on the sign and magnitude of the energy gap, as well as on the distance between the antenna chromophores and the reaction center, either of these function is more strongly expressed, or at least more obvious.

The subject of this article may be properly described as a funnel for collecting and feeding excitation energy into the reaction center. It is the biliprotein antenna systems of cyanobacteria and red algae, and of the cryptophytan algae. These pigments are only loosely attached to the photosynthetic membrane and water-soluble, which greatly facilitated their investigation and made especially the former two the hitherto probably best understood antennas on a molecular basis.

This review is focused on the molecular aspects of the process. For recent reviews on biliproteins written from various points of view and citing earlier literature, the reader is referred to the articles of: Bennett and Siegelman (1979), Berns (1971), Bogorad (1975), Chapman (1973), Gantt (1975, 1979), Glazer (1977), MacColl and Berns (1979), O'Carra and O'hEocha (1976), Rüdiger (1971, 1975, 1978, 1979), Scheer (1978, 1981), and Troxler (1975). Biliproteins containing structurally very similar chromophores, the phytochromes and phycochromes, are also involved as reaction center pigments in sensory transduction of green plants and many algae. For recent surveys on these subjects, see: Björn (1979), Hartmann and Haupt (1978), Lazaroff (1973),

1 Botanisches Institut der Universität München, Menzinger Straße 67, 8000 München 19, FRG

Table 1. Photosynthetic apparatus concerned with the light reactions in various organisms

	Organism type	Reaction centers Major pigments	Location	Antenna Major pigments	Location	Leading references
Anoxygenic photosynthesis	Photosynthetic bacteria (Nonsulfur and purple sulfur)	4 Bphe *a* or *b* 2 Bphe *a* or *b*	PS-Membrane (integral proteins)	Bchl *a* or *b*	PS-Membrane	Clayton and Sistrom (1978)
	Photosynthetic bacteria (Green sulfur)	4 Bphe *a* or *b* 2 Bphe *a* or *b*	PS-Membrane (integral proteins)	Bchl *c, d, e*	Chlorosomes (separate membrane-covered vesicles)	Clayton and Sistrom (1978)
Oxygenic photosynthesis	Cyanobacteria	1 or 2 Chl *a* (PSI) 1 or 2 Chl *a* (PSII)	PS-Membrane (integral proteins)	Chl *a* (PSI) Biliproteins (PSII)	Phycobilisomes (separate particles at the membrane)	this chapter
	Prochloron	1 or 2 Chl *a* (PSI) 1 or 2 Chl *a* (PSII)	PS-Membrane (integral proteins)	Chl *a*, Chl *b*	PS-membrane	Giddings et al. (1980)
	Red algae	1 or 2 Chl *a* (PSI) 1 or 2 Chl *a* (PSII)	PS-Membrane (integral proteins)	Chl *a*, biliproteins	Phycobilisomes	Giddings et al. (1980)
	Cryptophytan	1 or 2 Chl *a* (PSI) 1 or 2 Chl *a* (PSII)	PS-Membrane (integral proteins)	Chl *a, c,* Biliproteins	Intrathylacoidal space	Gantt (1979)
	Brown algae Diatoms	1 or 2 Chl *a* (PSI) 1 or 2 Chl *a* (PSII)	PS-Membrane (integral proteins)	Chl *a, c,* carotenoids	PS-membrane	Alberte et al. (1981)
	Dinoflagellates	1 or 2 Chl *a* (PSI) 1 or 2 Chl *a* (PSII)	PS-Membrane (integral proteins)	Chl *a, c,* carotenoids	PS-membrane	Song et al. (1977)
	Green algae	1 or 2 Chl *a* (PSI) 1 or 2 Chl *a* (PSII)	PS-Membrane (integral proteins)	Chl *a*, Chl *b*	PS-membrane	Govindjee (1975)
	Green plants	1 or 2 Chl *a* (PSI) 1 or 2 Chl *a* (PSII)	PS-Membrane (integral proteins	Chl *a*, Chl *b*	PS-membrane	Govindjee (1975)

Marme (1977), Mitrakos and Shropshire (19720' Mohr (1972), Pratt (1978), Rüdiger (1971, 1980), Scheer (1981), Smith (1975), and Smith and Kendrick (1976).

2 Morphology

In electron micrographs of the unicellular red algae, *Porphyridium cruentum*, Gantt and Conti described in 1965 a new particle of oblong shape (≈ 4 nm diameter), which was arranged in a rather regular fashion on the outside of the thylakoid membrane. Subsequent investigations by several research groups (Gantt 1979; Glazer et al. 1979; Koller et al. 1978; Wildman and Bowen 1974) revealed similar particles, although of varying size and arrangement, as a main characteristic of the photosynthetic apparatus of cyanobacteria and red algae.

Both classes of photosynthetic organisms owe their coloration to biliproteins, which had been shown by bichromatic action spectroscopy to be major light-harvesting pigments, feeding excitation energy mainly to photosystem II (Emerson 1958; Haxo 1960). The early suspicion that the phycobiliproteins are contained in these particles was confirmed after their isolation as integral entities and the analysis of their composition (Bryant et al. 1976; Gantt and Lipschultz 1974; Gantt et al. 1979; Glazer et al. 1979; Gray and Gantt 1975; Koller et al. 1978; Rigbi et al. 1980; Wanner and Köst 1980). They are almost entirely (Koller et al. 1978; Tandeau de Marsac and Cohen-Buzire 1977; Yamanaka et al. 1978) composed of phycobiliproteins, and are thus properly termed phycobilisomes (Gantt and Conti 1966).

The phycobilisomes contain three different types of biliproteins[2], the phycoerythrins (PE) absorbing in the range between 480 and 580 nm, the phycocyanins (PC) absorbing in the range between 570 and 630 nm, and the allophycocyanins (APC) absorbing in the range between 610 and 670 nm (Table 2). In addition, minor amounts of uncolored proteins have been reported by several workers (Koller et al. 1978; Tandeau de Marsac and Cohen-Bazire 1977; Yamanaka et al. 1978). The majority of the pigment content is PE and PC. The ratio between the two is variable within different species, and in spite of many exceptions, the blue PC's are predominant in the cyanobacteria ("blue algae"), and the red PE's are predominant in the red algae. The ratio between the two is often also variable within a given species in response to the environmental light quality, the relative proportion of the PC's absorbing red light being higher in red and lower in green light (see "chromatic adaptation"). The sizes and fine structures of the phycobilisomes vary accordingly, although a common construction principle is currently evolving.

The detailed investigation of the phycobilisomes revealed a striking morphology. From dissociation experiments, Gantt and co-workers arrived at a model for *P. cruentum* phycobilisomes in which an APC core in contact with the photosynthetic membrane is covered by a roughly hemispherical layer of PC, which in turn is covered by

2 *Abbreviations:* PC = Phycocyanin, PE = Phycoerythrin, APC = Allophycocyanin. The prefixes C–, R– and B– stand for *C*yanobacteria, *r*ed algae and *b*angiales, an order of the red algae. Chl = Chlorophyll

Table 2. Properties of phycobiliproteins and classification according to occurrence, spectra, and subunit composition

Type[a]	Occurence	λmax (nm) in the visible spectral range	Chromophores α–	β–	γ–Chain
APC–I	Cyanobacteria and red algae	656	1x*1a*	1x*1a*	?
– II, – III	Cyanobacteria and red algae	650	1x*1a*	1x*1a*	–
– B	Cyanobacteria and red algae	670	1x*1a*	1x*1a*	–
C–PC	Cyanobacteria	635[d],620,(590)[e]	1x*1a*	2x*1a*	–
PEC[b]	Cyanobacteria	590,568	1xPXB[g]	2x*1a*	–
R–PC	Red algae	620,555	1x*1a*	1x*1a*,1x*2*	–
Cr–PC	Cryptophytan algae	645,610 580 (and others)	1x*1a*[h] (and others)	1x*1a*, 1x*2* (?)	–
C–PE	Cyanobacteria	575[d],560,540	2x*2*	3–4x*2*	–
R–PE[c]	Red algae	568,540,498[g]	2x*2*	? x*2* ? xPUB	? xPUB[j] ? x*2*
b–PE	Red algae	575[d],565,540	2x*2*	4x*2*	–
B–PE	Red algae[j]	565,545,498[f]	2x*2*	4x*2*	2xPUB[j] 2x*2*
Cr–PE	Cryptophytan algae	545–565	? x*2*	? x*2*	–

a Prefixes according to their occurence: B = Brangiales, an order of the red algae; Cr = Cryptophytan algae; C = Cyanobacteria, R = Red algae
b Phycoerythrocyanin
c R–PE has been reported to be a glycochromoprotein (Chapman 1973; Raftery and O'hEocha 1965)
d Possibly an aggregate form (Brown et al. 1975; Zilinskas et al. 1978)
e Shoulder, resolved at low temperatures (Frackowiak et al. 1975; Friedrich et al. 1981, Gray and Gantt 1975; Scheer and Kufer 1977)

Protein structure		Size (kDalton)			
Monomers	aggregation number	α	β	γ	References
$\alpha_3\beta_3\gamma$	1	18	18	30	Gantt and Canaani (1980); Gysi and Zuber (1976); Zilinskas et al. (1978)
$\alpha\beta$	(1),3,6	16	18	–	Bennett and Bogorad (1971); Bogorad (1975); Brown et al. (1975); Cohen–Bazire et al. (1977); Gysi and Zuber (1974); Zilinkas et al. (1978)
$\alpha\beta$	(1),3,6	16	20	–	Glazer and Bryant (1975); Ley et al. (1977)
$\alpha\beta$	1, 3,6	16	20	–	See reviews cited on p. 7
$\alpha\beta$	3	17	20	–	Bryant et al. (1978)
$\alpha\beta$	3,6	18	20	–	Chapman et al. (1967); Gantt and Lipschultz (1974); Glazer and Hixson (1975)
$\alpha\alpha'\beta_2$	1	9,10	16	–	Glazer and Cohen-Bazire (1975); Jung et al. (1980) Mörschel and Wehrmeyer (1975)
$\alpha\beta$	1, 3,6	17	21	–	See reviews cited on p. 7
$\alpha_6\beta_6\gamma$	1	19	19	35	O'Carra (1970); O'Carra and O'hEocha (1976)
$\alpha\beta$	3	19	19	–	Gantt and Lipschultz (1974); O'Carra (1970)
$\alpha_6\beta_6\gamma$	1	19	19	35	Abad-Zapatero et al. (1977); Gantt and Lipschultz (1974); Glazer and Hixson (1977); Sweet et al. (1977); van der Velde (1973)
$\alpha\beta$	2	10	17	–	Glazer et al. (1971); MacColl and Berns (1979); MacColl et al. (1976); Mörschel and Wehrmeyer (1977)

f Shoulder, due to phycourobilin chromophores

g Chromophore of unknown structure type (λ_{max} = 600 nm after denaturation in 8 M urea, pH 3.0)

h In addition, chromophores of unknown structure have been described (λ_{max} = 690 nm after denaturation in 8 M urea, pH 2.0)

i Chromophore of the urobilin spectral type. The exact structure is unknown (λ_{max} = 498 nm after denaturation in 8 M urea, 1 M HCl)

j A pigment spectroscopically similar to B–PE has been described to occur in a marine cyanobacterium (Shimura and Fujita 1975)

PE. Phycobilisomes can be isolated intact in high ionic strength buffer both with and without parts of the thylacoid membrane still attached, and the dissociation has been studied both by fluorescence spectroscopy (Gantt and Zilinskas 1976; see below) and immunoelectron microscopy (Gantt and Zilinskas 1978; Gantt and Lipschultz 1977). The phycobilisomes of *P. cruentum* are rather large. Indications of a fine structure have been obtained only recently, including a small stalk which may function as an anchor to the membrane (Wanner and Köst 1980).

The sequential arrangement of the pigments, APC, PC, and PE, has been supported by the analysis of phycobilisomes from another red alga, *Rhodella violacea*, by the group of Wehrmeyer (Koller et al. 1978; Mörschel et al. 1977). Due to their smaller size and an almost planar shape it has been possible to get a more detailed insight into their morphology, which is schematically shown in Fig. 1. The inner core of the phycobilisome consists of three double-platelets of APC, arranged in a triagular fashion and attached edge on to the thylacoid membrane. This core is "garnished" with short rods, termed tripartite units due to their composition of three double-platelets stacked face to face like a stack of coins. The inner double-platelet facing the APC core is an aggregate of PC, the outer two are composed of PE. The individual APC and PC platelets are trimers of the α, β - monomers of these pigments (see below) having a ring-shaped structure with an inner hole (Mörschel et al. 1980). The double-platelets then are hexamers. The PE present in *R. violacea* is a B-type pigment (Table 2), in which the inner hole of the hexamer contains an additional polypeptide chain, thus making it an $\alpha_6 \beta_6 \gamma$ complex (Abad-Zapatero et al. 1977; Köst 1980; Sweet et al. 1977; Wehrmeyer 1980).

Both trimers and hexamers have been shown by in-vitro studies to be the aggregates formed predominantly and reversibly from manomers in solution (Berns and Morgenstern 1968; Chen and Berns 1978; Lee and Berns 1968; MacColl and Berns 1973; MacColl et al. 1971 a, b; 1974; Saito et al. 1976). The same or similar aggregates

Fig. 1. Phycobilisome model of *Rhodella violacea*, a unicellular red alga, according to Mörschel et al. (1977). The discs arranged in the central triangle facing the viewer are APC hexamers, each stack attached to this core consists of one PC hexamer ($\alpha_6\beta_6$) and two B-PE "monomers" of $\alpha_6\beta_6\gamma$-structure. The phycobilisomes are arranged in regular and densely packed fashion on the outer surface of the thylacoid membrane. The membrane is omitted in this figure, it would be at right angle to the plane of the paper with its trace at the "bottom" of the phycobilisome in the orientation shown

have also been suggested as the building blocs of phycobiliprotein crystals (Bryant et al. 1976; Dobler et al. 1972; Sweet et al. 1977). Some apparently confliction results between morphological, biochemical and biophysical investigations on biliprotein aggregates have been discussed recently Mörschel et al. (1980).

The phycobilisomes of cyanobacteria have been characterized in comparable detail only recently, due to isolation problems (Gantt et al. 1979; Gray and Gantt 1975; Gray et al. 1973; Rigbi et al. 1980; Yamanaka et al. 1978). Electron microscopy of three species has yielded basically the same fine structure, with the number of the central APC platelets or the lenght of the rods being somewhat variable (Glazer et al. 1979).

The phycobilisomes are probably products of a complex self-assembly process which does not only require the aggregation of identical biliproteins (homo-aggregation), but also of different types of biliproteins with each other (hetero-aggregation) and with the membrane, in a highly ordered and regulated fashion. The organizing principles have only very recently begun to emerge. Homo-aggregates larger than the hexamer and heteroaggregates have been isolated by several groups from partly dissociated phycobilisomes (Kessel et al. 1973; Koller et al. 1978; Mörschel et al. 1977, 1980a, b; Yamanaka et al. 1978; Gantt et al. 1979; Rigbi et al. 1980; Grabowski et al. 1981), but their reassociation was rarely observed and difficult to achieve in a reproducible way.

After the identification of colorless proteins as integral components of phycobilisomes, at least some of them have tentatively been related to an ordering function (Tandeau de Marsac and Cohen-Bazire 1977b). An indirect evidence to this is the finding, that crude dissociates of phycobilisomes containing larger fragments still bearing colorless proteins can be reassociated into functional phycobilisomes (Canaani et al. 1980; Katoh, private communication). The largest one of the colorless proteins (75–90 kDalton) is probably involved in the attachment of phycobilisomes to the photosynthetic membrane (see Gantt 1981). A protein of this size has recently been identified both in isolated phycobilisomes and in membranes from which the biliproteins had been dissociated (Redlinger and Gantt 1980). This protein is blue, however, with the fluorescence characteristics of an APC (–B?), and its relation to the colorless proteins is still unresolved (Gantt 1981). Irrespective of its coloration, an "anchor" protein has also been suggested from electron-micrographs of phycobilisomes from *P. cruentum* showing a footlike protrusion (Wanner and Köst 1980).

Three different smaller colorless proteins (nominally 33, 30 and 27 kDalton) have recently been isolated from *Synechococcus* 6301 phycobilisomes (Lundell et al. 1981). From reassociation experiments with isolated phycocyanin and one or more of the rather hydrophobic colorless proteins, specific functions for the latter have been suggested. The 33 and 30 kDalton species are involved in stacking hexamers into rods, which may grow to exceptional and unnatural lengths. The 27 kDalton protein is rather related to rod termination. Only small stacks are formed in its presence, even if the 30 and/or 33 kDalton proteins are present as well. A small colorless protein has also been found in PC–PE complexes isolated from *P. sordidum* phycobilisomes, where it appears to function as a linker between the different pigments (Lipschultz and Gantt 1981).

These findings do assign specific functions to at least some of the colorless proteins (group II). As an intriguing aspect of these results, one can imagine, that the self-assembly of phycobilisomes is controlled by the relative concentrations of the different biliproteins *and* colorless proteins, which in turn is controlled by their biosynthesis and degradation.

In many cyanobacteria, the amounts of the different biliproteins (and other photosynthesis pigments) are regulated by light (chromatic adaption, see below and Bogorad 1975), by nitrogen supply (Allen and Smith 1969) and sulfur compounds in the medium (Schmidt 1980).

On the phycobilisome level, these regulations have been shown to involve the composition of the colorless proteins (Tandeau de Marsac and Cohen-Bazire 1977b; Yamanaka and Glazer 1980), as well as the composition of the biliproteins and the phycobilisome-architecture (Siegelman 1980; Yamanaka and Glazer 1980). For biophysical studies, the contamination of biliproteins with colorless proteins in amounts varying with the isolation procedure, has to be considered as a potential complication, which e.g. may be involved in the conflicting results on size and shape of biliprotein homoaggregates.

The morphology of cyptophytan antennas is different from those of cyanobacteria and red algae. Phycobilisomes are absent, and the biliproteins rather appear to be locadet at the inner side of the thylacoid membrane (Gantt et al. 1971; Wehrmeyer 1970). The grana lamellae are wider spaced than in green plants, and filled with an electron-dense material assigned to biliproteins, for which only recently indications of a fine structure have been obtained (Wehrmeyer, pers. comm., 1979). For a recent survey on cryptophytan biliproteins, see Gantt 1979).

3 Energy Transfer

The sequential arrangement of pigments in the phycobilisomes from PE to PC to APC and further to the membrane containing chlorophyll *a* is in the proper order for a downhill energy transfer (Fig. 2), which is indeed their major function. An transfer efficiency of close to 100% from PE to chlorophyll *a* has been determined in intact algae, and

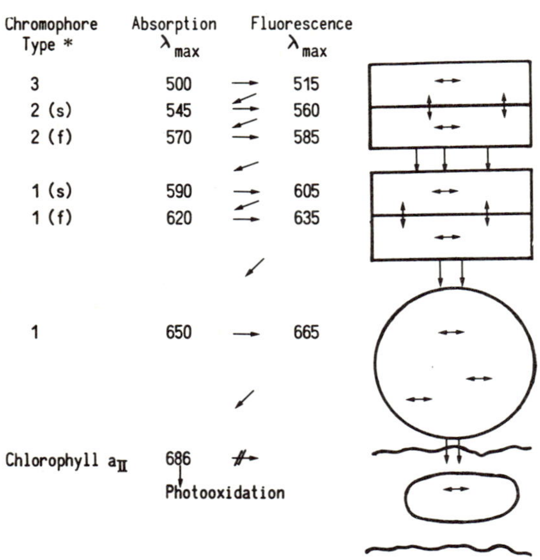

Fig. 2. Energy transfer in a phycobilisome. The *right side* is a diagrammatic representation of a phycobilisome of the type shown in Fig. 1, with one molecule of B–PE, one C–PC hexamer, one APC–hexamer, and the chlorophyll a_{II} in the photosynthetic membrane (*from top to bottom*). *On the left,* the chromophore types and their approximate absorption and emission maxima are given. The *arrows* correspond to the probable energy transfer directions. *See p. 18 for chromophore structures

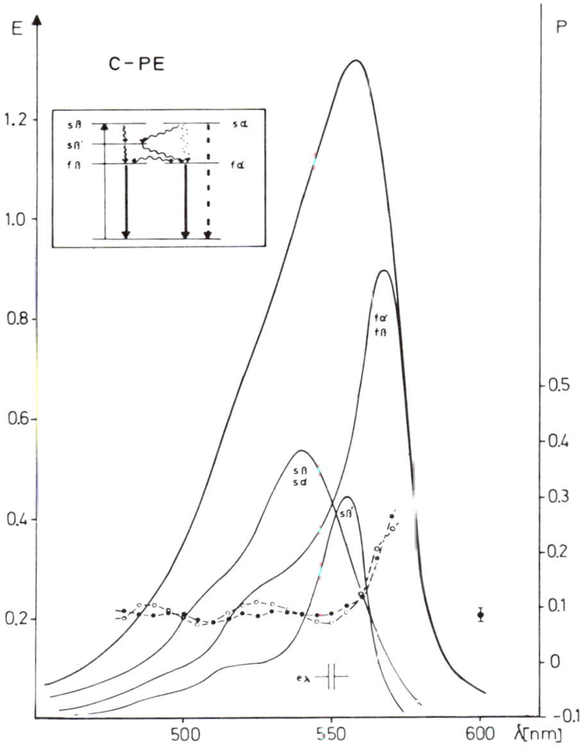

Fig. 3. Deconcolution of the long-wavelenth absorption band and the corresponding terms scheme of C–PE from *Pseudoanobaena* W 1173, according to Zickendraht et al. (1980). The deconvolution is based on absorption, fluorescence polarization data of the isolated α- and β-subunits, and of the αβ-monomers, as well as on an independent computer deconvolution of the absorption spectra The *open* and *full circles* are fluorescence polarization data (scale on the right)

from PE to APC in isolated phycobilisomes (Porter et al. 1978; Searle et al. 1978). This efficient energy transfer can be traced down to the subunits of the individual pigments. All biliproteins are highly fluorescent in their native state (see below), but their emission spectra are not symmetric to their long-wavelength absorption bands, and they have a low and, except for APC, varying degree of fluorescence polarization (Dale and Teale 1970; Grabowski and Gantt 1978a, b; Teale and Dale 1970; Zickendraht-Wendelstadt et al. 1980). Both results have been interpreted first by Teale and Dale (1970) and subsequently by others in terms of distinct and different chromophores in the biliproteins, with an efficient energy transfer between the chromophores, and only the ones lowest in energy fluorescing. On this basis, it is possible to deconcolute the absorption spectra of biliproteins and even their subunits into two (or more) components, arising from the fluorescing "f" and the sensitizing "s" chromophores, respectively (Fig. 3). This concept of different chromophores has later been supported by chemical evidence (see "the protein").

However, a distinction between "s" and "f" chromophores is purely phenomenological, since it is not a property of a chromophore per se, but rather of the aggregation state of the pigments. In all biliproteins and their functional aggregates, it is always the chromophores lowest in energy which fluoresce exclusively or at least predominantly. Biliproteins have 1 to 4 chromophores on the individual subunits, up to 40 chromophores in the monomer superstructures of B– or R –PE, or the APC, PC and PE hexamer building blocs of phycobilisomes, and between 900 and 2500 chromophores in a *P. cruentum* phycobilisome, as estimated from the data of Wanner and Köst (1980) and Gantt (1976)[3]. In building up the latter, the percentage of the "f" chromophores is constantly decreasing. Thus, the "f" chromophore of isolated PE becomes an "s" chromophore in PE–PC heteroaggregates (Koller et al. 1978; Grabowski et al. 1980), and in the phycobilisomes it is only the few APC chromophores which fluorescence (Grabowski and Gantt 1978a,b; Porter et al. 1978; Searle et al. 1978).

The energy transfer of the individual subunits, the monomers and the various aggregates of the pigments has been analyzed in terms of a weak coupling (Förster) process. From the absorption and fluorescence spectra critical distances for non ordered orientations have been calculated (see Grabowski and Gantt, 1978 for further references), which are considerably larger than the diameters of the subunits, allowing an efficient transfer. The spatial distribution of the chromophores in different isolated phycobiliproteins has been estimated based on the Förster mechanism and the Jablonski "active shere" approximation. They indicated a surface distribution (Dale and Teale 1970; Zickendraht et al. 1980), in agreement with chemical evidence (see below). The observed energy transfer times within individual pigments (Kobayashi et al. 1979) and phycobilisomes (Searle et al. 1978) are also cinsistent with and analyzed on the basis of this transfer mechanism. For phycobilisomes of *P. cruentum*, hopping times of 280 ± 40 ps, with an average of 28 jumps for the transfer from PE to PC, have been determined in agreement with the model described, when treated by the *Pearlstein* formalism (Grabowski and Gantt 1978b). The energy transfer has been investigated, too, for the individual subunits of a C–PE containing two and three chromophores, respectively, and the relative orientations of the dipoles determined (Zickendraht et al. 1980). The term scheme shown in Fig. 3 has been obtained from this work.

Strong coupling between chromophores seems to be less prominent in phycobiliproteins from cyanobacteria and red algae, and has in no case yet been shown conclusively. From CD data, exciton coupling has been implied for an APC (Gantt and Canaani 1980). The S-shaped CD bands for C–PC from *Pseudoanabaena spec.* W 1173 are indicative, too, of exciton splitting, but here a definite decision is difficult in view of the five different chromophores present (Langer et al. 1980).

An intermediate coupling, has finally been suggested in APC to account for the pronounced red-shift upon trimer formation without an accompanying CD-effect (McColl et al. 1980). In view of the increase of the oscillator-strenght of the long-wavelength band, a chemical change (conformation, protonation) may be considered as well.

By contrast, strong coupling between chromophores has been deduced mainly from CD data for a cryptophytan PC (Jung et al. 1980). This is supported by a fast component ($\leqslant 8$ ps) in the transient absorption of this pigment, assigned to an energy transfer process, as compared to 84 ps in a C–PC monomer (Kobayashi et al. 1979). This indicates again a certain separation of the cryptophytan biliproteins, which is further evidenced by their spectral diversity and the occurence of special chromophores (see below, and Gantt 1979).

3 The "monomers" of B–PE, R–PE and possibly one of the APC have the rather complex $\alpha_6 \beta_6 \gamma$-structure, in which the γ- subunit fills the inner hole of the torus-shaped trimers and hexamers typical for most biliproteins (see below)

From the morphological and energy transfer studies two different strategies for harvesting green and orange light, their conversion to excitation energy corresponding to red light quanta, and their funneling to the reaction centers, seem to have been followed during the evolution of biliprotein antennas. One is exemplified in the cyanobacteria, where each of the pigments present covers only a comparably narrow absorption range. To avoid gaps in the transfer chain and the absorption spectrum, the phycobiliproteins are arranged in intricate superstructures, the phycobilisomes. In the cryptophytan biliproteins, on the other hand, a miniature transfer chain has already evolved within each biliprotein, especially within the phycocyanins. Accordingly, a superstructure is not necessary, although the transfer from PE to Chl *a* would be facilitated by the Chl *c*'s present in these organisms as intermediate carriers. In the red algae, both concepts have been united, since they have biliproteins covering a broader range of energies organized in phycobilisomes.

4 Chromophore Structure

4.1 Chromophores Cleaved from Biliproteins

Only two chromophores are responsible for the broad range of absorption spectra of the majority of biliproteins (Table 2). Both are of the formerly[4] so-called IX *a* substitution type, characteristic for the mammalian bile pigments derived from heme cleavage at the methine bridge formerly designated "α", now C–5. Phycocyanobilin (*1*) is the blue chromophore of PC's and APC's. It is noteworthy that the same chromophore apears to be present in the phycochromes (Björn 1979; Ohad et al. 1979; Ohki and Fujita 1978; Scheibe 1972), and a very similar chromophore (18-vinyl instead of 18-ethyl) occurs in phytochrome(s) in the P_r - form (Grombein et al. 1975; Rüdiger 1980). Phycoerythrobilin (*2*) is the red chromophore of PE's. Some of the biliproteins contain additional chromophores. R– and B–PE's carry phycourobilin chromophores, for which structure *3* has been proposed (O'Carra and O'hEocha 1976). Phycoerythrocyanin has a red chromophore of unknown structure (Bryant et al. 1978), and at least two other chromophores have been proposed to occur in cryptophytan biliproteins (Glazer and Cohen-Bazire 1975; Jung et al. 1980; Mörschel and Wehrmeyer 1975). The evidence for these less-known chromophores comes from spectroscopic studies on the denatured pigments, which is useful for screening. If done under carefully controled conditions, chromophores other than *1* or *2* can be easily recognized. It should be pointed out, however, that the structures *1* and *2* have been strictly proven only for a few biliproteins, and that alterations, e.g., of the side chains or especially the second protein bond, may remain unnoticed in such studies.

All chromophores are covalently bound to their respective apoproteins. This prevented a direct examination by the common analytical tools with the exception of uv-vis spectroscopic technics, and required initial degradative steps. Chromophore cleavage,

4 A new IUPAC nomenclature of bile pigments has recently been agreed on (IUPAC 1979). For a survey of the older nomenclature systems, see Bonnett (1978)

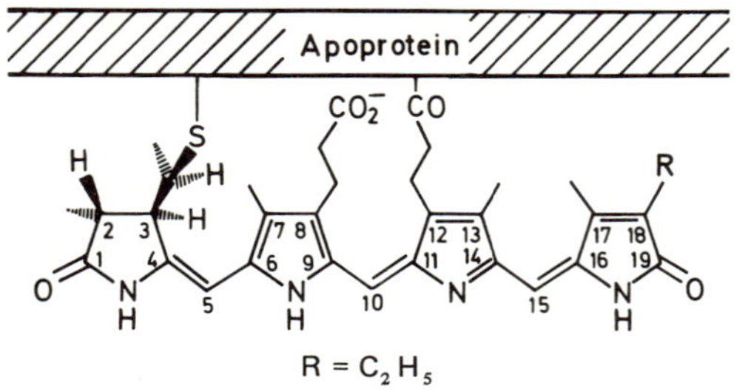

Structure 1

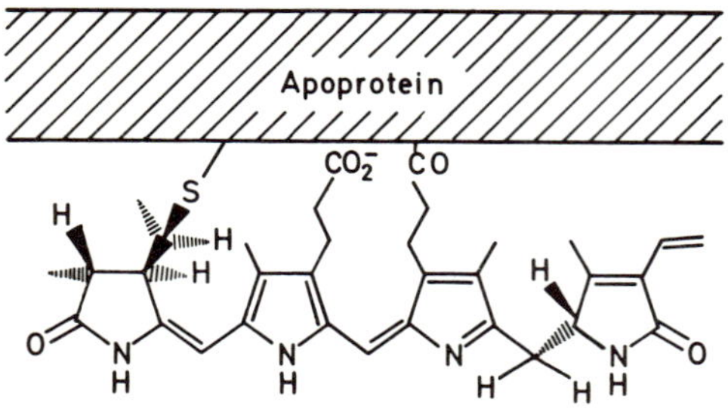

Structure 2

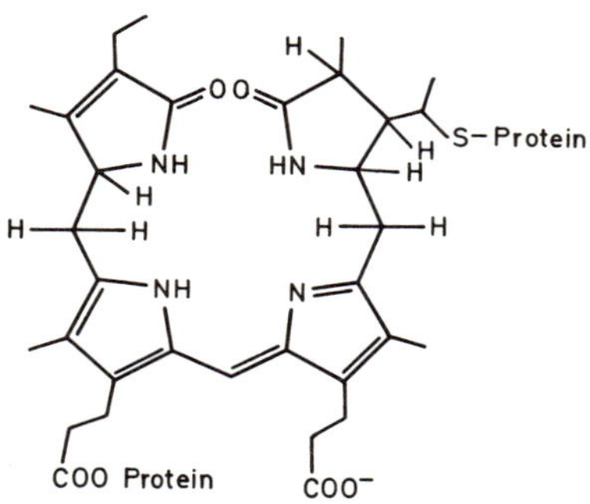

Structure 3

chromic acid degradation or proteolytic digestion of the peptide chains have been most useful. Recently, milder degradative techniques have been investigated, too.

The chromophore cleavage reactions yield different pigments depending on the method used. The temporary confusion concerning the nomenclature of the chromophores is mainly due to the different cleavage procedures yielding different products. For a discussion, the reader is referred to O'Carra (1980), Rüdoger (1971, 1975), and to the new IUPAC nomenclature on tetrapyrroles (1979). Most studies have focused on the 3-ethylidene bilins *4* and *5*, which can best be obtained by thermolysis in refluxing alcohols (Fujita and Hattori 1961; O'Carra and O'hEocha 1966; Fu et al. 1979), or by treatment with hydrogen bromide in trifluoracetic acid (Schram and Kroes 1971). The same products have been obtained recently by a flash-pyrolytic procedure (Fu et al. 1979). The structures *4* and *5* have been elucidated by chromic acid degradation (Rüdiger 1969) yielding succin- or maleimides carrying at least in principle the substituents of the parent pyrrolic rings (Rüdiger and O'Carra 1969; Rüdiger et al. 1967), and by ^{1}Hmr and and mass spectroscopy (Chapman et al. 1967; Cole et al. 1968; Crespi et

Structure 4

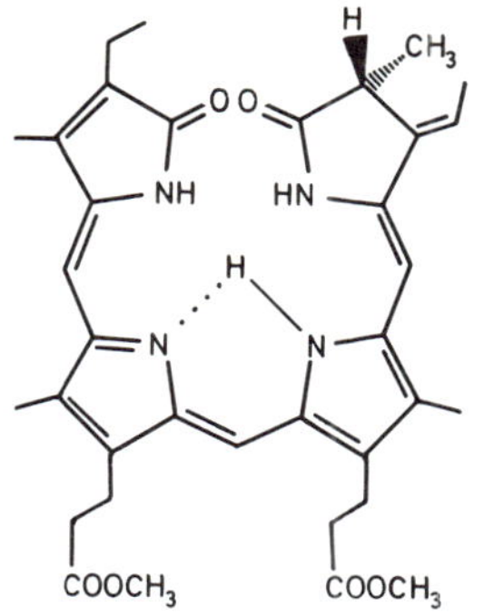

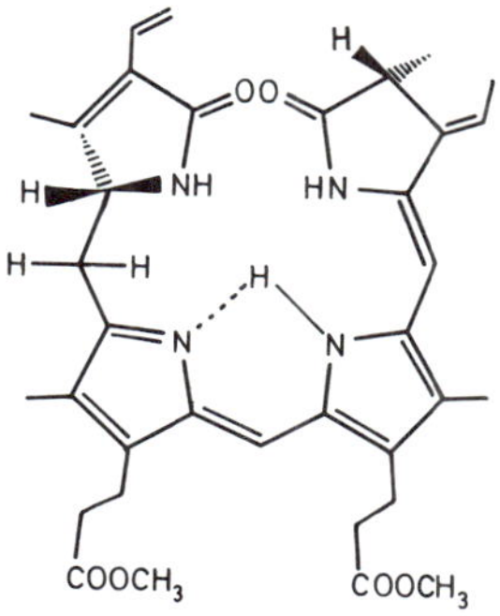

Structure 5

Structure 6

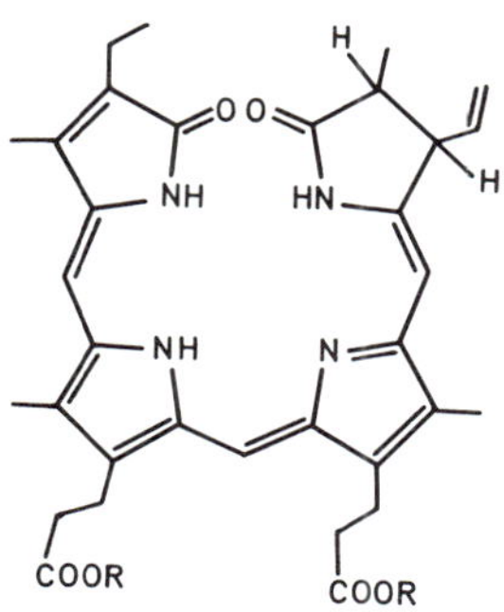

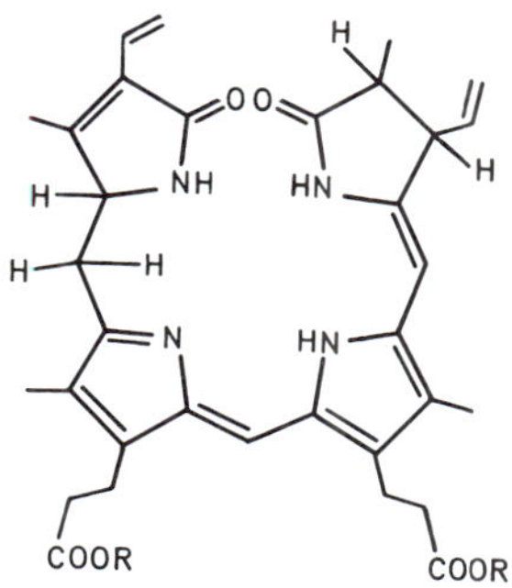

Structure 7

al. 1967, 1968). Both have been confirmed by total synthesis (Gossauer and Hinze 1978; Gossauer and Hirsch 1974; Gossauer and Weller 1978).

Cleavage in aqueous acid yields different pigments, to which structures *6* and *7* have tentatively been assigned. They contain the same conjugated system as do *1* and *2*, respectively, but the exact nature especially of the substitutent at C–3 has not yet been established. One of the possible products, the epimeric methanol adducts at C–3^1, has recently been synthesized (Gossauer et al. 1981). Products of this type probably derived from *4*, together with *4* itself, are excreted into the medium by a mutant of the blue-colored unicellular red algae, *Cyanidium caldarium* (Troxler et al. 1975; Beuhler et al. 1976).

The R configuration of *4* and *5* at the asymmetric C–2 has been arrived at by chromic acid degradation to an imide of known stereochemistry (Brockmann and Knobloch 1973), the 16R configuration by asymmetric synthesis (Gossauer and Weller 1978), and correlation with 4R, 16R urobilin (Cole et al. 1967). The cleavage of biliproteins to *4* (and probably *5* as well) yields a mixture of the Z and E isomers of the 3-ethylidene substituent, as shown by total synthesis (Weller and Gossauer 1980).

4.2 Chromophores Bound to the Protein

The conjugation systems of the chromophores still bound to the proteins have been investigated mainly by uv-vis spectroscopy. Although the spectra of biliproteins are strongly influenced by chromophore-protein interactions (see below), these can be completely abolished by unfolding the protein, e.g., by 8 M urea or guanidinium chloride (Köst et al. 1975; Kufer and Scheer, 1979; see there also for leading references to earlier work). Heat denaturation leads to complete uncoupling of the chromophores as well (Scheer and Kufer 1977), whereas mercurials (Pecci and Fujimori 1967, 1968; Erokhina and Krasnovskii 1968), pH changes (Erokhina and Kransnovskii 1968; Zickendraht-Wendelstadt 1980) and sodium dodecylsulfate give only partially uncoupled and/or further modified chromophores. The denatured pigments can be directly correlated to free bile pigments of known structure not only as the free bases, but also as the anions, zinc complexes and, preferably, the cations, the latter being most stable and having more intense and sharp absorption bands. By acid-base titration, the pK values for protonation and deprotonations are accessible as additional criteria (Grombein et al. 1975). By this means, *1* has been shown to have the same conjugation system as does *8* (Kufer and Scheer 1979), and *2* to have the conjugation system of *9* (Köst et al. 1975). The position of the long-wavelength maxima of bile pigments in the visible absorption range is roughly proportional to the length of the conjugation system, which is helpful in the investigation of the less-known chromophores (Köst et al. 1975).

Structure 8

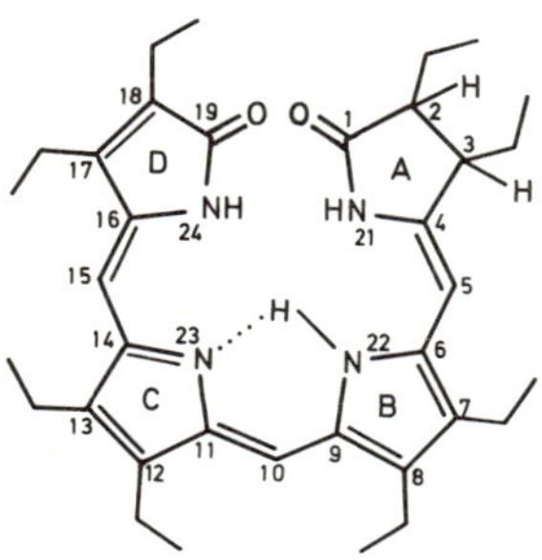

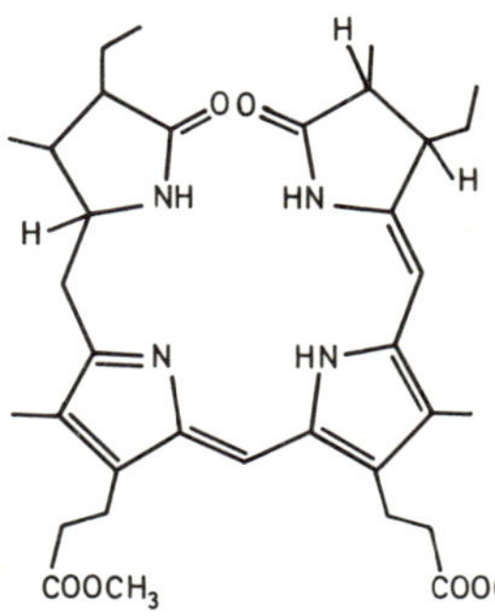

Structure 9

All bile pigment chromophores of phycobiliproteins are covalently bound to the proteins. A thioether bond joining ring A to a cystein residue has now been established as a characteristic feature of the biliproteins (formulas *1, 2*). This type of bond had already been proposed by *Fujiwara* in 1956 from the amino acid analysis of chromopeptides. Cystein as the binding amino acid could first be established for a peptide derived from B–PE of *Porphyridium cruentum* (Köst-Reyes at al. 1975). It has since been recognized as the binding amino acid for *all* chromophores in several PC's and PE's from sequencing studies (Brooks and Chapman 1971; Bryant et al. 1976; Byfield and Zuber 1972; Crespi and Smith 1970; Frank et al. 1978; Freidenreich et al. 1978; Harris and Berns 1975; Kililea and O'Carra 1968; Köst-Reyes and Köst 1979; Lagarias et al. 1979; Muckle et al. 1978; Troxler 1980; Williams et al. 1978; Zuber 1978) and a comparison of the number of free and bound SH residues (Glazer et al. 1979).

The chromophores are attached to the cystein residues via the 3-ethyl substituents as shown in formulas *1, 2* and *3*. Both the site of attachment at C–3^1 and the stereochemistry have been deduced from a modification of the chromic acid degradation and model studies on succinic imides, and from recent ^{1}Hmr investigations of bilipeptides.

Chromic acid degradation at ambient temperature leaves ring A attached to the protein, but the thioether is oxidized to a sulfone under the reaction conditions (Schoch et al. 1976). The latter can subsequently be eliminated unter very mild conditions with ammonia to yield ethylidene succinimide *5* (Klein et al. 1977). This sequence is stereospecific and allowed (in conjunction with the known 2R configuration (Brockmann and Knobloch 1973) the determination of the 2 R, 3 R, 3^1 R configurations of the three asymmetric C atoms at ring A (Klein and Rüdiger 1978; Lotter et al. 1977).

The thermal (Fujita and Hattori 1962; O'Carra and O'hEocha 1966; Fu et al. 1979) or acid/HBr catalyzed (Schram and Kroes 1971) cleavage of the chromophores from the protein proceeds by a similar mechanism to yield the well-known ethylidene bilins *4* and *5* from *1* and *2*, respectively. Under the reaction conditions, epimerization at C–2 (Gossauer and Weller 1978; Scheer and Bubenzer 1978) and isomerization of the $3,3^1$ - ethylidene substituent (Weller and Gossauer 1980) is possible. The 3-vinyl bilins *6* and *7* suggested as products from the acidic cleavage, can be rationalized by a different elimination mechanism (O'Carra et al. 1980). Other cleavage products may arise from further reactions of the 3-ethyldene, 3-vinyl and 18-vinyl groups (Brandlmeier et al. 1979; Klein and Rüdiger 1979; Weller and Gossauer 1980). Independent proof for the structure *1* and thioether bond in both PC from *Synechococcus* sp. 6701 and Phytochrom P_r comes from ^{1}Hmr spectroscopy of bilipeptides. Both yield very similar signals for the chromophore, with the exception of the ones originating from the C–18 substituent. In particular, the analysis of the signals originating from ring A-substituents support the thioether linkage (Lagarias et al. 1979; Lagarias and Rapoport 1980). In an extension of the studies on only one of the three different chromophore-regions (Lagarias et al. 1979), the remaining two have recently been investigated (Lagarias 1981). Surprisingly, one of them (β_2) appears to be differently bound, still via cystein, but possibly to ring D.

The evidence for additional protein-chromophore bonds is much less sound than for the thioether bridge. Chromic acid degradation has yielded conflicting results on the presence of a second protein bond to the propionic acid side chain at ring C (or B), involving most probably a serin residue (Brooks and Chapman 1972; O'Carra and O'hEocha 1976; Rüdiger 1971, 1975, 1979). Only one peptide containing such a bond has so far been identified (Muckle et al. 1978), indicating that in a given protein only part of the chromophores may contain a second bond of this type. Other additional bonds discussed are a ring A lactimester-aspartic acid junction (Crespi and Smith 1970), and a bond to glutamate (Brooks and Chapman 1972). No indications for such additional bonds have been obtained from ^{1}Hmr-studies of bilipeptides (Lagarias 1981; Lagarias et al. 1979). The problem in verifying such bonds is their lability as compared with the thioether bond to proteolytic or hydrolytic conditions, which requires less drastic degradation procedures. The chromate oxidation has been applied for this purpose (Rüdiger 1969, 1971, 1975), and recently two procedures cleaving selectively between rings A and B (Krauss and Scheer 1981) and between rings B and C (Kufer and Scheer 1981) have been applied to PC from *Spirulina platensis* (see Scheer, 1981, for discussion). The latter indicate, that bonds may be present in only part of the chromophores.

5 Noncovalent Protein Chromophore Interactions

Native phycobiliproteins are highly fluorescent (see Sect. 3), whereas the pigments denatured by urea or heat fluoresce with very small quantum yields (Kufer 1977; Kufer et al. 1980; Zickendraht-Wendelstadt 1980). Similar pronounced differences are observed in the absorption spectra (Fig. 4; Sect. 4.2) and the chemical reactivities (Table 3) of the native as compared to the denatured pigments. As can be shown, e.g., by SDS gel electrophoresis, the covalent bonds between chromophores and proteins are unchanged by denaturation. Moreover, the spectroscopic properties of the native pig-

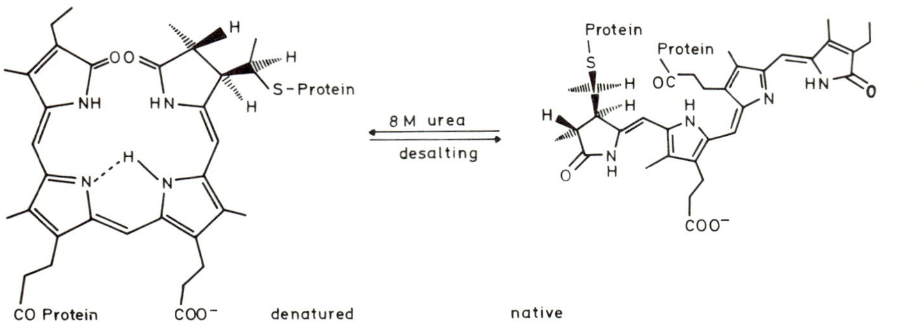

Structure 10a

Structure 10b

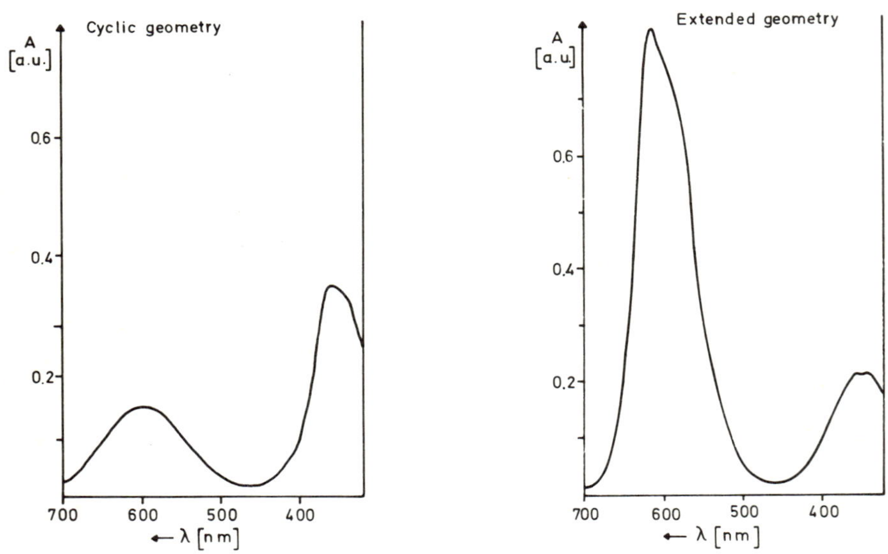

Fig. 4. Absorption spectra of denatured (*left*) and native (*right*) C–PC from *Spirulina platensis*, and the corresponding types of chromophore geometry. Both spectra correspond to identical pigment concentrations. The cyclic conformation is helical (see formula *11*). The extended conformation of the native state is chosen with respect to minimum steric hindrance of the substituents. It is probably uniformly twisted, too, as inferred from CD data (see text)

Table 3. Properties of native biliproteins, as compared to the properties of denatured biliproteins and free bilins

	Denatured biliproteins and free bilins	Native biliproteins	Ecological significance in the native pigments
Absorption spectra	Broad bands Near UV band: $\epsilon \approx 4 \cdot 10^4$ Visible band: $\epsilon \approx 2 \cdot 10^4$ λ_{max} determined by the chromophore only	Narrow bands Near UV band: $\epsilon \approx 2 \cdot 10^4$ Visible band: $\epsilon \approx 10^5$ λ_{max} influenced by the protein	Increased absorption, fine-tuning of λ_{max} However, only narrower spectral range covered by any given chromophore
Photochemistry	Radiationless deactivation predominant, small quantum yield of fluorescence and photochemical reactions	High quantum yield of fluorescence (phycobilins) or photochemical reactions (phytochrome), respectively	Decreased losses of excitation energy
Reactions: Complexation	Instantaneous complexation with Zn^{2+} at ambient temperature	Inert with Zn^{2+} at ambient temperature	Chromophore stabilization
Reduction	Complete reduction with dithionite ($c \geqslant 0.5$ mM)	Partial reduction only with dithionite ($c \geqslant 5$ mM)	
Oxidation	Oxidative bleaching within days at 4 °C	Stable for months at 4 °C	

ments can be restored fully and in good yield (Scheer and Kufer 1977) within the folding time of the protein (Bartholmes and Scheer 1980) if the concentration of the denaturing agent or the temperature is decreased.

The spectroscopic properties and reactivities of the denatured pigments are identical to those of free bile pigments of the proper structure, which has been commonly used in the structure analysis of biliproteins (see Sect. 4). This proves, on the other hand, that the properties of the denatured pigments can be accounted for completely by their proposed structures, and that the changes upon naturation are entirely due to noncovalent interactions between the chromophores and the proteins, and/or among the chromophores.

The ecological significance of these changes is obvious from Table 3. The oscillator strength, as a base for the efficiency of light absorption of biliprotein antenna systems, is enhanced. Radiationless deexcitation, corresponding to a waste of the absorbed energy, is decreased, and at the same time chemical and photochemical side reactions leading to the destruction of the sensitive bile pigment chromophores is impeded. The biliproteins thus present an excellent example of molecular ecology, e.g., the adaptation of photoreceptor molecule as unfit as an A-dihydrobilindion like *1* or *4*, to its function in an antenna pigment as efficient as phycocyanin. Recently, some progress has been made in understanding the principles involved, by a combination of chemical model studies, MO calculations, and denaturation-renaturation studies with the isolated pigments.

5.1 Topology of the Chromophore

The characteristics of the changes in the absorption spectra can be accounted for essentially by a change in the geometry of the chromophores. Bile pigments are at least principally flexible structures, and several research groups have treated theoretically the dependence of the absorption spectra on the molecular topology (Burke et al. 1971; Chae and Song 1975; Falk and Höllbacher 1978; Fuhrhop et al. 1974; Wagnière and Blauer 1976), and in some calculations also on a charge at or close to the π-system (Pasternak and Wagnière 1979; Sugimoto et al. 1976). The results agree in one fundamental aspect, that the ratio of the oscillator strengths of the two lowest electronic transitions,

$$Q = \frac{f_{vis}}{f_{near\ uv}}$$

is a rough measure of the shape of the molecule. Q is < 1 in cyclic porphyrin-type conformations, but becomes $\gg 1$ when the molecule is stretched to an extended conformation. On this basis, the spectroscopic properties of native and denatured phycocyanin can be rationalized by the former having an extended *(10b)*, the latter a cyclic conformation *(10a*, Fig. 4) (Scheer and Kufer 1977).

The theoretical calculations have been supported by the conformational analysis of bile pigments of the biliverdin type in solution and in the crystal. A cyclic-helical conformation was proposed first for optically active urobilin on the basis of the large cotton effects observed, which are typical for inherently dissymmetric chromophores (Moscowitz et al. 1964). If the two asymmetric centers at C–4 and C–16 have the same configuration, one of the two helices of opposite chirality is strongly favored due to steric hindrance. The effect of asymmetric centers in shifting the equilibrium between the two helices is principally possible, too, in other helical bilins. For denatured PC, an energy difference of $\Delta H = 0.1-0.2$ kcal/mol has been estimated from the steric hindrance arising from the asymmetric C–2, C–3, and C–3' (Scheer et al. 1979). In denatured PE, the contribution of C–16 is expected to counteract this effect. In agreement with this reasoning, denatured PC (Lehner and Scheer 1981), but not PE (Langer et al. 1980), has a pronounced optical activity.

On the other hand, optical activity can be induced, too, by a chiral environment. The solvent-induced circular dichroism (SICD) seems to be a safe indicator of helical conformations (Lehner et al. 1978, 1981). *All-syn, Z* biliverdin-dimethylester *(11)* gives a strong SICD effect (Lehner et al. 1978), but neither the *anti-E, syn-Z, Syn-Z* isomer *12* (Gossauer et al. 1980) nor the formyltripyrrinone *13* (Lehner et al. 1981) do. In solution *11* has a predominantly helical conformation (Lehner et al. 1978; Falk et al. 1978a), and a helical crystal structure (Lehner et al. 1978b; W.S. Sheldrick 1976). The energy barrier of 42 kJ/mol between the two helical forms allows a rapid interconversion at ambient temperatures (Lehner et al. 1979). In a chiral solvent, e.g., lactate esters, the equilibrium between the two helical forms is shifted. As both forms are expected to be strongly optically active, a slight shift of the equilibrium from K = 1 is already sufficient to give measurable CD signals. In another way of reasoning, the perturbation by the chiral solvent is transmitted through the entire molecule only in a helical structure. *12* has a more extended conformation (according to MO studies), with no interaction between rings A and D (Falk et al. 1978b), while *13* is planar, as has been shown by the X-ray structure of an analog (Cullen et al. 1978), and thus also lacks the principal requirements for a strong SICD-effect. Falk et al. (1981) have recently carried out force field calculations on bile pigments. For the Z, Z, Z-bilindion, they converge at a structure very similar to the crystal structure of biliverdin-dimethylester (Sheldrick 1976).

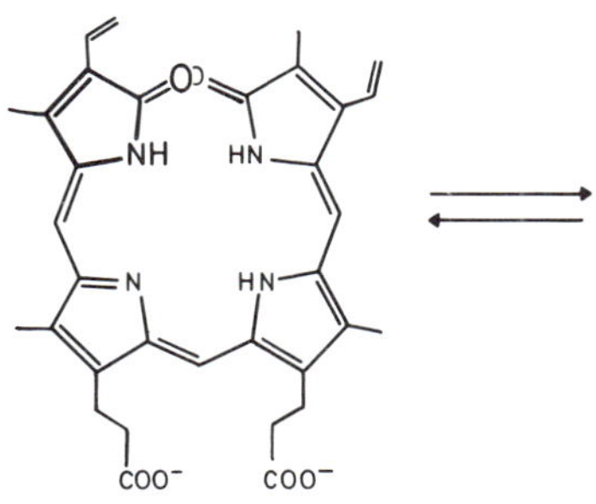

Structure 11a

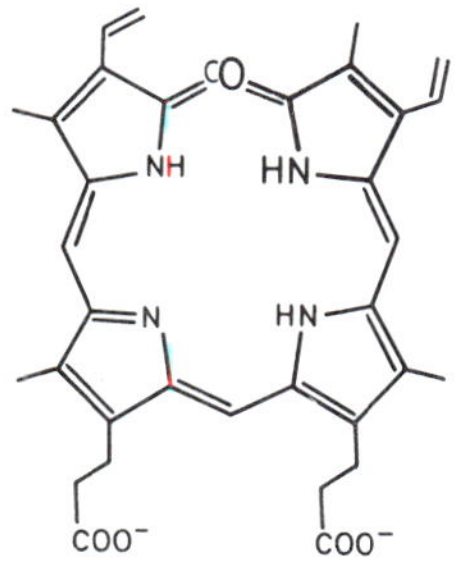

Structure 11b

Experimental data on bile-pigments with restricted conformational freedom have been provided from pigments bearing intramolecular bridges. A 21,24-methano-bilindion supports the assignment of cyclic-helical conformations to free bilindiones (Falk and Thirring 1981). Models for the extended conformations suggested for the native biliprotein-chromophores, have been provided by nature and biomimetic synthesis with isophorcabilin (*15*) and related polycyclic pigments (Bois-Choussy and Barbier 1978; Choussy and Barbier 1975). They are derivatives of the 3, 7, 12, 17-tetramethyl-8,13-divinyl-2, 18-dipropionyl-bilindion (*14*, "biliverdin IX γ")(Bois-Choussy and Barbier 1978; Choussy and Barbier 1975), and have been discovered during investigations on bile pigments from caterpillars and butterflies (*Lepidoptera*). Their importance for conformation analysis comes from their confinement to extended conformations by the additional rings, which are formed by intramolecular additions to the vinyl groups of *14*.

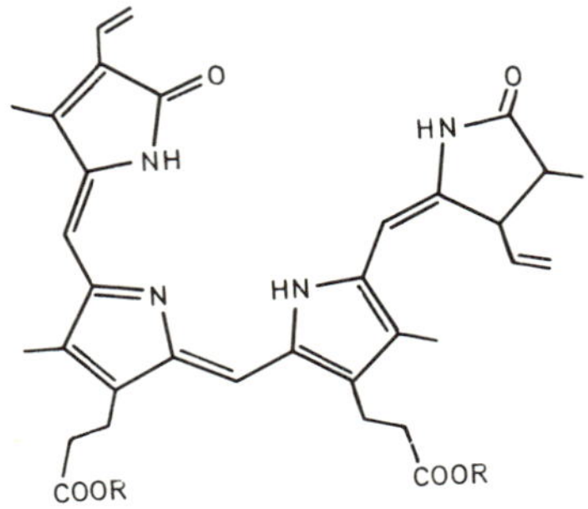

Structure 12

Structure 13

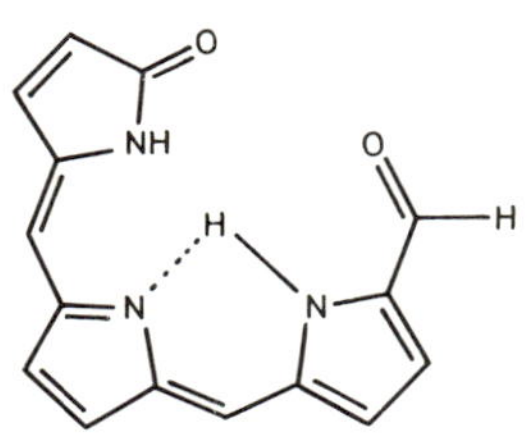

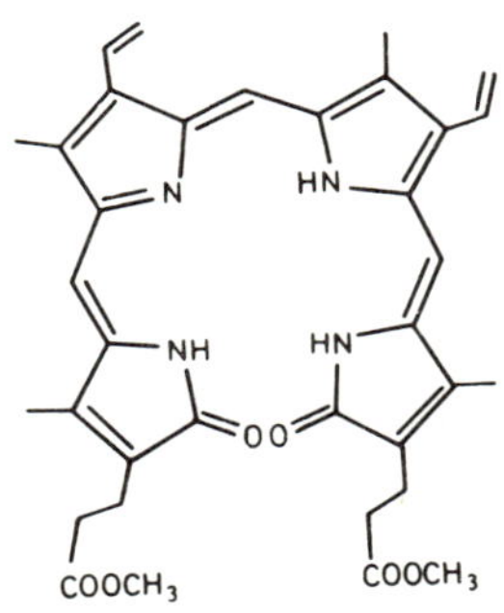

Structure 14

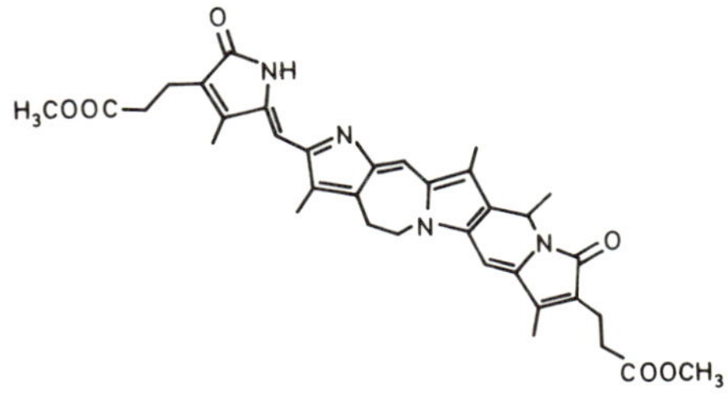

Structure 15

The isophorcabilin *15* has an intense visible absorption band and a very weak near absorption (Bois-Choussy and Barbier 1978; Brandlmeier et al. 1981), thus supporting the theory. Its spectrum is very similar to that of native PC, for which an extended conformation is therefore highly likely, too (Fig. 4) (Scheer and Kufer 1977). This extended conformation has to be brought about by the protein, but the details of the process are not yet understood. It should be pointed out, that extended conformers of bile pigments like *1* and *12* are less stable than their cyclic conformers (Falk and Grubmayr 1979; Scheer et al. 1979). The energy for unfolding the chromophores would then have to be provided by the protein (Kufer and Scheer 1979). In agreement with this reasoning, the energies of denaturation of various PC's determined recently by Chen and Berns (1978) are conspicuously low as compared with the respective energies of globular proteins of similar size (Knapp and Pace 1974; Salahuddin and Tanford 1970).

5.2 Conformational Mobility

Another factor of the protein seems to be a restriction of conformational mobility to the chromophore. Bile pigments of the biliverdin type have broad absorption spectra, which even at low temperatures show only little fine structure (Chae and Song 1976; Friedrich et al. 1981a,b; Gautron et al. 1976; Holzwarth et al. 1978; Petrier et al. 1979, 1980; Scheer and Kufer 1977). This has been rationalized as a superposition of the spectra several conformers with slightly different absorptions, which are in rapid equilibrium with each other (Lehner et al. 1978a; Scheer et al. 1977; Scheer and Kufer 1978). This interpretation has been substantiated recently by a careful fluorescence analysis. It could be shown that a solution of *11* contains at least two species, their equilibrium being solvent-dependent (Braslavsky et al. 1980; Tegmo-Larsson et al. 1980a, b). From the Q values of the excitation spectra, a cyclic and a more extended conformation have been proposed, the latter being favored in rigid solutions and especially in liposomes. Broad spectra are characteristic of most bile pigments, but a notable exception is the "purpurins", e.g., *13*. They have a double-peaked long-wavelength absorption, with each of the two components being narrow. *13* is planar (Cullen et al. 1978), and it has been suggested, that the two peaks arise from two comparably rigid, distinct forms, e.g., two tautomers, each fixed by a different type of intramolecular H-bond (Scheer et al. 1977).

Native biliproteins have broad absorption bands, too, in which some fine structure is generally obvious already at ambient temperature (see for examples: O'Carra and O'hEocha 1976) and becomes prominent at low temperatures (Frackowiak and Grabowski 1971; Frackowiak et al. 1975; Friedrich et al. 1980a,b; Gray et al. 1976; Scheer and Kufer, 1977; Zickendraht-Wendelstadt et al. 1980). Thus, their spectra are superpositions of different forms as well. There is, however, an important difference as compared with the denatured biliproteins or the free bile pigments with corresponding structures: The different forms are not in equilibrium, but rather correspond to different chromophores of the pigments in different environments well defined by the protein (see Scheer 1981; Zickendraht-Wendelstadt et al. 1980; Zuber 1978, for leading references). The situation is reminiscent the purpurins, but obviously for different reasons. As an example, C–PC contains three chromophores. Its long-wavelength absorption band is asymmetric with a distinct shoulder at shorter wavelength already noticeable at room temperature, which is split into two narrow components at low temperatures (Frackowiak et al. 1975; Friedrich et al. 1980a, b; Scheer and Kufer 1977). The assignment to individual chromophores is yet unclear. The renatured α– and β-subunits of C–PC from *Synechococcus spec.* (formerly *Anacystis nidulans*) absorb at 620 and 608 nm, respectively, indicating the α-subunit bearing the "f", the β-subunit bearing two "s"-type chromophores (Glazer et al. 1973). The superposition of the spectra of the two subunits does not yield that of the native protein, however. Such an additivity has been observed for a C–PE (Zickendraht et al. 1980). It is also indicated from the data of Binder et al. (1972) and unpublished data of the author for the renatured subunits of other C–PC's. In this case, the β-subunit bearing two chromophores absorbs at longer wavelength than the α-subunit bearing a single chromophore, which would correspond to the spectral shape of the native $\alpha\beta$-monomer. Irrespective of this conflicting interpretation is the assignment of the two compounds of the absorption spectrum of C–PC to different chromophores supported by photochemical hole-burning experiments (Friedrich et al. 1980a, b), by partial denaturation (Scheer and Kufer 1977) and chemical modification, e.g., with sodium dithionite (Kufer and Scheer 1979). In each case, differential effects on the two absorption bands are observed.

The bandwidth of the individual components is a narrow as in the purpurins (Fig. 5) (Scheer et al. 1977), or in the macrocyclic analogs, the chlorophylls. This is a good indication of a fixation of each of the chromophores, although possibly to different degrees. Similar results have been obtained for PE, although the analysis is more difficult due to the larger number of chromophores present. Reaction with sodium dithionite is preferential for the short-wavelength forms C–PE from *Pseudoanabaena* spec. W. 1173 (Kufer and Scheer 1979), and differential reactions have been observed too with organic mercury compounds (Pecci and Fujimori 1967, 1968). The subunits of this pigment have been analyzed in detail by fluorescence (Zickendraht-Wendelstadt et al.

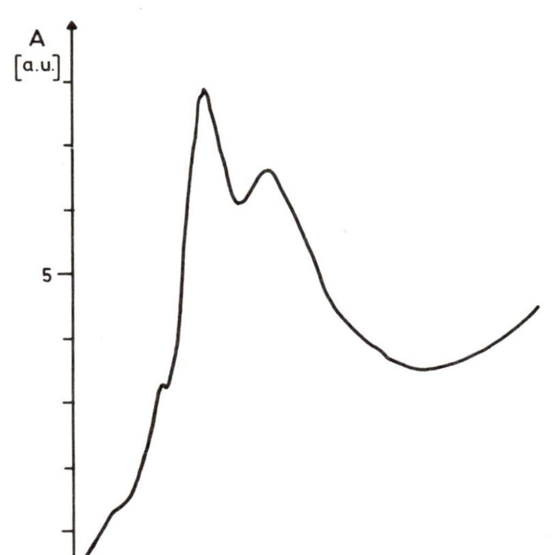

Fig. 5. Low-temperature spectrum of C–PC from *Spirulina platensis*, according to Scheer and Kufer (1977). The separation of the two bands corresponding to different chromophores is still increased at liquid helium temperatures. Only the long-wavelength peak is shown in Fig. 6

1980), and CD spectroscopy (Langer et al. 1980), and the results allow the distinction of all chromophores in the subunits, and of at least three of the five chromophores in the monomer.

The distinct environments of different chromophores in biliproteins is supported by analysis of the primary structure of chromopeptides and entire biliproteins. In all cases studied, there is a defined sequence to each chromophore, which is different for each chromophore in a given pigment, but similar for the corresponding chromophores in pigments from other organisms (Frank et al. 1978; Freidenreich et al. 1978; Otto et al. 1977; Troxler 1980; Zuber 1978).

A rigid fixation is necessary (although not sufficient, see below) to minimize radiationless decay of the excited states of biliproteins. This is accompanied by a narrowing of the absorption bands which is principally unfavorable to their antenna functions for two reasons: Only a narrow wavelength range is absorbed efficiently, and the overlap between emission bands of a sensitizing chromophore and the absorption of the next member of the Förster transfer chain becomes more crucial. Both effects are overcome by the development of chromophores absorbing at defined, closely spaced intervals, as realized most impressively in the PE's from red and cryptophytan algae, and in the phycobilisome superstructures (see Sect. 2).

Another necessary requirement for the suppression of radiationless processes is the suppression of photochemistry. This channel is important in the phytochromes and phycochromes (Björn 1979; Rüdiger 1980; Scheer 1981). Especially the results obtained recently in the latter case (see Sect. 6) indicate that small changes in the protein structure are already sufficient to open the photochemical channel, e.g., to increase radiationless transitions. Free bile pigments (Falk and Neufingerl 1979; Hudson and Smith 1975; Lightner 1977; Mac Donagh 1979; Manitto and Monti 1972), especially the A-Dihydrobilins (Kraus et al. 1979; Scheer and Krauss 1977; Scheer et al. 1977), are photolabile, but they react only with small quantum yields due to the competing radiationless deexcitation. It thus appears that the major effect of the protein is to decrease the latter processes, wile the change from efficient fluorescence to efficient photochemistry requires comparatively small modifications in the protein environment.

A subtle form of photochemistry, photoautomerization, is discussed as another pathway for rapid radiationless deexcitation of free bile pigments. Although at present no quantitative analysis has been done to assess its contribution in a certain pigment, examples are known from both bile pigments (Falk et al. 1979; Holzwarth et al. 1978) and other pigments (Schneider et al. 1979). Recently, photochemical hole-burning experiments have been performed on biliproteins (Friedrich et al. 1980a, b) which may ultimately help to answer this question, and give more insight into the protein-chromophore interaction. In this technic, a narrow, low-intensity laser beam is used to excite a glassy sample at low temperatures. Under certain conditions a narrow dip ("hole") in produced in the absorption band (Fig. 6) which may be accompanied by other holes, at longer wavelengths, and by a depression at the long-wavelength-side of the hole. The phenomenon is produced by a site-selective photochemical process, and the mechanism of this process has been studied in detail on small molecules. In hydroxyanthraquinones it could be identified as a hydrogen transfer process of a triplet state intermediate, which produces a thermolabile product absorbing at shorter wavelengths than the educt (Haarer et al. 1980), thus leaving a hole in the absorption envelope with its width deter-

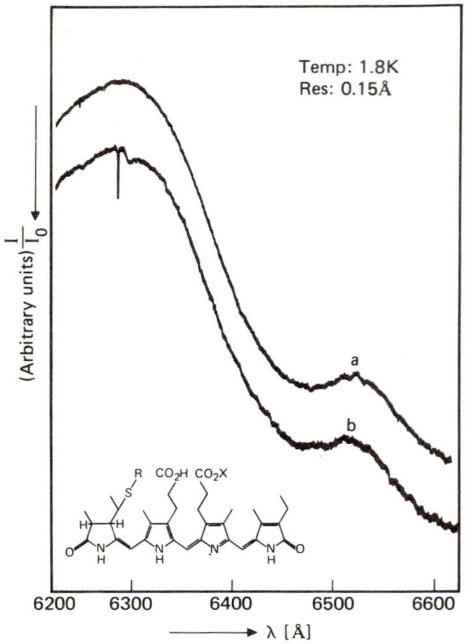

Fig. 6. Hole-burning experiment in C–PC from *Spirulina platensis*, according to Friedrich et al. (1980). Only the long-wavelength peak of the two visible bands of the low-temperature spectrum (Fig. 5) is shown in trace *a*. Trace *b* is obtained after 1 min irradiation with light (605 nm). The narrow hole has a half-width of 1.8 cm^{-1}, it has a small side band at longer wavelengths corresponding to a weak phonon coupling. Similar holes are obtained by irradiation of the APC band at 655 nm. In C–PE, additional narrow holes are produced, probably due to energy transfer between chromophores

mined by the relaxation kinetics. The situation is depicted in the term scheme (Fig. 7). Hole burning has been observed in APC, C–PC, C–PE, and in *8* as a model for the chromophores of the former two pigments, and the same mechanism (proton transfer from a triplet intermediate) has been suggested from the similar phenomenology. Bile pigment triplets have recently been studied by Land (1979).

Obviously, all three pigments can undergo low-temperature photochemistry. Processes of this type may account for the radiationless losses of isolated biliproteins, which are in the range of 20%–50% at ambient temperature (see Grabowski and Gantt 1978a) and do not change considerably at lower temperatures (Zickendraht-Wendelstadt et al. 1980). In the intact antenna, the energy transfer to Chl *a* is obviously much more rapid than photochemical processes, as transfer efficiencies of 99% have been determined experimentally. Further important information can be extracted from an analysis of the hole-burning features. In the case of the biliproteins, a small phonon coupling (corresponding to unfavorable Franck-Condon factors for radiationless decay) has been demonstrated, and the results for C–PE indicate a rapid energy transfer process among a highly ordered chromophore assembly. According to these results, the chromophores are well isolated from the protein matrix with regard to energy losses, but tightly coupled at the same time with regard to conformational restrictions.

An indirect form of protein-chromophore interactions is manifested in chromophore-chromophore interactions. There is increasing evidence for interactions of this type, mainly from CD spectra. S-shaped spectra or subspectra have been observed in APC's (Canaani and Gantt 1980), a cryptophytan PC (Jung et al. 1980), and a C-PE (Langer et al. 1980). A distinction between exciton coupling, or different chromophores with Cotton effects of opposite sign as the origin for these S-shaped bands is not trivial. The native *α*-subunit of C–PC from *Spirulina platensis* has a positive long-wavelength

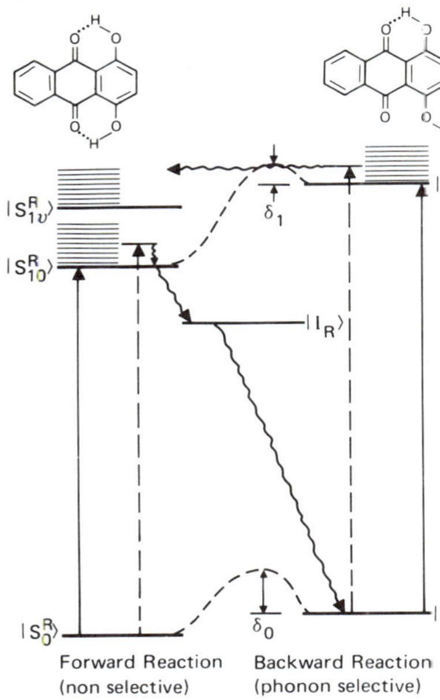

Fig. 7. Term scheme for 1,4-dihydroxyanthraquinone (*left*) and its photoproduct (*right*) according to Haarer et al. (1979). S_O and S_I are ground state and first excited singlet states, respectively. For the latter, the vibrational ground state S_{I0} and a vibrational state ($S_{I\gamma}$) as well as additional phonon sublevels are shown, The superscripts refer to the educt (*R*) and the product (*P*). I_R is an intermediate, probably a triplet, the formation of which is rate-determining for the photoreaction. A similar term scheme is suggested for the biliproteins from hole-burning experiments like the one shown in Fig. 5

CD-band (Lehner and Scheer 1981), whereas P_r bearing a structurally closely related chromophore of the same absolute configuration exhibits a negative effect (Brandlmeier et al. 1981). Since the CD-spectra of both pigments bearing only a single chromophore, are very similar in the denatured state, the opposite sign is due to chromophore-protein interactions of opposite chirality. It is then difficult to distinguish in pigments bearing up to six chromophores, between chromophore-protein interactions of opposite chirality on different chromophores, and S-shaped bands originating from exciton coupling. The former is supported from the correlation of the absorption maxima with the points of zero ellipticity in the CD spectra of APC, and controled denaturation experiments (similar decrease of the positive and negative extrema in C–PE).

Intermediate strength (viz. CD-incative) couplings between chromophores have been discussed for chromophores on different proteins in the APC-trimer (Mac Coll et al. 1980). The absorption of native APC is shifted from 620 to 650 nm and increased in intensity upon trimerization (Brown et al. 1975; Cohen-Bazire et al. 1977; Erokhina and Krasnovskii 1974; MacColl et al. 1981), similarly striking CD-changes, and less pronounced absorption changes have been observed for other biliproteins as well upon dissociation (Bennett and Bogorad 1971; Glazer et al. 1973; Gray and Gantt 1975). The CD changes observed recently upon controled dissociation of phycobilisomes have been interpreted on the same basis (Rigbi et al. 1980).

6 The Proteins

Protein structure and aggregation of biliproteins have been summarized recently in several reviews and shall be discussed only briefly in this context. One of the aspects studied in more detail is the phylogenetic relationship between the various biliproteins.

The family tree shown in Fig. 8 has been derived from sequenation studies and immunochemical investigations (Zuber 1978). The completed sequences of two C–PC's (Frank et al. 1978; Troxler 1980), an APC (Zuber et al. 1981) and of one subunit each of another C–PC (Freidenreich et al. 1978), principally support this picture, but there is evidence for very pronounced variations with sequenation studies of a marine cayno-bacterium *Agmenellum quadruplicatum* (Fox 1980). The completed sequences also indicate the presence of conservative as well as of more variable regions, the former being associated with the chromophore environments (see Muckle et al. 1977; Zuber 1980). This part may well be crucial for the efficiency and stability of the chromophores.

Due to conflicting earlier results (Berns 1967; Glazer et al. 1971), the cryptophytan biliproteins had not been included in Fig. 8. Recent immunological evidence supports a relationship of both cryptophytan PC and PE with rhodophytan PE, but neither with rhodophytan PC nor cyanobacterial pigments (MacColl et al. 1976). This is yet another aspect of the somewhat special character of these pigments, since in both cyanobacterial and rhodophytan pigments the immunology parallels the spectroscopic classification. But it also points to the increasing complexity of biliproteins emerging with an increasing number of pigments studied. Phycoerythrocyanin and R–PC are other examples of pigments with mixed chromophores, which could be correlated to the "common" pigments only by immunochemical methods (see Table 2 for references). Another example is the increasing number of biliproteins containing a γ-chain, which may have to be added to the family tree as another branch. Besides R– and B–PE, which form an $\alpha_6\beta_6\gamma$-structure, (Abad-Zapatero et al. 1977; Gantt and Lipschultz 1974; Sweet et al. 1977), APC–I with probably $\alpha_3\beta_3\gamma$-structure is the third example of this class (Zilinskas et al. 1978).

One of the most prominent properties of cyanobacterial and rhodophytan biliproteins is their pronounced aggregation. Especially the aggregation of PC's has been studied in great detail. Monomers, trimers, and hexamers are the most abundant species in solutions of the purified pigments (MacColl et al. 1971), but dimers (Iso et al. 1977) and tetramers (Neufeld and Riggs 1969) have been observed, too. The influence of various parameters including concentration (Lee and Berns 1968; MacColl et al. 1974), ionic strength (MacColl et al. 1971a), pH (Lee and Berns 1968; Saito 1976), temperature (MacColl et al. 1971a), aromatic compounds (MacColl and Berns 1973), substitution of [^{1}H] by [^{2}H] (Hattori et al. 1965; Lee and Berns 1968), inorganic and organic salts (Berns and Morgenstern 1968; MacColl et al. 1971b), biotope of the parent organisms (Chen and Berns 1979; Kao et al. 1973, 1975), and others (Chen and Berns 1978), have been investigated. A major contribution of hydrophobic interactions as the driving force for aggregation has been concluded.

The trimers and hexamers are the building blocs both of phycobilisomes (Gantt 1975; Glazer et al. 1979; Mörschel et al. 1980) and biliprotein crystals (Abad-Zapatero et al. 1977; Bryant et al. 1976; Dobler et al. 1972; Sweet et al. 1977). Accordingly, higher aggregates (Berns 1971; MacColl and Berns 1979) as well as hetero-aggregates (Grabowski et al. 1980; Koller et al. 1978; Rigbi et al. 1980) can be obtained from controled dissociation of the former. In view of the architecture of the phycobilisomes, the γ-chain in R– and B–PE as well as in APC–I may then serve as a further means to stabilize the trimeric and hexameric building bloc, forming a core around which the

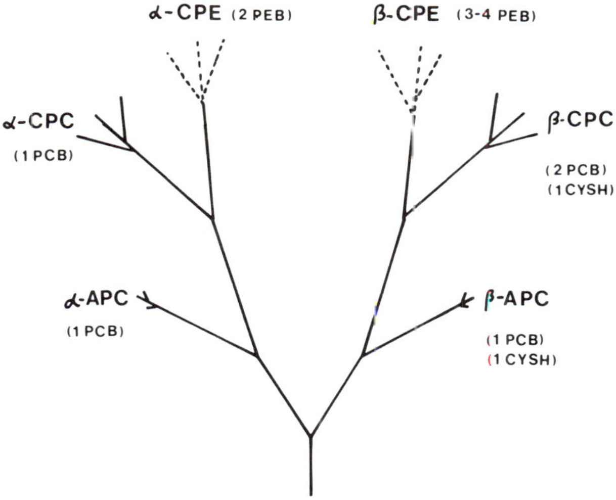

Fig. 8. Family tree of phycobiliproteins, according to Zuber (1978), as derived from sequenation studies and immunochemical relationsships. The abbreviations refer to the chromophores present (*PCB* = *1*, *PEB* = *2*) or to free cystein residues (*CYSH*)

α- and β-subunits are arranged. It is interesting in this respect, that PC-hexamers isolated from phycobilisomes do not show the expected "inner hole" in negatively stained electron micrographs, and may contain a colorless core-protein instead (Mörschel et al. 1980). The cryptophytan biliproteins do not form similar large and stable aggregates, corresponding to the absence of phycobilisomes in these organisms (see Gantt 1979).

7 Biosynthesis

The name biliprotein is derived from these pigments bearing chromophores structurally similar to the bile pigments of higher animals. The latter are derived from oxidative breakdown of the porphyrin macrocycle of hemes, which is apparently the only physiological way to preserve the iron for recycling (MacDonagh 1978; O'Carra 1975). From this point of view, the bile pigments are then only waste products of this process. By contrast, bile pigments in cyanobacteria, red and cryptophytan algae, are synthesized for the essential function of photosynthesis.

In spite of this teleological difference, the biosynthesis of the two types of bile pigments is surprisingly similar in several aspects. Both are derived from protoporphyrin (IX), in both cases the macrocycle is opened at the C–5 position (see Troxler 1976). Apparently, the mechanism of the ring opening is similar, too. One molecule of CO is released (Troxler 1972), and two molecules of oxygen yield each one oxygen atom to the bile pigments (Troxler et al. 1979). It is not yet clear, whether the educt for this ring opening is an iron porphyrin in the chromophore biosynthesis of biliproteins. Chemical model studies support a Mg-porphyrin as a precursor as well (Barrett 1967; Hudson and Smith 1975), and a tetrahydroporphyrin with an ethylidene group at the proper position is known with bacteriochlorophyll *b* (Scheer et al. 1974). Brown et al. (1980) have, however, recently demonstrated the incorporation of exogenous hemin into phycobilins, but not chlorophylls in *Cyanidium caldarium*, which strongly supports the "iron-pathway" in the former.

Nothing is yet known as to the point at which during biosynthesis the chromophore is linked to the protein. Facets to this problem are (1) the finding of various strains of *Cyanidium caldarium* which excrete after treatment with δ-aminoevulinic acid (ALA), the ethylidene bilin *4* into the medium, together with products derived from addition reactions at the ethylidene function (Troxler and Bogorad 1966; Troxler et al. 1978); (2) the hitherto unsucessful search for the free apoprotein of any of these pigments; and (3) the demonstration of reversible addition reactions to the 3-ethylidene bond of *4* (Beuhler et al. 1976; Gossauer et al. 1980; Klein and Rüdiger 1978). Taken together, they indicate that the bilin *4* may be the biosynthetic precursor, which is attached to the protein in one of the last steps.

The biosynthesis of biliproteins is effected by several nutritional factors, of which nitrogen (see Bogorad 1975), sulfur (Schmidt 1980) and especially light, are studied in detail. Light regulates the synthesis of both chlorophylls and biliproteins. The tetrapyrole skeletons of both pigments are derived from ALA and share a common part of their biosynthetic pathways, although possibly not their ALA pool. In a recent study, with mutants of *Cyanidium caldarium* defective in chlorophyll or biliprotein biosynthesis, Schneider and Bogorad (1979) obtained evidence for two different, but strongly interacting photoregulation systems. A protochlorophyll-type and a hemoprotein-type pigment, respectively, may be involved as photoreceptors. For leading references on light-dependent development in cyanobacteria and red algae the reader is referred to: Björn 1979, 1980; Bogorad 1975; Bogorad et al. 1975; Lazaroff 1973.

In many cyanobacteria and red algae, light-stimulated developmental responses have been observed, which suggest the presence of photoreversibly photochromic pigments as photoreceptors (see Björn 1979; Björn 1980; Bogorad 1975) (Table 4a). Chromatic adaptation is a widespread response, by which (among other effects) the composition of the biliprotein antenna is changed with the environmental light quality. The relative amount of green-light absorbing PE's is reversibly increased in green light and decreased in red light (Bogorad 1975; Lazaroff et al. 1973; Tandeau de Marsac 1977; Wagenmann 1977), with a concomitant change in phycobilisome composition (Siegelman 1980). Another light-mediated response is the induction of filamentous growth, e.g., in *Nostoc* (Ginsburg and Lazaroff 1973). Corresponding to these effects, the terms adaptochromes and phycomorphochromes have been suggested (Bogorad 1975). In analogy to phytochrome, the photoreversible photochromic sensory pigment of higher plants (Rüdiger 1980), the term phycochromes is often used synonymously for these receptors.

It my be useful to reserve the term phycochromes to photoreversibly photochromic pigments irrespective of any proven function as a photoreceptor for the present time (see Björn 1979, for a discussion). This distinction arose from the attempts to isolate adaptochromes and phycomorphochromes, e.g., the genuine photoreceptors. During this search, Scheibe (1972) first isolated a biliprotein fraction from photobleached *Tolypothrix tenuis*, which gave photoreversible absorption differences reminiscent of the action spectra for chromatic adaption in this species. The Björns have since characterized four different phycochromes (*a, b, c, d*) from various organisms, and purified to a different degree (Björn 1978; Björn and Björn 1976). All appear to be biliproteins, and thus require very tedious and extensive purification from the bulk pigments, with isoelectric focusing as the key step. In only one of the photochromic fractions, the absorption differences exceed a few percent, although the quantum yields are reasonable. Surprisingly, only two of these four phycochromes (Table 4b) can presently be correlated spectroscopically to a known light response in the parent organisms, which led to the aforementioned distinction of phyochromes (pigments with photoreversible photochromic properties) on one hand, and adaptochromes and phycomorphochromes (as photoceptors for the respective processes) on the other. To link these chemical and functional properties to a certain pigment, it may be necessary to obtain further information besides the spectral data. As one such possibility, a differential temperature effect on the forward and back reaction of the photoreversion process has been suggested recently (Ohad et al. 1979).

The need for a distinction becomes even more obvious from recent results on the photochemical properties of phycobiliproteins partially denatured with low concentrations of guanidinium chloride (Ohki and Fujita 1979) or a decrease in pH (Ohad et al. 1979, 1980) (Table 4c). By this treatment, APC and PC obtain photoreversible photochromic properties reminiscent of the phycochromes *a* and *c* respectively, of Björn and Björn (1976). According to all evidence, this is no artifact due to a co-isolated impurity. It rather appears that at least in these cases phycochromes are not separate pigments, but rather slightly denatured states of well-known biliprotein antenna pigments, or certain sub-populations there of. It is interesting in this respect that phycochrome *d* has been associated recently with another biliprotein, phycoerythrocyanin (Björn 1980). The denaturant-induced photochromicity of PC and APC decreases again at more severe denaturation conditions. Probably, the "tickling" of the protein looses the interactions with the protein sufficiently to open a photochemical channel, while internal conversion and destructive photochemistry of the pigments are still inhibited. More severe uncoupling (see Sect. 5) then favors the latter processes. Stepwise denaturation has been observed with PC *Spirulina platensis*, suggesting a distinct intermediate in the unfolding process (Scheer and Kufer 1977). Similar intermediates may be produced in the pigments isolated from *T. tenuis* (Ohki and Fujita 1979) and *F. diplosiphon* (Ohad et al. 1979, 1980). MacColl et al. (1980b) recently pointed out, that similar difference spectra as in the phycochrome reaction, are observed in the reversible dissociation of trimeric $(\alpha\beta)_3$ *APC II*. It would then be crucial to link the dissociation-reassociation unequivocally to the light reaction rather than to an artefact, e.g., due to sample heating.

8 Concluding Remarks

This review on phycobiliproteins is far from complete. This is not only due to the widespread and still growing interest in these pigments, which is prohibitive to the citing of all pertinent work, but also to the deliberate attempt of the author, to focus mainly on the chromophore protein interactions. These interactions in the native biliproteins are crucial for their functions, and only recently some general principles have been recog-

Table 4. Action maxima for light-stimulated developmental responses in whole organisms (a), for reversible absorption changes in biliprotein fractions (b), and for reversible absorption changes in partially denatured, purified biliproteins

	"green" form		"red" form		References
	λ_{max}	action	λ_{max}	action	
(a) Action spectra for Photoresponse (chromatic adaptation or photomorphose) in whole organisms					
Tolypothrix tenuis	541 550, 350	PE formation	641 660, 360	PC formation	Fujita and Hattori (1962) Diakoff and Scheibe (1973) Ohki and Fujita (1978)
Fremyella diplosiphon	550, 387, 540	PE formation	641, 463, 650, 360	PC formation	Vogelmann and Scheibe (1978) Haury and Bogorad (1977)
Nostoc muscorum	~550	Reversion	650	Induction of filamentous growth	Lazaroff and Schiff (1962)
Nostoc commune	520		640		Robinson and Miller (1970)
Cyanidium caldarium [c]	420, 550, 595	PBP formation	–	–	Nichols and Bogorad (1962) Schneider and Bogorad (1979)
(b) "Action spectra" for reversible absorption changes in algal extracts or enriched fractions of phycobiliproteins ("phycochromes")					
"Photoreversible Pigment" from photobleached *Tolypothrix tenuis*	~520	formation P_{650}	650	reversion	Scheibe (1972) Ohki and Fujita (1979a)
Phycochrome *a* from: *Phormidium luridum, Anabaenae, Tolypothrix distorta*	630	formation P_{580}	580	reversion	Björn and Björn (1978) Björn (1980)

Phycochrome *b* from *Tolypothrix distorta, Anabaenae* [b]	580	formation P_{500}	500	reversion	Björn and Björn (1976) Björn (1980)
Phycochrome *c* from *Nostoc muscorum*	630, 650[a]				Björn and Björn (1976) Björn (1980)
Phycochrome *d* from *Tolypothrix distorta Anabaenae*[b]	650	formation P_{650}	610 – 620	reversion	Björn (1978, 1980)
APC in phycobiliprotein mixtures obtained by isoelectric focusing (pI = 4.4.3) of extracts from *Fremvella diplosiphon* and *Nostoc muscorum*	645	formation P_{555}	560	reversion	Ohad et al. (1980)
(c) "Action spectra" for reversible absorption changes in partially denatured, purified biliproteins					
PC from *Tolypothrix tenuis* "tickled" with 0.5 M guanidinium	570	formation P_{630}	630	reversion	Ohki and Fujita (1979b)
APC 0.5 M guanidinium chloride	600	formation P_{650}	650	reversion	
APC from *Fremyella diplosiphon, Nostoc muscorum* and *Phormidium luridum* at pH 4	645	formation P_{555}	550 – 560	reversion	Ohad et al. (1980)

a Probably two different forms

b A correlation has recently been indicated between the occurence of phycochromes *b* and *d* and the presence of phycoerythrocyanin in the respective organism (G.S. Björn, to be published)

c Probably no photoreversibly photochromic pigment, although with an action spectrum similar to such pigments from *F. diplosiphon* nized. Most details are derived from spectroscopic and chemical studies of the chromophores. The protein environment is known only for a few pigments in its primary structure, and the amino acid residues responsible for the noncovalent interaction in the native pigments are still unknown. X-ray structure, as the most promising tool to study these interactions, has encountered unexpected difficulties, but may provide these data in the near future. Other major gaps in our present knowledge concern the last steps in the biosynthesis of biliproteins, the processes of biliprotein aggregation to phycobilisomes, and the details about the phycobilisome-membrane junction critical for energy transfer to the reaction centers. It should also be borne in mind that biliproteins have been studied in detail only from a small number of species.

Phycobiliproteins are a good example of molecular ecology, and they may help not only in understanding better the processes of photosynthesis, but also the interactions between proteins and small molecules in general. This, together with the fun to work with these brilliantly colored pigments, is expected to fill at least part of the gaps in the near future.

Acknowledgement. The cited work of the author was supported by the Deutsche Forschungsgemeinschaft, Bonn-Bad Godesberg.

Notes Added in Proof

This manuscript contains references acessible to the author until mid 1981. Some recent progress is added below. The reader is also referred to a review on phycobilisomes and phycobiliproteins by R. MacColl to appear in Photochem. Photobiol.

Energy Transfer

The transfer efficiencies from R–PE to R–PC (93%), R–PC to "bulk" APC (98%) and "bulk" APC to the "terminal" APC (96%) have been determined for a rhodophytan phycobilisome [Bekasova et al.; Biofizika **26**, 74 (1981)]. Several groups are investigating the energy transfer in biliproteins and phycobilisomes by picosecond time-resolved techniques. By comparison of low and high intensity excitation, $^1S-^1S$ annihilation has been suggested as a complication both in isolated biliproteins [Wong et al.; Photochem. Photobiol. **33**, 651 (1981); Doukas et al. ibid. **34**, 505 (1981)] and in phycobilisomes (Holzwarth et al., unpubl.). Attempts to analyze the complex fluorescence or ground-state recovery are ambiguous, since good fits can be obtained using expotentials only, but also including nonlinear exponents ($t^{-1/2}$) originating from a Förster mechanistic analysis (Holzwarth et al., unpublished; Hefferele, Nies, Wehrmeyer and Schneider, unpublished). Depolarization studies have been included in the latter study and also in denaturation work with C–PC in defined aggregation states (Hefferle, John, Scheer and Schneider, unpublished) to aid this analysis. Brody et al. (Biophys. J. **34**, 439 (1981) reported the puzzling result, that the 715 nm fluorescence (generally associated with chlorophyll) arose faster than that of biliproteins when exciting the latter, which may indicate alternative transfer schemes, e.g., via carotenoids (Szalontay and v.d. Ven, FEBS Lett. **131**, 155 (1981).

Structure of Phycobilisomes and Reconstitution of Functional Complexes

Kirilowski and Ohad (Proc. Natl. Acad. Sci. USA, in press) reported the reassembly of fully dissociated phycobilisome-membrane complexes, and Glick and Zilinskas (Plant Physiol., in press) demonstrated the importance of the structural peptides by reassembling phycobilisomes from the individual pigments in the presence of a colorless 29 kdalton peptide. The function and properties of the large structural peptide (75-90 kdalton) have been studied in some detail and the results seem to rationalize earlier conflicting results. This colored peptide is readily susceptible to bleaching and proteolysis [Lundell and Glazer; J. Biol. Chem. **256**, 12600 (1981); Lundell et al., J. Cell. Biol. **91**, 315 (1981); Ruschkowski, Ph. D. Thesis, Rutgers University, 1981] and is probably the genuine γ-subunit of APC–I, serving as the terminal energy donor to chlorophyll in these algae, and as (one of) the anchor-protein(s) of the phycobilisome to the membrane. The facile proteolysis of this peptide is reminiscent of other peripheral membrane proteins, e.g., some b-cytochromes. The functionality and structure of phycobilisomes during N-starvation has been studied by Yamanaka and Glazer [Arch. Microbiol. **124**, 39 (1980)]. The APC core and adjacent PC hexamers essential for energy transfer are kept functional until a very late stage of starvation, whereas the phycobilisome periphery including specific structural peptides are digested. Nies and Wehrmeyer [Arch. Microbiol. **129**, 374 (1981)] studied in some detail the phycobilisomes from *M. laminosus* and its dissociation products to arrive at a model very similar to that of the red algae, *R. violacea* shown in Fig. 1.

Isolated Biliproteins

The primary structure of a third C–PC (*C. caldarium*) solved by Troxler et al. [J. Biol. Chem. **256**, 12176 (1981)] and Offner et al. [ibid. p. 12167] supports the homology between pigments from different species. Murakami et al. [J. Biochem. **89**, 79 (1981)] studied the reversible aggregation of APC under varying environmental conditions. By curve-resolution of the absorption band, they concluded that one of the two chromophores is unaffected by this process, whereas the second chromophore is blue-shifted and decreased in intensity in the monomer [Mimuro et al., Arch. Biochem. Biophys. (in press)]. This change is similar to the change observed upon deprotonation of bile-pigment cations to the free bases (see text). Yu et al. [Plant. Physiol. **68**, 482 (1981)] found a variation in the chromophore composition indicating an exchange of phycourobilin with phycoerythrobilin chromophores in R–PE of two closely related species, and also in a single species by chromatic adaptation.

References

Abad-Zapatero, C., Fox, JL., Hackert, M.L.: Biochem. Biophys. Res. Comm. **78**, 266 (1977)
Abeliovich, A., Shilo, M.: Biochim. Biophys. Acta **283**, 483–491 (1972)
Adams, S.M., Kao O.H.W., Berns, D.S.: Plant Physiol. **64**, 525–527 (1979)
Alberte, R.S., Friedman, A.L., Gustafson, D.L., Rudnick, M.S., Lyman, H.: Biochim. Biophys. Acta **635**, 304–316 (1981)
Allen, M.M., Smith, A.J., Arch. Mikrobiol. **69**, 114–120 (1969)
Barrett, J.: Nature **215**, 733–35 (1967)
Bartholmes, P., Scheer, H.: in preparation
Bellin, J.S., Gergel, C.A.: Photochem. Photobiol. **13**, 399–409 (1971)
Bennett, A., Bogorad, L.: Biochemistry **10**, 3525 (1971)
Bennett, A., Siegelman, H.W.: In: The Porphyrins; Dolphin, D. (ed.), pp. 493. New York: Academic Press 1979
Berns, D.S.: In: Subunits in Biological Systems; Timasheff, S.N., Fasman, G.D. (eds.), Part A, pp. 105, New York: Dekker 1971
Berns, D.S.: Photochem. Photobiol. **24**, 117–139 (1976)

Berns, D.S., Morgenstern, A.: Arch. Biochem. Biophys. **123**, 640 (1980)
Beuhler, R.J., Pierce, R.C., Friedman, L., Siegelman, H.W.: J. Biol. Chem. **251**, 2405–2411 (1976)
Binder, A., Wilson, K., Zuber, H.: FEBS Lett. **20**, 111 (1972)
Björn, G.S.: Physiol. Plant. **42**, 321 (1978)
Björn, G.S.: Ph. D. Thesis (Coden: Lunbds/CNBFB–1009)/1–28/ (1980), University of Lund 1980
Björn, G.S., Björn, L.O.: Physiol. Plant. **26**, 297 (1976)
Björn, L.O.: Quart. Rev. Biophys. **12**, 1–25 (1979)
Blacha-Puller, M.: Dissertation, Technische Universität, Braunschweig 1979
Bogorad, L.: Annv. Rev. Plant Physiol. **26**, 369–401 (1975)
Bois-Choussy, M., Barbier, M.: Heterocycles **9**, 677–690 (1978)
Bonnett, R.: In: The Porphyrins; Dolphin, D. (ed.), Vol. I, p. 1, New York: Academic Press 1978
Boucher, L.J., Crespi, H.L., Katz, J.J.: Biochemistry **5**, 3796–3802 (1966)
Brandlmeier, T., Blos, I., Rüdiger, W.: In: "Photoreceptors and Plant Development"; de Greef, J. (ed.), p. 47–54, Antwerpen University Press, 1980
Brandlmeier, T., Scheer, H., Rüdiger, W.: Z. Naturforsch. **36c**, 431–439 (1981)
Braslavsky, S.E., Holzwarth, A.R., Langer, E., Lehner, H., Matthews, J.J., Schaffner, K.: I sr. J. Chem. **20**, 196–202 (1980)
Brockmann, H. jr., Knobloch, G.: Chem. Ber. **106**, 803 (1973)
Brooks, C., Chapman, D.J.: Phytochemistry **11**, 2663–2670 (1972)
Brown, A.S., Foster, J.A., Voynow, P.V., Franzblau, C., Troxler, R.F.: Biochemistry **14**, 3581 (1975)
Brown, S.B., King, R.F.G.J.: Biochem. J. **170**, 297–311 (1978)
Bryant, D.A., Glazer, A.N., Eiserling, F.A.: Arch. Microbiol. **110**, 61–75 (1976)
Bryant, D.A., Hixson, C.S., Glazer, A.N.: Biol. Chem. **253**, 220–225 (1978)
Burke, M.J., Pratt, D.C., Moscowitz, A.: Biochemistry **11**, 4025 (1972)
Byfield, P.G.H., Zuber, H.: FEBS Lett. **28**, 36–40 (1972)
Calzaferri, G., Gugger, H., Leutwyler, S.: Helv. Chim. Acta **59**, 1969 (1976)
Canaani, O.D., Gantt, E.: Biochemistry **19**, 2950–2956 (1980)
Canaani, O.D., Lipschultz, C.A., Gantt, E.: FEBS Lett. **115**, 225–229 (1980)
Chae, Q., Song, P.S.: J. Am. Chem. Soc. **97**, 4176 (1975)
Chapman, D.J.: In: The Biology of Blue-Green Algae; Whitton, B.A. (ed.), p. 162. Berkeley: University of California Press 1973
Chapman, D.J., Cole, W.J., Siegelman, H.W.: J. Am. Chem. Soc. **89**, 5976 (1967)
Chapman, D.J., Cole, W.J., Siegelman, H.W.: Biochem. J. **105**, 903–905 (1967)
Chen, C.H., Berns, D.S.: Biophys. Chem. **8**, 191–202 (1978)
Chen, C.H., Berns, D.S.: Phys. Chem. **82**, 2781–2787 (1978a)
Chen, S.S., Berns, D.S.: J. Membrane Biol. **47**, 113–127 (1979)
Choussy, M., Barbier, M.: Helv. Chim. Acta **58**, 2651–2661 (1975)
Clayton, R.K., Sistrom, W.R.: The Photosynthetic Bacteria. New York: Academic Press 1978
Cohen-Bazire, G., Béguin, S., Rimon, S., Glazer, A.N., Brown, D.M.: Arch. Microbiol. **111**, 225–238 (1977)
Cole, W.J., O'hEocha, C., Moscowitz, A., Krueger, W.R.: Eur. J. Biochem. **3**, 202–207 (1967)
Cole, W.J., Chapman, D.J., Siegelman, H.W.: Biochemistry **7**, 2929 (1968)
Crespi, H.L., Smith, U.H.: Phytochemistry **9**, 205–212 (1970)
Crespi, H.L., Boucher, L.J., Norman, G.D., Katz, J.J., Dougherty, R.C.: J. Am. Chem. Soc. **89**, 3642 (1967)
Crespi, H.L., Smith, U., Katz, J.J.: Biochemistry **7**, 2232 (1968)
Csatorday, K.: Biochim. Biophys. Acta **504**, 341 (1978)
Cullen, D.L., Meyer, E.F. jr., Eivazi, F., Smith, K.M.: J. C. S. Perkin II **1978**,259
Dale, R.E., Teale, F.W.J.: Photochem. Photobiol. **12**, 99 (1970)
Diakoff, S., Scheibe, J.: Plant Physiol. **51**, 382 (1973)
Dobler, S.D., Dover, K., Laves, K., Binder, A., Zuber, H.: J. Mol. Biol. **71**, 785–787 (1972)
Eder, J., Wagenmann, R., Rüdiger, W.: Immunochemistry **15**, 315–321 (1978)
Emerson, R.: Annu. Rev. Plant Physiol. **9**, 1–24 (1958)
Erokhina, L.G., Krasnovskii, A.A.: Biologya (Engl. transl.) **2**, 442–450 (1968)

Erokhina, L.G., Krasnovskii, A.A.: Molekulyarnaya Biologiya (Engl. transl.) **8**, 517–523 (1974)
Evstigneev, V.B., Bekasova, O.D.: Biofizika **17**, 997–1006 (1972)
Falk, H., Grubmayr, K.: Monatsh. Chem. **110**, 1237 (1979)
Falk, H., Höllbacher, G.: Monatsh. Chem. **109**, 1429 (1978)
Falk, H., Neufingerl, F.: Monatsh. Chem. **110**, 987 (1979)
Falk, H., Thirring, K.: Tetrahedron **37**, 761–766 (1981)
Falk, H., Grubmayr, K., Thirring, K.: Z. Naturforsch. **33b**, 924 (1978a)
Falk, H., Grubmayr, K., Haslinger, E., Schlederer, T., Thirring, K.: Monatsh. Chem. **109**, 1451–1473 (1978b)
Falk, H., Grubmayr, K., Neufingerl, F.: Monatsh. Chem. **110**, 1127 (1979)
Falk, H., Höllbacher, G., Hofer, O., Müller, N.: Monatsh. Chem. **112**, 391–403 (1981)
Frackowiak, D., Grabowski, J.: Photosynthetica **5**, 146 (1971)
Frackowiak, D., Skowron, A.: Photosynthetica **12**, 76 (1978)
Frackowiak, D., Fiksinski, K., Grabowski, J.: Photosynthetica **9**, 185 (1975)
Frackowiak, D., Grabowski, J., Manikowski, H.: Photosynthetica **10**, 204–207 (1976)
Frank, G., Sidler, W., Widmer, H., Zuber, H.: Hoppe-Seyler's Z. Physiol. Chem. **359**, 1491 (1978)
Freidenreich, P., Apell, G.S., Glazer, A.N.: J. Biol. Chem. **253**, 212–219 (1978)
Friedrich, J., Scheer, H., Zickendraht-Wendelstadt, B., Haarer, D.: J. Am. Chem. Soc. **103**, 1030–1035 (1981a)
Friedrich, J., Scheer, H., Zickendraht-Wendelstadt, B., Haarer, D.: J. Chem. Phys. **74**, 2260–2266 (1981b)
Fu, E., Friedman, L., Siegelman, H.W.: Biochem. J. **179**, 1 (1979)
Fuhrhop, J.-H., Wasser, P.K.W., Subramanian, J., Schrader, U.: Liebigs Ann. Chem. **1974**, 1450
Fujimori, E., Pecci, J.: Biochim. Biophys. Acta **221**, 132 (1970)
Fujita, Y., Hattori, A.: J. Biochem. (Tokyo) **51**, 89 (1962)
Fujita, Y., Hattori, A.: Plant Cell Physiol. **3**, 209–220 (1962)
Fujiwara, T.: Biochem. (Tokyo) **43**, 195–203 (1956)
Gantt, E.: Bio Science **25**, 781–788 (1975)
Gantt, E.: In: Biochemistry and Physiology of Protozoa; Levandowsky, M., Hutner, S.H. (eds.), 2nd ed., pp. 121–137. New York: Academic Press 1979
Gantt, E.: Ann. Rev. Plant Physiol. **32**, 327–347 (1981)
Gantt, E., Conti, S.F.: J. Cell. Biol. **29**, 423 (1966)
Gantt, E., Lipschultz, C.A.: Biochemistry **13**, 2960–66 (1974)
Gantt, E., Lipschultz, C.A.: J. Phycol. **13**, 185–192 (1977)
Gantt, E., Edwards, M.R., Provasoli, L.: J. Cell. Biol. **48**, 280–290 (1971)
Gantt, E., Lipschultz, C.A., Grabowski, J., Zimmermann, K.: Plant Physiol. **63**, 615–620 (1979)
Gautron, R., Jardon, P., Petrier, C., Choussy, M., Barbier, M., Vuillaume, M.: Experientia **32**, 1100 (1976)
Gendel, S., Ohad, I., Bogorad, L.: Plant Physiol. **64**, 786–790 (1979)
Ginsburg, R., Lazaroff, N.: J. Gen. Microbiol. **75**, 1–9 (1973)
Glazer, A.N.: Mol. Cell. Biochem. **18**, 125–140 (1977)
Glazer, A.N., Bryant, D.A.: Arch. Microbiol. **104**, 15–22 (1975)
Glazer, A.N., Cohen-Bazire, G.: Arch. Microbiol. **104**, 29–32 (1975)
Glazer, A.N., Hixson, C.S.: J. Biol. Chem. **250**, 5487 (1975)
Glazer, A.N., Hixson, C.S.: J. Biol. Chem. **252**, 32–42 (1977)
Glazer, A.N., Cohen-Bazire, G., Stanier, R.Y.: Arch. Microbiol. **80**, 1 (1971a)
Glazer, A.N., Cohen-Bazire, G., Stanier, R.Y.: Proc. Natl. Acad. Sci. USA **68**, 3005 (1971b)
Glazer, A.N., Fang, S., Brown, D.M.: J. Biol. Chem. **248**, 5679–5685 (1973)
Glazer, A.N., Hixson, C.S., De Lange, R.J.: Anal. Biochem. **92**, 489 (1979a)
Glazer, A.N., Williams, R.C., Yamanaka, G., Schachman, H.K.: Proc. Natl. Acad. Sci. USA **76**, 6161–6166 (1979b)
Gossauer, A., Hinze, R.-P.: J. Org. Chem. **43**, 283 (1978)
Gossauer, A., Hirsch, W.: Liebigs Ann. Chem. **1974**, 1496–1513 (1974)
Gossauer, A., Weller, J.-P.: J. Am. Chem. Soc. **100**, 5928 (1978)
Gossauer, A., Blacha-Puller, M., Zeisberg, R., Wray, V.: Liebigs Ann. Chem. **1981**, p. 342–346

Gossauer, A., Hinze, R.P., Kutschan, R.: Chem. Ber. **114**, 132–146 (1981)
Govindjee: Bioenergetics of Photosynthesis. New York: Academic Press 1975
Grabowski, J., Gantt, E.: Photochem. Photobiol. **28**, 39–45 (1978a)
Grabowski, J., Gantt, E.: Photochem. Photobiol. **28**, 47–54 (1978b)
Grabowski, J., Lipschultz, C.A., Gantt, E.: Plant Physiol., submitted for publication
Gray, B.H., Gantt, E.: Photochem. Photobiol. **21**, 121–128 (1975)
Gray, B.H., Lipschultz, C.A., Gantt, E.: J. Bacteriol. **116**, 471–478 (1973)
Gray, B.H., Cosner, J., Gantt, E.: Photochem. Photobiol. **24**, 299–302 (1976)
Grombein, S., Rüdiger, W., Zimmermann, H.: Hoppe-Seyler's Z. Physiol. Chem. **356**, 1709–1714 (1975)
Gysi, J., Zuber, H.: FEBS Lett. **48**, 209–213 (1974)
Gysi, J., Zuber, H.: FEBS Lett. **68**, 49–54 (1976)
Hackert, M.L., Abad-Zapatero, C., Stevens, S.E. jr., Fox, J.L.: J. Mol. Biol. **111**, 365–369 (1977)
Harnischfeger, G., Codd, G.A.: Biochim. Biophys. Acta **502**, 507–513 (1978)
Harris, J.U., Berns, D.S.: J. Mol. Evol. **5**, 153–163 (1975)
Hartmann, K.M., Haupt, W.: In: Biophysik; Hoppe, W., Lohmann, W., Markl, H., Ziegler, H. (eds.), p. 449. Berlin-Heidelberg-New York: Springer 1978
Hattori, A., Crespi, H.L., Katz, J.J.: Biochemistry **4**, 1225–1238 (1965)
Haury, J.F., Bogorad, L.: Plant Physiol. **60**, 835–839 (1977)
Haxo, F.T.: In: Comparative Biochemistry of Photoreactive Systems; Allen, M.B. (ed.), pp. 339–360. New York: Academic Press 1960
Holzwarth, A.R., Lehner, H., Braslavsky, S.E., Schaffner, K.: Liebigs Ann. Chem. **1978**, 2002
Hudson, M.F., Smith, K.M.: Chem. Soc. Rev. **4**, 363 (1975)
Ilani, A., Berns, D.S.: Biochem. Biophys. Res. Comm. **45**, 1423–1430 (1971)
Iso, N., Mizuno, H., Saito, T., Nitta, N., Yoshizaki, K.: Bull. Chem. Soc. Jap. **56**, 2892 (1977)
IUPAC: Nomenclature on Tetrapyrroles. Pure Appl. Chem. **51**, 2251 (1979)
Jung, J., Song, P.-S., Paxton, R.J., Edelstein, M.S., Swanson, R., Hazen, E.E. jr.: Biochemistry **19**, 24–32 (1980)
Kessel, M., MacColl. R., Berns,D.S., Edwards, M.R.: Can. J. Microbiol. **19**, 831–836 (1973)
Kililea, S.D., O'Carra, P.: Biochem. J. **110**, 14–15P (1968)
Klein, G., Rüdiger, W.: Liebigs Ann. Chem. **1978**, 267
Klein, G., Rüdiger, W.: Z. Naturforsch. **34c**, 192–195 (1979)
Klein, G., Grombein, S., Rüdiger, W.: Hoppe-Seyler's Z. Physiol. Chem. **358**, 1077–1079 (1977)
Knapp, J.A., Pace, C.N.: Biochemistry **13**, 1289–1294 (1974)
Kobayashi, T., Degenkolb, E.O., Behrson, R., Rentzepis, P.M., MacColl, R., Berns, D.S.: Biochemistry **18**, 5073 (1979)
Koller, K.P., Wehrmeyer, W., Moerschel, E.: Eur. J. Biochem. **91**, 57 (1978)
Köst, H.-P.: Private communication (1980)
Köst, H.-P., Rüdiger, W., Chapman, D.J.: Liebigs Ann. Chem. **1975**, 1582 (1975)
Köst-Reyes, E., Köst, H.-P.: Eur. J. Biochem. **102**, 83–91 (1979)
Köst-Reyes, E., Köst, H.-P., Rügider, W.: Liebigs Ann. Chem. **1975**, 1594
Kotera, A., Saito, T., Iso, N., Mizuno, H., Taki, N.: Bull. Chem. Soc. Jap. **48**, 1176–1179 (1975)
Krauss, C., Scheer, H.: Angew. Chem. **94**, in press (1982)
Krauss, C., Bubenzer, C., Scheer, H.: Photochem. Photobiol. **30**, 473–477 (1979)
Kufer, W.: Diplomarbeit, Universität München, 1977
Kufer, W., Scheer, H.: Hoppe-Seyler's Z. Physiol. Chem. **360**, 935–956 (1979)
Kufer, W., Scheer, H.: Angew. Chem. **94**, in press (1982)
Kufer, W., Holzwarth, A.R., Scheer, H.: to be published
Kumbar, M., MacColl, R.: Res. Commun. Chem. Pathol. Pharmacol. **11**, 627–637 (1975)
Lagarias, J.C., Glazer, A.N., Rapoport, H.: J. Am. Chem. Soc. **101**, 5030–5037 (1979)
Lagarias, J.C., Rapoport, H.: J. Am. Chem. Soc. **102**, 4821–4830 (1980)
Lagarias, J.C.: Private communication, in part presented at the Europ. Symp. on Photomorphogenesis, Bischofsmais, Bavaria 1981
Land, E.J.: Photochem. Photobiol. **29**, 483–487 (1979)
Langer, E., Lehner, H., Rüdiger, W., Zickendraht-Wendelstadt, B.: Z. Naturforsch. **35c**, 367–375 (1980)

Lazaroff, N.: In: The Biology of Blue-Green Algae; Carr, N.G., Whitton, B.A. (eds.), pp. 279–319 Oxford: Blackwell 1973
Lazaroff, N., Schiff, J.: Science **137**, 603–604 (1962)
Lee, J.J., Berns, D.S.: Biochem. J. **110**, 465 (1968)
Lehner, H., Scheer, H.: Unpublished (1981)
Lehner, H., Braslavsky, S.E., Schaffner, K.: Liebigs Ann. Chem. **1978**, 1990 (1978a)
Lehner, H., Braslavsky, S.E., Schaffner, K.: Angew. Chem. **90**, 1012–1013 (1978b)
Lehner, H., Riemer, W., Schaffner, K.: Liebigs Ann. Chem. **1979**, p. 1798
Lehner, H., Krauss, C., Scheer, H.: Z. Naturforsch. **36c**, 735–738 (1981)
Ley, A.C., Butler, W.L., Bryant, D.A., Glazer, A.N.: Plant Physiol. **59**, 974 (1977)
Lightner, D.A.: Photochem. Photobiol. **26**, 427 (1977)
Lipschultz, C.A., Gantt, E.: Biolchemistry **20**, 3371–3376 (1981)
Lotter, H., Klein, G., Rüdiger, W., Scheer, H.: Tetrahedron Lett. **1977**,2317
Lundell, D.J., Williams, R.C., Glazer, A,N.: J. Biol. Chem. **256**, 3580–3592 (1981)
MacColl, R., Berns, D.S.: Arch. Biochem. Biophys. **156**, 161–167 (1973)
MacColl, R., Berns, D.S.: Photochem. Photobiol. **27**, 343–349 (1978)
MacColl, R., Berns, D.S.: Trends Biochem. Sci. **4**, 44–47 (1979)
MacColl, R., Berns, D.S.: Biochem. Biophys. Res. Comm. **90**, 849–855 (1979)
MacColl, R., Berns, D.S., Koven, N.L.: Arch. Biochem. Biophys. **146**, 477–482 (1971a)
MacColl, R., Lee, J.J., Berns, D.S.: Biochem. J. **122**, 421–426 (1971b)
MacColl, R., Edwards, M.R., Mulks, M.H., Berns, D.S.: Biochem. J. **141**, 419–425 (1974)
MacColl, R., Berns, D.S., Gibbons, O.: Arch. Biochem. Biophys. **177**, 265–275 (1976)
MacColl, R., Csatorday, K., Berns, D.S., Traeger, E.: Biochemistry **19**, 2817–2820 (1980)
MacDonagh, A.F.: In: The Porphyrins; Dolphin, D. (ed.), Vol. VI, pp. 294–492. New York: Academic Press 1979
Manitto, P., Monti, D.: Experientia **28**, 379 (1972)
Marmé, D.: Annu. Rev. Plant Physiol. **28**, 173 (1977)
Mitrakos, K., Shropshire, W. jr.: Phytochrome. New York: Academic Press 1972
Mohr, H.: Lectures on Photomorphogenesis. Berlin-Heidelber-New York: Springer 1972
Mörschel, E., Wehrmeyer, W.: Arch. Mikrobiol. **105**, 153–158 (1975)
Mörschel, E., Wehrmeyer, W.: Arch. Microbiol. **113**, 83–89 (1977)
Mörschel, E., Koller, K.P., Wehrmeyer, W., Schneider, H.: Cytobiol. **16**, 118 (1977)
Mörschel, E., Koller, K.P., Wehrmeyer, W.: Arch. Microbiol. **125**, 43–52 (1980)
Moskowitz, A., Krueger, W.L., Kay, I.T., Skews, G., Bruckenstein, S.: Proc. Natl. Acad. Sci. USA **52**, 1190 (1964)
Muckle, G., Rüdiger, W.: Z. Naturforsch. **32c**, 957 (1977)
Muckle, G., Otto, J., Rüdiger, W.: Hoppe-Seyler's Z. Physiol. Chem. **359**, 345–355 (1978)
Neufeld, G.J., Riggs, A.F.: Biochim. Biophys. Acta **181**, 234 (1969)
Nichols, K.E., Bogorad, L.: Bot. Gaz. **124**, 85–93 (1962)
O'Carra, P.: Biochem. J. **119**, 2P–3P (1970)
O'Carra, P.: In: Porphyrins and Metalloporphyrins; Smith, K.M. (ed.), pp. 123–157. Amsterdam: Elsevier 1975
O'Carra, P., O'hEocha, C.: Phytochemistry **5**, 993 (1966)
O'Carra, P., O'hEocha, C.: In: Chemistry and Biochemistry of Plant Pigments; Goodwin, T.W. (ed.), 2 nd, pp. 328–376. New York: Academic Press 1976
Oesterhelt, D.: In: Energy Transformation of Biological Systems. Ciba Found. Symp. (New Ser.) **31**, 147–167 (1975)
Ohad, I., Clayton, R.K., Bogorad, L.: Proc. Natl. Acad. Sci. USA **76**, 5655–5659 (1979)
Ohad, I., Schneider, H.-J.A.W., Gendel, S., Bogorad, L.: Plant Physiol. **65**, 6–12 (1980)
O'hEocha, C., O'Carra, P.: J. Am. Chem. Soc. **83**, 1091–1093 (1961)
Ohki, K., Fujita, Y.: Plant Cell Physiol. **19**, 7 (1978)
Ohki, K., Fujita, Y.: Plant Cell Physiol. **20**, 1341–1348 (1979a)
Ohki, K., Fujita, Y.: Plant Cell Physiol. **20**, 483–490 (1979b)
Pasternak, R., Wagnière, G.: J. Am. Chem. Soc. **101**, 1662 (1979)

Pecci, J., Fujimori, E.: Biochim. Biophys. Acta **131**, 147–153 (1967)
Pecci, J., Fujimori, E.: Biochim. Biophys. Acta **154**, 332–341 (1968)
Petrier, C., Jardon, P., Dupuy, C., Gautron, R.: J. Chim. Phys. **76**, 97–103 (1979)
Petrier, C., Dupuy, C., Jardon, P., Gautron, R.: Tetrahedron Lett. **22**, 855–858 (1981)
Petrier, C., Kufer, W., Scheer, H., Gautron, R.: To be published
Porter, G., Tredwell, L.J., Searle, G.F.W., Barber, J.: Biochim. Biophys. Acta **501**, 232 (1978)
Pratt, L.H.: Photochem. Photobiol. **27**, 81 (1978)
Pullin, C.A., Brown, R.G., Evans, E.H.: FEBS Lett. **101**, 110–112 (1979)
Raftery, M.A., O'hEocha, C.: Biochem. J. **94**, 166 (1965)
Redlinger, T., Gantt, E.: Proc. 5 th. Int. Congr. Photosynth., Kassandra-Halkidiki, Greece
Rigbi, M., Rosinski, J., Siegelman, H.W., Clark, J.: Proc. Natl. Acad. Sci. USA **77**, 1961–1965 (1980)
Robinson, B.L., Miller, J.H.: Physiol. Plant **23**, 461–472 (1970)
Rüdiger, W.: Hoppe-Seyler's Z. Physiol. Chem. **350**, 1921 (1969)
Rüdiger, W.: Fortschr. Chem. Org. Naturst. **29**, 60 (1971)
Rüdiger, W.: Ber. Deutsch. Bot. Ges. **88**, 125 (1975)
Rüdiger, W.: In: Plant Growth and Light Perception; Deutsch, B., Deutsch, B.I., Gyldenholm, A.O. (eds.), pp. 53–74. University of Aarhus 1978
Rüdiger, W.: Ber. Deutsch. Bot. Ges. **92**, 413–426 (1979)
Rüdiger, W.: Struct. Bond. **40**, 101 (1980)
Rüdiger, W., O'Carra, P.: Eur. J. Biochem. **7**, 509 (1969)
Rüdiger, W., O'Carra, P., O'hEocha, C.: Nature **215**, 5109 (1967)
Rüdiger, W., Brandlmeier, T., Blos, I., Goassauer, A., Weller, J.P.: Naturforsch. **35c**, 763–769 (1980)
Saito, T., Iso, N., Mizuno, H., Kitamura, I.: Bull. Chem. Soc. Jap. **51**, 3471–3474 (1976)
Salahuddin, A., Tanford, C.: Biochemistry **9**, 1342–1347 (1970)
Scheer, H.: In: Plant Growth and Light Perception; Deutsch, B., Deutsch, B.I., Gyldenholm, A.O. (eds.), pp. 25–52. University of Aarhus 1978
Scheer, H.: Angew. Chem. **93**, 230–250 (1981). – Angew. Chem. Int. Ed. (english) **20**, 241–261 (1981)
Scheer, H., Bubenzer, C.: Unpublished
Scheer, H., Krauss, C.: Photochem. Photobiol. **25**, 311 (1977)
Scheer, H., Kufer, W.: Z. Naturforsch. **32c**, 513 (1977)
Scheer, H., Formanek, H., Rüdiger, W.: Z. Naturforsch. **34c**, 1085–1093 (1979)
Scheer, H., Svec, W.A., Cope, B.T., Studier, M.H., Scott, R.G., Katz, J.J.: J. Am. Chem. Soc. **96**, 3714 (1974)
Scheer, H., Linsenmeier, U., Krauss, C.: Hoppe-Seyler's Z. Physiol. Chem. **358**, 185 (1977)
Scheibe, J.: Science **176**, 1037–1039 (1972)
Schmidt, A., MacDonagh, A.F.: In: The Porphyrins; Dolphin, D. (ed.), Vol. VI, p. 258. New York: Academic Press 1978
Schmidt, A.: Private communication, in part presented at the Botaniker Tagung, Bochum 1980
Schneider, H.-J.A.W., Bogorad, L.: Z. Pflanzenphysiol. **94**, 449–459 (1979)
Schoch, S., Klein, G., Linsenmeier, U., Rüdiger, W.: Liebig Ann. Chem. **1976**, 549 (1976)
Schram, B.L., Kroes, H.H.: Eur. J. Biochem. **19**, 581 (1971)
Searle, G.F.W., Barber, J., Porter, G., Tredwell, C.J.: Biochim. Biophys. Acta **501**, 246 (1978)
Sheldrick, W.S.: J.C.S. Perkin II **1976**, 1457–1462
Shimura, S., Fujita, Y.: Mar. Biol. **31**, 121–128 (1975)
Siegelman, H.W.: Private communication, 1980
Smith, H.: Phytochrome and Photomorphogenesis. London: McGraw Hill 1975
Smith, H., Kendrick, R.E.: In: Chemistry and Biochemistry of Plant Pigments; Goodwin, T.W. (ed.), p. 378. New York: Academic Press 1976
Stoeckenius, W.: In: The Photosynthetic Bacteria; Clayton, R.R., Sistrom, W.R. (eds.), pp. 571–592. New York: Acadecmic Press 1978
Struckmeier, G., Thewaldt, U., Fuhrhop, J.-H.: J. Am. Chem. Soc. **98**, 278 (1976)
Sugimoto, T., Ishikawa, K., Suzuki, H.: J. Phys. Soc. Jap. **40**, 258 (1976)
Sweet, R.M., Fuchs, H.E., Fisher, R.G., Glazer, A.N.: J. Biol. Chem. **252**, 8258 (1977)

Tandeau de Marsac, N.: J. Bacteriol. **130**, 82 (1977a)

Tandeau de Marsac, N., Cohen-Bazire, G.: Proc. Natl. Acad. Sci. USA **74**, 1635 (1977b)

Teale, F.W.J., Dale, R.E.: Biochem. J. **116**, 161 (1970)

Tegmo-Larsson, I.M., Braslavsky, S.E., Nicolau, C., Schaffner, K.: Submitted for publication (1980a)

Tegmo-Larsson, I.M., Braslavsky, S.E., Schaffner, K.: Private communication (1980b)

Troxler, R.F.: Biochemistry **11**, 4235–4242 (1972)

Troxler, R.F.: In: Chemistry and Physiology of Bile Pigments; Berk, P.D., Berlin, N.I. (eds.), pp. 431–454. Washington, DC.: US Dept of Health DHEW Publ. No. (NIH) 77–1100, 1975

Troxler, R.: In preparation 1980

Troxler, R.: Private communication, 1980

Troxler, R.F., Bogorad, L.: Plant Physiol. **41**, 491–499 (1966)

Troxler, R.F., Brown, A.: Biochim. Biophys. Acta **215**, 503–511 (1970)

Troxler, R.F., Dokos, J.M.: Plant Physiol. **51**, 72–75 (1973)

Troxler, R.F., Brown, A.S., Köst, H.-P.: Eur. J. Biochem. **87**, 181 (1978)

Troxler, R.F., Brown, A.S., Brown, S.B.: J. Biol. Chem. **254**, 3411–3418 (1979)

Velde, H.H. v.d.: Biochim. Biophys. Acta **303**, 246 (1973)

Vernotte, C.: Photochem. Photobiol. **14**, 163 (1971)

Vogelmann, T.C., Scheibe, J.: Planta **143**, 233–239 (1978)

Wagenmann, R.: Dissertation, Universität München, 1977

Wagenière, G., Blauer, G.: J. Am. Chem. Soc. **98**, 7806 (1976)

Wanner, G., Köst, H.-P.: Protoplasma **102**, 97–109 (1980)

Wehrmeyer, W.: Arch. Microbiol. **71**, 367–383 (1970)

Wehrmeyer, W.: Private communication, 1979

Wehrmeyer, W.: Private communication, 1980

Weller, J.-P., Gossauer, A.: Chem. Ber. **113**, 1603–1611 (1980)

Wildman, R.B., Bowen, C.C.: J. Bacteriol. **117**, 866–881 (1974)

Williams, V.P., Glazer, A.N.: J. Biol. Chem. **253**, 202–211 (1978)

Yamanaka, G., Glazer, A.N.: Arch. Microbiol. **124**, 39–47 (1980)

Yamanaka, G., Glazer, A.N., Williams, R.C.: J. Biol. Chem. **253**, 8303–8310 (1978)

Zickendraht-Wendelstadt, B.: Dissertation, Universität München, 1980

Zickendraht-Wendelstadt, B., Friedrich, J., Rüdiger, W.: Photochem. Photobiol. **31**, 367–376 (1980)

Zilinskas, B.A., Zimmermann, B.K., Gantt, E.: Photochem. Photobiol. **27**, 587–595 (1978)

Zuber, H.: Ber. Deutsch. Bot. Ges. **91**, 459–474 (1978)

Chapter 3
Structure and Excitation Dynamics of Light-harvesting Protein Complexes

Godfrey S. Beddard[1] and Richard J. Cogdell[2]

1 Introduction

The light-absorbing pigments in photosynthetic organisms are not just randomly distributed throughout the hydrophobic interior of the photosynthetic membranes, but are combined with proteins forming very specific pigment-protein complexes. Moreover, these complexes are also highly organised within the membrane.

There are two major classes of pigment-protein complexes, those which fulfill a light-harvesting or antenna role and those which serve to "trap" the absorbed light energy, the "reaction centres." Light absorbed by the antenna complexes is transfered to the reaction centres where it is trapped in the primary photochemical redox reaction. The combination of the antenna complexes and the reaction centre (s) forms the so-called Photosynthetic Unit.

The light-harvesting pigment-protein complexes, which are the subject of this chapter, serve the same function as the reflector of a radiotelescope. In the radiotelescope there is only a small sensor device which converts the incident radio waves into an electrical signal. If there were no reflector, then the chance of a radio wave striking the detector would be very low. However, with the reflector to focus the radio waves captured over a large surface area onto the sensor, the probability of the sensor receiving a signal is greatly increased. In an analogous way, the antenna pigments serve to increase the effective cross-sectional area of absorbance of each reaction centre.

The major antenna pigments of plants, algae, and photosynthetic bacteria are the chlorophylls and bacteriochlorophylls (Fig. 1). However, there is also a range of accessory light-harvesting pigments, notably the carotenoids and the bile pigments (Fig. 2). When these accessory pigments absorb incident radiation they pass on the absorbed light energy to the chlorophyll pigments. In this way serve to extend the spectral range over which light may drive photosynthesis.

The general in-vivo organisation of both chlorophyll and bacteriochlorophyll has been excellently reviewed elsewhere [1, 2]. We therefore propose to deal with this topic in a different way. Rather than provide a comprehensive review of the structure and function of the full range of known antenna complexes found in photosynthetic organisms, we shall select a few of those complexes and discuss them in detail beginning

1 Davy Faraday Research Laboratory, Royal Institution, 21 Albemarle Street, London W1X 4BS, England
Present address: Department of Chemistry, University of Manchester, Manchester M13 9PL, England
2 Department of Botany, University of Glasgow, Glasgow G12800, Scotland

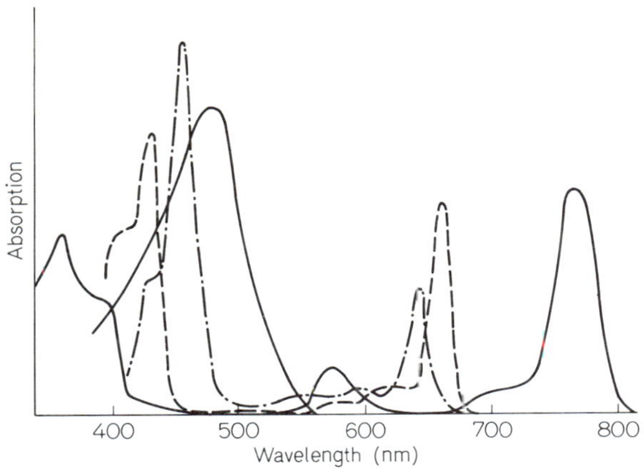

Fig. 1. Absorption spectra of various photosynthetic pigments ——— Bacteria chlorophyll; – – – – Chlorophyll *a*; – · – · – Chlorophyll *b*; ——— curve peaking at 490 nm, xanthophyll

with a general discussion of energy migration which is common to these organisms. We shall also briefly discuss the picosecond instruments used to measure fluroescence decays and excited state annihilation processes within antenna arrays.

1.1 General Discussion of Excitation Migration

The high efficiency with which energy is transported from the site of light absorption to the traps (the reaction centres) depends primarily on two factors. These are the ability photosynthetic pigments (porphyrins, carotenes, etc.) have for rapid interpigment transfer and the absence of any efficient quenchers other than the reaction centres.

When the pigments interact weakly, as is often the case, then dipole-dipole energy transfer can occur between molecules. The rate of transfer between two of these molecules, as defined by Förster [3], is proportional to (1) the angle between the emission and absorption dipoles, (2) the fluorescence rate constant and (3) the normalised area of overlap of the donor fluorescence and acceptor absorption spectra divided by the sixth power of the distance apart of the two molecules. In fact, Förster transfer has been most extensively used as a "spectroscopic ruler", i.e. as a probe of intermolecular distances [4]. If the acceptor (A) energy level is below that of the donor (D), then the energy transfer ($D \rightarrow A$) is more probable than the reverse transfer, simply because most of the fluorescence is at longer wavelengths than the corresponding absorption. If the energy levels of the donor and acceptor are very close and there is a large overlap between the absorption and emission spectra, energy can transfer to and fro in an incoherent manner between the molecules for as long as the excitation exists. If many similar molecules exist in the absence of traps, then the energy can migrate freely among them and decays with the same lifetime as from an ensemble of similar molecules in which no energy transfer occurs.

We shall consider two different cases in which Förster transfer occurs. The first is the most familiar and occurs in a homogeneous system where donor-donor transfer does not occur and the donor/acceptor ratio is $\ll 1$. Single step (i.e. irreversible) transfer occurs and the fluorescence intensity at time t, I(t) has the familiar form $I(t) = \exp(-t/\tau_0 - \gamma(t/\tau_0)^{1/2})$ where τ_0 is the donor lifetime and γ a constant dependent on

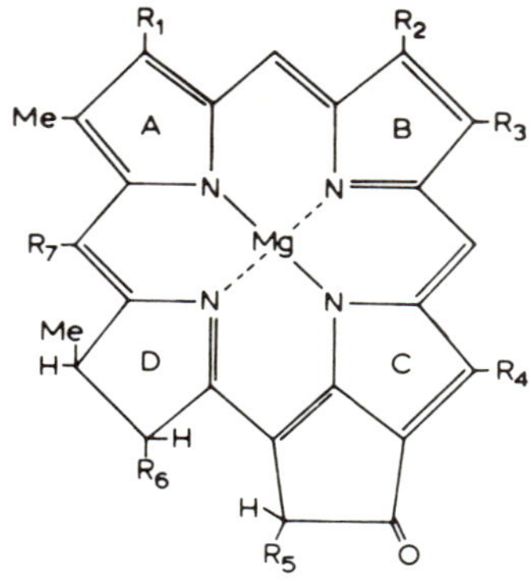

<table>
<tr><th></th><th>R_1</th><th>R_2</th><th>R_3</th><th>R_4</th><th>R_5</th><th>R_6</th><th>R_7</th></tr>
<tr><td>Chlorophyll a</td><td rowspan="4">$-CH=CH_2$</td><td>$-CH_3$</td><td rowspan="3">$-CH_2CH_3$</td><td rowspan="6">$-CH_3$</td><td rowspan="6">$-CO-O-CH_3$</td><td rowspan="2">Phytol Ester[b]</td><td rowspan="6">-H</td></tr>
<tr><td>— b</td><td>-CHO</td></tr>
<tr><td>— c_1</td><td rowspan="6">$-CH_3$</td><td rowspan="2">Free Acrylic Acid</td></tr>
<tr><td>— c_2</td><td>$-CH=CH_2$</td></tr>
<tr><td>Bacteriochlorophyll a</td><td rowspan="2">$-CO-CH_3$</td><td>$-CH_2CH_3$</td><td rowspan="2">Phytol[c] Ester[b]</td></tr>
<tr><td>— b</td><td>$=CHCH_3$</td></tr>
<tr><td>— c</td><td rowspan="3">$-\underset{OH}{CH}-CH_3$</td><td rowspan="3">Alkyl[a]</td><td rowspan="3">$-C_2H_5$</td><td rowspan="3">-H</td><td rowspan="3">Farnesol Ester[b]</td><td>$-CH_3$</td></tr>
<tr><td>— d</td><td>-H</td></tr>
<tr><td>— e</td><td>-CHO</td><td>$-CH_3$</td></tr>
</table>

A

Fig. 2. A Structures of various chlorophyll and bacteriochlorophyll derivatives occuring in photosynthetic organisms. **B** Structures of phycocyanobilin and phycoerythrobilin. **C** Structure of two photosynthetic carotenoids

the donor and acceptor characteristics and concentrations [3]. An example of such a system, of which many are known, is rhodamine quenching sodium fluorescein fluorescence in fluid solution.

The second case occurs where donor-donor transfer rate is high and the donor/acceptor ratio $\gg 1$. The energy now migrates among donors until an acceptor trap is found or fluorescence occurs. The equations describing this migration are not fully solved and, furthermore, this high donor to acceptor ratio is probably the condition under which energy migration occurs in the photosynthetic unit. However, in the limit of very rapid donor-donor transfer, all donors have an equal probability of being excited for all times of interest and the excited donor population P(t) decays exponentially in time [5]. The fluorescence intensity is $I(t) = \exp(-t/\tau - C_a \Sigma_i k_i t)$ where the acceptor concentration is C_a and the donor-acceptor transfer rates from a donor to all acceptors (i) is $\Sigma_i k_i$.

In the second case we have to make a further distinction about the acceptors, or traps. If on arrival at the trap – in the photosynthetic organism the reaction centre – there is only, say, a 10% probability of trapping, the energy will then be forced back into the bulk pigments (the donors) and repeated revisting of the trap occurs until,

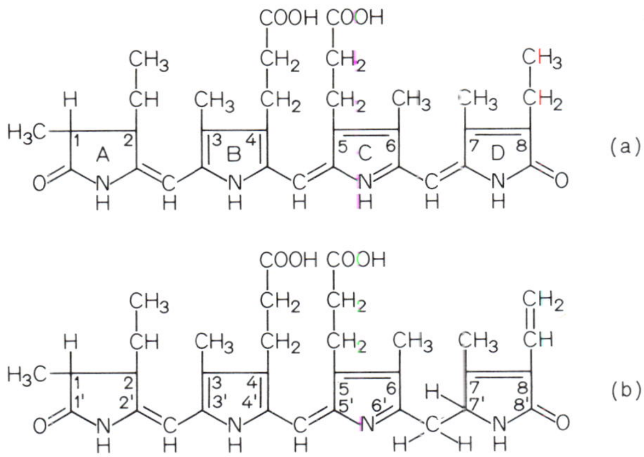

Structures of (a) phycocyanobilin and (b) phycoerythrobilin

Fig. 2B

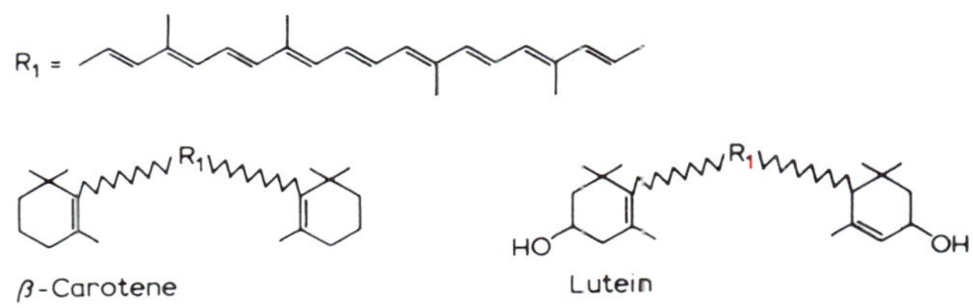

Fig. 2C

finally, this energy is irreversibly trapped. In this situation the trapping is limited by the trap concentration, i.e. reaction limited. Conversely, if trapping becomes irreversible on first arrival to the trap, then the rate of trapping is limited by diffusion. Two further consequences arise from these two limits, firstly, if reaction-controlled, the migration must be much faster than for the diffusion-controlled case if trapping is to occur during the singlet state lifetime. Secondly, if trapping is reaction limited exponential fluorescence kinetics are expected, but if diffusion controlled, this may not necessarily be so. At short times the variation in donor to trap energy transfer rates leads to a nonexponential part of the fluorescence decay. At longer times diffusion governs the fluorescence decay. An additional complicating effect also occurs at short times, this is the initial nondiffusive nature of the excitation migration and leads to a "diffusion coefficient" which varies with time.

Calculations on irregular two-dimensional lattices illustrate this complex fluorescence decay [5a]. Figure 3 shows some calculated fluorescence decays at different donor concentrations but at fixed trap (or quencher) concentration. The data were obtained by performing repeated random walks on an irregular lattice of 300 points, each of 10Å diameter among ~ 500,000 "empty" points. The walks were continued without restriction until fluorescence occurred or the trap was encountered. Intermolecular transfer rates were calculated assuming Förster energy transfer. If the mean square displacement of the random walk multiplied by the probability of fluorescence is plotted versus time as shown in Fig. 4, it is seen how small a distance the excitation

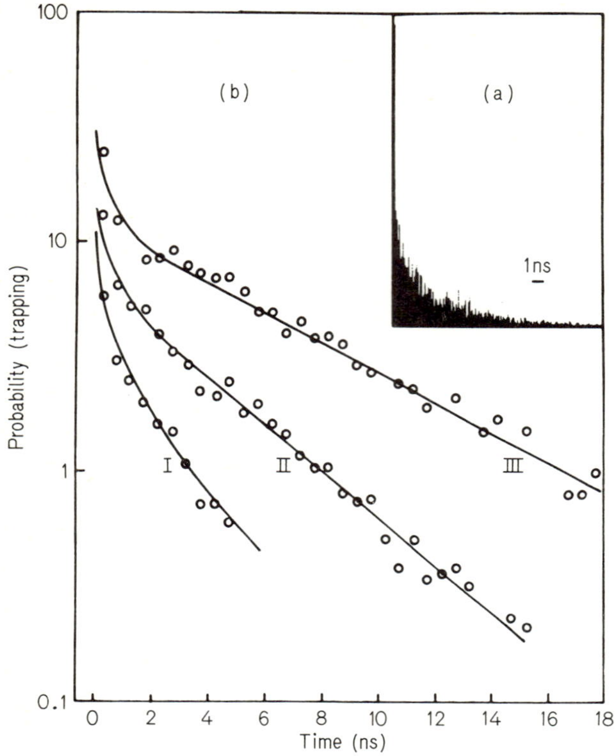

Fig. 3. Typical decay obtained by simulated energy migration; *(a)* composite of 3000 individual walks. Full time scale: 18 ns, intensity: arbitrary units. *(b)* Semilogarithmic plots of the probability of trapping as a function of time at three donor concentrations; *curve I*: 0.25 mol nm^{-2}; *curve II*: 0.1875 mol nm^{-2}; *curve III*: 0.125 mol nm^{-2}. Each curve is a composite of over 2000 data points

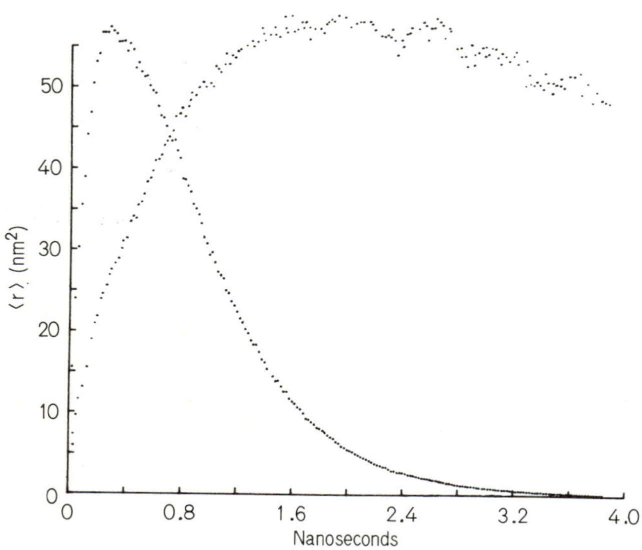

Fig. 4. Calculated mean square displacement of excitation migration by simulated energy migration. The packing density is 0.025 mol nm^{-2}. To illustrate the effect of fluorescence on the distance migrated by the excitation, the same displacement curve has been multipled by a 500 ps decay, curve peaking at 400 ps and by a 5 ns decay, curve peaking at approx. 1.6 ns. This latter curve has been scaled down by dividing by 5

moves from its original site before being trapped when fluorescence competes with trapping.

Calculations by Haan and Zwanzig [6] and Gochanour et al. [7] have shown how complex fluorescence decays may also be expected in three-dimensional systems. The effects are less pronounced than in two dimensions and are most noticeable at lower concentrations than those estimated for chlorophyll in the PSU. These transient non-exponential effects would be unlikely to last for more than 50 ps. This is perhaps unfortunate since fluorescence lifetime data cannot easily be used to distinguish between reaction-limited and diffusion-limited quenching processes.

The recently developed transient grating technique [8] could be used to determine the diffusion coefficient of the energy migration directly. In this experiment a picosecond laser pulse is split into two parts, then made to recombine to form interference fringes inside the sample, as in a "Young's Slit" experiment. A third part of the beam then probes the amplitude of this grating as it decreases, due to energy migration or on a much slower time scale (a few hundred picoseconds), by fluorescence. The grating is initially produced by molecules of the sample absorbing where the light interferes constructively, and light and dark bands of sinusoidal intensity produce a grating. As molecular events take place, such as bulk diffusion of the excited molecules by energy migration, the diffraction grating will smear out in space and time leading to a change in intensity of the monitoring beam that is diffracted off this transient grating. The manner in which the intensity of the diffracted beam decays in time is related to the molecular events. In this way, the diffusion coefficient could be measured directly, although at present no experiments appear to have been performed in vivo.

An alternative method would be to measure fluorescence yields and lifetimes with added quenchers present. Experiments have been performed for many years in which the reaction centres are closed artificially by light and chemical additions, and a linear "Stern-Volmer" relationship has been found between yield and trap concentrations, but there are still two unknowns, the rate of trapping and rate of energy migration or diffusion. If artificial traps could be introduced, whose rate of quenching with chlorophyll could be independently determined, then the rate of diffusion could be measured, and in experiments with active reaction centres the rate of trapping also.

We have considered, so far, what might happen in a very simple system and the picture that presents itself is of the excitation moving randomly from one pigment to another among the pigment-protein complexes until the reaction centre is discovered. Because the transfer rate depends upon mutual orientation of the dipoles, a preferential alignment can be used to increase the overall rate of transfer to the reaction centre. This effect has been well illustrated in model calculations by Seely [9] in a regular lattice, where changing orientation of only a few molecules, in crucial positions, radically changed the ability of the reaction centre to trap the excitation. Such advantageous orientations may aid transfer at the junction between two proteins, where the extra distance (and hence lower rate of transfer and higher probability of fluorescence) can be compensated for by fovourable orientation. Similarly, the use of pigments with slightly different absorption spectra can act in a number of ways to encourage energy to move to the reaction centre. For instance, short-wavelength absorbing pigments can be placed farthest from the reaction centre and, whilst energy can migrate among these pigments, once transfer to the next pigment type (at longer wavelength and lower energy) occurs the process has become, for all practical purposes, irreversible, compared to the forward reaction. Hence pigments are arranged functionally as a funnel, down which the energy runs to be trapped. Structures like this are seen in *Rhodopseudomonas sphaeroides* and in the phycobilisomes of lower algae [10-12].

Alternatively, pigments such as chl-b in chl-a/b proteins of green plants can be isolated from one another by chl-a molecules. The energy absorbed by the short-wavelength chl-b is almost always transferred in the first step to the chl-a lattice where it remains until trapping or fluorescence occurs. This type of structure allows energy on the chl-a lattice to visit (possibly) any one of many reaction centres connected to this pigment bed. In the funnel model, such as with phycobilisomes, only one reaction centre can be reached for each free antennae system, unless a group of reaction centres is connected at the end of the funnel, as with *Rps. sphaeroides*.

1.2 Coherence

Not all of the photosynthetic pigments interact weakly, as is shown by exciton splittings in the *Chlorobium limicola* and *Rps. sphaeroides* adsorption spectra. But this is a static, i.e. time-independent process, occurring when dimers or trimers, etc. are formed. In molecular solids, coherent energy transfer can occur, such as by Frenkel excitons, and the energy is delocalised. The exciton is described by a series of wave functions analogous to lattice modes in a crystal. Consequently, superposition of several modes will form a localised wave packet that propagates through the crystal which, since the excitation is coupled to the lattice, creates a distortion which travels adiabatically with the electronic excitation. All the various couplings between the lattice and the exciton have to be considered when describing the exciton. In the Frenkel limit, these result in a partial localisation of the excitation but still allows it to propagate as a coherent wave packet, provided that the phase relationship between states lasts for times long compared to the time for intermolecular energy transfer. If this phase relationship is lost rapidly, compared to the intermolecular transfer time, then the energy transfer and subsequent migration is incoherent. At room temperature, intra and intermolecular interactions destroy the phase memory of the excitation within a few picoseconds and energy transfer is incoherent; this is most probably the case in photosynthetic organisms, even when there is a strong coupling between the molecules as detected by absorption spectroscopy.

2 Experimental Methods

In the observation of the time evolution of the fluorescence from the antennae, the streak camera-Nd laser system and single photon counting have been widely used techniques. These instruments are described briefly below and block diagrams are shown in Figs. 5 and 6, respectively.

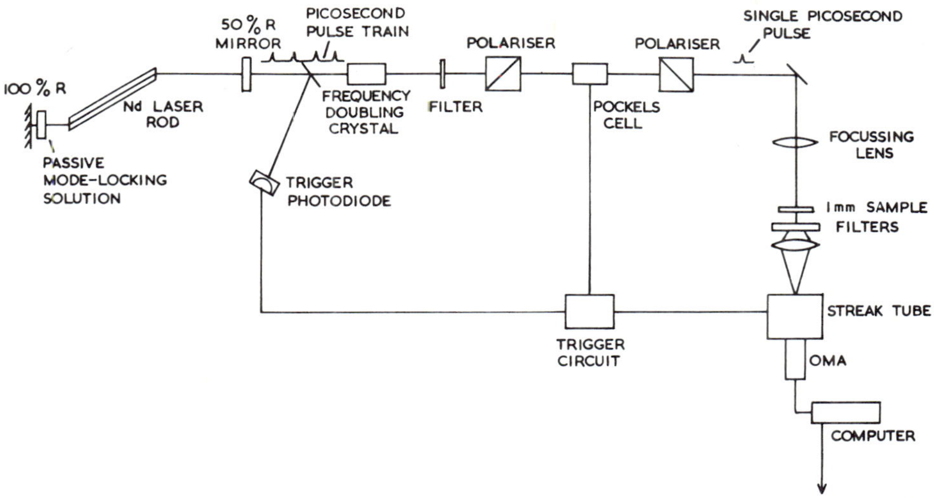

Fig. 5. Nd laser and streak camera set up for measuring fluorescence lifetimes

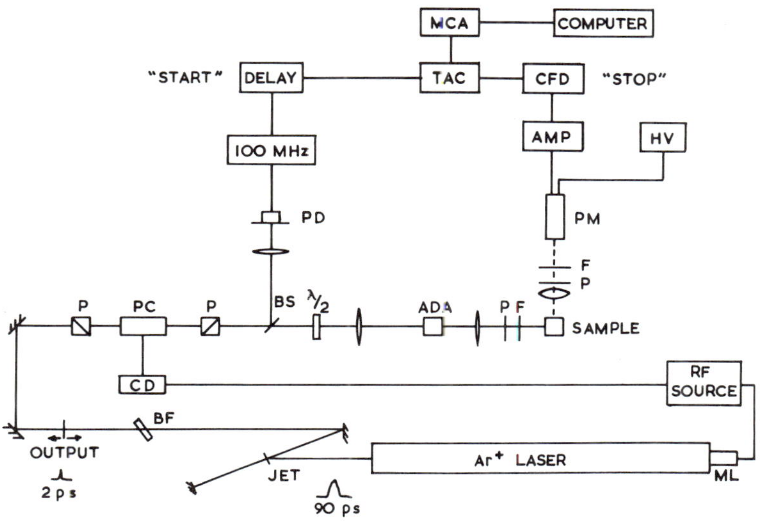

Fig. 6. Schematic of the sync-pump photon counting fluorescence lifetime instrument: *ML*, mode locker on the Ar^+ ion laser; *JET*, dye laser ethylene glycol jet stream; *BF*, biregringent wavelength tuning element; *out*, dye laser output mirror on a translation stage; *P*, polarisers; *PC*, pockels cell driven by pulses from the mode locker via *CD*, countdown logic; *BS*, beam splitter; *ADA*, frequency doubling crystal; *P*, polariser; *F*, filter; *PD*, timing photodiode; *PM*, photomultipler; *AMP*, Amplifier; *CFD*, and *100 MHz* are discriminators; *TAC*, time to amplitude converter

2.1 Streak Camera and Neodymium Laser

In the neodymium laser, the laser emission is at approximately 1064 nm, and the laser is mode-locked by a saturable absorber placed inside the laser cavity. This compound is bleached (becomes ~ 100% transmitting) by an intense burst of light from the laser rod; less intense bursts are absorbed and so are not amplified. The pulse is repeatedly amplified on its passage up and down the laser cavity until gain saturation occurs. Laser pulses are produced at the output mirror, one for each trip around the laser cavity, and a train of pulses typically of 5–8 ps duration are produced, each separated by the cavity round trip time, the intensity envelope of the pulses following that of the pumping flash lamps. The saturable absorber acts not only as a switch, but its rapid recovery from bleaching shortens the laser pulse and also locks the longitudinal modes in the laser cavity together since only those modes present when the switch is open (i.e. absorption saturated) are amplified. The 1064 nm is not useful for exciting chlorophyll *a* or bacteriochlorophyll *a*, so the laser output is frequency doubled to 530 nm, one pulse from the pulse train can also be selected by a Pockels cell, placed between crossed polarisers. In many Nd lasers, several spatial modes (transverse modes) exist across the laser beam leading to a non uniform intensity cross section, and several hot spots may be present. The pulse intensity can be high ~ 10^{16} photons/cm^2 /pulse and can lead to unexpected effects such as excited state annihilation which alters the fluorescence lifetime.

The fluorescence is collected by high f-number optics and focussed onto the entrance slit of the streak camera. In this device, time is represented by a linear displacement of electrons across a phosphor screen; rather like in an oscilloscope. In the streak camera, the electron source is a photocathode, excited by the laser pulse rather than an electron gun. To achieve high time resolution, the gain of the streak tube is kept low as this avoids electron-electron repulsion, which would spatially distort, and hence, temporarily distort the image. This image is then amplified and viewed by a photographic plate or by a vidicon-optical multichannel analyser (OMA) assembly.

Unlike most other instruments used to measure temporal information, the resolution of the streak camera becomes worse as the time scale is increased to longer times. This is because the resolution is governed by the width of the slit in front of the photocathode; this produces an image of the same horizontal length, irrespective of sweep speed. Thus, below 100 ps and down to 2 ps, the streak camera has superb resolution but, above 100 ps, the instrument function has to be taken into account, just as with photon counting which we describe next.

2.2 Single Photon Counting

Time-correlated photon counting is a repetitive technique, hence, many experiments are combined to produce one fluorescence decay (see Fig. 6). The time differences between exciting the sample and fluorescence being detected are repeatedly measured. This difference depends upon the residence time of the molecule in the excited state but, when collected and combined as an amplitude histogram, will represent the excited state decay. Conventional photon counting can be upgraded by using a picosecond light source such as from a synchronously pumped laser. The dye laser is pumped by a continuous train of 90 ps pulses from an acousto-optically mode-locked ion-laser. A standing acoustic wave, producing Debye-Sears diffraction of the laser light, modulates the loss in the ion laser cavity and thus mode-locks the laser. The mode-locked pulses are used to modulate the gain in a jet-stream dye laser having the same round trip time as pulse separation in the ion laser. When the dye laser cavity length is correctly adjusted, tunable wavelength pulses of less than 1 ps can be obtained. The pulses repetition frequency is typically 70 to 90 MHz and must be reduced by an electro-optic modulator or cavity dumper to a frequency of say 70 kHz to be useful for photon counting. The laser pulses are significantly less intense than from a solid state laser, being $\sim 10^9$ photons/pulse. Wavelengths in the 450 to 1000 nm range can be obtained using a series of laser dyes, and also frequency doubled for uv excitation.

3 Excited State Annihilation

As a result of using Nd or ruby lasers, excited state annihilation has been discovered in photosynthetic systems. These effects are discussed next.

One important effect, common to most antennae systems, is the ability of excited states to annihilate one another when, by using an extremely intense laser pulse, many molecules are excited in close proximity to one another. These effects have been observed in green plants, algae, and bacteria [13, 14, 15]. The excited singlet state (S_1) and also triplet state (T) can interact by the following reactions

$$S_1 + S_1 \longrightarrow S_o + S_1 ; S_1 + S_1 \longrightarrow T - T ; T + S_1 \longrightarrow T + T ; T + S_1 \longrightarrow S_o + T_n$$

where S_o is the ground electronic state and T_n a higher triplet. The excited states are generally those of chlorophyll molecules, in *Porphyridium cruentum* annihilation in the pigments of the phycobilisome is also possible. Either a single intense pulse of light or a pulse train can produce the effect and a severe decrease in fluorescence decay time (and fluorescence yield) occurs as a result of the quenching processes described above (see Fig. 7). When whole pulse trains are used, triplet states produced by annihilation, or directly by intersystem crossing, persist from pulse to pulse and produce additional quenching processes when later pulses excite the sample.

If we assume that the concentration of triplet states is small, the equation describing the singlet state population at a time = t is given by the equation

$$\frac{d(S_1)^*}{dt} = P(t) - k(S_1)^* - \gamma(S_1{}^*)^2 \qquad (1)$$

where k is the sum of all unimolecular loss processes from the $S_1{}^*$ singlet state and γ is the bimolecular annihilation rate. P(t) describes the evolution of the laser pulse and absorption cross section of the S_o state. If we assume P(t) is narrow in time compared to the singlet decay, as is in fact the case with a 5 ps laser pulse, then the relative fluorescence yield $\phi(I)/\phi(o)$ is given below for an initial laser intensity of I.

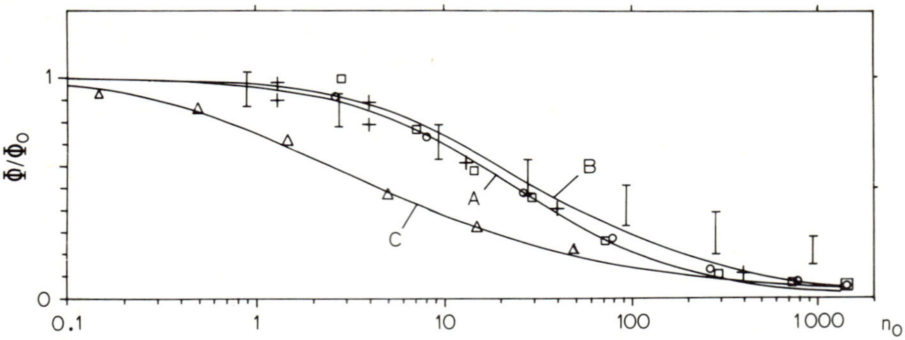

Fig. 7. Relative fluorescence yield Φ/Φ_0 versus average number of hits on the PSU, n_0. *Curve A:* Eq. (2) ($K = 2.510^9 s^{-1}$, $\gamma = 2.510^8 dm^3 m^{-1} s^{-1}$); *Curve B:* assuming transient diffusional terms are important in the quenching [17]; *Curve C:* poisson model of Mauzerall [82] Experimental points: △ Mauzerall [82], ○ Campillo et al. [63], + Geacintov and Breton [14] and FEBS Letts 69, 86 (1976), 735 and 683 nm, respectively; □ Monger et al. [30]

$$\phi(I)/\Phi(0) = \frac{k}{\gamma I} \log_e \left(1 + \frac{\gamma I}{k}\right) \qquad (2)$$

Figure 7 shows this equation plotted together with data from a number of different photosynthetic organisms. When deriving this equation it has been assumed that the excitation moves rapidly about the photosynthetic unit randomising the excitation on a time scale short compared to the annihilation. It is also assumed that the ground state depletion is negligible. This latter assumption may not be valid when high intensities, i.e. $> 10^{16}$ photons/pulse are used. A more detailed analysis of the annihilation process takes into account the various sizes of the PSU and distribution of chromophores; nevertheless the final results are still described quite closely by Eq. (2) [16].

The similarity of the quenching curve for each organism strongly suggests that there is a common quenching mechanism between them. Inferring a common pigment structure from the yield curves is not necessarily valid, however, since the fluorescence yield is an average property and several pigment structures could conceivably give indistinguishable quenching curves.

A more stringent test of the annihilation process is to measure the fluorescence decay curves directly using a streak camera. In molecular crystals, annihilation is observed and the decay follows that expected from Eq. (1), in photosynthetic organisms, where poorer signal to noise ratios are obtained, such a clear fit to a theoretical curve is not possible. In fact, it has been suggested that transient diffusional quenching kinetics [i.e. $I(t) = \exp(-at - bt\ 1/2)$] will fit the data in a more satisfactory manner [17] rather than a dual exponential expected from Eq. (1).

4 Anaerobic Photosynthetic Bacteria

When Bchl is extracted into organic solvents it shows a single absorption band in the near infrared (in 7 : 2 Acetone methanol the peak is ~ 772 nm). However, the absorption spectrum of a typical photosynthetic bacterium shows several peaks in the near infrared, usually three or even four, and moreover those absorption bands may be shifted by 100 nm or more to the red compared with Bchl in organic solvent. These adsorption bands reflect the presence of different antenna pigment-protein complexes and illustrate the strong influence the pigment-protein interactions have upon the absorption of the pigment.

The different light-harvesting complexes are usually identified by the position of their near infrared absorption bands. For example, the light-harvesting complex from *Rps. sphaeroides,* which absorbs at 800 and 850 nm, is called B800–850 (B for Bulk Bchl). We shall adhere to this convention.

Two light-harvesting complexes have received much more detailed study than the rest. One is the B800–850 complex from *Rps. sphaeroides*, and the other is the water-soluble Bchl complex from *Prosthecochloris aestararii*, strain 2k.

4.1 The B800–850 Light-harvesting Pigment-Protein Complex Isolated from Rps. sphaeroides

The most convenient preparation of the B800–850 complex involves solubilising the chromatophore membrane with the Zwitterionic detergent lauryl-dimethylamine-N-oxide (LDAO). The detergent-solubilised complex is then purified by a combination of sucrose gradient centrifugation and ammonium sulphate precipitation [18]. The purified B800–850 complex is hydrophobic and only water soluble in the presence of detergents.

Fig. 8 shows the room temperature absorption spectrum of the B800–850 complex from *Rps. sphaeroides* strain 2 · 4 · 1. As well as the absorption due to Bchl at 850, 800, 590, and 380 nm, the presence of a triad of absorption bands due to carotenoid can be seen between 440–530 nm.

The polypeptide composition of the B800–850 complex has been analysed by SDS polyacrylamide gel electrophoresis [18, 19]. When the complex is denatured and run on 10% gel, a single protein band is seen with an apparent molecular weight of

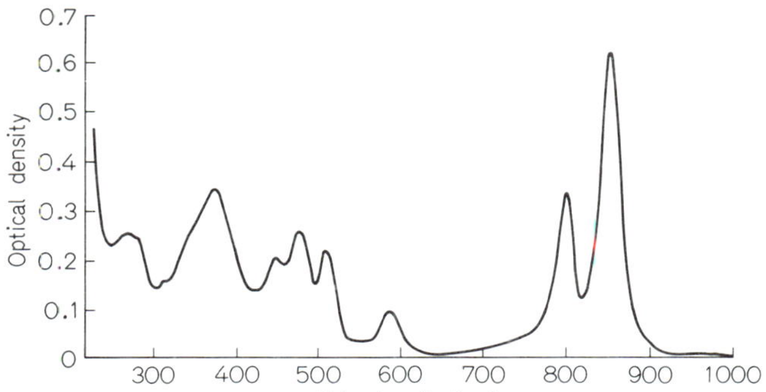

Fig. 8. Room temperature absorption spectrum of the B800–850 complex from *Rps. sphaeroides* strain 2 · 4 · 1. [From Biochim. Biophys. Acta **459**, 506 (1977)]

about 9,000. The undenatured complex runs together with its pigment in one or two bands in the 80–120 kD molecular weight region, the pattern obtained depending upon the concentration of SDS in the electrophoresis buffer [18, 20].

Sauer and Austin [21] carefully analysed the Bchl: protein ratio in this complex and concluded that it contained 3 Bchl per pair of 9,000 molecular weight polypeptides.

However, more recently, prompted by the finding of more than one type of polypeptide chain associated with the analogous B800–850 complexes from *Rps. capsulata* and *Rps. palustris* [22] the polypeptide composition of the B800–850 complex from *Rps. sphaeroides* has been reinvestigated [22, 23]. SDS polyacrylamide gel electrophoresis on gradient gels of higher acrylamide concentration of 11.5%–15% does resolve two low molecular weight polypeptides. Their separation is not very large, though, even at the higher percentage of acrylamide. In contrast, if the complex is analysed by isolectric focussing, then two bands present in approximately equal amounts are well resolved. The amino acid analysis of the complex as a whole and the two individual bands has been determined [23]. Both polypeptides are quite hydrophobic and both lack cysteine residues.

Cogdell and Crofts [20] investigated the pigment content of the B800–850 complex and found that it contained carotenoid and bacteriochlorophyll in the ratio of 1 carotenoid: 3 Bchl. The type of carotenoid present in the complex isolated from a range of different carotenoid containing mutants reflected the carotenoid composition of the parent organism, and no evidence for specificity was obtained. This information together with that obtained by Sauer and Austin [21] suggests that a minimal unit of the B800–850 complex consists of one each of the two polypeptides, 1 carotenoid and 3 Bchl. The in-vivo form of the complex is almost certainly an aggregate of this minimal unit. In Fig. 9 (a, b) A and B show the CD and fourth derivative spectra of the B800–850 complex in the near infrared region. The CD spectrum shows a large exciton-coupled band at 850 nm and a band more consistent with a monomer at 800 nm. Sauer and Austin [21] found that in preparations where the 800 nm had been destroyed the complex lost one of the three Bchls. In this case the 850–nm band still showed the same exciton-coupled CD spectrum at 850 nm. The fourth derivative spectrum also shows the presence of two transitions in the 850-nm Band and one transition in the 800 nm-band. Combining this information suggests that the 800-nm absorption band represents the third Bchl (which is more monomeric in character).

The carotenoid present in the complex also shows a strong induced CD spectrum, reflecting the specific binding of the carotenoid to the protein. The resonance Raman spectroscopy of the complex indicates that the carotenoid is probably in the all-trans configuration [83].

Bolt and Sauer [24] have studied the linear dichroism of the B800–850 complex. The complex was imbedded in a polyvinyl alcohol film and orientation was induced by stretching the hydrated film. They were able to deduce that the two exciton-coupled dipoles in the 850-nm band were tilted $\sim 68^{\circ}$ and 51° to the particle axis. The carotenoid transition dipole was tilted 58° to the particle axis. Further details of the mutual orientation of the pigments should be obtained when the fluorescence polarisation of complex has been described.

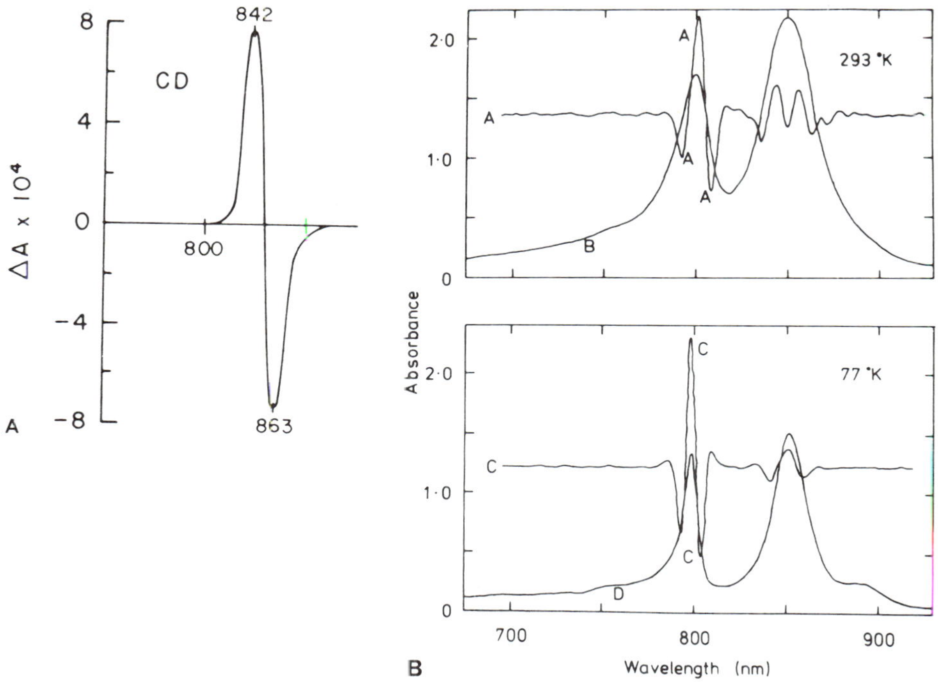

Fig. 9. A Circular dichroism spectrum of Bchl light-harvesting complex from *Rps. sphaeroides* solubilized in triton. [From Biochemistry **17**, 2011 (1978)]. **B** The near infrared absorption spectra of the B800–850 complex from strain 2 · 4 · 1 and their fourth derivatives. *Trace A* is the fourth derivative of the room temperature absorption spectrum of the B800–850 complex from strain 2 · 4 · 1. The complex was suspended in 50 mM Tris-HCl, pH 8.0, 1% lauryl-dimethyl-amine-N-oxide in a 1 cm pathlength cuvette. For both *trace A* and *trace C* the derivative intervals were 4.0, 3.5, 3.0 and 2.5 nm. *Trace C* is the fourth derivative of the 77k absorption spectrum. The complex was suspended in 60% glycerol/40% 50 mm Tris · HCl, pH 8.0/1% lauryldimethyl-amine-N-oxide in a 1 mm cuvette in a homemade dewar. The *vertical scale* for the fourth derivative curves is arbitrary, but the same in each case. The size of the peak in the fourth derivative depends upon the width of the absorption band, the narrower the band the larger and sharper the fourth derivative. Thus, the peak in the 800 nm band is larger than that in *trace A*. [From Bioch. Biophys. Acta **502**, 409 (1978)]

The emission spectrum of the B800–850 complex [25, 26] shows one major band at ~ 872 nm. The presence of the 800 nm emission indicates that energy transfer from the 800 nm to the 850 nm band is not 100% efficient.

The fluorescence may also be sensitized by light absorbed by the carotenoid. The relative efficiency of the carotenoid to bacteriochlorophyll energy transfer at the level of the excited singlet state is high (~ 85–100%) [25].

Sauer and Austin [21] measured the absolute yield of fluorescence of the B800–850 complex as ~ 25%. From this they predicted a fluorescence lifetime of 4–5 ns.

The current idea as to how the B800–850 complexes are inserted into the membrane is presented in Fig. 10. Two lines of evidence support this model. The first is derived from studies on the development of the photosynthetic apparatus and the second from studying exciton interactions within the intact photosynthetic unit.

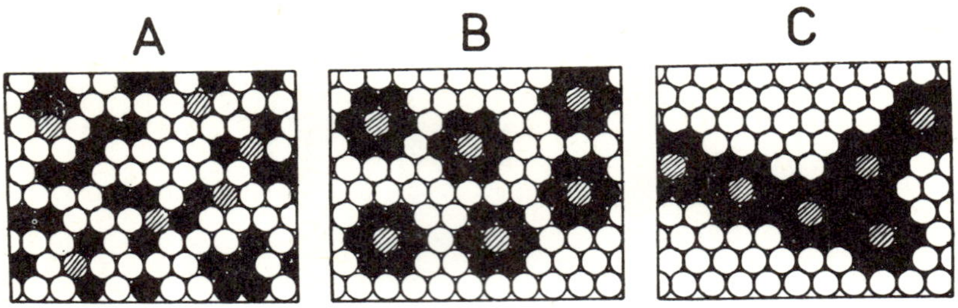

Fig. 10 A–C. Variants of the lake model of the photosynthetic apparatus. *Striped circle*, reaction centre; ● B870; ○ B800–B850. See the text for details. [From Biochim. Biophys. Acta **460**, 393 (1977)]

Rps. sphaeroides, like other purple nonsulphur photosynthetic bacteria, is able to grow either by respiration, in the presence of oxygen in the dark, or by photosynthesis, anaerobically in the light. When the cells are grown in the respiratory mode, then allowed to go anaerobic in the light, the photosynthetic apparatus is synthesized, largely de novo. If the cells are maintained at a low density in very high light, the major antenna pigment-protein complex synthesized is the B870. Moreover, the ratio of Bchl to reaction centre in this case is low (~ 20–30 : 1) [28] and relatively constant. Growth under low light condition induces the synthesis of the B800–850 component and results in the enlargement of the size of the photosynthetic unit, sometimes up to 300–350 Bchl. per reaction centre. These results suggest a more intimate relationship between the B870 complexes and the reaction centre than between the B800–850 complexes and the reaction centre.

4.2 Kinetic Studies

On the microsecond time scale Monger and Parson [10] have studied the fluorescence yield as a function of triplet quencher. In these double flash experiments an intense (20 ns) ruby laser pulse at 695 nm excited the antennae, and Bchl triplets are produced mainly by intersystem crossing. The concentration of triplet produced was measured by absorption spectroscopy and a second weak 1 μs long pulse interrogated the sample after 8 μs when the Bchl singlet states have decayed. The triplet ($t_{1/2}$ decay = 55 μs) quencher concentration was monitored as a function of fluorescence yield. Using these induced quenchers, the quencher concentration availiable was double that possible by closing the reaction center traps as had been done by Duysens [29]. A linear relationship between reciprocal yield and relative quencher concentration was observed in the carotenoidless R26 strain, indication that quenchers exist in a lake of antennae. A puddle model containing less than approximately seven connected units would give a curved line on the same axis.

In carotenoid containing strains, energy transfer from Bchl triplet to carotene occurs (to produce its triplet) in a few tens of nanoseconds [30]. Fluorescence quenching is still observed, however, and in parallel with an increase in carotene triplet formation. Furthermore, the carotene triplet is capable of quenching the fluorescence directly.

However, even when the carotene is present, the extent of quenching by the carotene is still much less than by Bchl triplets.

An insight into the topology of the antennae complex has been obtained for strain *Rps. sphaeroides Ga* grown anaerobically but with varying amounts of B800–850 and a constant ratio of B870 antennae to reaction centre. Except at very high exciting flash intensities, the amount of triplet carotene formed was independent of the B800–850 present, which indicates that most quenching occurs in the B870 complex. A rapid transfer of energy from the site of absorption (in B800–850) to the lower energy singlet states of B870 must be occurring. Similar effects have been observed in *Chromatium vinosum.* Figure 10 illustrates three possible lake models for the organisation of P870, B870, and B800–850, only Model C will explain the data. In this model P870 are connected together with B870 adjacent to this region. This organisation allows transfer from one reaction centre to another along a ribbon of antennae, the B800–850 pigments surround the other pigments.

Because of the closeness in energy between B850 and B870 singlet states, thermally assisted energy transfer (uphill transfer) from the B870 to B850 is possible at a rate not too much smaller than the forward reaction. Zankel and Clayton [27] studied the fluorescence from B870 relative to B850 when the reaction centre traps were open and closed. The largest change was observed when the reaction centre was open, indicating that equilibrium between two pigment beds is not established during the fluorescence lifetime because of the loss introduced by the reaction centre. The "uphill" transfer rate was calculated as ~ 1/2 the "downhill" (B850 ⟶ B870) rate and is comparable with the rate of trapping from B870. When many excitations arrive at B870 within a time short compared to the Bchl fluorescence decay time, the reaction centre will become saturated leaving the remaining excitations to either fluorescence, intersystem, cross to the triplet, annihilate one another, or return to the B850 pigments. One might therefore expect annihilation processes to be observable in the fluorescence from B850 pigments. Experiments with carotene containing antennae show that excited state annihilation does not occur in the B800–850 pigments to any great extent. We suggest the following tentative explanations for this effect. Immediately after excitation all the excited molecules are in the B800–850 Bchl singlet states. At this time the probability of annihilation in this pigment complex is highest. The excitation then rapidly enters B870 and the reaction centre is closed by the first few excitations to arrive at P870. The remainder now in B870 fluoresce, remigrate back to B850, or annihilate one another, or intersystem cross. The scheme below shows this in outline

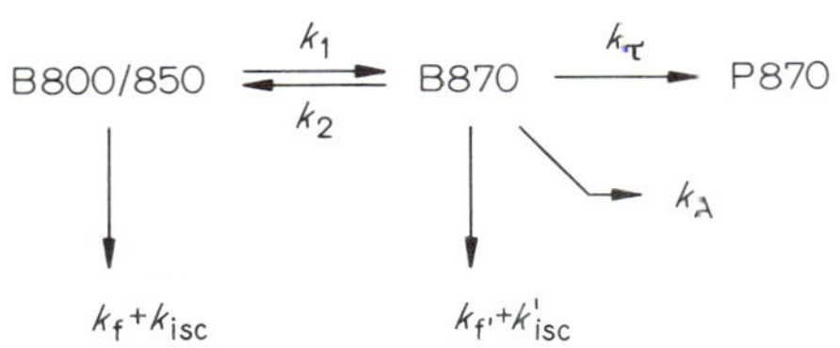

with the rate constants for fluorescence: k_f, $k_{f'}$, trapping k_T and annihilation, k_A and k_1 and k_2 are the rates into and out of B870 and B800–850. The larger size of B800–850 compared to B870, the faster rate $k_1 \sim 2k_2$, and possibly its topology help to keep annihilation low in B800–850. The topology and intermolecular distance between Bchl may be such that transfer to the B870 occurs after only a few excitation transfers, thus limiting migration in B800–850 pigments. The smaller size of B870 [28] will then cause an increase in the annihilation rate. Additionally, as excited state annihilation progress in B870, excitation is transferred preferentially into B870 at rate k_1 ($>k_2$) in an attempt to restore the equilibrium shown in the scheme above. This process will also be assisted by triplet quenchers produced in B870. Reactions such as $S^* + S^* \longrightarrow T + T$ produce Bchl triplets which cannot migrate on the time scale of the singlet state lifetime $< 10^{-8}$ s and provide additional quenchers by processes such as $S^* + T \longrightarrow T + T$ (or $T + S_o$). As the annihilation rate depends on the square of the excited state concentration, only a slight decrease in this concentration will drastically reduce the rate of annihilation relative to fluorescence or trapping by the reaction centre.

Using the scheme above, the fluorescence decay kinetics can also be derived. If initially we ignore the annihilation term, the decay at B800–850 consists of the sum of two exponential terms and B870 the difference of the same two exponentials, i. e.

$$I(f)_{800-850} \propto e^{-\lambda_1 t} + Ae^{-\lambda_2 t} \text{ and } I(f)_{870} \propto e^{-\lambda_1 t} - e^{-\lambda_2 t}$$

where $\lambda_2 > \lambda_1$ and λ_1, λ_2 and A depend upon the rate constants for fluorescence and interconversion between the two pigment beds. When the rate constants k_1 and k_2 are equal

$$I(t)_{800-850} = I(t)_{870} \propto e^{-\lambda_1 t}$$

and except at short times the decays are exponential and of the same lifetime. When annihilation is present, the loss processes in B870 become greater making λ_2 even larger, and the initial part of the fluorescence decreases more steeply.

Early measurements of the fluorescence decay of photosynthetic bacteria used the phase method whereby the continuous excitation beam is modulated at high frequency and the phase lag between the input beam and the modulated fluorescence measured. This method was applied by Merkelo et al. [31] and also by Borisov and Godik [32], and time resolutions of a few picoseconds have been claimed. Decays in the 7 to 70 ps range have been reported for a number of bacteria and PSI complexes of green plants. More recent measurements using streak cameras indicate that lifetimes in the 200 ps region exist for chromatophores or whole cells [33].

Using different strains of *Rhodoseodomonas sphaeroides,* the effect of increased absorption cross section and changes in quenching by the reaction centres have been studied as a function of light intensity [33]. At low excitation intensities the fluorescence decays appear exponential, but it is difficult to obtain accurate and reproducible decay profiles using streak cameras, particularly at lowish excitation energies ($\gtrsim 10^{13}$ photons/cm^2/pulse) so that the exact form of the decay has still to be confirmed. Notwithstanding, a comparison of different strains has been possible. In 2 · 4 · 1, the normal strain, and in Ga a green mutant, both have fluorescence decay times of 100±20 ps. In R26 which lacks carotenes, B800–850 decays in 300 ps, and PM8 which lacks a reaction centre decays in 100 ps. In each strain the mean fluorescence lifetime (or yield) versus exciting light intensity follows a curve described by Eq. (2), but the half-quenching intensities are displaced from one another. A combination of changes in cross section and lifetime can explain this phenomenon. The relative half-quenching intensities for PM8; 2 · 4 · 1; Ga; and R26 are 1; 2; 8; 16. Where the fluorescence lifetime is the same, as in 2 · 4 · 1 and Ga, the higher onset of annihilation is due to the four times smaller absorption (at 530 nm) of Ga. This is caused by the longer fluorescence decay time (due to lack of reaction centre, which permits a higher probability of "collisions" with other excited molecules, which lead to quenching. In the carotenoidless R26 strain, which also only contains B870 pigment complex, the data cannot so easily be explained by the combination of only two effects. The longish, 300 ps lifetime, would be expected to offset the smaller absorption to produce quenching similar to Ga, instead of which it is much higher. One possibility is a lower Bchl singlet-singlet migration rate.

In cells and chromatophores of *Rps. sphaeroides* strain 1760 – 1 a decay time of 200 ps was measured by Pashenko et al. [34a] also using a streak camera, and in *R. rubrum* Borisov [34b] measured fluorescence lifetimes of 1 · 1 to 1 · 5 ns from antennae complexes. These decay times were, however, found to be dependent upon the redox state of the reaction centres. In the active state, decay times of 30–60 ps were measured and a trapping rate of only $2 \times 10^{10} s^{-1}$ was calculated. When the reaction centre was closed, an efficient quenching was also observed.

4.3 The Water-soluble Bchl-a Antenna Complex from P. aestuarii, Strain 2K

This complex is discussed because it is the only one so far which has been crystallised, and its three-dimensional structure analysed by x-ray crystallography [35]. However, in some ways, it is rather atypical as a model for bacterial light-harvesting complexes. Firstly, it is water-soluble and most other types of antenna complexes are hydrophobic, integral membrane proteins. Secondly, it is more akin to a plant or algae antenna complex in that it has a larger number of pigment molecules present per polypeptide chain than the smaller more typical bacterial antenna complexes. *P. aestuarii* is a green sulphur photosynthetic bacterium. This group with their chlorobium vesicles have several similarities with the red algae and their phycobilisomes, and this may explain the presence of this rather unique water-soluble complex.

The existence of the Bchl-*a* protein was first demonstrated by Sybesma and Olson [36]. The Bchl-*a* complex is easily prepared in large amounts following cell disruption, ammonium sulphate precipitation, and DEAE cellulose chromatography. The pure protein may be crystallised from a solution 1 M in NaCl. The complex is a trimer. SDS polyacrylamide gel electrophoresis of pure preparations has been reported [37, 38]. Electrophoresis of the denatured complex reveals a single protein band with an apparent molecular weight of 40,000 [38]. It contains 354 amino acids. When the complex is subjected to isoelectric focussing, multiple bands are seen. These seem to be the result of either microheterogeneity or variable deamidation of external amide groups during the isoelectric focussing process. The N-terminal amino acid is alanine [39]. Each 40,000 molecular weight subunit contains 7 molecules of Bchl *a*. The polypeptide consists of a distorted hollow cylinder with the Bchl-*a* aggregate inside, one end of the cylinder is open but is covered in the trimer. The outside surfaces of the trimer consist of β-pleated protein sheet, and the side of the cylinder in contact with the adjacent units of short lengths of α-helix interspersed with regions of irregular conformation. The Bchl occupy a space of 45 · 35 · 15°A in which the average centre to centre distance of the porphine rings is 12Å for nearest neighbours, the rings of the Bchl lie at between 10° and 40° to one another. In the trimer the distance of closest approach of the Bchl in adjacent subunits is 24Å, the trimer is approximately an oblate sphaeroid of 83Å and 54Å along the long and short axes. The structure is more easily visualised with the help of Fig. 11.

There is no evidence for a chemical interaction between the Bchl, but there is solid evidence for an interaction between each Bchl *a* and the polypeptide backbone. In each Bchl *a* electron density maps indicate an interaction in the fifth position onto the Mg atom, i.e. hydrogen bonding and liganding, the latter possibly by histidine. The phytyl chain appears to be anchored to the protein by "hydrophobic" interactions. Because of these interactions, the Bchl are each held singly in the protein rather than with other

Fig. 11. The polypeptide backbone and the chlorophyll core of one subunit of the Bchl-*a* protein. The view is from the threefold axis, which is horizontal, towards the outer surface of the subunit. The Bchl. chromophores, *1* to *7*, lie on the surface of a sphaeroid and are shown in *heavy line*. The phytyl chains are omitted. [From Meth. Enzym. 69, 336 (1980)]

Bchl molecules and also the Bchl do not occur in regular arrays as is sometimes postulated. Circular dichroism spectra resulting from the interaction of the close Bchl molecules show six components arising from seven molecules [40]. The reasons for the arrangement of Bchl can be rationalised by consideration of the conditions necessary for energy transfer between pigments. Experiments in vitro on Chl *a* [41] have shown that Chl-Chl interactions lead to a quenching of the Chl excited singlet state, this process being almost 100% efficient when the molecules are at $\sim 10^{-2}$ (Ml^{-1}). The distance between molecules necessary to prevent quenching at this concentration, and the similar concentrations in vivo, has been found by computer simulation of random walks, prior to quenching by pairs of molecules, to be > 10Å [42]. The features of the Bchl protein complex thus fulfill this criterion, since the nearest distance is 12Å, but this distance is still near enough to allow rapid Bchl-Bchl energy transfer, and similarly the 24 Å to the adjacent subunit also allows fast transfer compared to the rate of fluorescence. Energy transfer through this trimer from the previous pigment complex in the antenna and to the next complex will lead to essentially no loss of excited state energy, an ideal situation for a light-collecting apparatus.

At room temperature the Bchl-*a* protein shows a rather simple absorption spectrum with a single broad band in the near IR centred at about 810 nm. However, as the temperature is lowered the spectral resolution increases, and at 5°K at least four different peaks (~ 790, ~ 805, ~ 814, ~ 825 nm) may be detected. Careful 4th and 8th derivate analysis and Gaussian curve fitting to the absorption spectrum have recognised at least six components, but it is not clear how these relate to the seven pigment molecules present in each subunit of the native complex [43].

The CD spectrum of the complex at 77°K shows three troughs and two peaks. This complicated pattern arises from strong exciton coupling between the seven Bchl-*a* molecules [44]. It is perhaps worthwhile noting that the CD spectrum can be simulated assuming six components, even though we now know that there are seven Bchl-*a* molecules present [38].

It is also interesting to note that this complex is also an exception in that it does not contanin any accessory light-harvesting pigment, such as carotenoids.

5 Lower Algae

In the lower algae two light-harvesting complexes, the peridinin-chlorophyll *a* protein from the dinoflagellate Glenodinium and the biliproteins (of which there are several different ones) of the red algae, have been particularly well studied.

5.1 The Perdinin-Chl *a* Protein from Glenodinium

This protein was first described a number of years ago by Bode and Hastings [45]. It is another of the water-soluble antenna complexes and this is probably why it has been so extensively studied.

The peridinin-Chl *a* complex is isolated and purified from broken cells by sephadex and DEAE cellulose chromatography [46a]. The complex isolated from Glenodinium

Table 1. Amino acid composition of purified peridinin-chlorophyll *a* proteins from two dinoflagellate algae (average integral ratios, mol^{-1})

	A. carterae pI 7.5	*Glenodinium sp.* pI 7.3
Trp	2	2
Lys	27	30
His	4	6
Arg	3	0
Cys	0	0
Asp	35	34
Thr	10	9
Ser	22	25
Glu	21	17
Pro	15	7
Gly	22	24
Ala	55	54
Val	30	19
Met	10	13
Ile	14	13
Leu	19	16
Tyr	9	11
Phe	8	11
Total residues per mol	306	291
MW without chromophores	31,800	30,600

is a dimer composed of two identical subunits of molecular weight ~ 31,000. SDS polyacrylamide gel electrophoresis of the denatured complex reveals only a single protein band; however, multiple forms are seen with isoelectric focussing. This multiplicity of bands probably arises from deamidation of glutamine and asparagine residues. The pI of the major band is 7.3. Table 1 shows the amino acid analysis of this complex. The complex consists of 291 amino acids per subunit and also does not contain any cysteine residues.

Analysis of the pigment content of this complex has shown that it consists of four molecules of peridinin and one molecule of Chl *a* per subunit. Figure 12A shows the room-temperature absorption spectra of the complex, the bands at 435 and 669 nm are due to Chl *a*, and the broad band at 476 nm to the peridinin. The fluorescence emission spectrum is shown in Fig. 12B. There is a single emission band centred at ~ 670 nm and the emission is from the Chl-*a* molecule. Light absorbed by the peridinin molecules sensitize the chlorophyll fluorescence with ~ 100% efficiency [46b]. The photosynthetic unit in these organisms is large and contains 600 chlorophyll molecules and, depending upon the family, the peridinin-chlorophyll protein contains from 30% to 50% of the chlorophyll. The a/c complex contains 20% to 40% chlorophyll and the light-harvesting chlorophyll *a* ~ 20%.

The fluorescence lifetimes are between 4 and 5 nanoseconds [42], similar to those for free chlorophyll. The surface of the protein is highly polar and the chromophores are accommodated in a hydrophobic cleft since concentrated potassium iodide (iodide is a heavy atom quencher for excited states) does not affect the fluorescence decay

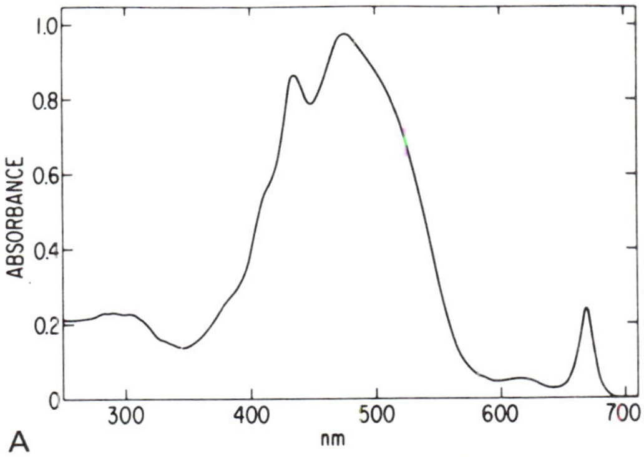

Fig. 12. A Absorption spectrum of the peridinin-chlorophyll-*a* protein, pI 7.5, at 20°C in 1 mM Tris-HCl buffer, pH 8.4. [From Brookhaven Symposia in Biology **28**, 162 (1976)]. **B** Fluorescence excitation and emission spectra of peridinin-chlorophyll-*a* protein, pI 7.5 Quantum-corrected excitation spectrum is shown from *260* to *660* nm. Emission spectra are uncorrected and are shown for 479-nm excitation (*upper*) and 330-nm excitation (*lower*). (Ibid.)

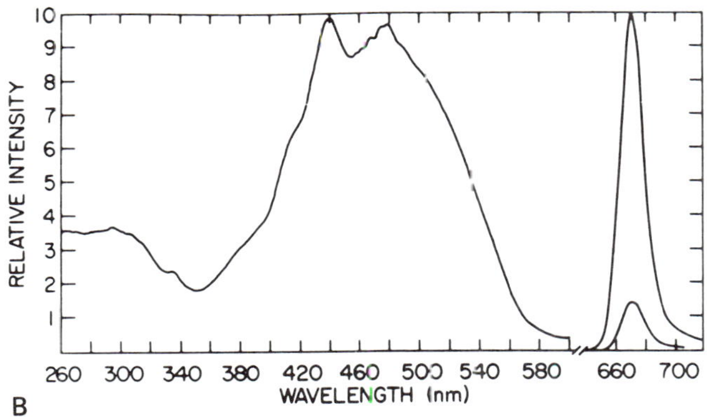

time. From circular dichroism and fluorescence polarisation spectra it is suggested [47] that the peridinin-chlorophyll protein has two peridinin molecules on either side of the chlorophyll macrocycle with ~ 12 Å between dimer pairs (see Fig. 13). Exciton resonance may enhance the peridinin singlet state lifetime, although still extremely short (estimated at < 1 ps), thereby making energy transfer more probable. With such a short singlet lifetime and ~ 8 Å for the Förster critical distance it is hard to see how Förster (dipole-dipole) energy transfer can occur with the ~ 100% efficiency with which energy is transferred to the chlorophyll. This problem exists for all carotenoid compounds where the singlet lifetime is known to be extremely short, yet transfer energy to chlorophyll or bacteriochlorophyll with high efficiency. Other interactions, such as exciton coupling, might occur when there is a strong interaction, due to proximity, of the molecules' excited states.

5.2 The Phycobiliproteins of the Red Algae

The Phycobiliproteins have probably received the most detailed biochemical analysis of all the types of antenna complexes so far studied, for two important reasons: firstly, the bile pigments are covalently linked to their apoproteins, and secondly, they are

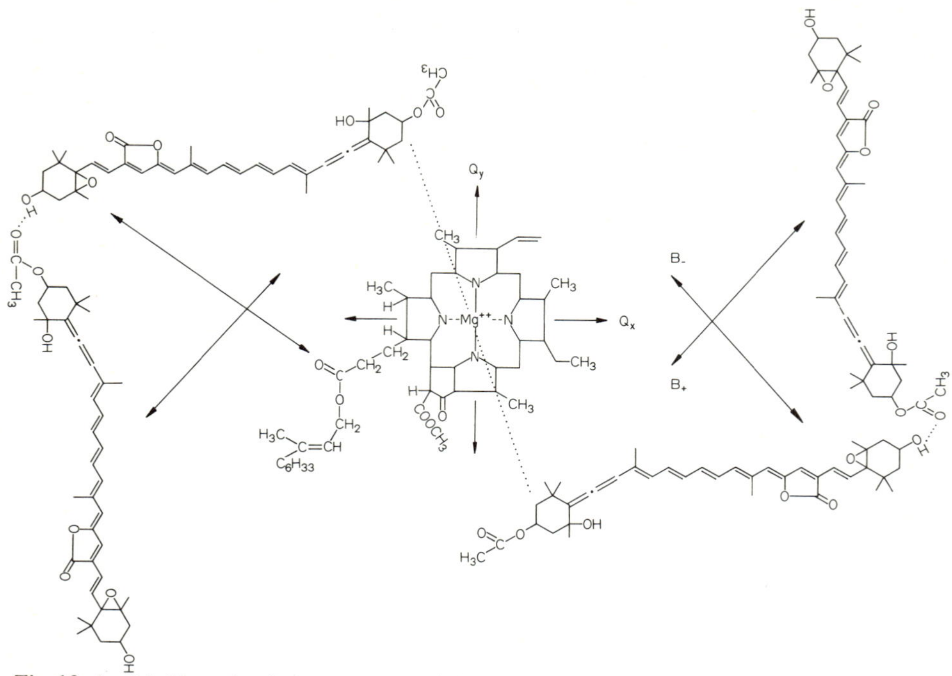

Fig. 13. A probable molecular arrangement of Chl *a* and peridinins based on relative orientations of transition moments (*double arrows*) of fluorescence Q_y, and excitation transitions B^+. [From Biochem. **15**, 4422 (1976)]

water-soluble proteins. Good reviews of the structure of a wide range of biliproteins have been presented by Borgorad [48] and Gantt [49], and the protein structure has also been reviewed [50].

The bili pigments are linear tetrapyrolles. There are three major classes of phycobiliproteins: the phycoerythrins, the phycocyanins, and the allophycocyanins. In general, they have a multisubunit structure composed of α and β subunits aggregated together, usually in the form of $\alpha_3\beta_3$ or $\alpha_6\beta_6$ complexes. Both types of subunit bind pigment.

The biliproteins are rather straightforward to isolate. Their preparation involves cell breakage and removal from the photosynthetic membrane by such treatments as phosphate buffer washes. They can be purified to homogeneity by conventional protein purification techniques. Table 2 shows that the three classes have differing absorption and fluorescence maxima. These differences in the fluorescence maxima allow almost unambigous monitoring of the fluorescence of one pigment, even in the presence of the others.

Table 2

Type	Abs. max. (nm)	Fl. max. (nm)	τ_R(PS)	$\tau_{1/e}$(PS)
Allophycocyanin	650	660	24	118
C- Phycocyanin	615	647	12	90
C-Phyoerythrin	565	575	*	70

* Appears with laser pulse

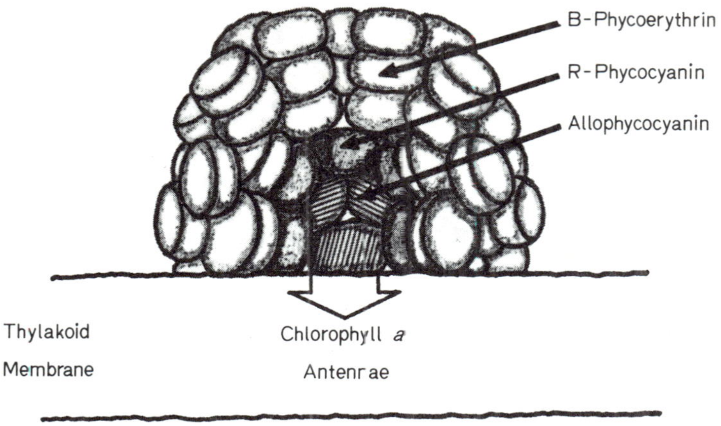

Fig. 14. Model of a phycobilisome as envisioned to exist in the red algae *Porphyridium cruentum*, based on Gantt et al. [Biochim. Biophys. Acta. **430**, 375 (1976)]

The biliproteins are organised into the so-colled phycobilisomes. These can be seen under the electron microscope as densely staining bodies, closely applied to the photosynthetic membrane. Figure 14 illustrates a model of phycobilisome. It has been postulated that the biliproteins are organised in the phycobilisomes, prolate hemiellipsoids measuring 300 Å high and 450 Å wide, so that the phycoerythrin forms the outer layer and the phycocyanin is between it and the allophycocyanin [51]. The allophycocyanin is thought to be in contact with the membrane and acts as a channel for electronic excitation to reach the chlorophyll in the thylakoid membrane. This model would favour energy transfer from the outer to the inner side.

5.3 Kinetic Studies

Picosecond spectroscopy using a streak camera has clearly demonstrated sequential energy transfer from the pigments phycoethrin → phycocyanin → allophycocyanin → chlorophyll [52, 53]. At 530 nm, the excitation wavelength, a large fraction of the incident light is absorbed by the phycoethyrin (see Table 2) and according to the proposed structure this pigment also forms the outside shell.

Both energy migration between similar pigments and transfer from one pigment type to another occurs. The consecutive layers of chromoprotein in the phycobilisome maximises the area for possible transfer from one layer to the next and also limits the energy migration in one type of protein which must inevitably lead to wasteful fluorescence. The fluorescence yield is consequently low and ~ 99% of the absorbed light is transferred to the chlorophylls.

The fluorescence from the phycoethyrin rises with the time resolution of the streak camera detection (< 5 ps) but decays in a complex manner with a mean decay time of 70 ps. None of the fluorescence from the other pigments rises promptly however, but takes a finite time determined by the rate of energy transfer from the previous chromoprotein. Table 2 also shows the fluorescence "rise times" τ_R and also the 1/e decay times for

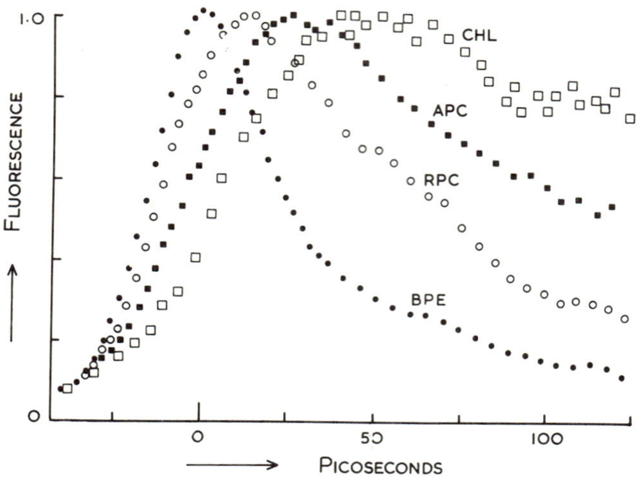

Fig. 15. Rise times of wavelength-resolved fluorescence from *P. cruentum*. *BPE*, B-phycoerythrin; *RPC*, R-phycocyanin; *APC*, allophycocyanin; *Chl*, chlorophyll *a*

the intact phycobilisome. In Fig. 15 the rise and decay of the fluorescence from each pigment can be seen. The chl rise time is 50 psec.

These fluorescence rise times and decays can be analysed in a sequential manner [52], the resulting equations describe the fluorescence as a difference of exponential terms, where the negative amplitude terms describe the rise of the fluorescence and the other terms the fluorescence decay. Porter et al. [52] described each intrinsic fluorescence decay by an exp $(-At^{1/2})$ decay law which was found empirically to describe the fluorescence. Their analysis demonstrated the consecutive nature of the energy transfer rather than being able to determine the energy transfer rates. As the pigments are on many different sites in the protein, of which they form a part, the fluorescence decay kinetics are probably best described by the sum of a number of exponential terms.

When the phycobilisome is isolated from the thylakoid membrane the allophycocyanin cannot transfer its energy to the chlorophyll and decays with its normal lifetime of 4 ns; this fluorescence however takes ~ 150 ps to reach its full intensity (see Fig. 16). As the decay time is long by comparison, this rise time reflects the rate of energy transfer from the R-phycocyanin. At high light intensities $> 10^{14}$ photons cm^{-2}, excited state annihilation is observed mainly in the allophycocyanin, as a decrease in fluorescence decay time. This decay is exponential up to ~ 50% quenching intensity after which a shorter component is observed. It has been suggested [52, 17] that the dynamics of the excitation motion are responsible for the nonexponential decay, as only at high excitation intensities, and hence a large excited state population, does quenching occur before diffusion spatially randomizes the excitations. The bimolecular quenching kinetics [Eq. (1)] predict a nonexponential fluorescence behaviour and this may also explain the fluorescence decay profiles. The lack of any significant quenching in the phycoethyrin pigment shows that the average rate of energy migration is here slower than the average rate of transfer to the phycocyanin, the next pigment.

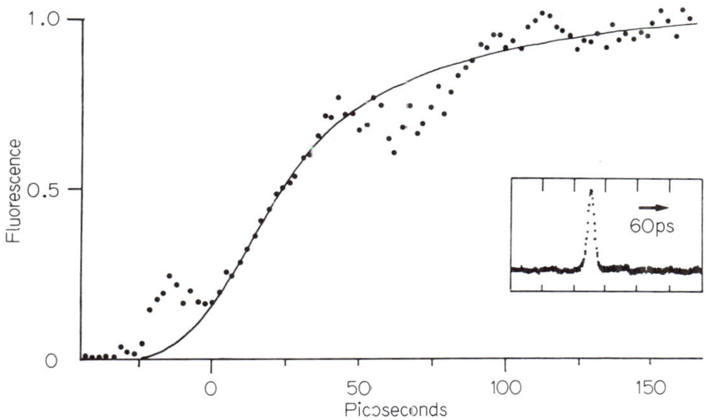

Fig. 16. The rise time of allophycocyanin fluorescence in phycobilisomes isolated from *Porphyridium cruentum*. Fluorescence was monitored above 645 nm and is expressed on a normalized scale, with the asymptote of the rise curve being taken as unity. Each experimental *point* represents an individual channel of the OMA memorly. The *solid line* represents the theoretical curve derived from a kinetic analysis of sequential energy transfer. The *inset* shows the laser pulse profile taken at the same streak rate. [From Ciba Foundation Symposium **61**, New Series, 257 (1979)]

6 Antenna Pigment-Protein Complexes from Higher Plants

Up to 1976 a rather simple picture of the higher plant photosynthetic unit was generally accepted [54]. If the chloroplast membrane was solubilized with SDS and fractionated by SDS polyacrylamide gel electrophoresis, in addition to free pigment two major chlorophyll pigment-protein complexes were resolved; a P700 Chl-*a* complex and a Chl-*a*/*b* protein. Thus the PSU was thought to consist of the P700 Chl-*a* protein, the Chl-*a*/*b* protein (the major antenna species) and a presumed PSII particle, which was assumed to be destroyed by this treatment.

However, more recently, less harsh isolation procedures have been employed, which strip very little of the pigments from the proteins, and a very much more complicated pattern of complexes has been discovered. Indeed, there is now an embarrassment of complexes and as yet there is no generally accepted model which accounts for all the types now discovered. For example see Markwell et al. [55]. However, it seems that there are probably two classes of antenna complexes in higher plants. The well-known Chl-*a*/*b* protein which may exist in multiple forms [57] and antenna complexes which are normally very intimately connected to the PSI and PSII particles [54]. The second class of complexes is as yet poorly described and so we shall concentrate on the Chl-*a*/*b* protein.

The Chl-*a*/*b* protein can be isolated from higher plants by treatment with SDS followed by purification with hydroxyl appatite chromotography and ammonium sulphate precipitation [58]. The Chl-*a*/*b* protein shows two major peaks in the red region of the spectrum at ~ 652 nm (Chl *b*) and ~ 670 nm (Chl *a*) (see Fig. 17), and fluorescence

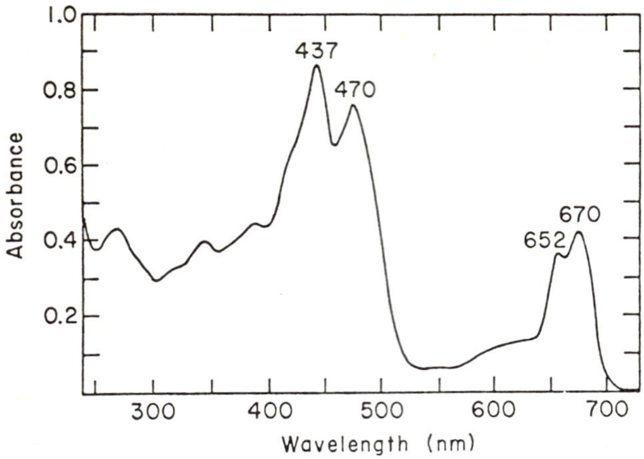

Fig. 17. Room temperature absorption spectrum of the light-harvesting chlorophyll-*a*/*b* protein in 50 mM Tris-HCl at pH 8. [From Meth. Enzym. **69**, 150 (1980)]

mainly from the Chl *a* with an emission peak at 685 nm at 77°K. The Chl-*a*/*b* ratio is normally in the range of 1 : 1 to 1 · 3 : 1 and the chlorophylls are bound to two polypeptides of 25–35,000. It has been calculated that there are about 3 Chl *a*, 3 Chl *b* and ~ 1 carotenoid present per 30 KD of protein [58]. It is probable that in vivo the complex exists in an aggregated form [56].

Considerable effort has been put into the measurement of the fluorescence decay times of pigment-protein complexes in chloroplasts. Initial experiments used nanosecond duration discharge lamps [59] and were only just able to time-resolve the fluorescence; more accurate measurements became possible when the mode-locked laser and streak camera were first used. In the first measurements using a streak camera Beddard et al. [60] detected rapid exponential decay curves of 108 ± 10 ps for Chlorella, 92 ± 10 ps for Porphyridium and 134 ± 10 ps for spinach chloroplasts. In similar experiments Shapiro et al. [61] and Kollman et al. [62] reported shorter lifetimes of 75 ± 10 ps for Anacystis. The possibility that singlet-singlet or singlet-triplet annihilation could shorten the fluorescence decay times was also considered [60] but was initially discounted when a tenfold decrease in excitation intensity revealed no change in the fluorescence kinetics. It was shown later, and may be seen in Fig. 7, that at the high excitation intensities used, ~ 10^{16} photons cm^{-2}, the decay time and fluorescence yield are almost independent of laser intensity.

In 1976 Campillo et al. [63] demonstrated that the fluorescence yield of Chlorella depends strongly upon excitation intensities greater than 10^{13} photons cm^{-2} when a single picosecond pulse was used. The dependence of the fluorescence lifetime on excitation intensity was established by Porter et al. [64] who also showed that some quenchers can last from one pulse to another when a train of picosecond pulses was used to excide the sample. Figure 18 shows the dramatic shortening in decay time due to annihilation. Chlorophyll triplet states are now though to be responsible for this extra quenching [65]. Triplet states had previously been suggested as quenchers on the microsecond time scale [66].

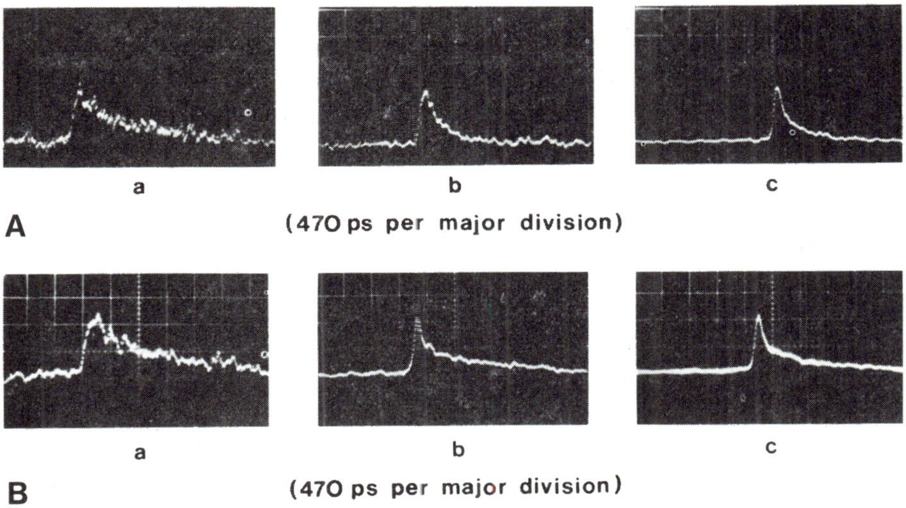

Fig. 18. A Fluorescence decay traces for dark-adapted *Chlorella* at excitation intensities of: *a* $3 \cdot 10^{13}$, *b* $5 \cdot 10^{14}$, *c* $8 \cdot 10^{15}$ photons/cm^2. **B** Fluorescence decay traces for pre-illuminated chlorella with DCMU, at excitation intensities of: *a* $3 \cdot 10^{13}$, *b* $5 \cdot 10^{14}$, *c* $8 \cdot 10^{15}$ photons/cm^2

These annihilation effects clearly demonstrate the necessity of using low repetition frequency and low energy pulses in lasers used to excite photosynthetic organisms. Usually $<10^{13}$ photons cm^{-2} are needed in a single pulse, and this pulse must have a smooth spatial distribution to prevent annihilation from occurring from hot spots in the beam. An other wise low energy beam may give lifetimes shorter than anticipated from a knowledge only of the total number of photons present.

Most groups are now in general agreement about the mean fluorescence decay times for Chlorella, Chloroplasts, PSII and PSI subchloroplast preparations at room temperature. These results come from measurements using streak cameras, single photon counting, and phase shift methods [15b, 67, 68]. The main contention is now on the exact form of the fluorescence decay profile. Experiments have also been performed with the reaction centres open and closed, and in the presence of divalent cations. In the dark adapted state the mean lifetimes are about 500 ps for chlorella, 450 ps for Chloroplasts and 110 ps for PSI subchloroplast particles. When the reaction centres are closed by DCMU and continuous illumination, the decay times is 1.5 ns in Chlorella and 1.3 ns in chloroplasts.

There have been differing results over the form of the fluorescence decay profile. Campillo and Shapiro [69] using a streak camera found exponential fluorescence decays at low excitation intensities which became nonexponential according to Eq. (1) as the excitation intensity increased. Harris et al. [64], who also used a streak camera, found that at high intensities (10^{15} photons cm^{-2}) the decay was fitted to an equation of the form $e^{-Kt^{1/2}}$ after 100 ps of the maximum intensity but with an initial short decay 1/e of 32 ps. At lower intensities a similar equation, $e^{-K't-Kt^{1/2}}$, was used to analyse the whole fluorescence decay where the rate constant K' is small compared to K. The decays of Chlorella and chloroplasts have been described in this way.

Using single photon counting either with discharge lamps [70] or sync-pump lasers [64], a bi-exponential fluorescence decay was observed for chloroplasts and Chlorella. The date were best fitted by a sum of two exponential terms as shown in Fig. 19, rather than terms involving $t^{1/2}$. The reasons for the discrepancy between decays measured by the streak camera and photon counting may rest with the techniques themselves. Although the streak camera can have better time resolution than photon counting, it is difficult to obtain accurate and reproducible decay profiles at the low excitation intensities needed to measure the fluorescence. Computer signal averaging of the streak camera data is easily performed, but there are difficulties associated with averaging data of varying amplitudes and position on the OMA vidicon. This latter effect occurs as a result of time jitter in the camera trigger circuits. Except at the faster time scales ($<$ 100 ps decay times approximately), as illustrated by the data on Porphyridium which could not be obtained by photon counting, the photon counting techniques, if properly used, provide a more reliable representation of the decay profile even when convolution with the excitation profile is required.

We wish to distinguish clearly between the different causes of nonexponentiality in fluorescence decays and suggest what kind of information can be obtained in each case. Deviations from exponentiality can arise from two distinct causes: (1) an intrinsic time dependence of the emission probability, the type of nonexponentiality can give information on the mechanism of the energy transfer or migration, and (2) inhomogeneity in the emitting species. This inhomogeneity can arise from trivial effects such as differences between cells, or many reflect different quenching probabilities in different regions of the cell and thus give structural information. Additionally, nonexponentiality can be experimentally induced such as by excited state annihilation, reabsorption of the fluorescence [68], or by stimulated emission. It has been proposed that this latter effect might be the cause of the intensity-dependent fluorescence decay times [72], although no lasing from any in vivo system has been observed [73] and would be hard to achieve because of the efficient transport of energy to the reaction centres.

A physical justification for the intrinsic time dependence of the quenching type (1) can be found in the nondiffusional nature of the excitation migration at short times [7] as described in Sect. 1.1. Recent calculations show that this effect should be measurable but at the expected chlorophyll concentration in vivo, assuming a three-dimensional organisation should last not longer than 50 ps [7]. In two-dimensional systems this time dependence can last for as long as 500 ps [5a].

Nonexponential decays arising from structural inhomogeneity can have several forms: (1) fluorescence from PSI as well as PSII since the PSI subchloroplast particles also emit at 685 nm, and (2) contributions from excitations near open and closed traps, i.e. on the state of the primary donors and acceptors. As the fluorescence yield of PSII pigments depends upon the reaction centre state, inhomogeneity may be introduced depending upon the ratio of open to closed traps, since the excitation near to these traps will have different lifetimes. If there are a number of distinct sites (open and closed traps), and the excitation is able to visit them all during its lifetime, then single exponential decays are expected and the sample is homogeneous on the time scale of the fluorescence decay.

The lifetimes of Chlorella, PSI particles and Chloroplasts with various amounts of closed traps, taken from [67] are shown in Table 3. The Chloroplasts are clearly an

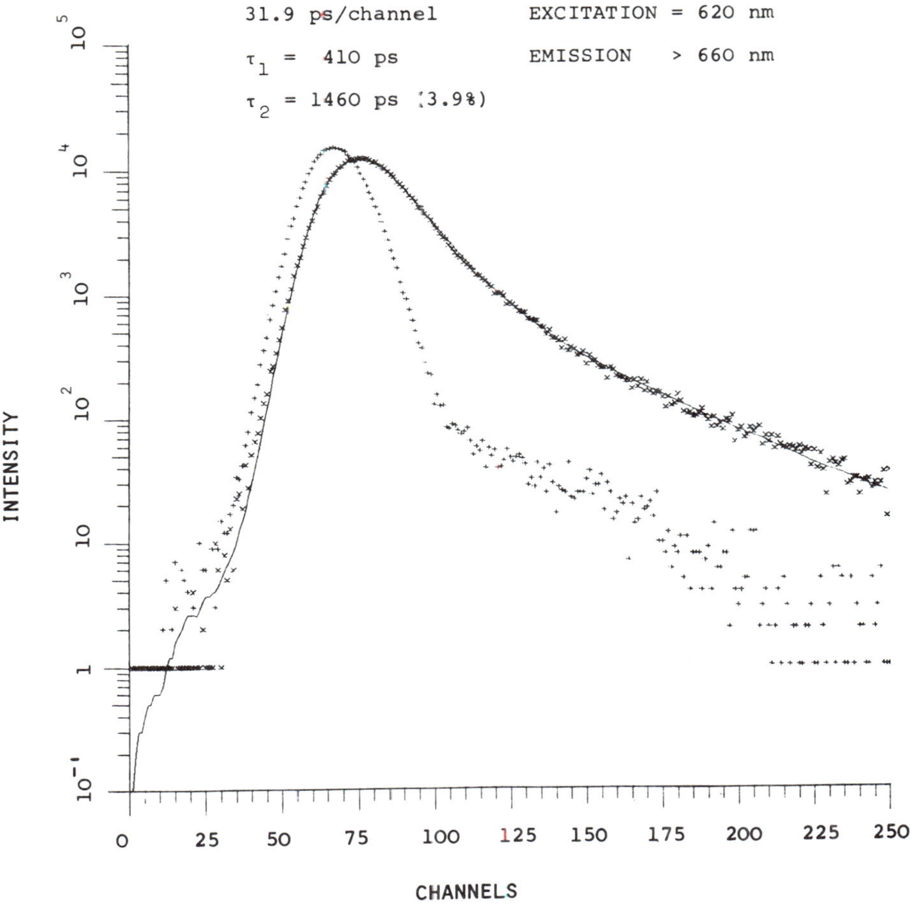

Fig. 19. Fluorescence emission of dark-adapted Chloroplasts at 31.9 ps/channel. The fluorescence is shown together with the calculated decay (*solid line*). The instrument function is also shown as the narrow curve of crosses (+)

example of the heterogeneous case (2) above, as two decay times are observed under all conditions. The shorter decay time is not due to PSI emission, as the data could not be fitted with an ~ 100 ps component in any proportion. The result is then due to the presence side by side of reaction centres in the open (435 ps) and closed (1–4 ns) states. This is supported by the observation that as more traps are closed by illumination and DCMU, only the percentage of the longer component and not its lifetime changes. The mean decay times are also in good agreement with the data of Moya [68] who used a phase shift method and measured the mean lifetime directly. The fluorescence lifetimes are consistent with an isolated unit model since if the excitation could visit both 400 ps and 1.4 ns antennae, an exponential lifetime continuously varying between ~ 0.4 and 1.4 ns would be observed as the traps are closed. The model envisaged consists of light harvesting pigments which can transfer energy to many photosyn-

Table 3. Characteristics of the fluorescence decay of chlorophyll in vivo

Sample	τ_1(ps)	τ_2(ps)	$\%\tau_{2_{\tau_1}}$	ϕcalc	τ_{mean}(ps)
Chlorella, dark adapted	492	–	–	0.025	–
Chloroplasts					
Dark adapted	413	1,463	3.8	.023	453
Light adapted + DCMU	453	1,328	9.9	.028	540
Light adapted + DCMU + Mg^{2+}	462	1,342	36.6	.039	784
SLV fraction (PSI particles)	113	1,192	3–9	.0056	–

thetic units, which are located in potential energy wells with the traps at the minima. Thus, once an excitation is in the vicinity of the trap it cannot return to the light-harvesting pigment and to another trap [67]. Later experiments by Searle et al. [74], using barley chloroplasts and streak camera detection, confirmed the two exponential decays described above. Chl-*a*/*b* complexes extracted by digitonin (F III) have a 4 ns fluorescence lifetime and chlorophyll *a* protein complexes prepared with SDS have a 4.5–5 ns lifetime at room temperature, indicating that little concentration quenching occurs in the antennae and little quenching in the absence of the reaction centre. The lifetime of Chl *a* in solution is 6.5 ns.

At low temperatures (77°K) a PSI emission at 735 nm can be detected; at room temperature this is so weak that it is buried under the tail of the 685 nm emission. When the 685 and 695 nm emission yields from the chlorophyll *a*/*b* complex are quenched by annihilation, the lifetime of the 735 nm emission remains unchanged at 1.5 ns. This fluorescence is thought to reflect energy transferred to PSI, but is not PSI antennae emission [75].

As the temperature is lowered from –60° to –196° this lifetime and yield both increase and a special pigment C705 has been proposed to give rise to the 735-nm emission [75]. Additionally, a 150 ps rise time in the 735-nm fluorescence has been detected, this rise time is much slower than the instrumental response with which the 685 nm fluorescence appeared [76]. Campillo et al. [76] suggest that this may be due to a few chlorophylls in the light-harvesting complex being associated with PSI, while the rest are only loosely associated and consequently transfer to PSI takes a comparatively long time, i.e. 150 ps. In similar experiments in the authors' laboratory, however, no rise time in this 735 fluorescence was detected [77]. If, as proposed, population of the C705 species depends on transfer from (and hence decay time of) PSI antennae pigments, any shortening of this decay time will result in a more rapid rise time in the C705 fluorescence. The results from both laboratories [76, 77] have been rechecked and confirmed as being different. The discrepancy that still exists may then lie in the differences in laser excitation intensities, giving differing PSI lifetimes, or possibly in sample preparation and handling, etc. The fluorescence yield curves at 685 nm and 735 nm have an identical shape as a function of pulse energy after single picosecond pulse excitation [78]. In contrast to this, when a microsecond pulse is used, the 735 nm fluorescence is quenched far more than the 685 nm and indicates a preferential buildup of quenchers, possible triplet states of chlorophyll or carotenes, in the PSI antennae rather than PSII antennae.

The decrease in fluorescence lifetime and yield, at room temperature, as a result of intense excitation, has demonstrated the presence of singlet-singlet and singlet-triplet quenching [15a, 15b]. Even when single laser pulses are used, triplets can be produced via S + S ⟶ S + T processes but singlet-singlet interaction usually dominates the quenching. Excellent agreement is found between the measured relative fluorescence yield and $\Phi(2)/\Phi(a)$ as a function of excitation energy, as described by Eq. (2). This is shown in Fig. 7. Data taken down to 21 K also follow quenching curves of this form [78]. The bimolecular rate constant for singlet-singlet quenching γ [Eq. (1)] is calculated to have values of 5 to 15 $10^{-9}\ cm^3 s^{-1}$. Lower bounds for diffusion of the excitation of ~ 200Å and a diffusion coefficient of $> 10^{-3} cm^3 s^{-1}$ were obtained [78] from an analysis of the quenching curves. These values are similar to ones obtained in organic molecular crystals [79].

According to the model of Paillotin et al. [80] two to four photosynthetic units, each of approximately 300 chlorophyll molecules, are involved in the excitation migration. Because of the large diffusion coefficient, migration in this model will be limited only by the singlet lifetime, and quenching due to annihilation is thought to begin when one photon is absorbed for every 2000 to 3000 chlorophyll molecules. In a model consisting of isolated puddles of photosynthetic units of ~ 300 chlorophyll molecules, higher excitation intensities would be required for the onset of annihilation. In the latter case the Poisson statistical approach of Mauzerall [81] would be appropriate but, as was pointed out above, the data is well fitted by a kinetic [17, 80] rather than a statistical model. The Poisson curves predict a much more rapid decrease in fluorescence yield with energy than is actually observed (Fig. 7).

The excited state annihilation data support a lake model of excitation migration where there is free diffusion of the excited singlet state and migration is limited only by the excited state lifetime. Recent fluorescence lifetime data however support a model in which excitation is captured near to open or closed reaction centres and cannot travel from one to another. One possible explanation of this apparent contradiction has been presented [67] and is that a common pool of antennae molecules feed energy into the vicinity of PSII or PSI, where lower singlet energy levels or geometrical restrictions on the extent of energy migration (such as in the phycobilisomes) and the presence of trapping can prevent efficient back transfer into the bulk pigment. This model would account for two fluorescence lifetimes and also for excited state annihilation being best described by a lake model.

References

1. Thornber, J.P., Alberte, R.S., Hunter, F.A., Shiozawa, J.A., Kan, K-S.: Brookhaven Symposia in Biology **28**, 132–148 (1977)
2. Thornber, J.P., Trosper, T.L., Strouse, C.E.: In: The Photosynthetic Bacteria; Clayton, R.K., Sistrom, W.R. (eds.), pp. 133–180. New York – London: Plenum Press 1979
3. Förster, Th.: Ann. Physik **62**, 55 (1948)
4. Stryer, L., Haugland, P.: Proc. Natl. Acad. Sci. (USA) **58**, 719 (1967)
5. Huber, D.: Phys. Rev. B. **20**, 2307 (1979)
   a) Altmann, J., Beddard, G., Porter, G.: Chem. Phys. Lett. **58**, 54 (1978)
   b) Altmann, J., Beddard, G., Porter, G.: Ciba Foundation 61, New Series, p. 191. Amsterdam: Elsevier North Holland 1979

6. Haan, S., Zwanzig, R.: J. Chem. Phys. **68**, 1879 (1978)
7. Gochanour, C., Anderson, H., Fayer, M.: J. Chem. Phys. **70**, 4254 (1979)
8. Phillion, D., Kuizenga, D., Siegman, A.: Appl. Phys. Lett. **27**, 85 (1975)
9. Seely, G.: J. Theor. Biol. **40**, 189 (1973)
10. Monger, T.G., Parson, W.W.: Biochim. Biophys. Acta **460**, 393–407 (1977)
11. Gantt, E., Lipschultz, C.A., Zilinskas, B.A.: Brookhaven Symp. in Biology **28**, 347–357 (1977)
12. Porter, G., Tredwell, C., Searle, G., Barber, J.: Biochim. Biophys. Acta **501**, 232 (1978)
13. Porter, G., Synowiec, J., Tredwell, C.: Biochim. Biophys. Acta **459**, 329 (1977)
14. Geacintov, N., Breton, J.: Biophys. J. **17**, 1 (1977)
15. a) Campillo, A., Kollman, V., Shapiro, S.: Science **193**, 227 (1976)
    b) Tredwell, C., Synowiec, J., Searle, G., Porter, G., Barber, J.: Photochem. Photobiol. **28**, 1013 (1978)
16. Paillotin, G., Swenberg, C.: Chlorophyll Organization and Energy Transfer in Photosynthesis. Ciba Foundation 61, New Series, p. 201. Amsterdam: Elsevier North Holland 1979
17. Beddard, G.S., Porter, G.: Biochim. Biophys. Acta **283**, 492–504 (1972)
18. Clayton, R.K., Clayton, B.J.: Biochem. Biophys. Acta **283**, 492–504 (1972)
19. a) Fraker, P.J., Kaplan, S.: J. Biol. Chem. **247**, 2732–2737 (1972)
    b) Huang, J.W., Kaplan, S.: Biochim. Biophys. Acta **307**, 332–342 (1973)
20. Cogdell, R.J., Crofts, A.R.: Biochim. Biophys. Acta **502**, 409–416 (1978)
21. Sauer, K., Austin, L.A.: Biochemistry **17**, 2011–2019 (1978)
22. Drews, G., Feick, R., Schumacher, A., Firsow, N.: In' Preceedings of the IVth International Congress on Photosynthetic Research; Hall, D.O., Goodwin, T.W., Coombs, J. (eds.), pp. 83–93. London and Colchester: The Biochemical Society 1978
23. Cogdell, R.J., Lindsay, J.G., MacDonald, W., Reid, G.P.: Biochem. Soc. Trans. **7**, 184–187 (1979)
24. a) Cogdell, R.J., Lindsay, J.G., Reid, G.P., Webster, G.D.: Biochim. Biophys. Acta **591**, 312–320 (1980)
    b) Bolt, J., Sauer, K.: Biochim. Biophys. Acta **546**, 54–63 (1979)
25. Cogdell, R.J., Hipkins, M.F., MacDonald, W., Truscott, T.G.: Biochim. Biophys. Acta **634**, 191–202 (1981)
26. Broglie, R.M., Hunter, C.N., Delepelaire, P., Neiderman, R.A., Chua, N.H., Clayton, R.K.: Natl. Acad. Sci. USA **77**, 87–91 (1980)
27. Zankel, K.L., Clayton, R.K.: Photochem. Photobiol. **9**, 7–15 (1969)
28. Aagaard, J., Sistrom, W.R.: Photochem. Photobiol. **15**, 209–225 (1972)
29. Vredenberg, W., Duysens, L.: Nature 4865, p. 355 (1963)
30. Monger, T.G., Cogdell, R.J., Parson, W.W.: Biochim. Biophy. Acta **449**, 136 (1970)
31. Merkelo, H., Hartman S., Mar, T., Singhal, G., Govindjee: Science **164**, 301 (1969)
32. Borisov, A.Yu., Godik, V.: J. Bioenergetics **3**, 211 (1972)
33. Campillo, A., Hyer, R., Monger, T.G., Parson, W.W., Shapiro, S.: Proc. Natl. Acad. Sci. USA **74**, 1997 (1977)
34. a) Pashenko, V., Kononenko, J., Protasov, S., Rubin, A., Uspenskaya, N.: Biochim. Biophys. Acta **461**, 403 (1977)
    b) Godik, A., Borisov, A.: FEBS Lett. **82**, 355 (1977)
35. Fenna, R.E., Matthews, B.W.: Brookhaven Symposium Biology **28**, 170–182 (1977)
36. Sybesma, C., Olson, J.M.: Proc. Natl. Acad. Sci. USA **49**, 248–253 (1963)
37. Olson, J.M., in: The Photosynthetic Bacteria. Clayton, R.K., Sistrom, W.R. (eds.), pp. 161–178. New York – London: Plenum Press 1979
38. Olson, J.M., Shaw, E.K., Englberger, F.M.: Biochem. J. **159**, 769–774 (1976)
39. Thornber, J.P., Olson, J.M.: Biochemistry 7, 2262 (1968)
40. Olson, J., Ke, B., Thompson, K.: Biochim. Biophys. Acta **430**, 524 (1976)
41. Beddard, G., Carlin, S., Porter, G.: Chem. Phys. Lett. **43**, 27 (1976)
42. Beddard, G., Porter, G.: Nature **260**, 366 (1976)
43. a) Olson, J.M., Ke, B., Thompson, K.H.: Biochim. Biophys. Acta **430**, 524–537 (1976)
    b) Whitten, W.B., Naim, J.A., Pearlstein, R.M.: Biochim. Biophys. Acta **503**, 251–262 (1978)
44. Phillipson, K.D., Sauer, K.: Biochemistry **11**, 1880–1885 (1972)

45. Bode, V., Hastings, J.: Arch. Biochem. Biophys. **103**, 488 (1965)
46. a) Siegleman, H., Kyrcia, J., Haxo, T.: Brookhaven Symp. **28**, 162 (1978)
    b) Song, P.–S., Koka, P., Prezelin, B., Haxo, F.: Biochemistry **15**, 4422 (1970)
47. Koka, P., Song, P.–S.: Biochim. Biophy. Acta **495**, 220 (1977)
48. Bogorad, L.: Annu. Rev. Plant Physiol. **26**, 369 (1975)
49. Gantt, E.: Photochem. Photobiol. **26**, 685 (1977)
50. a) Williams, V., Glazer, A.: J. Biol. Chem. **253**, 202 (1978)
    b) Freidenrich, P., Aptell, G., Glazer, A.N.: ibid. **253**, 212 (1978)
    c) Bryant, D.A., Hixson, C.S., Glazer, A.N.: ibid. **253**, 220 (1978)
51. Gantt, E., Lipschultz, C., Zilinskas, B.: Biochim. Biophys. Acta **375** (1976)
52. Porter, G., Tredwell, C., Searle, G., Barber, J.: Biochim. Biophys. Acta **501**, 232 (1978)
53. Searle, G., Barber, J., Porter, G., Tredwell, C.: Biochem. Biophys. Acta **501**, 246 (1978)
54. Thornber, J., Alberte, R., Hunter, F., Shiozawa, J., Kan, K.–S.: Brookhaven Symposium in Biology **28**, 132 (1977)
55. Markwell, J.P., Thornber, J.P., Boggs, R.T.: Proc. Natl. Acad. Sci. USA **76**, 1233–35 (1979)
56. Henriques, F., Park, R.: Biophys. Res. Commun. **81**, 1113 (1978)
57. Mullet, J., Burke, J., Arntzen, C.: Palnt Physiol. **65**, 814 (1980)
58. Kan, K.–S., Thornber, J.: Plant Physiol. **57**, 47 (1976)
59. Brody, S., Rabinowich, F.: Science **125**, 555 (1957)
60. Beddard, G., Porter, G., Tredwell, C., Barber, J.: Nature **258**, 166 (1975)
61. Shapiro, S., Kollman, V., Campillo, A.: FEBS Lett. **54**, 358 (1975)
62. Kollman, V., Shapiro, S., Campillo, A.: Biochem. Biophys. Res. Commun. **63**, 917 (1975)
63. Campillo, A., Shapiro, S., Kollman, V., Winn, K., Hyer, R.: Biophys. J. **16**, 93 (1976)
64. a) Porter, G., Synowiec, J., Tredwell, C.: Biochim. Biophys. Acta **459**, 329 (1977)
    b) Harris, L., Porter, G., Synowiec, J., Tredwell, C., Barber, J.: Biochim. Biophys. Acta **449**, 329 (1976)
65. Breton, J., Geacintoy, N., Swenberg, D.: Biochim. Biophys. Acta **548**, 616 (1980)
66. Zankel, R.: Biochim. Biophys. Acta **325**, 138 (1973)
67. a) Beddard, G., Fleming, G., Porter, G., Searle, G., Synowiec, J.: Biochim. Biophys. Acta **545**, 165 (1979)
    b) Beddard, G., Fleming, G., Synowiec, G., Porter, G.: Biochem. J. **6**, 1385 (1978)
    c) Beddard, G., Fleming, G., Porter, G., Tredwell, C.: Picosecond phenomena. In: Advances in chemical Physics Vol. **4**, p. 140; Shank, C , Ippen, E., Shapiro, S. (eds.). Berlin – Heidelberg – New York: Springer 1978
68. Moya, F., Govindjee, Vernotte, C., Briantais, J.: FEBS lett. **75**, 13 (1977)
69. Campillo, A., Shapiro, S.: Photochem. Photobiol. **28**, 975 (1978)
70. Sauer, K., Brewington, G.: Proceedings Int. Congress on Photosynthesis. Hall, D., Coombs, J., Goodwin, T. (eds.), p. 409. London: Biochem. Soc. 1977
71. Birks, J.: Photophysics of Aromatic Molecules. pp. 92–93. New York: Wiley 1970
72. Hindman, J., Kugel, R., Svirmickas, A., Katz, J.: Chem. Phys. Lett. **53**, 197 (1978)
73. Swenberg, C., Geacintov, N., Breton, J.: Photochem. Photobiol. **28**, 999 (1978)
74. Searle, G., Tredwell, C., Barber, J., Porter, G.: Biochim. Biophys. Acta **545**, 496 (1979)
75. Satoh, K., Butler, W.: Biochim. Biophys. Acta **502**, 103 (1978)
76. Campillo, A., Shapiro, S., Geacintov, N., Swenberg, C.: FEBS Lett. **83**, 316 (1977)
77. Butler, W., Tredwell, C., Malkin, C., Barber, J.: Biochim. Biophys. Acta **545**, 309 (1979)
78. Geacintov, N., Breton, J., Swenberg, C., Paillotin, G.: Photochem. Photobiol. **26**, 619 (1 977)
79. Swenberg, C., Geacintov, N.: Exciton interactions in organic solvents. In: Organic Molecular Photophysics; Birks, J. (ed.). New York: Wiley 1973
80. Paillotin, G., Swenberg, G., Breton, J., Geacintov, N.: Biophys. J. **25**, 513 (1979)
81. Mauzerall, D.: Biophys. J. **16**, 87 (1976) and J. Phys. Chem. **80**, 2306 (1976)
82. Lutz, M., Cogdell, R.J.: Unpublished observation

Chapter 4
Photooxidation of the Reaction Center Chlorophylls and Structural Properties of Photosynthetic Reaction Centers*

A.J. Hoff[1]

Abbreviations and Symbols

A	primary acceptor of photosystem 1 as measured by ESR
A, a	hyperfine splitting
ATP	adenosine triphosphate
A_1, A_2	intermediary acceptor of photosystem 1 as measured optically
B	primary acceptor of photosystem 1 as measured by ESR
Bchl	bacteriochlorophyll
Bph	bacteriopheophytin
C.	*Chromatium*
CD	circular dichroism
Chl	chlorophyll
cyt	cytochrome
D	a low-temperature donor to $P680^+$
D	zero field parameter of triplet state
$D_{a,b}$	electron transfer energy distribution
DMF	dimethyl formamide
Δ	Stokes shift parameter
ΔA	absorption difference
ΔCD	circular dichroism difference
ΔH	peak-to-peak line width of derivative ESR spectrum
ΔLD	linear dichroism difference
$\Delta\epsilon$	differential extinction coefficient
δE	$E_a - E_b$
$\delta E^{\ddagger}$	activation energy of electron transfer
E	zero field parameter of triplet state
$E_{a,b}$	redox energy levels
E_h	redox potential
E_m	redox midpoint potential
ENDOR	electron nuclear double resonance
ESR	electron spin resonance
ET	electron transfer
ϵ_1	extinction coefficient for left-handed circularly polarized light
ϵ_r	same for right-handed circular polarization
G	Gauss
g	electronic g-value
H	"heavy" subunit of RC protein of *Rps. sphaeroides* R–26
hfs	hyperfine splitting
I	intermediary acceptor in bacterial photosynthesis
J	exciton interaction; exchange interaction
k	Boltzmann's constant
$k_{a,b}$	force constants defining curvature of potential well
k_c	rate of charge separation
$k_{x,y,z}$	decay rates of triplet sublevels in zero magnetic field
LD	linear dichroism
LDAO	lauryl dimethylamine oxide
LM	sub-RC particle of *Rps. sphaeroides* R–26 consisting of "light" and "medium" subunit
m_s	magnetic quantum number
NAD	nicotinamide adenine dinucleotide
NADP	nicotinamide adenine dinucleotide phosphate
NMR	nuclear magnetic resonance
P, P860, P870, P960	primary donor in bacterial photosynthesis
P^*	exited state of the primary donor
P^F	the transient state P^+I^- of the bacterial reaction center
P^R	state $P^TI\ X^-$ of the bacterial reaction center
P^T	triplet state of the bacterial primary donor
p	polarization ratio

* See also *Notes Added in Proof* on pp. 322-326

1 Biophysics Department, Huygens Laboratory, University of Leiden, Leiden 2405, The Netherlands

pH	$-\log[H^+]$
pK	pH at which $[AH]/[A^+] = 1$
Pheo	pheophytin
PS 1	photosystem 1
PS 2	photosystem 2
P680	primary donor of PS 2
P700	primary donor of PS 1
Q	primary acceptor of photosystem 2
Q_x, Q_y	electronic transition moments
$Q.Fe^{2+}$	quinone-iron primary acceptor complex in bacterial photosynthesis
Q-band	microwave frequency of about 36 GHz
R	distance between two sites
R.	*Rhodospirillum*
RC	reaction center
RPM	radical pair mechanism
Rps.	*Rhodopseudomonas*
S I	signal I, ESR signal of $P700^+$
SDS	sodium dodecyl sulfate
T	temperature
T_{ab}	tunneling matrix element
T_c	transition temperature in tunneling theory
$T_{1,0,-1}$	triplet levels in high magnetic field
$t_{\frac{1}{2}}$	decay half time
t_m	transition dipole moment
THF	tetrahydrofuran
TMPD	N-tetramethyl-p-phenylene diamine
UQ_1, UQ_2	primary and secondary ubiquinone functioning as primary and secondary acceptor in *Rps. sphaeroides*, respectively
W_{ab}	electron transfer rate
X	primary acceptor in bacterial photosynthesis; intermediary acceptor of photosystem 1 as measured by ESR
X-band	microwave frequency of about 9 GHz
ZFR	zero field resonance

1 Introduction

The central issue of photosynthesis is the primary photochemical event, the charge separation process. With a high efficiency the energy of a light quantum is converted into chemical free energy confined in the charged donor-acceptor pair, D^+A^-. From the chemical point of view this is a most unusual process. Commonly, photochemistry is carried out with high-energy ultraviolet light. Reactions utilizing light of wavelengths between 400 nm and 900 nm are few and have a low quantum yield. Yet, this is precisely the region of wavelength of interest to anyone who intends to put solar energy to use on a technological scale. Nature, in the course of evolution, has found a solution to the problem of efficient solar energy conversion and it seems appropriate to look to her for guidance in the difficult problem of how to engineer a way to harness the Sun's energy as efficiently, cheaply and nonpollutingly as Nature does. It is clear then that understanding the mechanisms of the primary photoact of photosynthesis, $DA \xrightarrow{h\nu} D^+A^-$, is our most important task. In the last 20 odd years, starting with the pioneering application of absorption difference spectroscopy by Duysens [1], much progress has been made in the elucidation of the primary photochemistry of photosynthesis. In the last decade a great variety of techniques has augmented and supplemented classical optical difference spectroscopy. As a result of the concerted efforts of relatively few groups of investigators, we are now in a position to identify a good number of the primary reactants that hitherto went under the disguise of labels as P, Q, X, etc.

In this chapter I will first trace the lines of research that have enabled the indentification of the primary components. The main results will be reviewed in some detail, finally a tentative sketch will be given of the molecular mechanism leading to charge separation.

A list of recent review papers pertaining to the primary processes is given in the bibliography. Of course, the act of charge separation, albeit of primordial importance, is but one in the sequence of chemical reactions that lead to the overall result of photosynthesis; i.e., the fixation of CO_2 to carbohydrate as summarized by van Niel's equation [2]: $CO_2 + 2H_2A \rightarrow (CH_2O) + H_2O + 2A$, where H_2A stands for water in plant photosynthesis, and for a number of reduced substrates (as H_2S, or organic molecules) in bacterial photosynthesis. These reactions, as well as the process of light capture and resonant transfer of energy to the phototrap, are not touched upon here. A number of excellent books have recently appeared where extensive up-to-date information on these and related subjects can be found (see bibliography).

1.1 Energetics of Photosynthesis

As much of the discussion in this chapter will be focused on the redox properties of the primary reactants, it is helpful to summarize the chain of the primary photoreaction(s) and subsequent dark reactions on a redox energy scale. For plant photosynthesis this is the so-called Z-scheme (Fig. 1.1a), for bacterial photosyntheses an analogous scheme can be drawn (Fig. 1.1b). The Z-scheme comprises two photoreactions: photosystem 1 (PS 1) and photosystem 2 (PS 2), which are connected by a series of dark redox reactions. The ultimate electron donor is water, the final electron acceptor is oxidized nicotinamide adenine dinucleotide ($NADP^+$), which after reduction is used in a cyclic series of reactions, known as the Calvin cycle, to reduce CO_2 to carbohydrate. A number of dark reactants have been identified (see Fig. 1 and legend). The primary donor and acceptor of PS 1 are commonly labeled P700 and X, respectively, that of PS 2, P680 and Q, respectively. P stands for pigment, the number refers to the wavelength in nm at which in absorption difference spectroscopy a bleaching is observed. The effect of the two photoreactions is the generation of a compound of sufficiently high reductive power to reduce $NADP^+$. The minimal redox span needed ist +0.82 eV (the midpoint redox potential, E_m, of the couple $H_2O/2H^+ + O + e$) – (–0.32 ev) (the E_m of the couple $NADP^+/NADP$) = 1.14 eV. The energy contained in the two quanta (680 and 700 nm) is 3.61 eV, so that the overall efficiency is 32%. Part of the loss is recaptured by the synthesis along the chain of probably two molecules of adenosine triphosphate (ATP), the universal energy carrier in the cell (see for reviews section III of ref. [2]). This amounts to 2 x 0.165 = 0.31 eV, so that the efficiency becomes 40%. Another way of assessing the economics of the two photoreactions is the observation that for one molecule of oxygen liberated, four electrons are needed. This means that both PS 1 and PS 2 should turn over four times, i.e., eight quanta are used. This amounts to an overall efficiency of 37%, since the net reaction of photosynthesis $CO_2 + H_2O \rightarrow \text{carbohydrate} + O_2$ requires 120 kCal/mol of free energy and each quantum of 680 nm is equivalent to 41 kCal/mol. Apparently, the loss of photoenergy is partly due to the necessity to create energy barriers in order to prevent rapid back reactions. A thermodynamic argument has been given by Duysens [3], which was later amplified by others [4, 5], that the accumulation of long-lived photoproducts at low light intensity necessitates a loss of about 30%. One should be cautioned, however, that this reasoning is not applicable to the primary photoact as such, since its products undergo extremely fast back or forward reactions. This means that the photoexcited primary donor and the first electron acceptor may have comparable redox levels (see e.g., a discussion by Parson [6]).

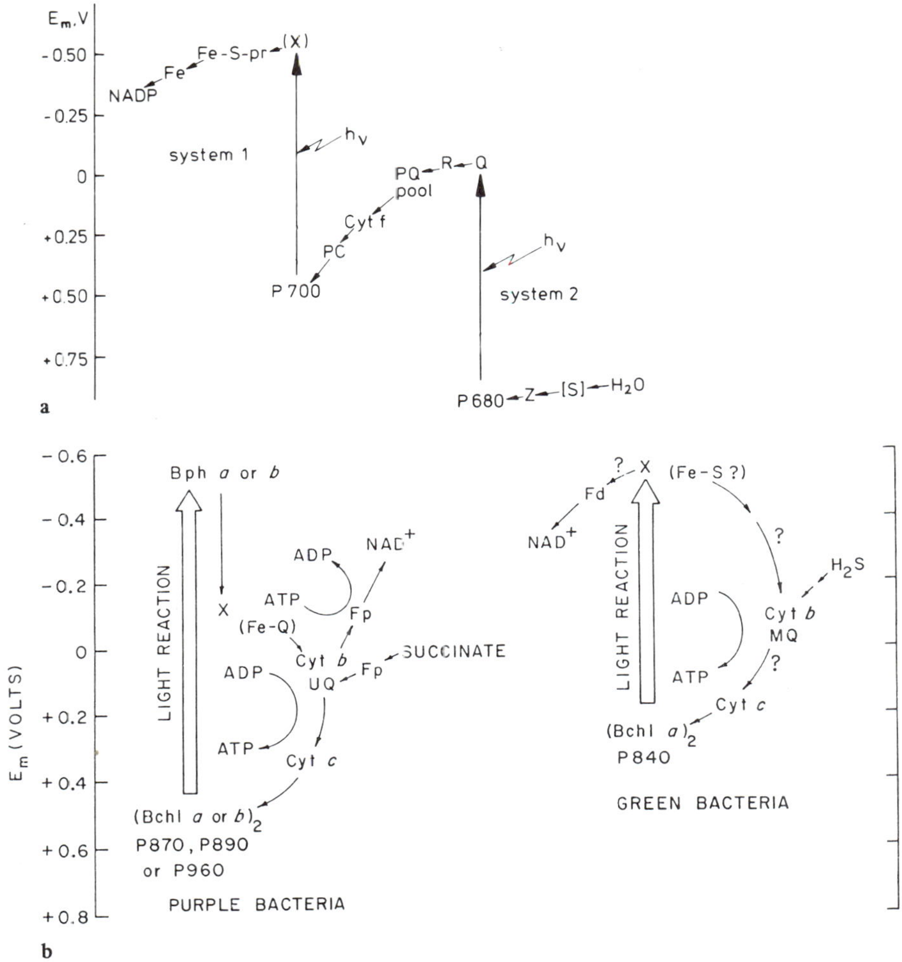

Fig. 1.1. **a** The Z-scheme. Location of electron carriers of photosystems 1 and 2 on a redox potential scale. **b** Same for the bacterial photosystem. (From [18])

Bacterial photosynthesis is considerably less complex than that of algae and higher plants and requires less energy. Only one photoreaction is needed to create a redox span of 0.6 to 0.7 V between the E_m (0.45 V) of the primary donor and the E_m(−0.15 to −0.20 V) of the primary acceptor of the purple photosynthetic bacteria that can use inorganic sulfur in one or another form as electron donor, the Chromatiaceae, and those that use organic compounds for this purpose, the Rhodospirillaceae. For green photosynthetic bacteria (the Chlorobiaceae) the primary acceptor has a much lower mid-point potential of about −0.5 V, which reductive power is sufficiently high to directly reduce oxidized nicotinamide adenine dinucleotide (NAD^+) to NADH (E_m (pH 7.0) = −0.32 V), which in turn is used in a cyclic series of reactions to reduce CO_2. For the purple bacteria the reductive power created by the light reaction is too small to reduce NAD^+. It is used to produce ATP which in turn is used to reduce NAD^+ in a process called reverse electron flow (see e.g., [7]). The purple bacteria have phototraps with near-infrared absorption maxima ranging from 860 nm to 960 nm, the green bacteria use light around 840 nm. The E_m of the couple $NAD^+/NADH$ is −0.32 V, that of the primary donor for green bacteria is +0.24 V at pH 7.0, so that the efficiency for direct reduction of NAD^+ is about 40%. The overall efficiency for purple bacteria is somewhat less easily calculated as the detailed energy balance of reverse electron flow coupled to cyclic electron transport is not fully known. As discussed in Sect. 2, the species that use light around 860 nm generate reduced electron acceptors with an E_m of about −0.20 V, those that use 960 nm light have a primary acceptor with an E_m of about −0.15 V. The E_m of the primary electron donors ranges between +0.45 and +0.50 Volt so that the redox span is about 0.65 V, and the efficiency for 870 nm light (1.35 eV)[2] is 48%, and for 930 nm light (1.26 eV)† 51%. It may be observed that the efficiencies of both plant and bacterial photosynthesis, as calculated above, are rather similar, viz. 40% to 50%. However, if one takes the redox span between the ultimate electron donor [e.g., succinate (E_m succinate/fumarate ~ 0.0 V) for purple bacteria] and the primary acceptor, bacterial photosynthesis would seem to be much less efficient than plant photosynthesis.

1.2 Chlorophylls, Quinones and Related Molecules

Anticipating on the data reviewed below, a number of structural formulas of chemical compounds that play a role in primary reactions are listed in Fig. 1.2.

2 The Photosystem of Purple Bacteria[3]

2.1 Optical Investigations

2.1.1 Absorption Difference Spectroscopy

The first indication that a long-wavelength form of Bchl is involved in the primary photoact came from work on whole cells of *Chromatium vinosum* and *Rhodospirillum*

---

2 By the Franck-Condon principle the usable quantum energy is that of the 0−0 transition, which is about 0.1 eV lower than the energy corresponding to the peak of the absorption band. Including in the calculation the light-induced membrane potential (about 100 mV) will raise the efficiency somewhat

3 It is outside the scope of this chapter to describe the organization of the photosynthetic apparatus. Suffice it to observe that it is located inside a membranal system. A number of antenna pigments absorbing at various wavelengths in the visible region collect light and transfer the photon energy to the phototrap by resonant energy transfer. Among the antenna pigments Bchl *a* (and in plants Chl *a* and Chl *b*) plays a dominant role. We will limit this review to purple photosynthetic bacteria. Up to very recently, the green bacterial photosystem was largely unstudied. For a series of recent investigations, see *Notes Added in Proof*, p. 322.

rubrum by Duysens [1, 8, 9], who discovered with absorption difference (ΔA) spectroscopy small light-induced absorptivity changes at 870 nm. Goedheer [10], and Fuhrop and Mauzerall [11], compared the ΔA spectrum of chemically oxidized Bchl in vitro and that of chemically oxidized chromatophores[4] of *Rps. sphaeroides* and *R. rubrum* confirming the earlier conclusion [9] that the light-induced ΔA spectrum was caused by oxidation of one or more Bchl molecules with a red absorption maximum around 870 nm. In Fig. 2.1 a recent ΔA spectrum of chromatophores of *R. rubrum* is displayed, together with the corresponding absorption spectrum of the chromatophores. The main features of the ΔA spectrum are a strong bleaching at 865 nm, with corresponding bleachings at 605 and 385 nm, and an apparent wavelength shift of an absorption peak at 800 nm. The blue-green region of the ΔA spectrum is compared with the difference spectrum of oxidized *minus* neutral Bchl *a* in methanol in Fig. 2.2. It is seen that the two agree remarkably well, indicating that most of the absorption changes of the phototrap in the blue-green region are due to oxidation of a bacteriochlorophyll species. The far-red region is somewhat more difficult to interpret, as the $Bchl^+$ – Bchl spectrum shows a strong bleaching at 770 nm instead of 865 nm. The shift toward the red of the bleaching measured in vivo is usually taken to indicate that the phototrap pigment has a strong interaction with one or a few chlorophyllous species [14, 15]. Note that also the antenna pigments strongly absorb around 870 nm. This bathochromic shift of the antenna pigments has been similarly attributed to pigment-pigment interaction, but we will see later that this interpretation of the far-red wavelength shifts is not necessarily correct (Sect. 3.1.1).

From the early work on chemically oxidized phototraps and in-vivo oxidation of Bchl it is clear that Bchl is involved in the primary donor complex. More precise data on the phototrap were obtained with traps isolated from the intracellular membrane with detergents [16, 17]. Notably, work on the carotenoidless mutant R–26 of *Rps. sphaeroides* resulted in the isolation of highly purified phototraps, the reaction centers (RC). Later, also from other species RC preparations of comparable purity have been prepared [301, 302] (see for a review [18]). It appears, that the RC of practically all species studied is composed of a protein of a molecular weight of about 100 kD. By treatment with the detergent SDS, it can be decomposed in three (in *Rps. gelatinosa* two, [303]), subunits of comparable molecular weight. The protein contains four Bchl molecules, two bacteriopheophytin (Bph) molecules, and one or two ubiquinones. Mild treatment with a "cocktail" of detergents and chaotropic agents causes the RC to split in two parts, one of which contains two subunits. This heavier particle is still photoactive, it consists of residues of unusual low average polarity. These units secure the photoreactive part into the membrane. An absorption spectrum of RC of *Rps. sphaeroides* R–26 and its corresponding ΔA spectrum are displayed in Fig. 2.3.

It is seen that the red part of the absorption spectrum is now resolved into three distinct bands: the 865 nm band is, on the basis of its bleaching in the absorption difference spectrum, attributed to the primary donor, the 800 nm band to the associated Bchl's and the 780 nm band to the two Bph's. The bands in the green are partly due to

4 Chromatophores are membrane vesicles containing the "complete" photosynthetic apparatus, that are obtained by disrupting the bacterial cell wall by sonication or forced flow through a narrow orifice (press treatment after French)

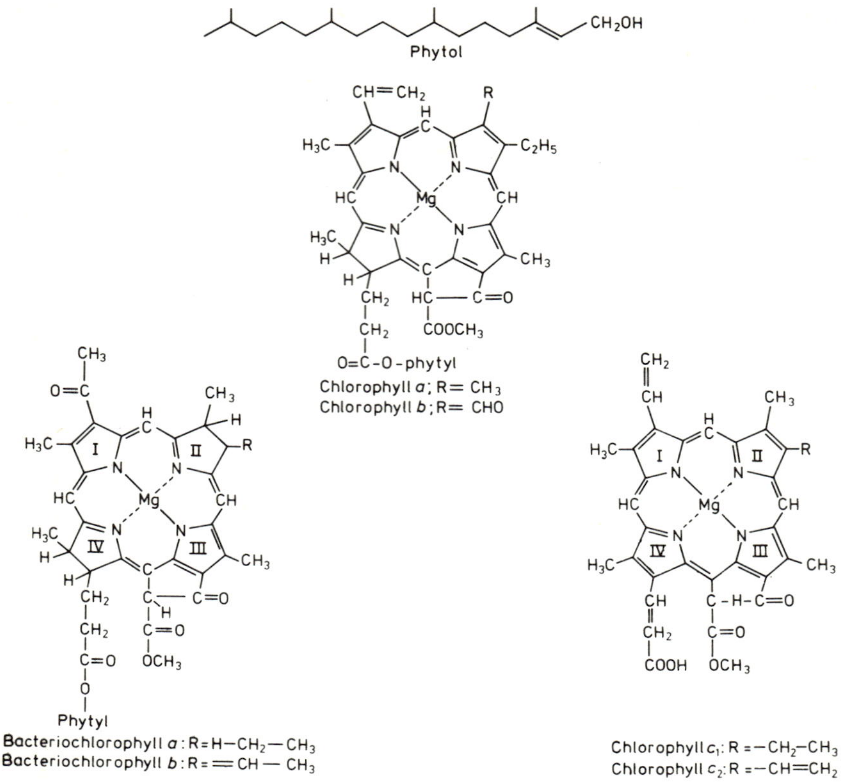

Fig. 1.2. Chemical formulas of chlorophylls, quinones and carotenoids

Bchl (near 600 nm) and partly to the Bph's (near 530 nm). The Soret region around 400 nm shows overlap of all chlorophyllous pigments, and is for this reason not very informative. The ΔA spectrum shows the already mentioned almost complete bleaching of the 865 nm band, an apparent blue shift of a few nm at 800 nm[5] and 365 nm, and a partial bleaching and a shift at 600 nm. A differentail extinction coefficient $\Delta\epsilon = 90\ mM^{-1}\ cm^{-1}$ has been obtained for P865 in chromatophores of *R. rubrum* [19], and $\Delta\epsilon = 112\ mM^{-1}\ cm^{-1}$ for RC of *Rps. sphaeroides* R–26 [20].

A striking feature of the ΔA spectrum is the appearance of a relatively strong band at 1245 nm. This band is not present in the ΔA spectrum of $Bchl^+$ – Bchl. A theoretical calculation of the latter spectrum [21] did predict a ΔA band at about this wavelength but with smaller oscillator strength than the experimental ΔA band. The oscillator strength of the bands at 800 and 865 nm are about equal (based on a Gaussian approxi-

5 For a shift of a Gaussian absorption band, one can write $x/\Delta\lambda = (2 \ln 2)^{-1/2} \left(\frac{\text{arctanh}\ a/x}{a/x}\right)^{1/2}$, where $\Delta\lambda$ is the width at half height of the absorption band, x the measured peak-to-peak width of the ΔA spectrum, and "a" the band shift. For $0 < a/\Delta\lambda < 0.5$, $x/\Delta\lambda \sim (2 \ln 2)^{-1/2}$ (it varies between 0.85 and 0.90). Thus, the value of x, 25 nm, from the ΔA spectrum of Fig. 2.3 is consistent with a shift of ~6 nm (Fig. 2.3 *bottom*) together with $\Delta\lambda$ (800 nm) ~ 30 nm. However, in Sect. 2.1.4 other interpretations of the ΔA spectrum around 800 nm are discussed

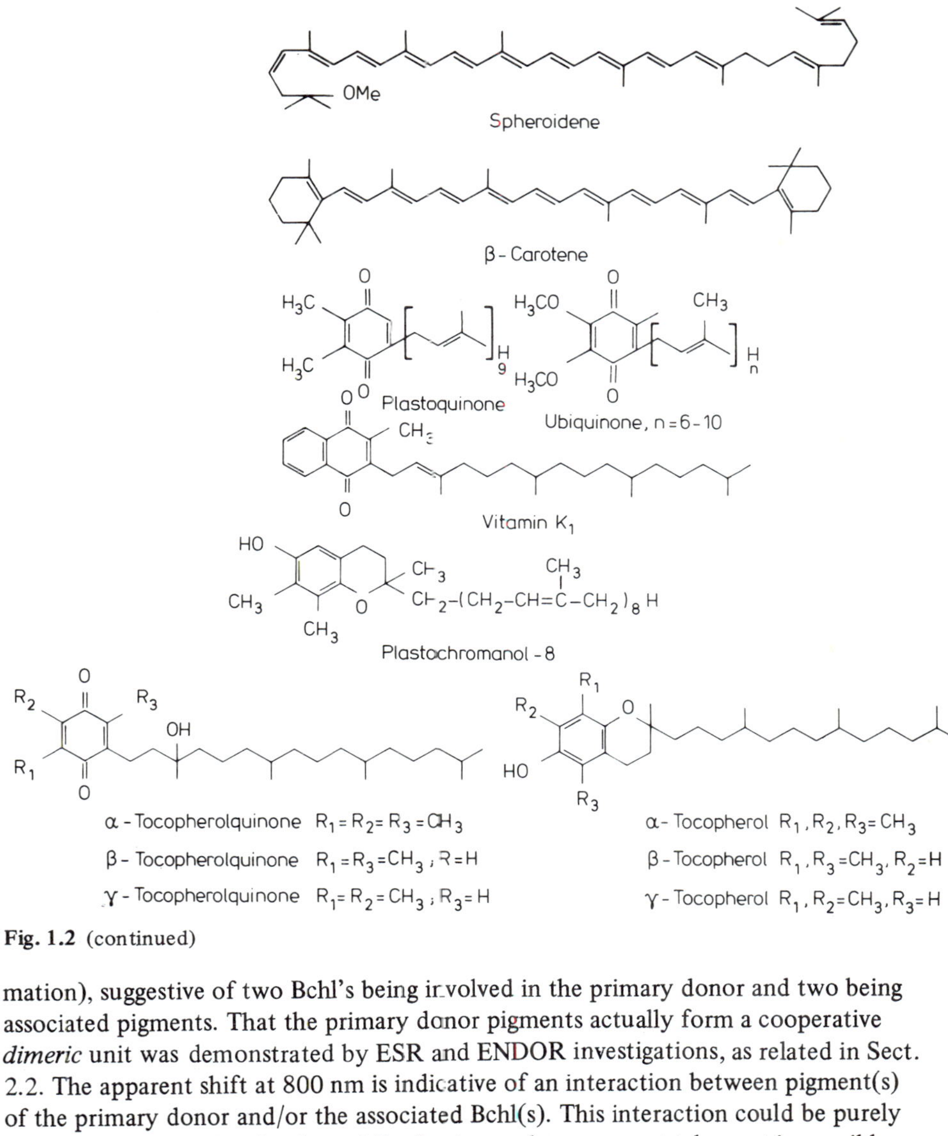

Fig. 1.2 (continued)

mation), suggestive of two Bchl's being involved in the primary donor and two being associated pigments. That the primary donor pigments actually form a cooperative *dimeric* unit was demonstrated by ESR and ENDOR investigations, as related in Sect. 2.2. The apparent shift at 800 nm is indicative of an interaction between pigment(s) of the primary donor and/or the associated Bchl(s). This interaction could be purely electrostatic, or excitonic: the oxidized primary donor cannot take part in possible exciton[6] interaction anymore, which would result in rather drastic changes in the absorp-

6 An exciton state is a collective excitation of an assembly of molecules having such a strong resonant coupling that it is not possible to excite one molecule individually. The exciton model was first developed by Davydov [31] and extended by Kasha and co-workers [32–34]. The source of the coupling is in electrostatic interaction between transition charge distributions, which is usually approximeted by interaction between transition dipoles. These interactions result in the splitting of the energy level into a number of new levels equal to the number of equivalent interacting molecules, i.e., for a dimer two levels result (the Davydov components) which are split by the exciton interaction J. Spectral properties of exciton states have been calculated by Tinoco [35, 36]

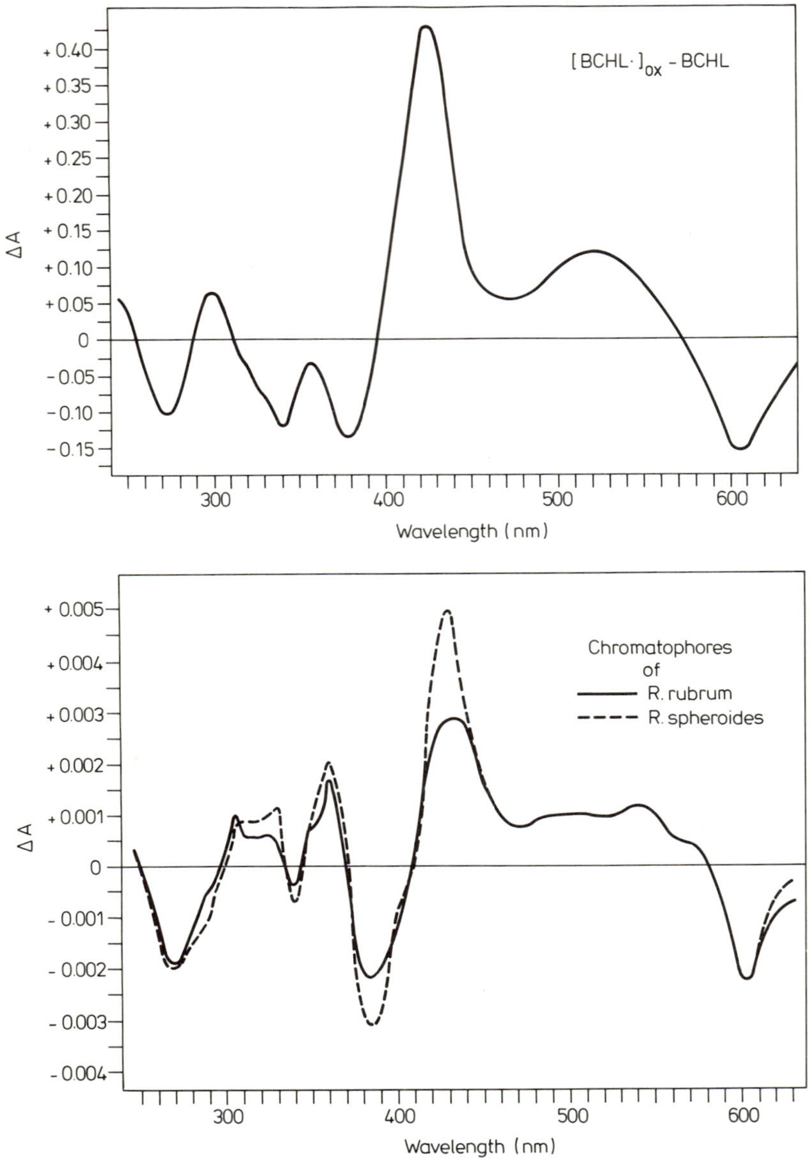

Fig. 2.1. Difference absorption spectrum of the oxidation of Bchl in methanol by I_2 (*top*) and of the photooxidation of chromatophores of *R. rubrum* and *Rps. sphaeroides (bottom)*. (From [12])

tion spectrum of the donor and the associated pigments.

The analysis of the ΔA spectrum is aided by circular and linear dichroism studies. A review of the results obtained with these methods is postponed to the next section.

The redox midpoint potential of the primary donor in chromatophores and reaction centers of the purple bacteria lies in the range +390 to +500 mV (pH 7.0) [18, 22].

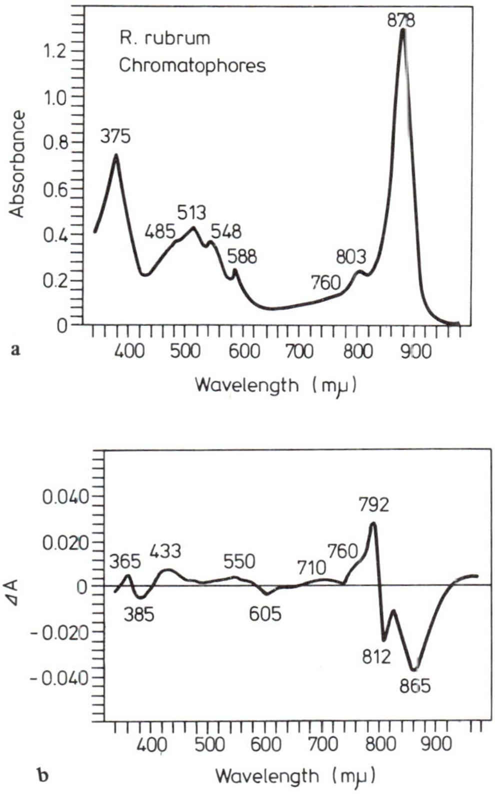

Fig. 2.2. **a** Absorption spectrum and **b** absorption difference spectrum light-*minus*-dark of chromatophores of *R. rubrum*. (From [13])

The E_m of the donor in the green bacterium *Chlorobium thiosulphatophilum* is somewhat lower, viz E_m = +220 to +250 mV (pH 7.3–6.8) [23, 24]. These E_m values compare to the E_m of the couple Bchl/Bchl$^+$ in methanol of +270 mV [11].

The discussion up to this point has been limited to Bchl *a*, and Bchl *a* containing species. Recently, much work has been devoted to Bchl *b* and Bchl *b* containing organisms (the purple bacteria *Rps. viridis* and *Thiocapsa pfennigii*). In Bchl *b* the ethyl group on ring II is replaced by an ethylidene group (Fig. 1.2). This causes the red absorption bands to shift to longer wavelength and, because of the greater spread in wavelength, the various bands in RC preparations of *Rps. viridis* are better resolved. The near-infrared spectrum shows maxima at 790, 810, 830, 850, and 960 nm [25, 26, 27] which are resolved at low temperatures. The 790 nm band is ascribed to Bph *b*, the bands at 850 and 960 are bleached when the primary donor is oxidized. The other bands belong to associated Bchl *b*. Upon oxidation of P960, a band at 1310 nm appears, which is analogous to the 1245 nm band for Bchl *a* containing species. The molar differential extinction coefficient for oxidation of P960 in *Rps. viridis* was determined to be $\Delta\epsilon = 123 \pm 25\ mM^{-1}\ cm^{-1}$ [27]. At low temperature (30 K) the Bchl-*b* bands at 600 nm are resolved in two bands at 601 and 606.5 nm, whereas in this spectral region the two Bph-*b* molecules then show up at 531 and 543 nm [27].

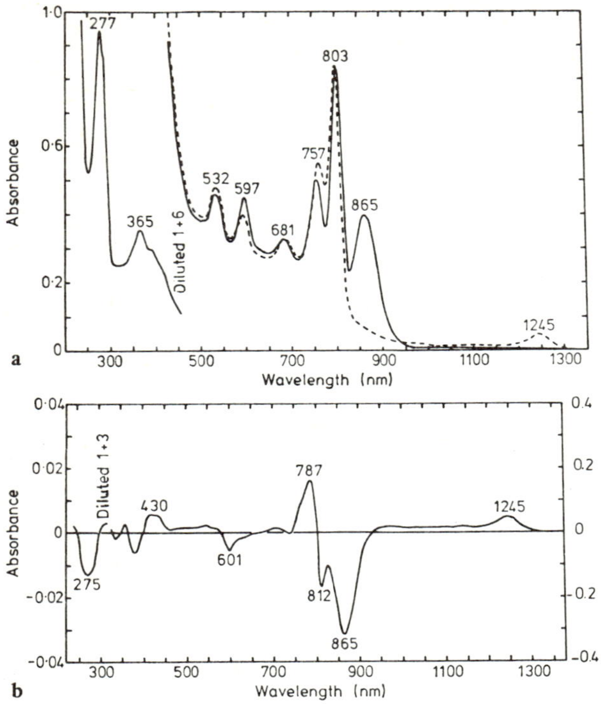

Fig. 2.3. a Absorption spectrum of reaction centers of *Rps. sphaeroides* R–26 ——— dark, – – – – – under illumination by 800 nm actinic light. **b** Absorption difference spectrum light *-minus*-dark. The left hand scale refers to the UV; the right hand scale to the visible and infrared. (After [287])

From the optical experiments reviewed above, from redox potentiometry and from chemical analysis of unmasked phototraps [28, 29] and of purified RC [30] it is abundantly clear that the primary donor consists of bacteriochlorophyll. The pigment(s) are identical in chemical structure to the antenna Bchl and the RC-associated Bchl as evidenced by optical and chromatographical analysis of the extracted pigments. Also, the ESR and ENDOR properties of oxidized extracted RC Bchl are identical to those of oxidized antenna Bchl (G. Feher, pers. comm.). Thus, one must conclude that the special donor property of the primary donor complex is due to its geometrical structure, interaction with other pigments and possible protein or exogeneous ligands, and to the presence of an acceptor species in direct proximity to the donor complex. The assessment of the complex structure of the reaction center aggregate is the subject of the following sections.

2.1.2 Spectroscopic Nomenclature of Bchl

Before discussing the circular dichroism, CD, and linear dichroism, LD, of the reaction center it will be helpful to introduce some terminology [37, 38]. In Fig. 2.4a the Bchl

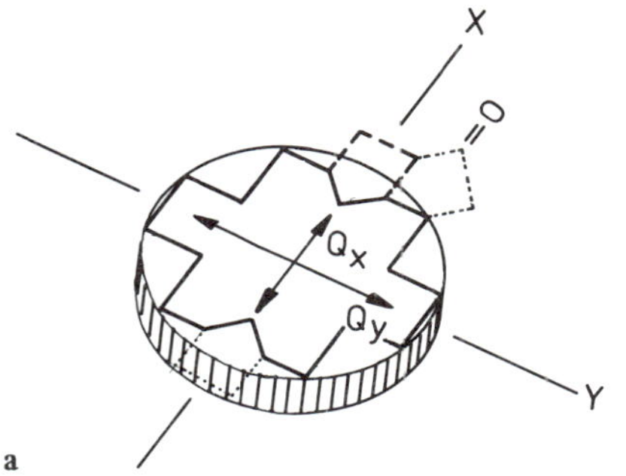

a

Fig. 2.4. a The chlorophyll π-electron system. The molecule is represented as an *oblong box*. The *heavy drawn lines* represent the unsaturated ring system of Bchl. The *heavy dashed line* stands for the unsaturated ring II of Chl *a, b*. The *thin dashed lines* represent the saturated rings IV and V. **b** Absorption spectra of Chl *a* and Bchl *a* in ether solution for equimolar concentrations. The polarization of the bands (x or y) corresponds to the direction of the transition moment indicated in **a**. (From [288])

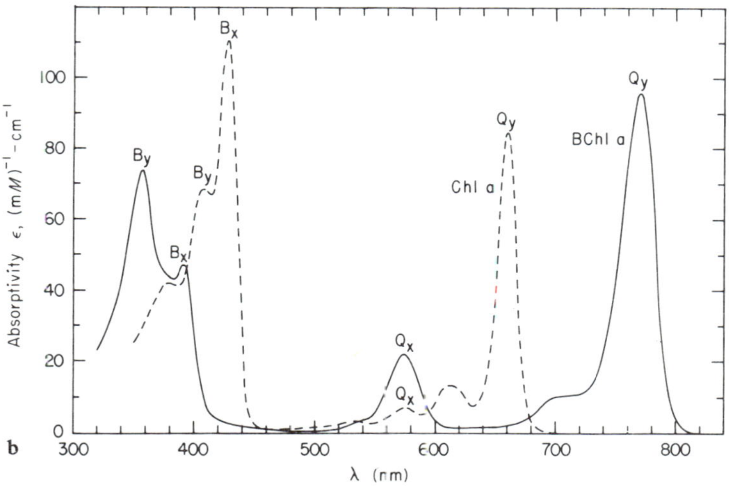

b

molecule is schematically depicted. The heavy lines indicate the conjugated pathway. The molecule can be regarded as an oblong box filled with π-electrons. The electronic transitions are polarized in the plane of the molecule (it is easy to change the charge distribution along the long and short axis of the box, but not perpendicular to the flat sides of the box). By convention, the transition along the long axis of the unsaturated pyrrole rings is called the Q_y transition dipole moment (tm), that polarized along the short axis, the Q_x tm. The Q_y and Q_x tm of Bchl *a* in vivo lie in the range of 800 to 870 nm, and around 600 nm, respectively (Fig. 2.4b). Both are S_0–S_1 transitions. The absorption band in the blue (Soret band or B transition) consists of overlapping S_0–S_2 transitions of opposite polarity, i.e., it is almost circularly polarized in the plane of the molecule. The transitions of Bph are shifted somewhat to the blue; apart from this the electronic spectrum is very similar to that of Bchl, and the bands are also classified into Q_x, Q_y . . . . . . bands.

The linear dichroism corresponding to Q_y and Q_x tm will have opposite sign. The same is true for the CD bands associated with these transitions, but the reason for this is not as trivial as for the LD bands [39, 40].

If two Bchl molecules are coupled by exciton interaction, the excited electronic energy levels are split. The oscillator strength of the transitions to the two levels depends on the relative orientation [33, 34]: for parallel tm, the transition to the lower level is forbidden, for colinear tm, the transition to the upper level is forbidden. When the tm of the monomers make an angle, the transitions to the two exciton levels are both permitted and polarized perpendicular to each other, with relative oscillator strength dependent on the angle.

2.1.3 Circular Dichroism

Circular dichroism (CD) is the difference between the absorption of light of left- and right-handed circular polarization: CD = $\epsilon_1 - \epsilon_r$. The difference is caused by intrinsic asymmetry of the molecule (optical activity) or by externally induced asymmetries, e.g., by exciton coupling between two or more chromophores [35, 36]. CD studies of the bacterial photosystem were first performed by Dratz et al. on chromatophores of *Rps. sphaeroides* and *R. rubrum* [41]. They were later complemented by studies on chromatophores and reaction center preparations of various species [42–44].

The red region of the CD spectrum of RC preparations of Bchl *a* containing species all show four characteristic bands (Fig. 2.5): a weak negative band at 750 nm, a strong narrow positive band at 795 nm, a strong narrow negative band at 810 nm, and a broad, rather weak, band at about 860 nm (the wavelength figures vary slightly from species to species). Upon oxidation the strong negative band at 810 nm and the broad positive band nm are largely bleached, whereas the band at 795 nm shows a small increase. The ΔCD spectrum (oxidized-*minus*-reduced) displayed in Fig. 2.6 shows a narrow positive peak at 810 nm and a broad negative band at 860 nm. The above features are reproduced in the CD and ΔCD spectrum of *Rps. viridis* at longer wavelengths (Fig. 2.6). The

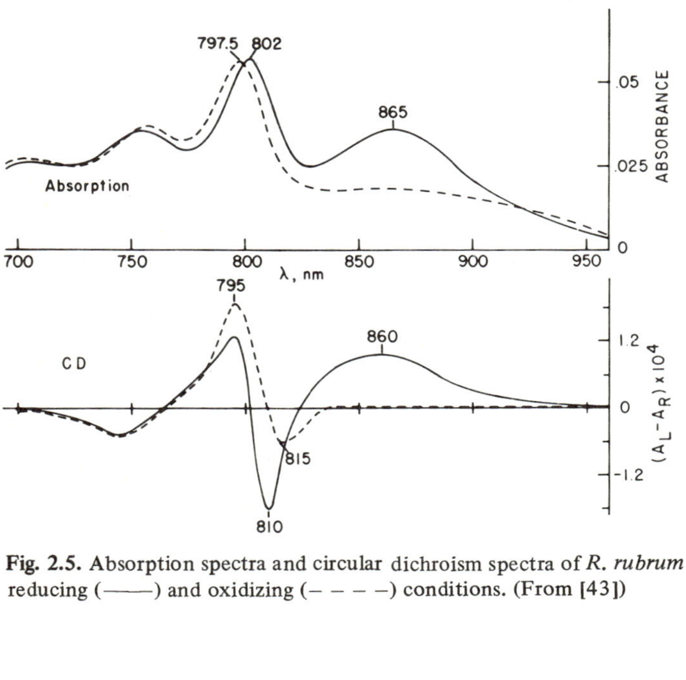

Fig. 2.5. Absorption spectra and circular dichroism spectra of *R. rubrum* reaction centers under reducing (——) and oxidizing (– – – –) conditions. (From [43])

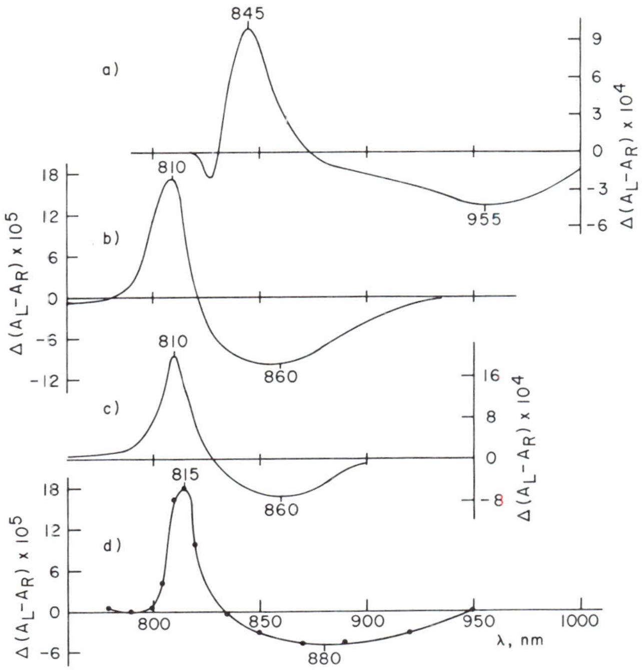

Fig. 2.6. Oxidized-*minus*-reduced CD difference spectra of reaction center preparations of *a Rps. viridis, b R. rubrum, c Rps. sphaeroides, d C. vinosum.* (From [43])

4-banded structure of the red region of the CD spectra was interpreted to arise from Bph (795) and three [42] or four [43, 44] Bchl molecules of which those absorbing at 803 nm interact, leading to CD bands of opposite sign as a result of exciton splitting [36]. A shoulder at 810 nm observed in the B800 absorption band by Feher [17] supports the notion that this band is a composite.*

The CD spectrum of *Rps. viridis* under reducing conditions was measured by Shuvalov et al. [25] who found that illumination at a redox potential of −450 mV resulted in a strong decrease of the positive CD band at 827 nm (this band corresponds to the 795 nm band in a Bchl *a* containing species). Illumination at low redox potentials causes reduction of one of the Bph-*b* molecules (Bph−790). As will be discussed later, this molecule is an intermediary electron acceptor in the primary reaction $(Bchl)_2$ Bph $Q.Fe^{2+} \xrightarrow[3\ ps]{h\nu} (Bchl)_2^+\ Bph^-Q.Fe^{2+} \xrightarrow{200\ ps} (Bchl)_2^+\ Bph\ Q^-.Fe^{2+}$ where $Q.Fe^{2+}$ (a quin-one-iron-complex) is the "primary" electron acceptor. Thus, oxidation of $(Bchl)_2$ causes disappearance of the negative 847 nm CD band, whereas reduction of Bph−790 causes the disappearance of the positive 827 nm band. In addition to these CD spectra, Shuvalov et al. [45] measured linear dichroism of RC of *R. rubrum* and of *Rps. viridis* [289] by the method of photoselection. On the basis of their CD and LD spectra, the authors offered a model for the organization of the pigments in the reaction center unit. Before discussing this model, however, we will review the results of LD spectroscopy obtained by other authors on Bchl *a* containing species.

* Asterisks refer to *Notes Added in Proof*, (see pp. 322-326)

2.1.4 Linear Dichroism

By definition linear dichroism (LD) is the difference in absorbancy of a sample when monitored with a measuring beam of two mutually perpendicular directions of polarization: LD = $A_{//} - A_{\perp}$, where $A_{//}$ and $A_{\perp}$ are the absorbance of the sample for the two directions of polarization (unspecified as yet with respect to the laboratory coordinate frame).

The value of LD is zero, when the absorbing pigments are oriented randomly, or when the distribution of their transition moment is spherically symmetric. Thus, to obtain LD $\neq$ 0 one has to orient the pigments in some way, or one has to introduce an asymmetry by selecting a particular slice out of a random distribution by producing photochemistry with a beam of actinic light (usually polarized, although this is not necessary per se): the method of photoselection. The latter method is fairly old (its principle and general formulation were already discussed in detail in 1935 by Jablonski [46]), but it has been used only recently in the study of photosynthetic complexes. An excellent introduction to this technique is given by Albrecht [47].

Orientation of the pigments has been achieved in a variety of ways: by spreading or drying preparations on glass slides, by applying a magnetic or an electric field, or by a velocity (flow) gradient. The brush-spreading technique depends on asymmetry in the shape of the cells or membrane system containing the pigments; it does not yield good results with RC preparations. Drying occasions a flattening of membranes and thus good orientation is achieved even for spherically symmetric systems as chromatophores, etc. The orientation of photosynthetic membranes (chloroplasts, bacterial cells) by a magnetic field (discovered by Geacintov et al. [48, 49]) and by an electric field (first employed by Gagliano et al. [50]) is perhaps the most reliable technique, although it is difficult to obtain fully oriented samples. Lately, it has been found that RC in gelatin films exhibit very good orientation upon stretching [51, 290] (Fig. 2.7a).

Bchl *a* Containing Species. LD in photosynthetic bacteria was first observed by Morita and Miyasaki [52] employing both the flow and the drying technique on whole cells of *Rps. palustris* (in this bacterium the membranes are stacked in regular arrays of lamellae; this is in contrast to bacteria such as *Rps. sphaeroides* and *R. rubrum* in which the pigmented membranes form spherical sacs, see, e.g., ref. [53]). They observed opposite signs of the dichroism for the 590 nm, and for the 800 nm and 870 nm absorption peaks, in line with the assignment of these bands to the Q_x and Q_y transitions of Bchl. A similar study on whole cells of *Rps. palustris* and *Rps. sphaeroides* was carried out by Breton [54], using magnetic field orientation. In both these studies, the dichroism observed is almost solely due to antenna pigments, obscuring information on reaction center pigment orientation.

Penna et al. [55] measured LD of an RC preparation of *Rps. sphaeroides* R–26 using the brush-spreading technique. Although the orientation was rather poor, they nevertheless were able to see relatively strong positive bands at 860 and 795 nm, and negative bands at 600 and 750 nm. At low temperature (77 K) the weak negative band at 810 nm was more pronounced and a weak, split, positive band at 540 nm appeared. The 750 nm band was assigned to the Q_y transition of Bph, the bands at 540 nm to its Q_x transition.

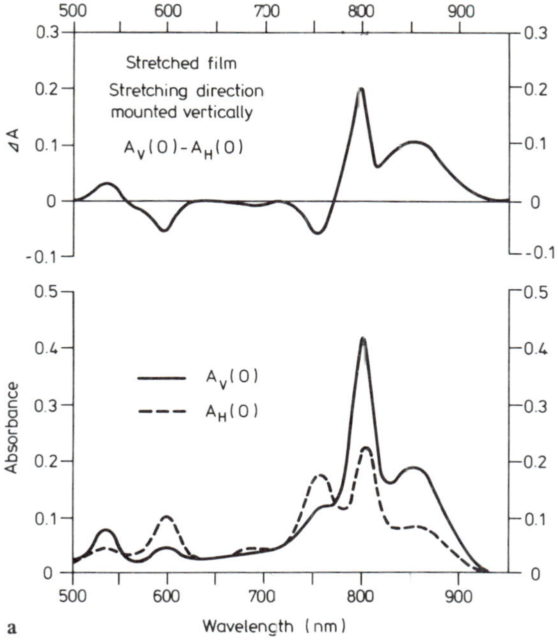

a

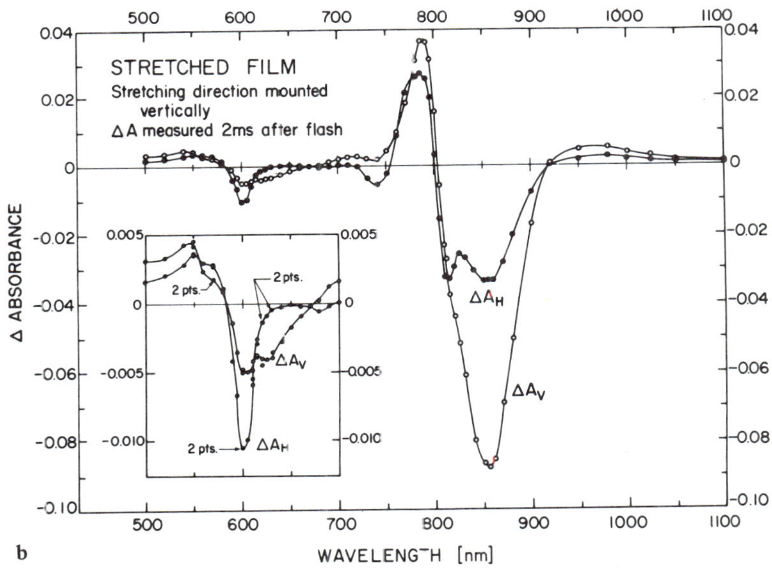

b

Fig. 2.7. a The polarized absorption and linear dichroism spectra of RC of *Rps. sphaeroides* R–26 in a stretched gelatin film. $A_V(O)$, direction of polarization parallel to the film plane and the stretching direction; $A_h(O)$, direction of polarization parallel to the film plane and perpendicular to the stretching direction. **b** The polarized ΔA spectrum 2 ms after a vertically polarized actinic flash that was incident on the vertical film plane under an angle of 45°. A_V, as in **a**; A_h, direction of polarization makes an angle of 45° with the film plane and is perpendicular to the stretching direction. (From [290])

The positive 795 nm and negative 810 nm LD bands associated with B800 were interpreted to arise from a pair of interacting Bchl molecules. At 175 K the 860 nm LD band showed a shoulder at 883 nm, at 77 K this band was shifted to 890 nm and appeared to be split. However, later work by Vermeglio and Clayton [56] makes it prob-

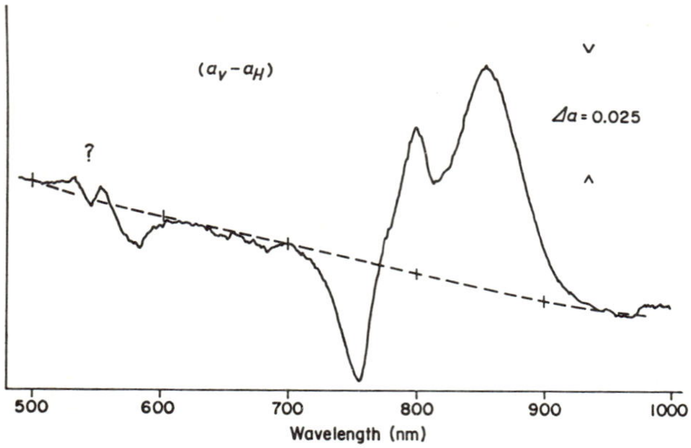

Fig. 2.8. Linear dichroism spectrum of chloroiridate-treated chromatophores of *Rps. sphaeroides* which were oriented by drying on a glass plate. $a_V - a_h$ denotes the difference in absorption between light polarized with the electric vector parallel to the plate and under an angle of 60° with the normal of the plate. (From [56])

able that the shoulder and splitting were artefacts arising from peaks in the Xenon lamp spectrum. The latter authors carried out a careful LD study of chromatophores of *Rps. sphaeroides* oriented by drying on a glass slide. The chromatophores were treated with the strong oxidant K_2IrCl_6 to irreversibly oxidize most of the antenna Bchl. The LD spectrum is reproduced in Fig. 2.8.

Denoting by A_V and A_h the absorbance for the analyzing beam polarized vertically and horizontally, repectively, Vermeglio and Clayton defined the dichroic ratio by A_V/A_h. They measured the response of this ratio to actinic light with the vertical plane of the slide making an angle of 30° with the analyzing beam and found that $\Delta A_V/\Delta A_h$ was appreciably smaller than unity at 600, 765, and 810 nm, but appreciably larger than unity for 790, 860, and 1250 nm. In other words, $\Delta LD = (= \Delta A_V - \Delta A_h) < 0$ for the first series, and $\Delta LD > 0$ for the latter series.

The bands mentioned correspond to bands measured in the ΔA spectrum. If the 790 and 810 nm bands from this spectrum are due to a blue shift of a single band of the (hypothetical) B800 pigment, then one would expect a ΔLD of equal sign. Thus, the opposite signs found experimentally indicate that the ΔA bands at 790 and 810 are *not* due to a blue shift of a single band.

Vermeglio and Clayton explained the ΔA spectral features around 810 nm by assumming that the monomeric Q_Y transition moments of the primary donor, the $(Bchl)_2$ dimer, are nonparallel, thus giving rise to exciton components at 860 nm (Q_Y) and 800 nm (Q_X). Upon oxidation, the exciton partners are bleached, and a band at 790 nm corresponding to the Q_Y band of the nonoxidized monomer appears (from resonance Raman experiments [291] it was concluded that the hopping rate of the hole over the two Bchl is less than $10^{13}\,s^{-1}$, slow on the optical time scale of $\sim 10^{-14}$ s; however, on a time scale corresponding to hyperfine interactions, $\sim 10^{-7}$ s, the hole is distributed on both molecules; cf. Sect. 2.2). A different explanation of the ΔA spectrum around 810 nm is given by Shuvalov et al. [45], who performed a photoselection study of

R. rubrum reaction centers by measuring $\Delta A_{//}$ and $\Delta A_{\perp}$ as a function of wavelength for polarized actinic light, where $A_{//}$ and $A_{\perp}$ refer to polarization of the measuring beam parallel and perpendicular to the actinic beam, respectively.

The $\Delta A_{//} + \Delta A_{\perp}$ spectrum (this spectrum is close to the $\Delta A = \Delta A_{//} + 2\Delta A_{\perp}$ spectrum for the values of p encountered)[7] showed similar width of the 790 and 810 nm bands, suggesting to the authors that these bands are due to an exciton splitting of B800, and that the P860 band (shifted to 890 nm at 77 K) represents one of the exciton partners of the primary donor dimer, the other having negligible oscillator strength because of colinear alignment of the Q_y transition moments of the monomeric components. From the oscillator strengths of the 790 and 810 nm bands an angle of 45° is calculated between the Q_y transition dipoles of the two B800 components.

The above explanation suffered from the defect that a single exciton band at 870 nm should exhibit a p-value of about 0.5 and not 0.22 as found by Shuvalov et al. [45], and Mar and Gingras [57]. However, Vermeglio et al. [58], in a photoselection study of *Rps. sphaeroides* chromatophores, found p-values approaching 0.5 for P870. The spectral features found by Vermeglio et al. [58] are essentially the same as those of Shuvalov et al. [45]. The lower value of p obtained by the latter authors and by Mar and Gingras [57] were attributed by Vermeglio et al. [58] to rather long period of excitation employed by these authors, causing saturation of a slower decaying component of P860 (this argument assumes identical behavior of *Rps. sphaeroides* and *R. rubrum*).

From an analysis of the ΔLD spectra around 800 nm upon excitation at 900 nm at temperatures from 160 to 290 K, Vermeglio et al. [58] arrived at the conclusion that the ΔA bands at about 790 and 810 nm result from the superposition of a band shift at 800 nm with polarization of $p = +0.23$ and a bleaching of a band at 805 nm, polarized perpendicularly to P870. This was taken to support the hypothesis of Vermeglio and Clayton [56] that the 805 and 870 nm bands arise from the excitonically coupled components of the primary donor.

The work by Vermeglio et al. [58] was extended by Rafferty, Clayton, and Vermeglio [51, 290, 292, 293] who in a series of papers combined LD and ΔLD measurements on RC oriented in stretched gelatin films [51, 290, 292] and chromatophores dried on a glass slide [293] with photoselection data [58, 292, 293]. They were able to deduce the angles between transition dipole moments (tm) of RC pigments and "a" symmetry axis of the RC, the angles between the tm of pigments and the normal to the chromatophore membrane and the angles between the tm of the pigments. As these studies make full use of the power of polarization spectroscopy, they will be reviewed here in some detail.

RC of *Rps. sphaeroides* R–26 show planar orientation when embedded in a gelatin film, i.e., an axis of symmetry of the RC lies in the plane of the film with random orientation. Upon stretching hydrated films (up to three times), the RC orient themselves such that the axis of symmetry lies predominantly in the direction of stretching. It is assumed that the tm of the various RC pigments are distributed randomly about

7 $p = (|\Delta A_{//}| - |\Delta A_{\perp}|) / (\Delta A_{//}| + |\Delta A_{\perp}|)$. For a single transition in randomly distributed pigments p is related to the angle α between the excitation transition moment and the transition moment detected by the measuring beam by $p = (3\cos^2 \alpha - 1) / (\cos^2 \alpha + 3)$, p varying from +0.5 ($\alpha = 0°$) to −0.33 ($\alpha = 90°$)

the axis of symmetry. The film was mounted vertically, the measuring beam horizontal and making an angle θ with the plane of the film ($\theta = 60^o$ or 45°).

As a first result it was found that the 600 nm band of the primary donor was split into components at 600 and 630 nm in a ΔA spectrum taken with the direction of polarization of the measuring beam vertical (ΔA_v spectrum). In the ΔA_h spectrum (polarization of the measuring beam horizontal), the 630 nm band is almost absent. Conversely, in the infrared there were two bands (at 810 and 850 nm) in the ΔA_h spectrum, and but one bleaching at 850 nm in the ΔA_v spectrum. The authors assign these bands to the exciton components of the two strongly coupled Bchl's of the primary donor (see Table 2.1 for nomenclature and assignment).

On the basis of earlier work on oriented layers [294–296] the authors derived a set of equations relating the dichroic ratio to the angle α between a tm and an axis of symmetry of the RC, taking into account refraction in the film and deviations from perfect orientation. It was found that in unstretched films a fraction $1-f_1 = 0.3$ is oriented randomly, whereas in stretched films a fraction $1-f_2 = 0.15$ was oriented with planar symmetry as in unstretched films. In the latter film, at most 10% of the RC was oriented

Table 2.1. Band positions of reaction center pigments in ΔA and ΔLD spectra

Rps. sphaeroides[a]	Wavelength of transition[b] (nm)	Type	Designation[d]	Pigment
(Bchl *a*)	870–850	Q_y	BY_1	$(Bchl)_2$
	812	Q_y	BY_2	$(Bchl)_2$ or associated Bchl
	800	Q_y	B800	associated Bchl
	630	Q_x	BX_1	$(Bchl)_2$
	600	Q_x	BX_2	$(Bchl)_2$
	760	Q_y	$PY_{1,2}$	BpH (2x)
	545	Q_x	PX_1	Bph ≡ I[c]
	535	Q_x	PX_2	Bph[c]
Rps. viridis	980	Q_y	BY_1	$(Bchl)_2$
(Bchl *b*)	850	Q_y	BY_2	$(Bchl)_2$ or associated Bchl
	830	Q_y		associated Bchl
	805	Q_y		$(Bchl)^+$
	650			Bph^-
	605	Q_x	$BX_{1(2)}$	$(Bchl)_2$
	600	Q_y		associated Bchl
	800–790	Q_y	$PY_{1,2}$	Bph
	546	Q_x	PX_1	Bph ≡ I

a Typical representative of Bchl *a* containing species; others may have slightly different band positions
b Approximate position, actual positions depend on medium and on temperature
c Split only at low temperatures (150 K)
d The designation and the assignment of the absorbance changes around 800 nm depend on the interpretation (exciton component of the primary donor or shift of accessory Bchl)

randomly. This model, which assumes domains of unoriented, "unstretched" material in a for the rest perfectly oriented stretched film, was tested on samples with different degrees of stretching, with good results.

The angles α were calculated from the dichroic ratios in the limit of perfect orientation for the four dimer components BY_1, BY_2, BX_1, BX_2 and for the bacteriopheophytin bands $PX_{1,2}$ and $PY_{1,2}$. The values of α are subject to the assumption that the angle δ between the axis of symmetry of the RC and the direction of stretching or the plane of the unstretched film is equal to zero. However, δ can be as large as 22.5^0 [290]. For this upper limit of δ, values of $\alpha_{measured}$ around 40^0 give "true values" of α that are lower by about 5^0, but for $\alpha_{measured} < 22.5^0$, α_{true} (the angel between tm and axis of symmetry) can be anywhere between 0 and 22.5^0 [this may apply for α (BX_1, BX_2)].

From the relative intensity of the presumed exciton components of the primary donor it was concluded that the Q_y tm of the Bchl's are predominantly colinear and roughly parallel, whereas the Q_x tm lie predominantly side by side and are also roughly parallel. Low temperature (150 K) photoselection measurements [58, 292, 293] yielded the angles γ between one tm (here BY_1) and other tm of interest, and between the 600 and 630 nm bands (PX_1 and PX_2). The basic assumptions were: (1) the tm in the RC have fixed orientation relative to one another, (2) the induced absorption changes are associated with a single, linear tm. A correction was applied for imperfect photoselection by depolarization of the excitation beam in the sample and excitation within a band consisting of more than one transition [293]. Using spherical triangulation the photoselection data were combined with the LD measurements on RC in stretched gelatin films to give the orientation (in spherical coordinates) of all tm with respect to the axis of symmetry of the RC.

The sign ambiguities resulting from the application of the cosine law of spherical triangulation could be partially resolved by the requirement of internal consistency ("compatibility"), taking all tm and the measured angles α and γ. However, the values for the angles ϕ between BY_1 and BX_1, BX_2 are subject to the assumption that $\delta \ll 22.5^0$, as are the angles between BX_1, BX_2 and all four tm of the Bph's [292]. This means that as long as δ is not known accurately, the relative orientation of these tm must be regarded with some caution.

Finally, ΔLD measurements on chromatophores oriented by drying on a glass slide were combined with the photoselection data to arrive at the polar coordinates of the tem of the primary donor with respect to the normal to the chromatophore membrane [293]. Here, the basic assumption were: (1) the membrane planes are precisely parallel to the glass slide, (2) there is no long-range lateral order in the plane of the chromatophores, (3) the angle of a particular tm to the membrane plane is the same for each RC or antenna pigment, (4) the measured absorbance is associated with a single linear tm. The first assumption is justified by the observation that the dichroic ration is independent of layer thickness, i.e., there is no accumulation of misorientation.

The 630 nm transition noted in the ΔA_V spectrum of RC in stretched gelatin films was also found in oriented chromatophores, demonstrating that it is not an artefact produced by film preparation. The polar coordinates of the Bph molecules with respect to the normal to the membrane were found by transforming the stretched film data with the aid of a set of equations determined by a best fit transformation of the four dimer tm from stretched film to oriented membrane coordinates [293].

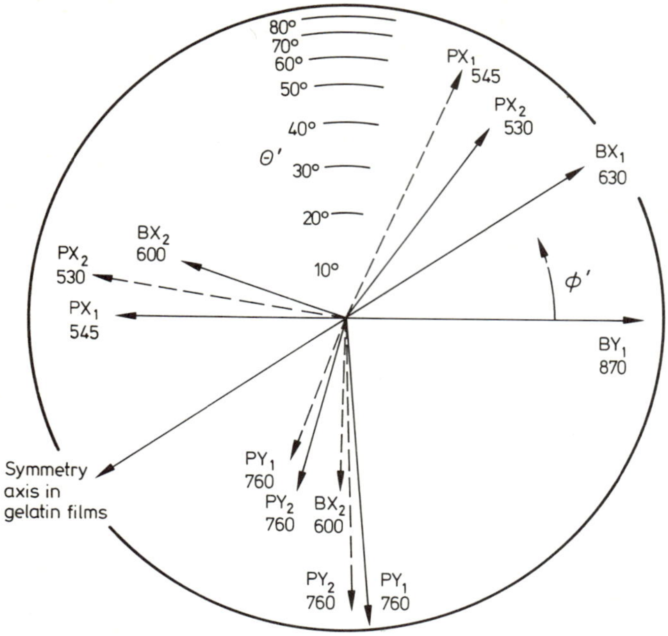

Fig. 2.9. Polar view of the orientations of optical transition moments of *Rps. sphaeroides* R–26 in a spherical coordinate system with the pole set normal to the plane of the chromatophore membrane. The transition moments are identified in Table 2.2. Alternative orientations are indicated by *solid* and *dashed arrows.* For either of the two alternative orientations of the BX_2 transitions, the two alternative sets of orientations for the PY_1, PY_2, PX_1 and PX_2 transitions are allowed. An equally acceptable picture would be a mirror image with ϕ' replaced by $-\phi'$ for all transition moments. (From [293])

The orientations of all tm are summarized in the polar diagram of Fig. 2.9. where the pole is the normal to the chromatophore membrane. Note that there are several alternative solutions, and that an equally acceptable picture is the mirror image obtained by taking $\phi = -\phi$ for all tm. In addition to the data depicted in Fig. 2.9, it was found that Q_y tm of the antenna Bchl are predominantly parallel to the membrane (as is BY_1) whereas the Q_x tm make an angle of 55–63° to the membrane plane (as does BX_2). This arrangement might facilitate energy transfer to the primary donor.

Very recently another method to orient particles was introduced [304] making use of pressed polyacryl amide gels. Dichroic spectra of *Rps. sphaeroides* R–26 reaction centers were almost identical to those of Rafferty and Clayton [304, 305]. However, the interpretation, based on a calculated extrapolation to perfect orientation, is rather different. As the corrected LD spectra are very different from the measured spectra, it is not clear what confidence limits must be placed on the reported values of the angle between the tm and the long axis of the RC.*

Although the model of Vermeglio and Clayton [56], Vermeglio et al. [58], and Rafferty and Clayton [290, 293] is appealingly simple, the difference in line width between the two exciton components (14 nm for the 805 and 44 nm for the 870 nm band) poses a problem, because to first approximation simple exciton theory predicts

* See *Notes Added in Proof*, p. 322

the same width for the exciton components (roughly equal to $1/\sqrt{2}$ times the monomer width). Suggestions offered to explain the difference [58] are as yet speculative. The competing model of Shuvalov et al. [25, 45], illustrated in Fig. 2.11 for the Bchl *b* containing species *Rps. viridis* (see below), has the merit that the proposed exciton components of B800 have the same width, but it is considerably more complex. Both models can explain the general features of ΔCD spectra (P870 oxidized *minus* P870 reduced) equally well, but for the disappearance of the positive CD peak at 827 nm in *Rps. viridis* upon reduction of the Bph intermediary electron acceptor the Vermeglio-Clayton model needs additional assumptions (always assuming that the structure of the reaction center in Bchl *a* and Bchl *b* containing species is similar).

Rps. viridis. *Rps. viridis* is an attractive organism to study with polarization spectroscopy because the absorption bands are better resolved than for Bchl *a* containing species, and because its cylindrical shape and the configuration of its cytoplasmic membranes make it possible to orient cells in a strong (20 kG) magnetic field. The long axis of the cylinder is then perpendicular to the direction of the field. Another advantage is that the states P^+I and $P\,I^-$ can be easily photo-induced at room temperature by adjusting the redox potential to +350 mV and lower than −400 mV, respectively.

Paillotin et al. [297] have carried out LD and ΔLD measurements on oriented *Rps. viridis* cells in the states P I, P^+I, and $P\,I^-$. The ΔLD data were obtained by monitoring ΔA_h and ΔA_v after a 66 ms pulse of blue, unpolarized light (h, v is horizontal and vertical with respect to the horizontal magnetic field, respectively).

The authors introduced order parameters assuming axial orientation and an axially symmetric distribution of pigments in the membrane. Expressions were derived relating ΔA_h and ΔA_v to $\Delta A_{//}$ and $\Delta A_{\perp}$, the absorption changes relative to the plane of the membrane. The order parameter S_i for the individual pigments is given by $S_i = \frac{1}{2}(3\overline{\cos^2}\,\phi_i - 1)$ with ϕ_i the angle between the ith tm and the normal to the membrane. The order parameter S defines the orientation of the cells by $S = \frac{1}{2}(3\overline{\cos^2}\,\theta - 1)$ with θ the angle between the plane of the membrane and the orientation axis. S is for all transitions the same, it can be increased by photoselection (excitation at 600 nm). From various experiments at different wavelength of observation it was found that $S = 0.32 \pm 0.03$ and $S_{970} < -0.45$. The values of S (20 kG) is close to the theoretical limit for perfectly oriented cylindrical membranes. The value of S_{970} indicates that the Q_y tm of the primary donor makes an angle of less than 10° with the plane of the membrane.

From S and the measured ΔA_h, ΔA_v spectra $\Delta A_{//}$ and $\Delta A_{\perp}$ spectra can be constructed. It was found that $\Delta A_{\perp} \sim 0$ for the 970 nm band, and that at 850 nm, $\Delta A_{\perp} > \Delta A_{//}$. This was taken as evidence that the 850 nm band is the exciton partner of the 970 nm band. In agreement with this interpretation, the LD at 850 nm is bleached to a large extent when the sample is illuminated with blue light to oxidize P970.

The 850 nm tm is tilted out of plane, with angle larger than 54°. Around 805 nm the absorbance changes are mainly parallel to the plane (angle smaller than 25°), suggesting that this tm is due to monomeric Bchl in the oxidized dimer. A band at 605 nm in the $\Delta A_{//}$ spectrum was tentatively attributed to the Q_x tm of the dimer.

From the relative intensity of the 970 and 850 nm bands it was concluded that the angle β between the Q_y transitions of the dimer components is $\beta = 35^o$, using the expression $A850/A970 = (1 + \cos\beta)/(1 - \cos\beta)$. Because the dimer 970 nm and the monomer 805 nm transitions make angles of less than 10° and 25°, respectively, with the plane of the membrane, it was concluded that the Q_y tm of the two Bchl are symmetrically arranged with respect to the plane, making an angle of about 18° with the plane.

Besides the appearance of the monomer-like band at 805 nm and the bleaching at 850 upon photooxidation of P970, there are features between 815 and 840 nm in the $\Delta A_{//}$ and $\Delta A_{\perp}$ spectra which are attributed to band shifts of the two associated Bchl absorbing at 830 nm. The fact that the zero crossing in these regions is different for the two ΔA spectra, together with the finding that, although the average angle of the 830 nm tm with the plane is less than 25°, the amplitude of the shift is nevertheless larger in the $\Delta A_{\perp}$ than in the $\Delta A_{//}$ spectrum, led the authors to the conclusion that the two associated Bchl's do not have the same orientation with respect to the plane, and that they shift is opposite direction. The best fit was obtained for the Q_y tm of one Bchl lying flat in the membrane and shifting from short to long wavelengths, whereas the Q_y tm of the other Bchl makes an angle of 50° with the membrane and shifts from long to short wavelengths with a three times larger shift amplitude.

When the cells are prepared in the state P $I^-$by strong illumination at low redox potential, half of the negative band in the LD spectrum at 800 nm bleaches, indicating that only one of the Bph bleaches. Formation of I^-shifts only *one* of the associated Bchl, because only in the $\Delta A_{//}$ spectrum, and not in the $\Delta A_{\perp}$ spectrum, occurs an absorption shift for both the Q_x and Q_y tm (at 600 and 830 nm, respectively).

In the $\Delta A_{\perp}$ spectrum, reduction of I not only entails a bleaching of the Bph bands, but also a bleaching at 845 and 605 nm. This is interpreted by the authors as a rearrangement of the dimer components such that they become more nearly parallel, whence one of the exciton bands decreases in amplitude. It should be noted that this proposal is a necessary consequence of the interpretation of the 850 band as the exciton partner of P970, because if the bleaching was due to a shift, then also the 970, should shift upon I^-formation, which it manifestly does not. On the other hand, if the 850 nm band feature in the P^+I ΔA spectrum is due to a shift of an accessory Bchl as proposed in [289], then the apparent bleaching at 845 nm can be interpreted to be part of a shift and one does not need to take recourse to a rearrangement of the dimer components.

An interpretation of the dichroic spectra of *Rps. viridis* in this vein has been offered by Shuvalov et al. [289]. These authors carried out CD, ΔA, and ΔLD measurements on RC of *Rps. viridis* in 80% glycerol at low temperature (100 K) using the photoselection technique. They studied the effects of photooxidation of P980 (the red shift of 10 nm is caused by the low temperature) and photoreduction of Bph (P790). Similarly to the results of Paillotin et al. [297], $P980^+$ formation entailed a bleaching at 980 nm in the ΔA and CD spectra, a decrease of the rotational and dipole strengths of the 850 nm band, and while the 790 nm band did not change the 830 nm band shifted slightly to 829 nm (Fig. 2.10. B). Photoreduction of I at $E_h < -300$ mV caused a bleaching in the ΔA and CD spectra at 790 and 546 nm, a decrease of the rotational and dipole strength and a blue shift at 830 nm, but the 850 and 980 nm bands were, apart from a small shift from 982 $\rightarrow$ 877 nm, preserved. All CD bands were attributed to interacting pigments, as the intrinsic CD of chlorophyllous pigments is very low. A minor band in the ΔA spectrum at 814 nm, which was not present in the CD spectrum, was attributed to a contaminant and disregarded. A weak band at 807 nm, which ap-

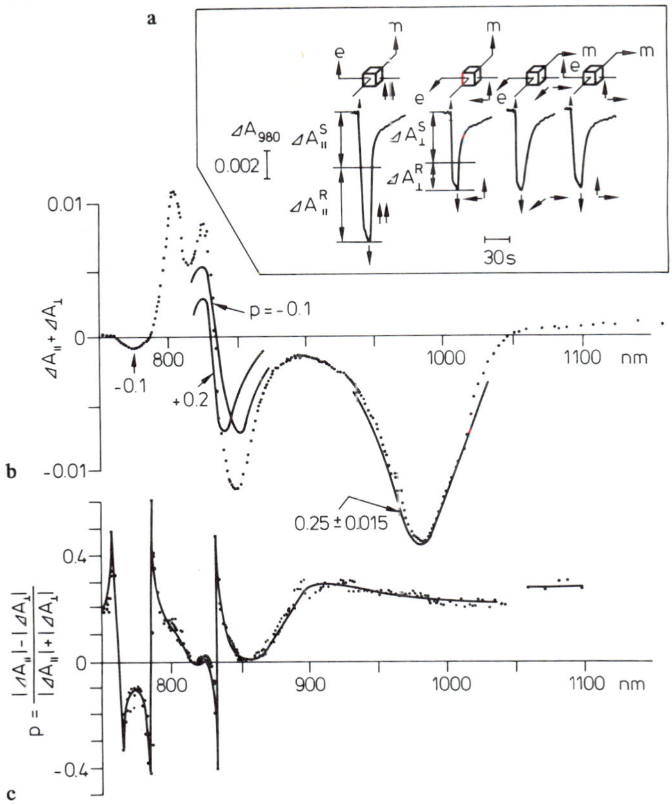

Fig. 2.10 a–c. Linear dichroism of absorbance changes induced by illumination at 980 nm of *Rps. viridis* RC at 100 K and E_h = 250 mV. **a** Kinetics of the absorbance changes (ΔA *980*) at various mutual orientations of the E-vectors of the measuring (*m*) and exciting (*e*) light. Note the rapid (ΔA^R) and slow (ΔA^S) decay components. **b** Spectrum of $\Delta A_{//} + \Delta A_{\perp}$ [this spectrum corresponds to ± 5% with the $\Delta A (= \Delta A_{//} + 2\Delta A_{\perp})$ spectrum for unpolarized light]. *Solid curves* indicate the resolved components of the 830–850 nm bands (see text). **c** Spectrum of the polarization $p = (\Delta A_{//} - \Delta A_{\perp})/(\Delta A_{//} + \Delta A_{\perp})$. (From [289])

peared in the ΔA spectrum upon photooxidation of P980, was tentatively assigned to monomeric Bchl *b*.

The dichroic measurements were presented as spectra of the polarization $p = (\Delta A_{//} - \Delta A_{\perp})/(\Delta A_{//} + \Delta A_{\perp})$, where p was calculated from a mixture of rapid and slowly decaying ΔA components after flash excitation (Fig. 2.10. A, C). The polarization spectra showed a number of singularities (sharp features where p changes sign) between 750 and 770, 770 and 800, and 820–870 nm. The singularities are due to the presence of at least two overlapping bands with different polarization. For the latter singularity, one of the bands, centered at 825 nm with p = +0.2, is though to stem from the 830 nm band which is blue-shifted and decreased in amplitude by P^+ formation, the other, also at 825 nm but with p = −0.1, from the 850 nm band which similarly undergoes a blue shift and a decrease in amplitude.

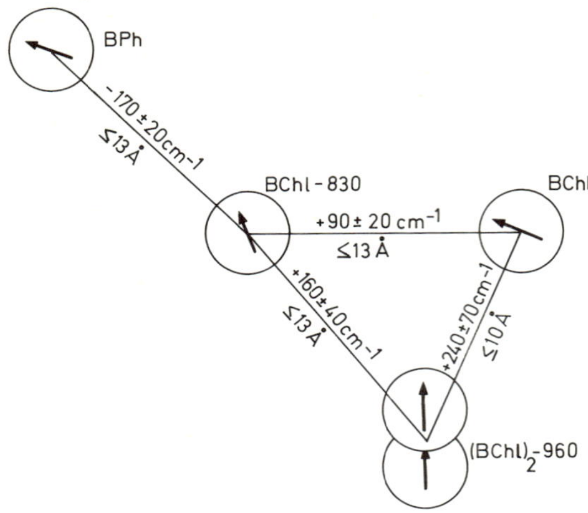

Fig. 2.11. Proposed arrangement of pigment molecules in RC of *Rps. viridis.* The *arrows* show the mutual orientation of the Q_y transition moments. The pairwise interaction in cm^{-1} between the transition dipoles, and the distance in A between the centers of the chromophores are indicated. (From [289])

Shuvalov et al. [289] explained their results in terms of five interacting chromophores, to wit the $(Bchl)_2$ dimer, the two accessory Bchl *b* pigments, and one Bph *b* pigment (Fig. 2.11). The 980 nm band is assigned to (Bchl $b)_2$. The strong red shift from the monomer absorption band position (798 nm for Bchl *b*) indicates that the pigments are strongly coupled and form an exciton complex. The 850 nm band is assigned to one accessory Bchl *b* that interacts with (Bchl *b*) and a Bchl *b* absorbing at 830 nm, which in turn interacts with the dimer and the intermediary acceptor, Bph *b*, absorbing at 790 and 546 nm. Upon photooxidation of P980 or photoreduction of P790, a redistribution of oscillator strengths and band positions occurs, giving rise to the various observed bleachings and blue shifts. The sum of the oscillator strenghts of the five pigments, however, should remain constant and correspond to the sum of the dipole strengths for the five isolated chromophores according to the rule $\Sigma_A \nu_A \mu^2{}_{OA} = \Sigma_i \nu_a \mu^2{}_{ioa}$ where ν_A, ν_a and μ_{OA}, μ_{ioa} are the frequencies and the tm of the transition O → A in the aggregate and o → a in the ith chromophore, respectively. The dipole strength of the isolated monomer tm is 30 and 19 debye^2 for Bchl *b* and Bph *b*, respectively.

From Table 2.2 it is seen that this rule holds good for the state P.Bph, viz. $\Sigma_A = 1.73 \times 10^6$ debye^2.cm^{-1} and $\Sigma_i = 1.74 \times 10^6$ debye^2.cm^{-1}. The sum of the oscillator strength of the 850 and 830 nm band is not far from twice the monomer value, the difference being partly due to intensity borrowing from the Bph, whereas the oscillator strength of the dimer is also close to that of twice the monomer value, suggesting that the companion exciton band is largely absent. Hence, according to this interpretation, the transition moments of the two Bchl in the dimer are parallel and colinear.

For the state P.$^+$Bph, the agreement to the sum rule is less good. In addition to the bleaching of the 980 nm band, also the 830 and 850 nm bands show a decrease in dipole strength, whereas the Bph band is unchanged. On the basis of the resonance Raman experiments of Lutz and Kleo [291], one would expect that oxidation of P980 would give rise to a band of monomer intensity. In view of the small intensity of the 807 nm band which appears upon P$^+$ formation, it is doubtful that this band accounts for all of

Table 2.2. Dipole strengths of transitions for various states of *Rps. viridis* reaction centers at 100 K (after [289])

Peak of band in the state P.Bph (nm)	Dipole strength (debye^2) P.Bph	$P.^{+}Bph$	$P.Bph^{-}$
982	67 ± 5	–	67 (977 nm)
850	17 ± 2	13 ± 2[a] (825 nm)	17 (850 nm)
833	60 ± 5	54 ± 5[a] (830 nm)	51 ± 5[a] (820 nm)
790	10 ± 2[b]	10 ± 2 (790 nm)	–

a Light-induced shift contributions are subtracted
b The photobleached part of the 790 nm transition

the 30 debye^2 of the monomer, whereas the missing dipole strength does not show up in the other pigments either.

From the strength of the interactions between the various pigments corresponding to the above interpretation of their data, Shuvalov et al. [289] estimated the distances between the tm. The angles a between the tm of P980 and the other tm followed from the values of p obtained for excitation at 980 nm and corrected for the presence of the slow decay component. The results are summarized in Fig. 2.11.

In reviewing the contrasting interpretations of the ΔA and dichroism data on photosynthetic bacteria by the Russian workers and the Saclay/Cornell groups, an attempt was made to bring out the strong and the weak points. At present, however, it does not seem possible to make a definite choice. For reasons of simplicity one may prefer the point of view of the latter workers, but Nature does not always choose the simplest solution. Clearly, more experimentation is needed to decide upon the important question of the geometrical configuration of the reaction center pigments.*

2.1.5 Nano- and Picosecond Spectroscopy

A breakthrough in the study of photooxidation of the primary bacterial donor came about by the advent of very fast absorption difference spectroscopy. Especially the new technique of ps spectroscopy (see for reviews refs. [59–62]) has already yielded a wealth of exciting results. As a first application, in 1973 it was established that the bleaching at 860 nm had a rise time of less than 6 ps [63] (the most recent $t_{1/2}$ value of the rise time is 3 ps [64], whereas the 1245 nm absorption characteristic of $P860^{+}$ had a rise time of less than 10 ps [65]).

Parson et al. [66] applied ns spectroscopy to RC of *Rps. sphaeroides* in the state $P X^{-}$, in which forward electron transport is blocked. They found two distinct transients, one (PF) was generated during the actinic flash (half width 20 ns) and decayed in about 10 ns to another state, P^{R}, which had a lifetime of about 6 μs at room temperature. The ΔA spectrum of PF showed a bleaching in the Bchl a absorption bands near 380, 600, and 870 nm, and in addition broad bands near 420 and 680 nm, to-

* See *Notes Added in Proof*, p. 322

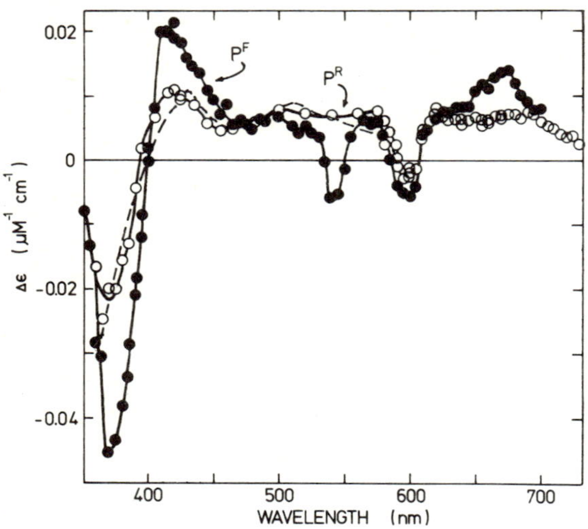

Fig. 2.12. Flash-induced difference spectra for the formation of P^F (●) and P^R (○) in reaction centers of *Rps. sphaeroides* R – 26 at room temperature. *Dashed line*: difference spectrum of Bchl $a \rightarrow$ Bchl a^T in ethanol-pyridine (7 : 2 v/v) scaled to the spectrum of P^R at 420 nm. (After refs. [66, 67, 71])

gether with a bleaching of bands near 540 and 760 nm (Fig. 2.12). The ΔA spectrum of P^R did not show the latter bleachings and had much less intensity at 360 and 680 nm (Fig. 2.12); it resembled the spectrum of Bchl *a* in the excited triplet state as measured in alcoholic solution [65]. The quantum yield of P^F was close to 1.0 at room temperature and at 15 K, that of P^R was only about 0.1 at room temperature, but it increased to 1.0 at temperatures below 20 K. The decay half time of P^R also increased with lower temperatures, to a value of 120 μs at 160 K. At 4.2 K it was biphasic, the two components having half times of 84 and 540 μs and amplitudes in a ratio of two to one, respectively [67].

Although short-lived, P^F coud not be due to an excited state of the primary donor, since the lifetime of P860* should be of the order of 20 ps in the state P^*X^-[68, 69]. It was tentatively assigned to the intermediary state P^+I^-, I being an intermediate acceptor. The short lifetime of P^F invited the application of ps spectroscopy.

In 1975 two groups [70, 71] independently published ΔA spectra of RC of dark-adapted *Rps. sphaeroides* R–26 taken at 13 or 20 ps after a light flash at 530 nm of about 5 ps duration. This spectrum exhibited the same characteristic bands as the ΔA spectrum called P^F, measured under reducing conditions on a nanosecond time scale [65]. The lifetime, $t_{1/2}$, of P^F under nonreducing conditions ranges between 180 and 240 ps, it decays to the state P^+X^-without development of P^R.

A composite of the ΔA spectrum of the in-vivo reduction of Bph *a* to Bph a^-, and the ΔA spectrum of the in-vivo oxidation of P860 and P860$^+$, resembled the ΔA spectrum of P^F [72]. Especially the bleachings at 542 and 760 nm observed for P^F are well reproduced in the composite spectrum if a red shift of some 10 nm is applied to the in-vitro spectrum. Such red shifts are generally found when comparing in-vitro to in-vivo spectra.

The question whether one, or both, of the Bph *a*'s, present in the RC, are reduced in the primary act was resolved by measurements of low-temperature absorption spectra of RC [73]. Absorption spectra of RC at 4 K showed the 535–540 nm absorption band of Bph *a* resolved into components with maxima at 532 and 544 nm. Apparently, only the latter Bph *a* is reduced in the state P^F.

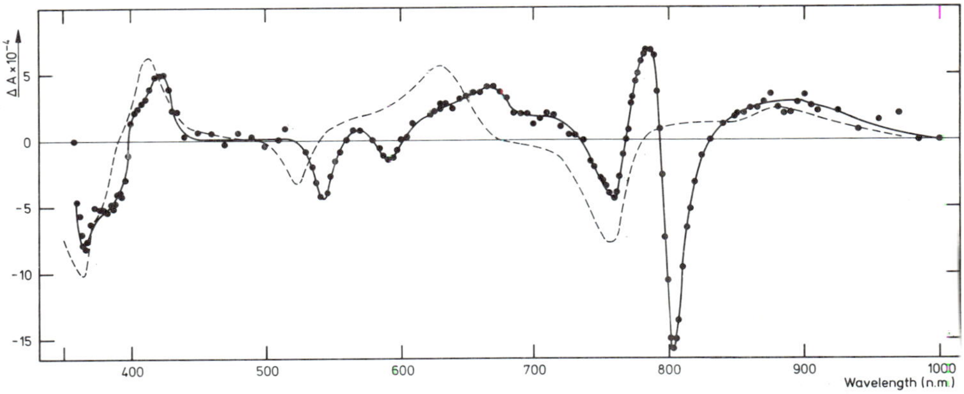

Fig. 2.13. Absorption difference spectrum of reaction centers of *C. vinosum* in which the intermediate acceptor I is reduced by illumination at $-15\ ^{\circ}C$ in the presence of the reductants dithionite and TMPD. *Dashed line*: ΔA spectrum for Bph *a* → Bph a^- in vitro (from [72]). The ΔA bands at 600 and 800 nm are attributed to a band shift of the accessory Bchl 800 pigment(s) caused by the reduction of I. (From [77])

Very accurate difference spectra of $I \rightarrow I^-$ were measured by exploiting the property of *C. vinosum* and *C. minutissimum* RC to retain a cyt c_{552} capable of electron donation to $P860^+$ [74–76]. If RC in the state P I X^- are illuminated by strong light at temperatures of $-15\ ^{\circ}C$ [77] or $-70\ ^{\circ}C$ [76], the cyt *c* irreversibly reduces P^+, so that eventually the state P I^-X^- appears.

A difference spectrum generated in this way is displayed in Fig. 2.13. Assuming that the ΔA band at 542 nm is solely due to reduction of one Bph *a*, Van Grondelle et al. [77] calculated a differential extinction coefficient for Bph *a* reduction of $\Delta\epsilon = 25\ mM^{-1}\ cm^{-1}$, in close agreement with the value $\Delta\epsilon = 21\ mM^{-1}\ cm^{-1}$ at 530 nm obtained for the in-vitro reduction of Bph *a* [72]. Thus, the notion that one Bph-*a* molecule is an intermediate acceptor in the primary charge separation is now amply substantiated. That it is also the *first* acceptor is demonstrated by ps experiments of Netzel et al. [78] who measured the bleaching of the 1235 nm band of $P860^+$ in *C. vinosum* in the states P I X^- and P I^-X^-. In the former case, the band was bleached within 10 ps, in the latter no bleaching was observed. Hence, oxidation of P and reduction of I are complementary, and there is no other acceptor on a ps time scale between P860 and Bph *a*.*

Virtually all the aspects of the picosecond ΔA spectrum of Bchl *a* containing species were reproduced in the ps ΔA spectrum of the Bchl *b* containing *Rps. viridis* [79]. The better spectral resolution around 830 nm (which is homologous to the 800 nm band in Bchl *a* containing species) made it possible to distinguish between the 810, 830, and 850 nm components.

The spectrum of $P^+I^-X^-$ showed bleachings at 380 and 545 nm and the appearance of a broad band at 680 nm, which features also accompany the reduction of Bph *b* to Bph b^- [80]. Thus, the intermediate I in *Rps. viridis* is most probably Bph *b*. The state P^R showed very broad bands near 450 and 560 nm, similar to the spectrum of P^R in Bchl *a* containing species. A noteworthy difference is the longer decay time of P^R in *Rps. viridis,* viz. 55 μs compared to 5 μs for *Rps. sphaeroides.* Another difference is that there appears to be no triplet energy transfer from $(\text{Bchl}\ b)_2^T$ to an RC carotenoid.

---

* See *Notes Added in Proof*, pp. 322, 323

The similarity in kinetics of the 850 and 960 nm bands supports the idea that they form the exciton components of the (Bchl *b*)$_2$ dimer. As is also borne out by the CD experiments of Shuvalov et al. [25] (Sect. 2.1.3), the ΔA band at 830 nm is profoundly affected by the conversion of I to I^-. This argues for rather strong excitonic coupling between I (one of the Bph *b*'s) and one of the associated Bchl *b*'s, but the absence of a ΔA band at about 1050 nm upon reduction of I [79, 80], and the results of ESR and ENDOR measurements on I^-([81, 82] and Sect. 2.2.3), make it improbable that this Bchl *b* shares the unpaired electron with I.

The increase in the $\Delta A(P^+ - P)$ spectrum at 810 nm could be due to the monomeric Bchl *b* that is liberated from exciton interaction upon oxidation of P; the effect of the reduction of I on this band to an electrochromic shift.*

Although most results of ps spectroscopy are now satisfactorily explained by the introduction of the intermediate I, there remain a number of puzzling observations. These include a 30 ps component in the increase of absorbance at 800 nm in RC of *Rps. sphaeroides* [71] and a corresponding 40 ps component at 810 nm for RC of *Rps. viridis* [79]. In RC of *R. rubrum*, excited at 800 nm, an initial absorption increase at 748 nm was followed by a bleaching with $t_{1/2}$ of 30 ps [83]. The 30 ps component was attributed to Bchl a^- as an intermediate step between the generation of the radical pair $P870^+$ Bph a^-. However, the width of the excitation flash (15 ps) would have made two-quantum processes quite likely, as discussed by Holten et al. [79], whereas the slow time response of the system (15 ps) was not particularly suited to extract information in the 0–30 ps domain. Alternatively, the 30 ps changes may be related to a reorganization of pigment geometry upon formation of P^+Bph^-[79].*

As this chapter is concerned with the photooxidation of chlorophyll, the fate of the electron after it passes from Bph *a* or Bph *b* to the "primary acceptor" X will not be discussed. Suffice it to say that X is a ubiquinone or menaquinone complexed to a high-spin (S = 2) Fe^{2+} atom. This iron may serve as a connection between the one-electron "primary" acceptor X and a second quinone which serves as a two-electron gate to the tertiary acceptor. Recent reviews discussing secondary electron transfer are found in refs. [84–87].

2.2 ESR and ENDOR

Charge separation leads to a radical pair and one would expect that the photoinduced paramagnetism would be detectable by electron spin resonance (ESR) spectrometry, Indeed, as early as 1956, Commoner et al. [88] found a light-induced ESR signal in a suspension of the photosynthetic bacterium *Rhodospirillum rubrum.* At first it was believed [89] that this ESR signal was not due to the primary reactions. However, in the following decades work by several other groups [19, 90, 91] has now firmly established that the light-induced ESR signal arises from the oxidized primary donor $P860^+$.

In the next section, this evidence will be briefly reviewed and the crucial electron nuclear double resonance (ENDOR) experiments that established the dimeric nature of P860 will be discussed.

2.2.1 Characteristics of the ESR Signal of $P860^+$

Upon illumination, whole cells or subcellular preparations of photosynthetic bacteria give rise to an almost pure Gaussian ESR line of width ΔH = 9.4 ± 0.2 G (data for

* See *Notes Added in Proof,* p. 323

R. rubrum [92, 93]. The width for other Bchl *a* containing species fall within the limits of error. However, the Bchl *b* containing species *Rps. viridis* and *Thiocapsa pfennigii* show a Gaussian line of width $\Delta H = 11.8 \pm 0.2$ G [80, 81].

For all species the g-value is $g = 2.0026 \pm 0.001$ [e.g., 87]. For *R. rubrum* it was shown to be almost isotropic [93]. The microwave power saturation behavior is characteristic for inhomogeneously broadened lines [94]. Deuteration leads to a narrowing by a factor 2.4 at X-band (9 GHz) and 2.0 at Q-band (35 GHz) [90, 92, 93]. ^{25}Mg ($I = 5/2$) enriched preparation show no broadening within the limits of errors hfs (^{25}Mg < 0.3 G in ^{2}H–*R. rubrum* [94,95]). Thus, the major part of the line width is due to hyperfine interaction with protons, with a small contribution from the pyrrole nitrogens ^{14}N ($I = 1$) and g-anisotropy [93].

The evidence that the ESR signal is due to the oxidized primary donor can be summarized as follows:

1. The signal is formed in a 1 : 1 stoichiometry with the light-induced bleaching at 860 nm [19, 91], viz. (number of spins, ESR) / (number of molecules, ΔA) = $= 1.25 \pm 0.25$ [19] or 1.07 ± 0.15 [91].
2. The rise time of the ESR signal after illumination by a light flash of saturating intensity at room temperature is instrument-limited (less than 1 μs; Gast, P., unpublished results).
3. The ESR signal is reversibly photoinduced at temperatures down to 1.5 K [90].
4. The ESR signal and the bleaching at 860 nm have identical decay kinetics, either measured simultaneously at room temperature [91, 96, 97], or in different samples at temperatures down to 1.5 K [90, 98].
5. The ESR signal and the bleaching at 860 nm have identical action spectra [99, 100].
6. Both the ESR signal and the bleaching can be induced by chemical oxidation, e.g., by ferricyanide (see refs. [101–105] for reviews).
7. The ESR signal is absent in mutants of *Rps. sphaeroides* that lack the reaction center [106].

Although neither of the items 1–7 is conclusive, together they present compelling evidence that the ESR signal is indeed due to the oxidized primary donor P 860$^+$. As optical evidence strongly suggested that P860 is a chlorophyllous species, it was of basic interest to make a detailed comparison of the ESR signal of P860$^+$ with that of oxidized monomeric Bchl in vitro. From this work [90, 93, 98] it transpired that the shape, the g-anisotropy and the g-values are identical. Deuteration led in both cases to line narrowing, but not to the same extent (at 9 GHz, a factor of 2.4 and 1.8 for RC and Bchl a^+, respectively). The line width, however, of protonated Bchl a^+($\Delta H = 13.0 \pm 0.2$ G [93]) is larger by a factor of 1.4 than that of reaction center preparations. Except for this deviation, also the ESR results supported the identification of the primary donor as a chlorophyllous species.

The narrowing by a factor of 1.4 of the ESR line of a large number of bacterial species, and of algae and higher plant preparations, compared to that of oxidized monomeric chlorophyll in vitro, suggested to Norris et al. [107] that the primary donor consists of two identical molecules on which the unpaired spin is completely delocalized. One then can easily show [107, 108] that for a large number of nuclei that have a hyperfine interaction with the unpaired spin the line width is narrowed by a factor of

$$\frac{\Delta H \text{ monomer}}{\Delta H \text{ dimer}} = \left(\frac{(\Sigma_i n_i I_i(I_i + 1) A_i^2)_{\text{monomer}}}{(\Sigma_i n_i I_i(I_i + 1) A_i^2)_{\text{dimer}}}\right)^{\frac{1}{2}} = \left(\frac{\Sigma_i n_i A_i^2}{\Sigma_i 2n_i(A_i/2)^2}\right)^{\frac{1}{2}} = \sqrt{2}.$$

Norris et al. [107] also studied in-vitro preparations of chlorophyll under various conditions. Except for aggregates of hydrated Chl *a* no photoinducable ESR signals were found, although in several samples the Chl-*a* molecules were present as dimeric or oligomeric species. The aggregates showed a very narrow ($\Delta H \sim 1$ G) ESR line. On the basis of this evidence, the authors proposed that the primary donor of photosystems consisted of a hydrated chlorophyll dimer, the liganding of water to the central magnesium of one molecule and to one of the keto groups of the other being essential for photooxidation. In the aggregates of hydrated Chl *a*, the unpaired electrons would then be delocalized on a large number of molecules via the H_2O bridges (or more probably by π-overlap of the molecules held together by H_2O in a favorable position), leading to a very narrow ESR line. Detailed models of the proposed "special pair" of various symmetries and stoichiometries of hydration have been drawn up (see for reviews refs. [14, 109, 110]). These largely speculative models are discussed elsewhere in this volume.

2.2.2 ENDOR of the Primary Donor

The proposal of Norris et al. that the primary donor is made up of two (bacterio) chlorophyll molecules was tested experimentally by electron nuclear double resonance (ENDOR) experiments [111–115]. This technique, which pairs the sensitivity of ESR with the resolving power of NMR, made it possible to extract information on hyperfine splitting (hfs) constants not obtainable from the broad, unstructured ESR line. (See for short introductions to the ENDOR technique refs. [116–118]). In Fig. 2.14 the low-temperature ENDOR spectrum of the light-induced ESR signal of *Rps. sphaeroides* chromatophores is compared to the spectrum of $Bchl^+$ in vitro. It is immediately seen that the hfs constants of the chromatophores are half those of the corresponding values of monomeric $Bchl^+$, just as expected for the unpaired electron of the oxidized primary donor to be fully delocalized over two Bchl molecules, i.e., P860 is a dimer.

Studies on model compounds, and on compounds in which part of the protons were substituted by deuterium atoms, permitted the assignment of the lines seen in Fig. 2.14, and of several other lines seen in the ENDOR spectrum of $Bchl^+$ in liquid solution (Fig. 2.15) [119]. In Table 2.3 the hfs constants determined for Bchl a,b^+ and for the oxidized primary donor of in-vivo samples are collected. Note, that the values for oxidized *Rps. viridis* are by no means half of those found for Bchl b^+, in keeping with the slight difference in line width of the ESR line of both preparations. Yet, the primary donor of *Rps. viridis* most probably is a dimer, because when oxidized it shows the characteristic absorption band in the infrared (at 1310 nm for *Rps. viridis*). Also, the zero field splitting parameters of its triplet state deviate from the monomeric values in a similar way as for Bchl *a* containing species. The broad ESR line of $P960^+$ (*Rps. viridis*) might be explained by a change in the dihedral angles between the β-protons and the molecular plane going from the monomer to the dimer [82], giving rise to enhanced hyperfine couplings of at least some β-protons in the dimer.

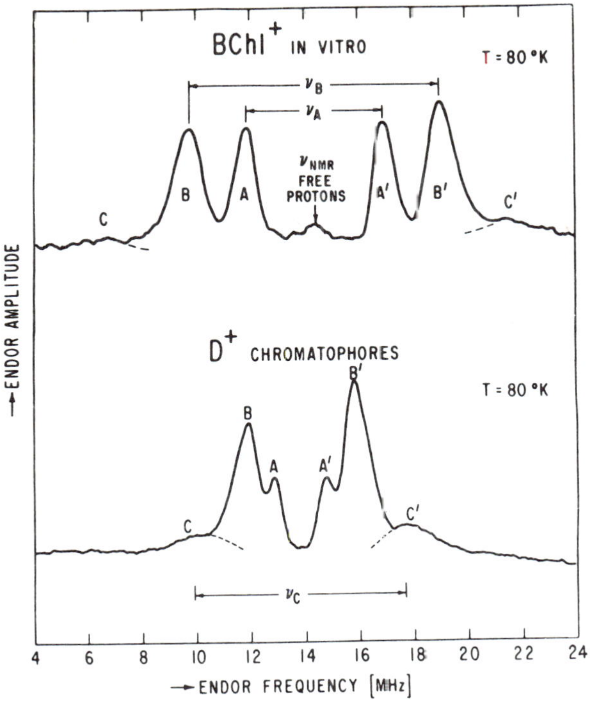

Fig. 2.14. Comparison of ENDOR spectra from Bchl a^+ in vitro (*top*) and chromatophores of *Rps. sphaeroides* R–26 (*bottom*). The hyperfine splitting ν_A and ν_B for Bchl a^+ in vitro and P860$^+$ are 5.0 ±0.1, 9.2 ±0.2, and 2.0 ±0.1, 4.2 ±0.2 MHz, respectively. The hyperfine splitting ν_C is approximately 16 and 8 MHz for Bchl a^+ and P860$^-$, respectively (From [114])

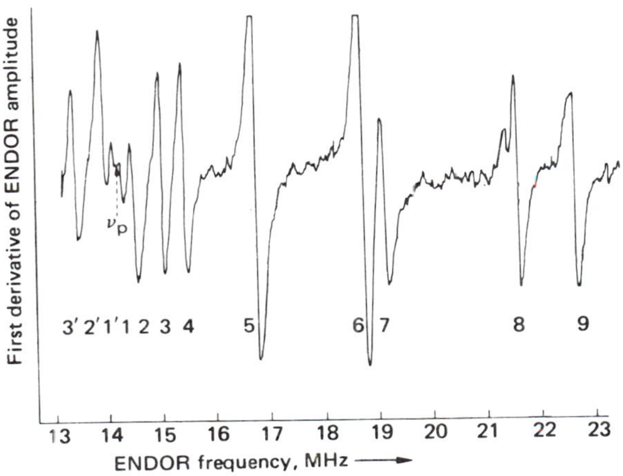

Fig. 2.15. Proton ENDOR spectra of Bchl a^+ in liquid solution. Lines *5* and *6* correspond to lines A' and B', and lines *7–9* to line C' of Fig. 2.14. Lines *5* and *6* are attributed to the methyl groups of ring I and III, respectively. Lines *7–9* arise from the β-protons of the saturated rings II and IV. Lines *1–4* are due to the bridge methine protons, the C–10 proton and/or unspecified γ-proton of the various side groups. ν_P is the free proton frequency. (From [119])

Table 2.3. Hyperfine coupling constants in MHz (2.8 MHz = 1 G) of the cation of Bchl *a* and Bchl *b* in solid state matrices and of the primary donor of photosynthetic bacteria

	Ring I CH_3	Ring III CH_3	Rings II, IV β-protons	Methine protons	Ethylidene-CH_3	References
Bchl a^+	5.0	9.2	~16	0.3, 1.2[b]		114
$P860^+$ (*Rps. sphaeroides*)	2.0	4.2	~ 8			114
(*R. rubrum*)	2.2	4.8	~ 7			115
[a]Bchl b^+	4.8	9.0	~12.5	1.1	9.0	82
[a]$P960^+$ (*Rps. viridis*)	2.8	4.2	9.2, 11.2		5.3	82

a Tentative assignments
b Measured at 2.1 K (A.J. Hoff, unpublished results, tentative assignment)

2.2.3 ESR and ENDOR of the Reduced Intermediary Acceptor, I^-

Obviously, the study of the electron accepting species is germane to the subject of chlorophyll oxidation. In the previous sections, optical investigations of the intermediary acceptor were discussed. In this section, the important features of ESR and ENDOR studies on I^- will be summarized.

Shuvalov et al. [74] and Tiede et al. [75] exploited the ability of *Chromatium* species (*C. minutissimum* and *C. vinosum*, respectively) to donate electrons from a cyt c_{553} molecule to the oxidized primary donor at low temperatures [120, 121] to build up a considerable concentration of the state cyt $c^+P\,I^-X^-$ by strong illumination of RC in which X was reduced. Subchromatophore particles of *Chromatium* thus prepared exhibited strong ESR signals. In *C. minutissimum* at 273 K, a 12.5 ± 0.5 G wide line appeared at g = 2.0025 ± 0.0005 [74]. In *C. vinosum*, the ESR spectrum at 8 K around g = 2 consisted of a singlet ($\Delta H \sim 15$ G at $g \sim 2.003$) and a doublet centered at g = 2.003 with splitting of about 65 G [75]. The doublet could only be observed at temperatures below 15 K, it was difficult to saturate at 8 K; the narrow signal was easily saturated.

Concurrently with the appearance of the above ESR signals, the intensity of the low-temperature light-induced triplet ESR spectrum (which is thought to stem from the back reaction $P^+I^-X^- \rightarrow P^T I\,X^-$, see Sect. 2.3) was decreased [75, 122]. This corroborates the findings that the yield at 273 K of a carotenoid triplet which originates from the light-induced reaction center triplet was decreased for RC in the state $P\,I^-X^-$ [74]. A tentative interpretation of the above observations is that the narrow singlet ESR line stems from the reduced intermediate I^-, whereas the doublet is due to exchange and/or dipolar coupling between I^- and X^-, the ubiquinone-iron complex, or between I^- and the iron alone.

The line width of the narrow signal of I^- in RC preparations of various bacterial species ranges between 12 and 15 G [74–76, 81, 82, 123–125], its g-value is 2.0036 ± ± 0.0002 [81, 82, 125]. These values are typical for the anions of Bchl *a,b* and Bph *a,b*. On the basis of the line width or the g-value it is not possible to discriminate between Bchl or Bph as candidates for I.

From the optical difference spectra (see Sect. 2.1) it was concluded that reduction of one of the Bph present in the RC samples caused changes in bands due to the Bchl molecules not involved in the primary donor. This observation suggested that I^- might be a complex of $(\text{Bph} - \text{Bchl})^-$ with the unpaired electron "delocalized" on the Bph and one or more of the Bchl molecules. One of the members of the complex might then have an exchange interaction with X^-. If the delocalization is a hopping, slow in the ESR time scale, then this could account for the appearance of two distinctly different ESR signals upon reduction of I. Alternatively, I^- could be a single reduced Bph which in part of the RC is in a different environment, distinguished by a different exchange interaction or dipolar coupling with X^-. The effect on the Bchl absorption at 800 nm would then be electrochromic. This view is supported by recent experiments by Okamura et al. [256], which suggest that under conditions of low redox potential and strong illumination X may be doubly reduced (see below).

More light was shed on the twin nature of the I^- ESR signal by investigations with the Bchl *b* containing bacterium *Rps. viridis* [122, 123]. Since the photon energy at the wavelength at which the phototrap of this species absorbs is only 1.26 eV compared to about 1.42 eV for Bchl *a* containing species, one might expect that the intermediary acceptor also has a higher redox potential. Otherwise, the back reaction $P^+I^- \rightarrow P^*I$ would have an enhanced probability, which would impair the quantum efficiency of the charge separation/stabilization process. Thus, in *Rps. viridis*, which species just as *C. vinosum* also contains a low-temperature cyt c_{553} donor to P^+, the E_m of I might be high enough to come into the range of chemical reduction. This turned out to be the case. Using the light-induced triplet state $P^T I X^-$ as an indicator of the redox state of I prior to illumination., ESR redox titration of P^T revealed a low potential $E_m =$ $= -400 \pm 25$ mV (pH 10.8) [122], which was taken to be the E_m of the couple I/I^-. At -500 mV practically no triplet signal was detected, indicating that then the RC were in the state $P I^- X^-$, for which photochemistry is blocked.

Interestingly, the mid-point potential of the primary acceptor X in *Rps. viridis* at pH 7.8 and higher is -150 mV [122], which is rather similar to the corresponding values in Bchl-*a* species (see Table 2.1). This leads to a redox span ΔE_m between the primary donor (Bchl *a*/Bchl a_2^+ at +500 mV) and the primary acceptor X/X^- of 650 mV, regardless of incident photon energy. This 650 mV seems to be the free energy difference necessary for secondary electron transport, that is shared by all photosynthetic purple bacteria.

The value of $E_m = -400$ mV for the couple I/I^- is reasonably close to the half-wave potentials found for elctrochemical reduction of Bph *b* in organic solvents (-500 mV in DMF, -560 mV in CH_2Cl_2) but far from the half-wave potential of Bchl-*b* reduction in DMF (< -700 mV) [81].

Similar work has been carried out on another Bchl *b* containing species, *Thiocapsa pfennigii* [124]. As in *Rps. viridis,* the donor signal consists of a single Gaussian line, $\Delta H = 13$ G at $g = 2.0025$, i.e., a line width close to the monomer value. The I^- signal, however, when trapped at 200 K resembled much that of *C. vinosum*, i.e., a broad split doublet, $\Delta H \sim 68$ G, centered at $g = 2.003$, that at higher temperatures (above 17 K) coalesced to a singlet with $\Delta H = 16$ G at 20 K and line shape between Lorentzian and Gaussian. Thus, the special properties of I in *Rps. viridis* seem to be specific to this species, and not to Bchl *b* containing bacteria in general. The redox mid-point potential of X/X^- in *T. pfennigii* was -130 mV ($n = 1$) above $pK = 6.5$, that of P^+/P was +490 mV (pH 7.0), so that the redox span between these couples is $\Delta E_m = 620$ mV, again very similar to that encountered in other species.

After it was found that I^- could be trapped in RC which contained a cyt *c* donor to P860, a similar technique was developed for RC of *Rps. sphaeroides* which do not contain a cyt *c* complement [126]. Addition of horse-heart cyt *c* and illumination at 300 K in the presence of dithionite gave an ESR signal that was similar to the narrow line found for SDS-RC of *Rps. viridis.* When in the RC the native acceptor ubiquinone is extracted and substituted by a naphtaquinone, the above procedure gives rise to a combination of the split and the narrow signal [256]. On the basis of the kinetics of formation it was suggested that the narrow signal originates from RC in which the primary acceptor is doubly reduced and therefore diamagnetic, whereas the split signal is due to magnetic interaction between I^- and the singly reduced primary acceptor.

In an effort to identify the narrow ESR signal of I^-, Feher et al. [125] performed a low temperature ENDOR study of I^- generated in RC of *Rps. sphaeroides* R-26 that were complemented with horse-heart cyt *c*. The ENDOR spectrum showed a single transition with a hyperfine splitting of 8–9 MHz (2.9–3.2 G). The shape and the position of the high-frequency ENDOR line of this transition were similar to that of the corresponding ENDOR line of Bph a^- and Bchl a^- in vitro. Fajer et al. [81, 82] performed ENDOR studies on the anion radicals Bph a,b^- and Bchl a,b^-, and on RC of *Rps. viridis* in the state P I^-X^- at high (liquid state) and low temperatures (frozen solutions). At -100 °C *Rps. viridis* RC showed ENDOR resonances with hyperfine splitting of 0.3, 2.8 and 3.2 (± 0.1) G. This compares with 0.7, 2.6 and 3.0 G for Bchl b^-, and 3.1 G for Bph b^-, at -140 °C in THF glassy matrices.

It is clear that the ESR and ENDOR parameters (line width, g-values, hyperfine splitting) do not permit to choose between Bph^- and $Bchl^-$ as the participants in I^-, but they do indicate that the species involved certainly is a *monomer* and not a dimer. This is in full agreement with the optical data which strongly suggest that only *one* molecule is involved. The identification of this molecule as a Bph, as concluded from the optical difference spectra, is consistent with the ESR and ENDOR data, so that there seems to be little reason to doubt this assignment.

2.3 The Triplet State of the Primary Donor

The discovery of the triplet state in reaction centers of *Rps. sphaeroides* under reducing conditions by Dutton et al. [121] furnished another probe of the structure of the primary donor. As its mode of formation is so intimately linked to primary charge separation, a brief discussion of its properties in this chapter seems in order (see for more extensive reviews refs. [87, 127, R21].

The bacterial triplet state shows strong spin polarization [128] of an unusual pattern [129, R21]. This pattern suggests that the triplet is generated via a back reaction between the oxidized primary donor and a reduced acceptor [129].

The reaction center triplet as detected by ESR is most probably identical to the state P^R detected by ns spectroscopy (Sect. 2.1.3), which state is also thought to result from the back reaction $P^+I^- \rightarrow P^TI$. The discrepancy between the apparent low-temperature decay time of the ESR triplet and P^R (~5 μs [128] and 110 μs [66] respectively) was recently solved by a redetermination of the decay rates corresponding to the canonical directions, k_x, k_y, and k_z, from the ESR spectrum by flash ESR spectroscopy [130], in which the effects of spin-lattice relaxation and stimulated emission were taken into account. An average decay half time (which obtains at temperatures where spin-lattice relaxation is fast) of 140 μs (± 20%) was found, whereas the individual rates k_x, k_y, and k_z corresponded nicely to the rates deduced from the biphasic decay of P^R at 4.2 K [67].

The notion that the RC triplet state is generated by the reaction $P\,I\,X^- \xrightarrow{h\nu} P^+I^-X^- \rightarrow P^TI\,X^-$ is considerably strengthened by the finding that its quantum yield is appreciably lower when the triplet state is photo-induced in a magnetic field [131, 132]. This behavior is expected when the triplet is formed via the radical pair P^+I^-. This pair can exist in two virtual states, the singlet state $(P^+I^-)^S$, where the two unpaired electrons are antiparallel, or the triplet state $(P^+I^-)^T$ with parallel spins. Upon recombination of the electron on I^- and the hole on P^+, either the singlet state P I or P^* is formed, or the triplet state P^TI. The virtual states $(P^+I^-)^S$ and $(P^+I^-)^T$ are not stable: the radical pair oscillates between them with an angular frequency ω that depends on the local and external magnetic fields experienced by the unpaired electrons. The probability of interconversion is proportional to $(E_T - E_S)^{-2}$, E_T and E_S being the energy levels of a triplet substate and the radical pair singlet state, respectively. When a strong magnetic field is applied, the $T_{\pm1}$ levels, corresponding to the magnetic quantum numbers $m_S = \pm 1$, are pulled out of the region where they can be populated from the state $(P^+I^-)^S$, as they are separated from the $m_S = 0$ and the singlet level by the Zeeman energy. In a high magnetic field, where the $T_{\pm1}$ levels are not populated, the probability to recombine from $(P^+I^-)^T$ to P^TI is lower, and the triplet yield is decreased. Although the details of the radical pair mechanism (RPM) are complicated by the possible occurrence of exchange interactions and recombination to the excited singlet state P^*I and renewed charge separation [130–135, 306–309], the magnetic effect on the triplet yield is qualitatively in good agreement with the expected effect as calculated from the RPM theory.

Another check on the RPM as mode of RC triplet formation was carried out by studying the effect of an external magnetic field on the fluorescence of cells of *Rps. sphaeroides* under reducing conditions at 1.2 K [136]. The magnitude of the effect was quantitatively in good agreement with the calculated value if it was assumed that the triplet state is formed via the RPM but not for formation via intersystem crossing.

Triplet states are characterized by their zero field splitting parameters D and E, which are determined by the electronic wave function and which express the way the two unpaired electrons are interacting via dipolar coupling, and by the sublevel decay rates which also depend on the electronic wave function. Thus, determination of D and E and of the molelucar decay rates gives insight into the electron configuration of the primary donor in its triplet state and, by inference, into its ground state geometrical structure. D and E values and the decay rates can be determined from ESR spectra and, more accurately, from ESR in zero field (zero field resonance, ZFR) [137–138].

For many bacterial species the values of D and E are about 20% lower than the corresponding values of monomeric Bchl in the triplet state. As is dicussed elsewhere in this volume, using a simple model, Clarke et al. [139] have related the values of D, E and the decay rates to the geometrical structure of the dimeric donor. The applicability of this model, however, is controversial as it does not incorporate charge transfer contributions from dimer triplet states of the form $Bchl^+ . Bchl^-$ or $(Bchl)_2^{+} . Bph^{-}$, which may have a profound influence on the values of D, E and the decay rates [310]. Taking the values of the triplet decay rates of the primary donor as determined by optical spectroscopy, by flash ESR spectroscopy or by ZFR [138, 140] together with the values of D and E, it is not possible to arrive at consistent results [140–142]. The deviating values of the decay rates cited in one report on ZFR of *R. rubrum* and *Rps. sphaeroides* [143] which are not in conflict with the simple exciton model, are incorrect [373, R23]. Whatever the virtues of the exciton model are (see also refs. [87, R23] and R.H. Clarke, this vol.), it is clear that the triplet state is an important probe of the structure of the primary donor, and future refinements of model calculation and further experimentation will certainly contribute to our knowledge of the structure of P860.*

---

* See *Notes Added in Proof*, p. 323

3 The Plant Photosystems

3.1 Optical Investigations of the Primary Donor of Photosystems 1 and 2

Photosynthesis in plants is considerably more complex than in photosynthetic bacteria, owing to the presence of two photosystems working in series. It is possible, however, by judiciously applying electron transport inhibitors, artificial electron donors and acceptors, and/or different wavelengths of excitation, to study in intact systems (algae, chloroplasts of higher plants) the response of the two photosystems to actinic light separately. Another avenue is the physical separation of PS 1 and PS 2. This has been achieved in a variety of ways, for instance by separation of the extended (stroma) and stacked (grana) lamellae of chloroplasts and subsequent detergent treatment (for reviews see refs. [144–148]). The resulting subchloroplast particles contain only one of the two photosystems, and are enriched in its reaction center with respect to antenna chlorophyll. In Table 3.1 a list of such particles is given. It has as yet not been possible, however, to isolate a reaction center comparable to the bacterial ones, that is devoid of antenna chlorophylls. This has considerably hampered the optical studies, which is one of the reasons why our knowledge of the photooxidation of the primary donor of PS 1 (P700) and PS 2 (P680) is much less detailed than that of the bacterial primary donor. In the following section, the optical investigations are reviewed followed by a discussion of the ESR and ENDOR work. It will be seen that there are many similarities between the bacterial and plant systems, but that there does not seem to be a one-to-one correspondence. In other words, although the main structures of the donor complexes are similar, yet Nature has found different ways to meet the different requirements set by the charge separation process in bacterial and plant photosynthesis.

Table 3.1. Photosystems 1 and 2 subchloroplast fragments

	Reaction center (P700 or P680) : chlorophyll	References
P700 fragments		
D-144 (Digitonin)	1 : 200	149
TSF-1 (Triton)	1 : 100	150
HP700 (Triton)	1 : 30	151
P-144 (French press)	1 : 100	152
LDAO PS 1 fragments	1 : 30	153
P700-Chl *a*-protein, higher plants (Triton)	1 : 40	154
P700-Chl *a*-protein, blue-green algae (Triton)	1 : 40	155
Digitonin PS 1 fragments	1 : 100	156
Triton-LDAO-SDS	1 : 20	161
P680 fragments		
F II	1 : 20–70	147, 157
(French press, dextran-glycol)	P700/P680 ~6	158
TSF-2a	1 : 40–80	159, 160

3.1.1 Absorption Difference Spectroscopy of P700

P700, the primary electron donor of photosystem 1, was discovered by Kok [162, 163] through application of ΔA spectroscopy on chloroplasts and algae. Photooxidiation of P700 is characterized by a bleaching at 703 nm and 430 nm, together with an apparent blue shift of a band at 675 to 680 nm. A recent ΔA spectrum of $P700^{+} - P700$ is shown in Fig. 3.1a. Apart from the ubiquitous red shift, it resembles closely the ΔA spectrum of the oxidation of Chl *a* in vitro [164] (Fig. 3.1b). The feature at 690 nm has been interpreted to be partly due to an electrochromic effect of the oxidation of P on associated pigments [165], partly to a band shift associated with the reduction of (one of) the primary acceptor(s) [166, 167].

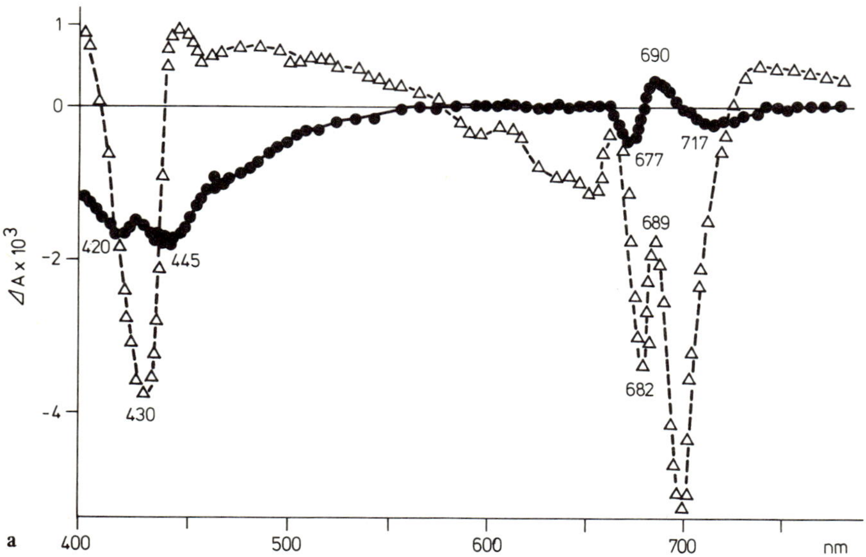

a

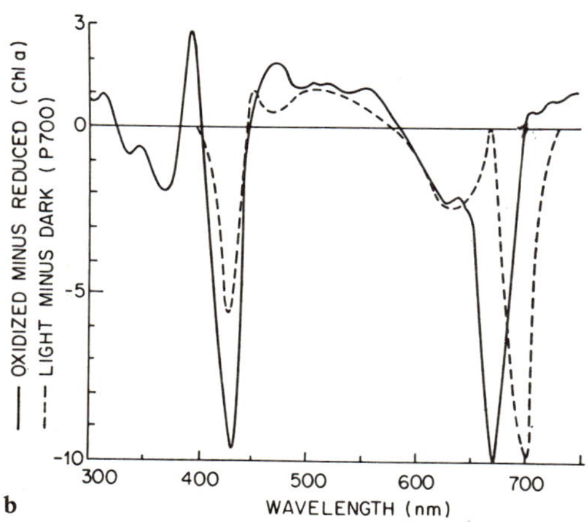

b

Fig. 3.1. a Absorption difference spectrum of the primary donor P700 (Δ) and the "primary" acceptor P430 (●) in photosystem 1 subchloroplast particles. (From [166] as redrawn in [84]). **b** Absorption difference spectrum of the oxidation of Chl *a* in vitro. Dashed line: ΔA spectrum of chloroplasts. (From [164])

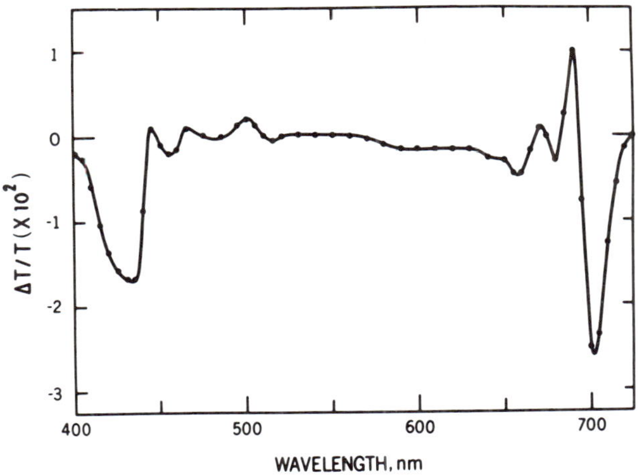

Fig. 3.2. Low temperature (86 K) ΔA spectrum of a PS 1 particle from the blue-green alga *Anabaena* (B. Ke, as reproduced in [168])

The feature at 690 nm, apparent as a trough in the room-temperature spectrum but much more pronounced in the low-temperature spectrum (Fig. 3.2) where it appears as a positive band, might also be due to the appearance of a monomer band upon bleaching of two exciton bands at 683 and 697.5 nm associated with a Chl *a* dimer (Fig. 3.3) [169]. According to this interpretation, which receives some support from ΔCD spectroscopy (Sect. 3.1.3), the band shift at 680 nm is caused by the partly overlapping positive ΔA band of the monomer and the negative ΔA of the bleached exciton band at 683 nm. Recent results of Schaffernicht and Junge [311], however, make it less probable that the 683–695 double band is due to exciton splitting of the dimer. Instead, it is attributed to the disappearance of a relatively wide dimer band, the appearance of a monomer band at usual width and half the oscillator strength of the dimer band and a small shift of antenna absorption due to local electrochromism. In addition, the red shift of P700 is not attributed to exciton interaction within the dimer, as the monomer is also red-shifted according to their interpretation.

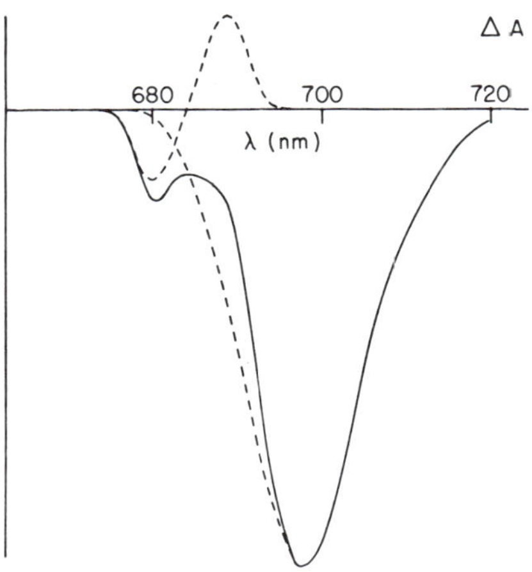

Fig. 3.3. A deconvolution of the absorption difference spectrum of subchloroplast particles into exciton components. The apparent band shift at 684.5 nm is though to be composed of a short-wavelength exciton component of P700 at about 684 nm and the appearance of a band at 686 nm owing to the absorbance of one unoxidized monomer in P700⁺. (From [169])

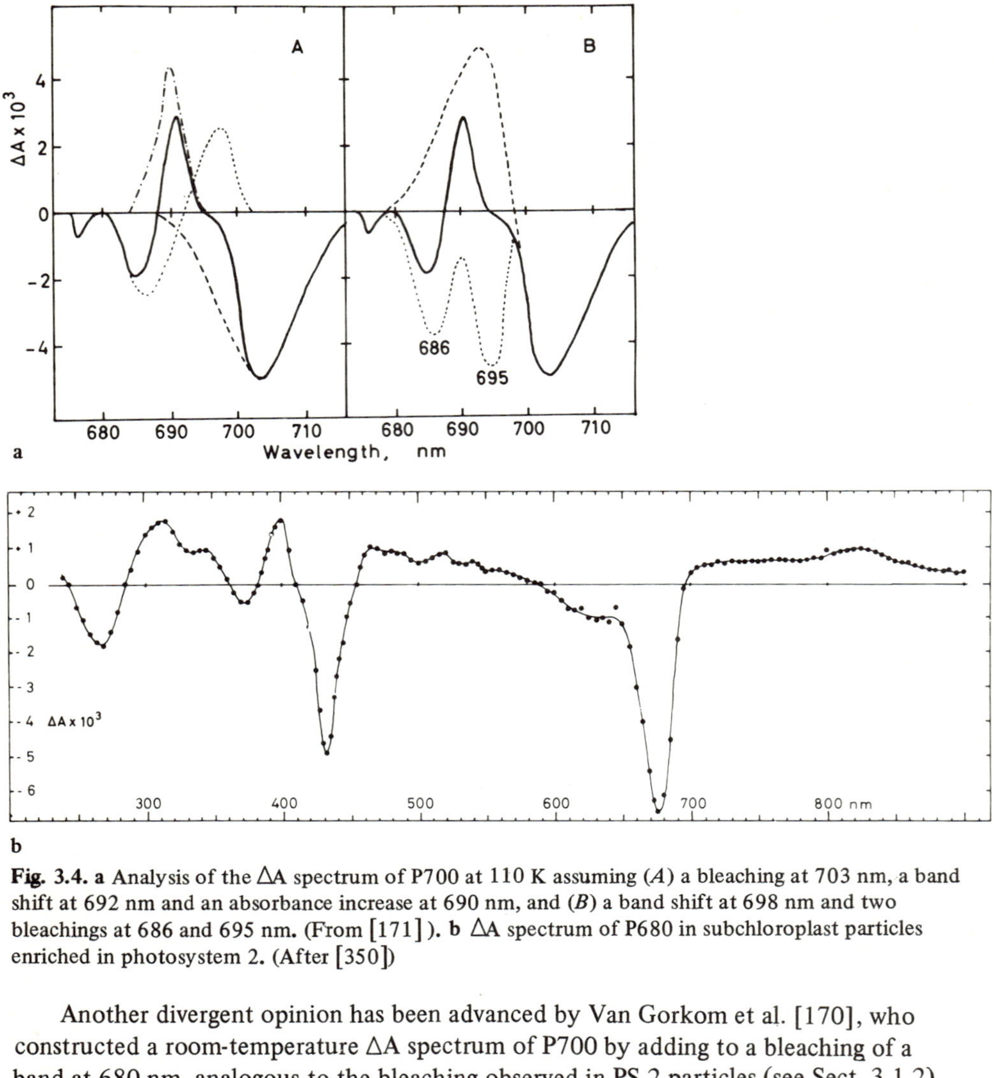

Fig. 3.4. a Analysis of the ΔA spectrum of P700 at 110 K assuming (*A*) a bleaching at 703 nm, a band shift at 692 nm and an absorbance increase at 690 nm, and (*B*) a band shift at 698 nm and two bleachings at 686 and 695 nm. (From [171]). **b** ΔA spectrum of P680 in subchloroplast particles enriched in photosystem 2. (After [350])

Another divergent opinion has been advanced by Van Gorkom et al. [170], who constructed a room-temperature ΔA spectrum of P700 by adding to a bleaching of a band at 680 nm, analogous to the bleaching observed in PS 2 particles (see Sect. 3.1.2), a hyposochromic band shift of about 12 nm at 700 nm. This shift is somewhat larger than the red shift at 690 nm proposed in the more generally accepted interpretation.

The ΔA spectrum of P700 at 100 K has been analyzed in both ways [171]. Assuming a bleaching at 700 nm with a red shift at 690 nm, the experimental ΔA spectrum could only be reproduced by adding a positive band at 690 nm (Fig. 3.4 **a** A); assuming a blue shift at 700 nm a two-banded bleaching at 680 nm resulted (Fig. 3.4 **a** B), which possibly reflects the disappearance of two exciton bands which are resolved at low temperature (see, however, Schaffernicht and Junge [311]).

The quantum yield of P700 formation approaches unity for excitation at 703 nm [172]; it can be induced at cryogenic temperatures down to 15 K [173]. The rise time of P700 is less than 20 ns [174]. The differential molar extinction coefficient is

64 mM^{-1} cm^{-1} for PS 1 particles and 70 mM^{-1} cm^{-1} for Triton-fractionated PS 1 particles [168] as measured by oxidation of the electron donor TMPD. For intact chloroplasts a value of 67 mM^{-1} cm^{-1} was found by relating the absorbance changes associated with the reduction of $P700^+$ after a series of closely spaced flashes to the number of electrons generated per flash as measured by the oxygen yield [175].

The E_m of P700 oxidation is controversial. Kok [176] obtained +420 mV for acetone-extracted chloroplasts, Ke et al. [150] found +470 mV for Triton PS 1 particles (TSF-1) but did not obtain a true end point of the titration. Calvin and Androes [177] obtained +420 mV by monitoring the ESR signal due to $P700^+$ in chloroplast fragments, while Knaff and Malkin [178] found +520 mV for digitonin PS 1 particles (D–144). Evans and co-workers [179–181] measured the E_m in Triton PS 1 particles (HP700) by optical and ESR methods and found a value of +375 mV when the ability of P700 to reduce the primary acceptor (ferredoxin center A) was monitored by low-temperature ESR. They attributed the higher values around +450 mV to artefacts from bulk chlorophyll oxidation.*

The shift of the wavelength of maximal bleaching (703 nm) compared to the red absorption band of Chl *a* (660–670 nm depending on the solvent [182]) has been interpreted to arise from exciton interaction between the components of a dimer. Calculation using the formalism developed by Hochstrasser and Kasha [183] yields an absorption maximum for a hydrated dimeric Chl *a* species halfway between that of the monomer and that of an infinite array (743 nm, as found for Chl $a \cdot H_2O$ aggregates) [184–187]. However, a model system composed of two Chl *a* molecules held rigidly together by two ester bridges [a so-called bis (Chl *a*) cyclophane] does not show a wavelength shift of the absorption maximum [188], so that it is questionable whether the observed shift is solely due to $\pi - \pi$ interactions. Recently, an attempt was made to include environmental effects in an exciton calculation for a P700 type dimer [312]. It was found that a polar "solvent" stabilizes charge transfer states of the dimer when significant overlap between the π-orbitals of the monomers is assumed. For monomers in a stacked configuration, with the central Mg atom translated by 8 Å [313], a red shift conforming to the observed red shift of P700 was obtained. For a Chl-*a* cyclophane, with the Mg atoms on top of each other, the allowed transition would be expected to be blue-shifted. Further evidence that it is not necessary to invoke aggregation of chromophores as the origin of the shift from the absorption maximum in organic solvents was provided by Rafferty et al. [341] who demonstrated by photooxidation of antenna Bchl in situ that monomeric Bchl in the antenna complex absorbs at 852 nm compared to 770 nm in acetone/methanol. Antenna Chl *a*-protein complexes also show a shift toward longer wavelengths [145, 148] which has similarly been attributed to $\pi - \pi$ interaction between neighboring pigment molecules. However, resonant Raman studies [189–191] yield quite different spectra for in vitro Chl-*a* dimers and for the Chl *a*-protein complexes. It was suggested that the Chl-*a* molecules are distributed in the enveloping proteins in a similar seemingly random way as the Bchl-*a* protein isolated from the green bacterium *Prosthecochloris aestuarii* of which the structure has been resolved by X-ray cristallography [192, 193]. Therefore, much more detailed exciton calculations seem to be required to obtain a reliable calculated absorption spectrum for antenna complexes [187].

---

* See *Notes Added in Proof*, p. 323

The hypothesis that P700 is a dimer is supported by CD and LD experiments to which we will turn in the next section. Also ESR and ENDOR investigations provide strong indications that P700 is a Chl *a* dimer analogous to the Bchl *a* dimer that makes up the bacterial primary donor (see Sect. 3.2.1 and 3.2.2).*

3.1.2 Absorption Difference Spectroscopy of P680

In 1969 Döring et al. [194] found in chloroplasts a rapidly decaying ($t_{1/2} \sim 200\ \mu s$) absorption change near 680 nm which was attributed to the photooxidation of the primary donor of PS 2. Its absorption difference spectrum in the red is composed of a large bleaching at 680 nm and a smaller one at about 635 nm. The species giving rise to the absorption change is generally labeled P680 to conform to the nomenclature for other primary donors, although sometimes the older name Chl a_{II} [194] is still used. The fast decaying changes are not induced by far-red (> 720 nm) illumination which is known to excite only PS 1, it is inhibited by the PS 2 inhibitor DCMU and its ΔA spectrum is identical to that of subchloroplast particles which lack PS 1 [157, 170, 194, 195]. Figure 3.4b shows a recent ΔA spectrum of P680 obtained in subchloroplast fragments [170]. Note the broad increase at 820 nm characteristic for oxidized Chl *a* [164]. Using faster response ΔA spectroscopy, Gläser et al. [196] found in chloroplasts a decay of about 35 μs. The 200 μs and the 35 μs decays are now attributed to rereduction of $P680^+$ by the primary acceptor Q^- and by a secondary donor, Z, on the oxidizing side of P680, respectively. Still faster rereduction of P680 was found by Van Best and Mathis [197] who, employing a ΔA spectrometer with nanosecond response time, found at 820 nm an absorption change which decayed with $t_{1/2} \simeq 30$ ns. Upon treatment with the reductant NH_2OH the decay became slower and biphasic with half times of $t_{1/2} \sim 30\ \mu s$ (major phase) and $t_{1/2} \sim 200\ \mu s$ (minor phase). Apparently, these components correspond to those found by Gläser et al. [196]; the use of a repetitive flash technique by these authors presumable converts the oxidizing side of PS 2 into a state rather similar to that obtained by inhibition with NH_2OH. The 30 ns reduction of P680 as measured at 820 nm is attributed to rapid electron donation by a secondary donor of PS 2 on the oxidizing side of P680 [197, 314].

At liquid nitrogen temperatures the decay of P680 is slowed to a few ms [198, 199]. Mathis and Vermeglio [199] followed in chloroplasts the change at 820 nm upon flash excitation and found a decay with a half time of 3 ms. This decay is reversible for about 75%, it probably results from the back reaction between $P680^+$ and the reduced primary acceptor, Q^-. The nonreversible part is due to competitive electron donation by cyt b_{559}; successive flashes reduce progressively the amplitude of the absorption change at 820 nm. The latter reaction is suppressed when cyt b_{559} is preoxidized with ferricyanide. In subchloroplast particles thus treated, one may observe at room temperature absorption changes analogous to the P680 ΔA spectrum ([170, 200] and Sect. 3.1.3). However, the changes are observed under continous illumination and may originate from photo-enhanced oxidation of bulk chlorophyll by ferricyanide, or by oxidation of a chlorophyllous alternative electron donor to P680 (see Sect. 3.1.3).

Recently, the decay of P680 was studied over the temperature range 300–30 K in TSF-IIa particles enriched in cyt b_{559} and PS 2 reaction centers [315]. At room tem-

---

* See *Notes Added in Proof*, p. 323

perature the decay was multiphasic, containing a major rapid phase which became progressively slower with lower temperature until at 200 K and below the decay was exponential with a temperatur-independent half time of 1.25 ms. From the ΔA spectrum it was concluded that the major decay phase represents charge recombination between $P680^+$ and Q^-, the reaction showing the temperature dependence typical of vibronically coupled tunneling (Sect. 4).

The reduction of $P680^+$ can also be followed by monitoring the prompt fluorescence (as $P680^+$ is a quencher) or the delayed (probably recombinational) fluorescence, which, at least at room temperature, originates mainly from PS 2. Den Haan et al. [316] found at 77 K a fast (30 μs) rise in fluorescence attributed to reduction of the quencher $P680^+$, but this component does not show up in low-temperature ΔA kinetics [171]. For details of related work on (delayed) fluorescence the reader is referred to Chap. III, 14 of ref. B2, Chap. 6 of ref. B3, and refs. B19, 201.

Van Gorkom et al. [170] have attempted to construct the ΔA spectrum of P680 from the spectra of an oxidized Chl-*a* monomer and a neutral Chl-*a* monomer, with the underlying assumptions that on an optical time scale the positive charge is localized on only one constituent of a Chl-*a* dimer and that exciton interaction between the two Chl-*a* molecules is destroyed by oxidation. The absorption spectrum of the dimer $(\mathrm{Chl}\ a)_2$ was taken from Sauer et al. [202] who studied the monomer-dimer equilibrium in Chl *a* in CCl_4. The agreement of the calculated and the experimental ΔA spectrum is rather good, but one should be cautioned that "dimer" spectra are sensitive to liganding. As it is farily certain that the primary donor Chl-*a* complex is hydrated, its absorption spectrum may in reality look more like the in vitro "P700" spectrum published by Fong and Koester [203], or like the spectra of synthetic dimers of chlorophyllous pigments obtained by Boxer and Closs [204] and Wasielewski et al. [205] in water-saturated nonpolar solvents. If such dimer spectra, which show a maximum absorption in the red close to 700 nm, are used for the construction, the agreement between calculated and experimental P680 ΔA spectra is much less good.

A quite different opinion is voiced by Davis et al. [317], who compared the ΔA spectrum of monomeric Chl a/Chl a^+ in CH_2Cl_2/25% THF to the ΔA spectrum of P680 in vitro. They found good agreement, attributing the shift of the bleaching at 680 nm in vivo to 670 nm in vitro to a difference in coordination of the central Mg atom, which effect would also be responsible for the change in line width of the ESR line of $P680^+$ compared to that of Chl a^+ (see Sect. 3.2.2). The authors contend that P680 might very well be a ligated Chl-*a monomer*. This is in line with the fact that ligation is important for the redox properties of Chl *a*. It is, for example, well known that hydrated Chl *a* is readily photooxidized, whereas "dry" Chl *a* is not [107, 209]. The stoichiometry of hydration is at present controversial (see refs. [14, 210]). There is evidence that in vitro an oxidized dimeric chlorophyll-dihydrate, $(\mathrm{Chl}\,a \cdot 2\,H_2O)_2^+$, may be capable of oxidizing water [211]. In vivo, however, this reaction does not seem to take place, because at the water-splitting side of P680 a complicated sequence of events, involving storage of four oxidative equivalents, leads to oxygen evolution (see, e.g., Chap. II, 8 of ref. B2 and Chap. 8 of ref. B3). The situation is complicated by the fact that the redox mid-point potential for P680 oxidation is not known. The early report that it is somewhat higher than +850 mV [258] is based on ESR experiments the validity of which has been rendered uncertain by later work (Sect. 3.2.2). As $P680^+$ must be capable of splitting water, its E_m (pH 7) must lie above +820 mV. Since there are a number of intermediates between water and P680, it may be appreciably higher.

3.1.3 Circular and Linear Dichroism of Photosystems 1 and 2

The fact that purified reaction centers of the plant photosystems have as yet not been prepared has presented a serious obstacle to the characterization of P700 and P680 and associated pigments by CD and LD. Most of the applications of dichroic techniques on plant material (chloroplasts, subchloroplast particles, algae) are concerned with antenna pigment orientation with respect to the thylakoid membrane. A useful survey of this work, which touches only tangentially the present subject, is given by Gregory [212].

One way to circumvent the difficulties arising from the presence of large amounts of bulk chlorophyll in even the most highly purified subchloroplast particles is the application of CD difference spectroscopy, as carried out by Philipson et al. [169]. The authors compared the ΔA and ΔCD (light *minus* dark) spectra of HP700 PS 1 particles (Fig. 3.5). The ΔCD spectrum consisted of a negative band at 688 nm and an equally strong positive band at 697.5 nm. Such a conservative CD spectrum (the rotational strength summed over the bands equals zero) is characteristic of exciton interaction between two pigment molecules [36]. Usually the CD intensity arising from exciton in-

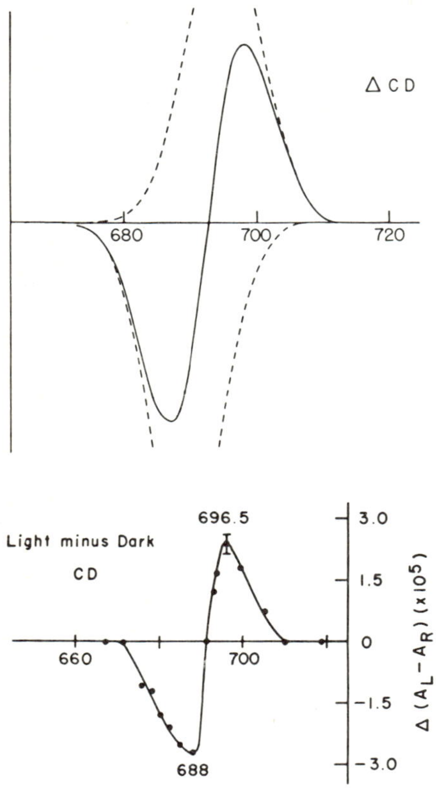

Fig. 3.5. Circular dichroism difference spectrum of the same preparation as in Fig. 3.3 and its deconvolution into exciton components centered at 689 (–) and 695 (+) nm (*top*). (From [169])

teraction is much stronger than that for the individual pigments. Hence, the destruction of the P700 exciton by oxidation would produce negligible CD, and the ΔCD spectrum consists predominantly of the P700 CD, reversed in sign (not counting possible interactions with closely associated pigments). Assuming strong excitonic coupling in P700, the ΔA spectrum could be satisfactorily explained (Fig. 3.3).

From the analysis presented by Philipson et al. [169] it follows that socalled P700 is a bleaching, and not part of a band shift as proposed by Van Gorkom et al. [170]. There are, however, a number of ambiguities that would deserve closer examination. For instance, the bandwidths of the exciton components of the absorption spectrum are rather dissimilar, the wavelengths of maximal oscillator and rotational strength of the ΔA and ΔCD spectrum are not related to each other, etc. There is clearly a need to quantify the exciton concept in ΔCD and ΔA measurements. In this respect, it would be helpful to carry out ΔCD experiments at low temperature as the various ΔA bands are then much better resolved (see Fig. 3.2).

Linear dichroism measurements suffer from the same obstacle as CD studies, i.e., the presence of large amounts of bulk chlorophyll. As for CD, this difficulty has been partly overcome by performing light *minus* dark difference spectroscopy. As discussed in Sects. 2.1.3 and 2.1.4, orientation was achieved by magnetic field or by flow techniques, or the necessary asymmetry was introduced by photoselection.

From the LD spectrum of oriented chloroplasts and subchloroplast PS 1 particles, it was concluded that antenna pigments absorbing beyond 680 nm are highly oriented with their Q_y transition moment parallel to the thylakoid membrane [213–215]. This finding was substantiated by measurements of the polarization of fluorescence in oriented samples [216, 217], it confirmed older measurements employing the polarizing microscope (see [212] for references). Employing difference spectroscopy in flashing light on PS 1 particles, it was found that the dichroic ratio $\Delta A_{/\!/}/\Delta A_{\perp}$ (for the experimental configuration see Fig. 3.6) for absorbance changes at 700 nm was 2.3–2.5 [218–220], which value is even higher than that found for antenna pigments ($\Delta A_{/\!/}/\Delta A_{\perp}$ = 1.6 [219]). A similar high value was found for the 820 nm band associated with the radical cation of P700. The high dichroic ratio indicated that also P700 is oriented almost parallel to the thylakoid membrane, as would be expected for efficient energy transfer from the long-wavelength antenna to P700. Interestingly, for the 650–670 nm region of the ΔA spectrum, much smaller dichroic ratios were found, e.g., for 660 nm $\Delta A_{/\!/}/\Delta A_{\perp} = 0.4$. The absorbance changes in this region are not due to electrochromic effects as in the experiments 10^{-6} M valinomycin was added to make the decay of these effects faster than the time resolution of the dichrograph. Assuming that the effects on ΔA bands around 660 nm are due to an optical transition of P700, the low dichroism was explained by two hypotheses: the 660 nm band could be the excitonic partner of the P700 band, with its transition moment polarized perpendicular to Q_y of P700, or it could be due to a slightly allowed Q_x transition of P700, shifted by exciton interaction. In either case, the planes of the tetrapyrrole rings of the members of the dimer cannot be *both* oriented parallel to the membrane.

In a series of parallel researches, Junge et al. [221–225] employed the technique of flash photoselection on chloroplasts and on PS 1 particles. In principle, this technique yields the same results as ΔLD spectroscopy (it also suffers from the danger of similar artefacts as the latter method). There is, however, the ambiguity that excitation

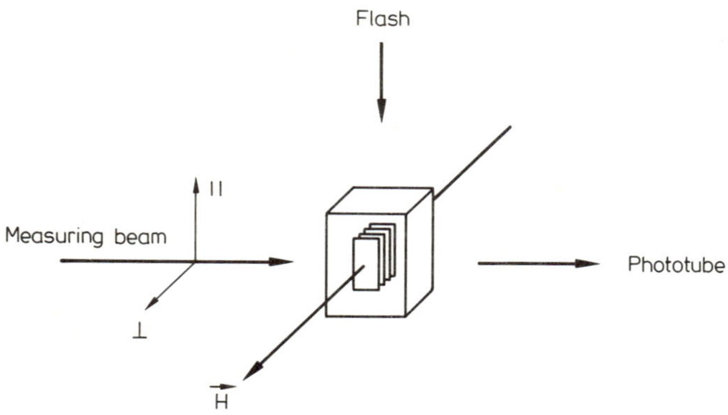

Fig. 3.6. The arrangement for linear dichroism measurements of chloroplast membranes oriented in a magnetic field. (From [215])

in an array of antenna pigments leads to extensive depolarization of the polarized excitation beam when not all the antenna transition moments are parallel.

Since there are many transfer steps between absorption of a light quantum by the antennae and its trapping by P700, and because it is improbable that these antennae are arrayed linearly, it is reasonable to assume that the excitation of P700 is circularly degenerated.

Junge and Eckhof [221, 222] found a dichroic ration $A_{//}/A_{\perp}$ (here $A_{//}$ and $A_{\perp}$ are the absorptions parallel and pependicular to the direction of polarization of the exciting light, respectively, the measuring and exciting beams being perpendicular to each other) for excitation at wavelengths greater than 690 nm of 1.16 at both 705 and 420 nm. This value is close to the value 4/3 expected for an absorption dipole parallel to a circularly degenerate antenna system [222]. They estimated that the plane of P700 makes an angle of about 25° to the membrane.

The above studies were refined in subsequent work [223–225], where excitation took place at 724 nm. Again the dichroism at 701 nm was close to 4/3; for PS 1 particles with P700 : Chl = 1 : 40, $A_{//}/A_{\perp} \sim 1.3$. Apparently, also at this long wavelength of excitation transfer of energy to P700, or absorption by P700 itself, takes place in a circularly degenerate way. Assuming circularly degenerate transfer of energy, a relation was derived between the number of antenna pigments and the angle β which their red transition moments make with the Q_y axis of the P700 dimer. At least two antenna molecules are required, for which then $\beta = 90°$. For four antennae, $\beta = 60°$, for six, $\beta = 55°$. As deconvolution of the absorption curve of the PS 1 particles does not permit more than eight chlorophylls absorbing at 700 nm and beyond, the last value of β constitutes a lower limit. The alternative case, i.e., the P700 dimer is the only species absorbing at 700 nm and beyond, leads to the interesting conclusion that the Q_y transition moments of the dimer constituents are perpendicular to each other. This is at variance with the postulated models for P700, viz. the Fong model in which the Q_y axes are inclined by 60° [226], and Shipman's model in which they are parallel [227].

Although the authors could as yet not discriminate between the two possibilities for want of knowledge of the number of antenna chlorophylls absorbing at 700 nm, they favored the circular degeneracy of the dimer itself because excitation of carotene molecules at 490 nm also resulted in a dichroic ration at 701 nm close to 4/3. The carotenes themselves are almost colinear, as evidenced by the high dichroic ratio (1.9) measured at 490 nm for excitation in the carotene band [223, 224], so that they would have to be inclined at roughly 60° to the red transition moment of a dimer which is not intrinsically degenerate. The authors considered it unlikely that both the antenna chlorophylls and the carotenes would show such a special orientation toward the red transition moment of P700.

One might ask whether it would be possible to derive the angle between the Q_x transition moments of the dimer, and consequently the angle between the planes of the two tetrapyrrole rings, from the dichroic ratio measured at 430 nm for excitation at 724 nm. This is a nontrivial task, as the Soret band is composed of overlapping bands of orthogonal polarization. Junge et al. [225] assumed circularly polarized absorption by P700 (one needs not discriminate between the two cases discussed above) and derived formulas relating the observed dichroic ratio (1.21) to the angles of inclination of the Q_x transition moments of the components of the dimer to the plane made up by the Q_y transitions Taking into account possible exciton interaction between the Q_x transition moments, they were able to delineate a field of possible configurations for the x-axes. It was found that if one x-axis is parallel to the plane of excitation, then the other x-axis is inclined to this plane at an angle between 43° and 29°, whereas for both axes having equal inclination the angle is 29°.

The above values can be used to deduce the angle ϕ between the porphyrin planes if the angle between the Q_y moments is known. If they are perpendicular then ϕ is between 28° and 40°. If one of the Q_x is strongly inclined, ϕ is between 29° and 43° irrespective of the angle between the Q_y's. If both Q_x's are about equally inclined, Shipman's model (Q_y's make an angle of 0° or 180°) yields ϕ 0° or $\phi = 50–58°$, whereas Fong's model (Q_y's make an angle of 60° or 120°) requires $\phi = 20–30°$ or 35–40°. As both models postulate essentially parallel porphyrin planes, either both models have to be modified for nonparallel rings, or the first model is right, but then there must be at least two antenna pigments with their Q_y transitions oriented almost perpendicular to the Q_y of P700. As Junge et al. [225] note, this is rather unfavorable for energy transfer.

It will be noted that the above discussion cannot directly be transferred to the bacterial photosystem. As discussed in Sect. 2.1.4, LD measurements point to almost parallel porphyrin rings for P860, whereas the calculations based on triplet parameters are ambiguous (Sect. 2.2.4). In view of the rather different aggregation configurations possible for Bchl *a* dimers compared to those for Chl *a* dimers, this need not be surprising. The study of recently synthesized, well-defined cyclophanes [185] might shed some light on the divergencies to be expected for the optical properties of the plant and the bacterial primary donor.

Note also that the interpretations of the dichroic studies on P700 all depend on the assumption that P700 is a bleaching. If it is a band shift as postulated by Van Gorkom et al. [170], then the conclusions advanced above are invalid.

As yet, only one report on ΔLD studies of PS 2 has appeared. Mathis et al. [228] measured the dichroic ratio at 825 nm in oriented chloroplasts for flash excitation at 600 nm. It was shown earlier [199] that the flash-induced absorbance increase at 825 nm decays biphasically, the fast phase being due to the reduction of $P680^+$ and the slow phase arising from the reduction of $P700^+$. The dichroic ratio for both phases was found to be similar, viz. 1.32 and 1.40 (± 0.05), respectively. The authors concluded that the Q_y transition moment of P680 is oriented parallel to the thylakoid membrane.

3.2 ESR and ENDOR

3.2.1 $P700^+$

Light-induced ESR signals in plant photosynthetic material were first observed by Commoner et al. [229]. Subsequent work [88, 230–232] revealed two signals, a narrow one, called Signal I (S I) and a broad one, called Signal II. The latter is not due to primary reactants and will not be considered further (see for a review on Signal II Chap. 3 of ref. [87]). Signal I arises from the oxidation of the primary donor of PS 1 as evidenced by the following:

1. Its E_m is similar to that of P700 as measured optically, viz. +430 to +520 mV [150, 177–179] (as suggested by Evans and co-workers [179–181], this value may be artifically high due to oxidation of bulk chlorophyll, which yields an ESR line similar in width and shape to S I),
2. S I and P700 have identical decay kinetics under various conditions as measured simultaneously optically and with ESR [233],
3. S I and P700 are produced in stoichiometric amounts [234, 235]. S I can be photoinduced at temperatures down to 1.2 K [236], its rise kinetics are complex, due to spin polarization effects [237–239].

Signal I is composed of a Gaussian line at $g = 2.0025 \pm 0.0001$ of width $\Delta H = 7.2 \pm 0.1$ G at room temperature [240]. It has no discernable hyperfine structure; its saturation behavior is characteristic of an inhomogeneously broadened line. Upon deuteration the line width is appreciably narrowed [92, 107], indicating that the width arises mainly from proton hf interactions.

The reduction of the width of S I in all species investigated (algae, higher plants) by a factor of 1.4 compared to the width of monomeric Chl a^+ (10.2 G) suggested to Norris et al. [107] that $P700^+$ is composed of an oxidized Chl-*a* dimer, in which the unpaired electron is fully delocalized (cf. Sect. 2.2.1). Support for this hypothesis came from experiments on ^{13}C-enriched algal samples, in which S I showed a broadening by a factor of 1.3 compared to monomeric ^{13}C-Chl a^+. A factor of 1.26 was computed from the non-Gaussian line shape of ^{13}C-Chl a^+ [241].*

As for the bacterial primary donor, proof of the dimer hypothesis was sought by ENDOR [112–115]. In Fig. 3.7 ENDOR spectra of Chl a^+ in vivo and in vitro are compared. The strong lines of the algal specimen are assigned to methyl groups on the basis of spectra obtained in vitro for selectively deuterated Chl a^+ and derivatives [112, 115]. If this assignment is correct (a somewhat different ENDOR spectrum showing one more set of strong lines was obtained for S I in PS 1 particles [87]), it follows from the hfs as measured for the methyl groups in vitro and in vivo that two Chl-*a* molecules participate in electron delocalization in $P700^+$.*

* See *Notes Added in Proof*, p. 323

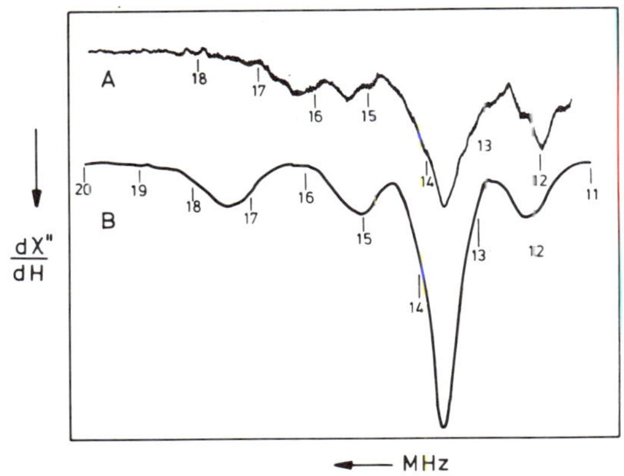

Fig. 3.7. ENDOR spectra at 97 K of the alga *C. vulgaris* oxidized by $K_3Fe(CN)_6$ (A) and of Chl a^+ *in vitro* (B). (After [115])

The photooxidation of P700 is reversible at room temperature, but at lower temperatures and moderate redox potentials it becomes gradually irreversible, until at temperatures below 15 K less than 10% of $P700^+$ is reduced in the dark after preceding illumination. This reduction is caused by a back reaction between $P700^+$ and the electron stored on one of the "primary" acceptors of PS 1. These acceptors, called, X, B, and A, act in series and are composed of iron-sulfur proteins as judged from their ESR spectrum in the reduced state (see for recent reviews on the acceptor of PS 1 refs. [84, 87, 242, 243]). Apparently, at temperatures below 15 K, the electron on either A or B cannot return to the oxidized primary donor. However, when A and B are both reduced before illumination at low temperatures (at redox potentials below –610 mV) the photoejected electron cannot leave X^- for A or B, and a back reaction occurs with $t_{1/2}$ of about 0.8 s [244]. Thus, at these low redox potentials, low temperature photooxidation of P700 as measured optically and by ESR is almost completely reversible [173, 245–250].

3.2.2 $P680^+$

It is well documented that Signal I arises from the photooxidation of P700. At room temperature, no other signals from primary reactants are found at g = 2.0. The absence of an ESR signal from the oxidized donor of PS 2, $P680^+$, is explained by very fast ($< 1\ \mu s$) reduction of $P680^+$ by a secondary donor [197, 255]. At much lower temperatures, $P680^+$ was shown to be reduced by a non-physiological donor, cyt b_{559} [256]. At sufficiently high redox potentials (above 375 mV) this cytochrome is oxidized, and one would except to observe an ESR signal from $P680^+$. Malkin and Bearden [257] found at 77 K an irreversible light-induced ESR signal similar in shape to S I on top of the signal from oxidized $P700^+$, with mid-point potential E_m = +477 mV [258]. It was attributed to $P680^+$, but subsequent ΔA studies by Lozier and Butler [165] carried out under similar conditions failed to show an irreversible bleaching at 680 nm. It was concluded that at 77 K and $E_h > 500$ mV, $P680^+$ is reduced in competitive reactions by a donor other than cyt b_{559} and by the back reaction $P680^+Q^- \rightarrow P680\ Q$.

Apparently, the signal found by Malkin and Bearden [257] arises from the oxidized alternative donor to $P680^+$, labeled D^+. Judging from the spectral characteristics (width, line shape and g-value, all similar to S I), D is a chlorophyllous species, probably a Chl-*a* dimer [257].

Subsequent experiments with flash illumination under oxidizing conditions at 77 K and lower temperatures [259, 260] revealed a transient ESR signal at g = 2.002, that decayed with $t_{1/2}$ = 5.0 ms, independence of temperature. A plot of the amplitude of the signal against magnetic field yielded a Gaussian ESR line of width 8 G, suggesting that also P 860 is a chlorophyll dimer (see, however, Sect. 3.1.2 and ref. [317]). The transient was partially reversible at redox potentials above −150 mV, the amplitude decreased with increasing number of flashes. The reversible part of the decay was attributed to the back reaction between $P680^+$ and Q^-, the nonreversible part to irreversible electron donation by D.

Another way to slow down the reduction of $P680^+$ is the preparation of PS 2 particles [67, 200]. In these highly purified particles no P700 activity is left. Upon illumination at room temperature an ESR difference spectrum (light *minus* dark) was observed that resembled S I with regard to line shape and line width. A similar spectrum was obtained by oxidation with 1 mM ferricyanide. The decay kinetics of the light-induced ESR signal were identical to those of Q^-as monitored optically at 320 nm. The ESR signal was attributed to $P680^+$ and it was concluded that P680 is a dimeric chlorophyll complex. The possibility, however, that these ESR signals (and the above mentioned transients) originate from photostimulated oxidation of bulk chlorophyll is not rigorously excluded [261]. Clearly, there is a need to quantitate the so-called $P680^+$ ESR signals. ENDOR experiment, which potentially could discriminate between an ensemble of oxidized bulk chlorophyll aggregates, an oxidized dimeric primary donor, and oxidized monomeric Chl *a*, have not been carried out so far, owing to the weak signal intensity and/or the transitory character of the signals.

3.3 The Intermediary Acceptors of Photosystems 1 and 2

3.3.1 Photosystem 1

In this section we will discuss the new evidence concerning the presence of an *intermediary* acceptor between P700 and X, the earliest ferredoxin type acceptor of PS 1.

The first evidence for an intermediate came from the work by Sauer et al. [251], who studied the decay rate of P700 and P430, the optical ΔA band associated with one of the primary ferredoxin type acceptors of PS 1 [252], following a light flash on TSF-1 PS 1 particles under reducing conditions. They found two intermediaries between P700 and P430, one of which was assigned to X, and the other to a more primary acceptor A_1. From studies of spin polarization effects of the light-induced ESR signal of $P700^+$ as a function of orientation of the chloroplasts, it was concluded that there is a chlorophyllous species acting as electron acceptor situated between P700 and X [239, 253]. Because in the ΔA spectrum in the region where $P430^-$accumulates (Fig. 3.1a) the difference peak of pheophytin a^-(Pheo a^-) is lacking since the absorption spectra of many PS 1 particles lack the Pheo *a* band at 540 nm [145], and because the E_m of the couple Pheo *a*/Pheo a^-(−0.64 V) is not negative enough [254], one would surmise that the postulated intermediate is Chl *a* (the E_m of Chl *a*/Chl $a^-$is −0.9 V [254]).

Recently, Shuvalov et al. [266, 267] have published a ΔA spectrum of a rapidly (1.3 ms) decaying absorption change after flash excitation at 5 K of PS 1 particles under strongly reducing conditions. The spectrum was attributed to the reduction of a Chl-*a* dimer functioning as an intermediate. The spectrum, however, looks rather similar to the difference spectrum expected for P700 → $P700^T$ formation, so that it seems not excluded that it evolves from a back reaction between a reduced intermediate and $P700^+$. The more so, because an earlier spectral change with a decay time of 10 ns was found after 60 ps flash excitation under similar conditions.*

Baltimore and Malkin [318] obtained a flash-induced ΔA spectrum of A_1 of LDAO-PS 1 particles in which the ferredoxin type acceptors were inactivated by heat treatment, while retaining a functional PS 1 reaction center with respect to P700 oxidation. The absorbance differences decayed with a half time at 25 °C of about 7 μs, similar to the half time of the back reaction of $P700^+$ and A_1^- observed in PS 1 particles from which the ferrodoxin type acceptors were removed by SDS treatment [319]. Subtraction of the ΔA spectrum of P700 from the measured ΔA spectrum gave the ΔA spectrum of A_1/A_1^-. It is characterized by bleachings at 375 and 410 nm, and an increase at 460 nm. These features are also found in the ΔA spectra of both Chl *a*/Chl a^- and Pheo *a*/Pheo a^-[254]. However, no significant changes between 500 and 540 nm were observed, strongly suggesting that A_1 is Chl *a* and not Pheo *a* [254].

The above experiments were complemented by ESR measurements on samples in which A_1^- was accumulated [320, 321]. Baltimore and Malkin [320] observed a near-Gaussian line at g = 2.003, with ΔH ~ 13 G, which values suggest a chlorophyllous monomeric anion [254]. Heathcote and Evans [321] observed in digitonin-PS 1 particles, in which A_1^- was accumulated by photoreduction under strongly reducing conditions, an asymmetric line with some unresolved structure at g = 2.0037 with ΔH = = 13.3 G. The addition of the detergent Triton X–100 produced a symmetric ESR line of similar width at the same g-value. Why the digitonin particles show an asymmetric line shape is not clear; it may be produced by magnetic interaction with the other paramagnetic compounds in the preparation. The g-value reported in ref. [321] is out of line with that of Chl a^-in vitro (2.0029, [254]) and that reported in ref. [320]. Possibly, this shift is also produced by magnetic (exchange) interaction.

3.3.2 Photosystem 2

The so-called "primary" acceptor of PS 2, Q, has been identified by optical spectroscopy as a plastoquinone molecule [206, 207]. To date no ESR signal of Q^-has been observed, perhaps due to excessive broadening arising from interaction with a transition metal.*

Although Q is the first relatively stable species reduced by P680, it appears not to be the first one. Recently, a ΔA spectrum of PS 2 particles at low redox potential ($E_h < -200$ mV) was obtained showing bleachings at 518 and 545 nm, and a broad absorption band at 450 nm, that were attributed to Pheo *a* reduction [208]. The presence of an intermediate analogous to the Bph-*a* monomer in the bacterial photosystem is also indicated by the observation of a magnetic field-induced increase of fluorescence in *Chlorella* under reducing conditions at 120 K [135]. This effect is analogous to the magnetic field-induced depression of the yield of the reaction center triplet in bacteria

* See *Notes Added in Proof*, p. 323

[131, 132], which effect is thought to stem from the influence of a magnetic field on the back reaction $P^{+}I^{-} \rightarrow P^{T}I$, where I is the intermediary electron acceptor [R24].

The reduction potential of Pheo *a* (~ −0.64 V in vitro [254], −0.61 in vivo [322]) is roughly midway between that of P680* [0.8 V (E_m of Chl *a*/Chl a^+) − 1.8 V ($h\nu$ at 680 nm) = ~ −1.0 V] and plastoquinone {two E_m values at pH = 7.6 of +35(25) mV and −240(−270) mV as measured by the absorption change of the indicator pigment at 518 nm [323] and (between parenthesis) by fluorescence induction [324] in chloroplasts; E_m = −130 mV in digitonin PS 2 particles at pK = 8.9, ΔE_m = −60 mV/pH unit [325]}. This has been taken to support its assignment as intermediary acceptor of PS 2 [254], but one must be cautioned in comparing the equilibrium redox potentials of isolated compounds in vitro and in vivo to those of electron carriers in vivo under conditions of actual electron transport [6, 307], (Chap. 3 of [B4].

Photoreduction at 295 K of PS 2 in TSF-IIa particles under reducing conditions (−450 mV) results in an ESR signal with characteristics analogous to those of the bacterial intermediate $I^- \equiv Bph^-$, viz. g = 2.0033 ± 0.0003, ΔH = 12.6 ± 0.3 G [326, 327]. Photoreduction at 220 K results in a mixed ESR signal. At 6 K, a singlet *and* a doublet is observed, the doublet disappearing above 15 K. When Q was extracted from lyophilized TSF-IIa particles with organic solvents, and the sample photoreduced at 220 K, at 6 K only the singlet was visible. These observations are very similar to those made in the bacterial photosystem under conditions where I^- is trapped (Sect. 2.2.3). The doublet ESR signal is analogously thought to result from exchange interaction between Pheo a^- and Q^-.

3.3.3 Triplet States

If intermediary acceptors analogous to Bph in the bacterial photosystem are located in PS 1 and PS 2, one might ask why there is no easily observable triplet state in these photosystems. The triplet states that have been observed by ESR in high magnetic field and in zero field [128, 262–265] are weak and, with one possible exception [128], do not show unusual polarization. Quite possibly, they are located in antenna pigments Recently, however, a triplet showing similar polarization as the bacterial triplet has been observed in chloroplasts and PS 1 particles in which the acceptors A, B, and X had been reduced by strong illumination during freezing in the prescence of dithionite [298]. Its zero field parameters, however, are indistinguishable from those of the triplet state of monomeric Chl *a* in solution. Accepting that P700 is a dimer and that the Q_y and Q_x transition moments of the monomer components are not parallel (Sect. 3.3.1), this observation is not easily reconciled with the notion of strong exciton coupling between the two Chl-*a* molecules.*

4 Structure of the Bacterial Primary Donor-Acceptor Complex

A review of the photooxidation of chlorophyll in photosynthesis is not complete without a discussion of the charge-separation process. We need a more or less accurate description of the physical basis of this process before we can attempt to assemble the

---

* See *Notes Added in Proof,* p. 323, 324

accumulated structural and kinetic data into a blueprint of primary photochemistry. Only recently, such a description has been presented by Hopfield [268, 269], Jortner [270], and Kuznetsov et al. [300]. The merits of these electron transfer (ET) theories and their relevance to charge separation and subsequent dark reactions in photosynthesis are extensively discussed by Blankenship and Parson in Chap. 3 of ref. [B4]. Although it may as yet be premature to apply the ET theories quantitatively, they provide a framework of considerable heuristic value. Therefore, in this section we will give a resume of Hopfield's version of ET, which embodies in a transparent way the principal features of current thinking about ET. The more complex treatment by Jortner gives at low temperatures a better quantitative description, but as its main results are functionally similar to those of Hopfield, the reader is referred to the original paper, the discussion in Chap. 3 of [B8] and refs. [R19, R20] for more details.

Application of simple ET theory gives approximate figures for the distances between reactants in the electron transport chain. Taking these results and using the geometrical data as reviewed in the preceding sections, a tentative sketch will be drawn of the arrangement of the donor-acceptor complex in the bacterial photosynthetic membrane. For the plant photosystem, the data are not yet sufficient to warrant a similar enterprise.

4.1 Electron Transfer Rates

In the classical description an electron transfer reaction takes place via an activated complex of the reaction partners. When during the lifetime of the activated state electron transfer has a high probability, the reaction is said to be *adiabatic* [B6, B7]. When it has a low probability, the reaction is *nonadiabatic* (diabatic). The latter case presumably prevails in photosynthetic electron transport: the interaction energy between the reactants is much smaller than the activation energy.

The activation of the electron donor to a state from which ET can proceed requires the expenditure of energy. This energy is in general not matched by the energy recovered when in the electron acceptor the nuclear configuration relaxes from its excited state just after ET to its new equilibrium state, corresponding to a minimum in the total energy including the effects of the extra electron. The energy difference between the "excited" and the relaxed nuclear configuration needs to be dissipated by molecular and lattice vibrations. This will be more difficult the larger the surplus energy is. Hence, the rate of the ET reaction must depend on the "activation" energy of both the donor and the acceptor. Moreover, in the activated state the molecule will as a rule be distorted, which means that the electron potential surface in this state is shifted with respect to that of the equilibrium state. Thus, in addition to the activation energy one then has to expend the vibrational "Franck-Condon" energy, both for the donor and the acceptor molecule.

Figure 4.1 illustrates the above considerations. Electron transfer is assumed to occur between two fixed sites, *a* and *b*. Each site is characterized by two potential energy curves of the form $\frac{1}{2}$ kx^2, taken along the transfer coordinate x (which might be viewed as a particular vibration) from site *a* to site *b*. One of the potential curves of site *a*, Φ_a, represents the initial equilibrium state with the electron on *a*, the other curve $\Phi^{\neq}_a$ representing the activated state of *a* and similarly for the acceptor site *b*. The equilibrium

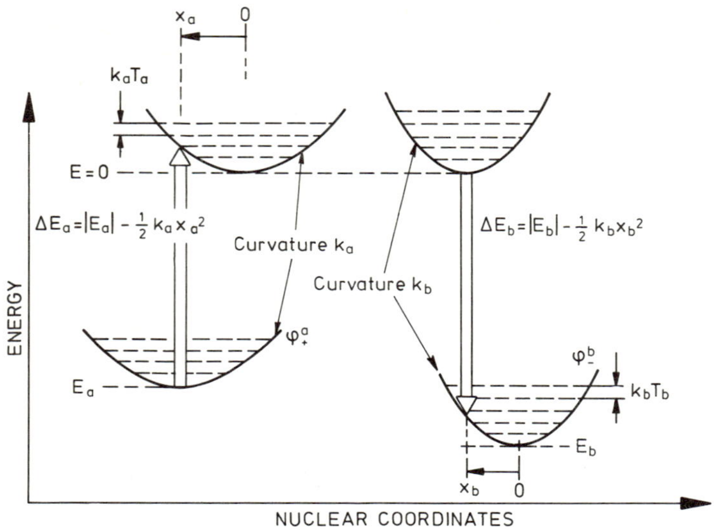

Fig. 4.1. Electron transfer by tunneling. The *left side* represents electron removal, the *right side* electron insertion. For further details see text. (Adapted from [269])

position of curve $\Phi_a^{\ddagger}$ is shifted with respect to that of Φ_a. Electron transfer is visualized as a resonant electron hopping, homologous with the Förster-type resonant transfer of excitation energy. Thus, site *a* undergoes the electron removal transition $\Phi_a \rightarrow \Phi^{\ddagger}{}_a$ and, at the same time, at site *b* an electron is added by the transition $\Phi^{\ddagger}{}_b \rightarrow \Phi_b$. Note that in this picture the activated complex $(ab)^{\ddagger}$ is represented by the two independently activated sites $a^{\ddagger}b^{\ddagger}$, i.e., there can be no coherent addition of vibrational modes which include *both* sites. The states $\Phi^{\ddagger}{}_a$ and $\Phi^{\ddagger}{}_b$ are arbitrarily set at zero energy, the states Φ_a and Φ_b lie at the redox potentials of molecules *a* and *b*, E_a and E_b, respectively. There is a weak coupling between sites *a* and *b*, characterized by an interaction matrix element T_{ab} resulting from overlap of the wave function of sites *a* and *b* at the intersection of initial state Φ_a and the final state Φ_b:

$$T_{ab} = \int \Phi_a(n)\Phi_b(m)\, V(r)\, \Phi_a(n-1)\Phi_b(m+1)\, dr \tag{1}$$

where $\Phi_a(n)$ and $\Phi_b(m)$ represent the highest occupied molecular orbital of sites *a* and *b* before ET and $\Phi_a(n-1)$, $\Phi_b(m+1)$ similarly after ET. V(r) is the interaction potential between *a* and *b*.

In the original Hopfield theory [268], the probability of findings the electron in Φ_a at the coordinate x is given by a classical Gaussian distribution, which results in an energy distribution for the electron removal transition $\Phi_a \rightarrow \Phi_a^{\ddagger}$

$$D_a(E) = \left(\frac{1}{2\pi kTk_ax_a{}^2}\right)^{\frac{1}{2}} \exp\left(\frac{-(E - E_a + \frac{1}{2}\, k_ax_a{}^2)^2}{2kTk_ax_a{}^2}\right) \tag{2}$$

and an analogous expression for the energy distribution $D_b(E)$ of the electron insertion transition $\Phi^{\ddagger}_b \rightarrow \Phi_b$ [8].

The rate of electron transfer, W_{ab}, is proportional to the square of the matrix element T_{ab}, and to the convolution integral of the functions $D_a(E)$ and $D_b(E)$. For high temperature, when the distribution over the vibrational levels approches a classical Boltzmann distribution,

$$W_{ab} = \frac{2\pi}{\hbar} T_{ab}{}^2 (4\pi\Delta kT)^{-\frac{1}{2}} \exp\left(-\frac{(E_a - E_b - \Delta)^2}{4\Delta kT}\right) \quad (3)$$

where $\Delta = \frac{1}{2} k_a x_a{}^2 + \frac{1}{2} k_b x_b{}^2$. At low temperature, a quantum mechanical correction for the distribution over the vibrational levels is introduced. For $T \rightarrow 0$, the denominator in the exponential then becomes $2\Delta kT_c$, where $T_c = \hbar\omega$ is a quantum of vibrational energy (neglecting differences between sites *a* and *b*), with corresponding changes in the normalization factor.

The essential features of Hopfield's theory are: (1) electron transfer takes place between two *bound* states, (2) the distance between the sites is not temperature-dependent, (3) transfer is strictly nonadiabatic, i.e., $T_{ab} \ll E_a, E_b$, (4) the process is thermally activated beyond $T = T_c/2$, (5) the sites are assumed to be vibrationally independent. This last assumption will not hold at very low temperature, where low-frequency phonons predominate. Then, the sites are coupled by lattice vibrations and their vibrations can be excited collectively. This effect is incorporated in Jortner's theory, it leads to a smaller estimate for the exchange matrix element T_{ab} if the low-temperature limit of the transfer rate is used for its evaluation.

T_{ab} is exponentially related to distance. An approximate expression is given by Hopfield [268], based on a calculation of p-orbital overlap for carbon atoms in π-bonding configuration of two aromatic rings in edge-to-edge contact

$$T_{ab} \sim 2.7 (N_a N_b)^{-\frac{1}{2}} \exp(-0.72\, R) \quad (4)$$

(T_{ab} in eV), where R is the distance between the edge atoms in Ångstrom. The factor $(N_a N_b)^{-\frac{1}{2}}$ takes into account electron distribution over the π-system of aromatic molecules at sites *a* and *b* consisting of N_a and N_b atoms, respectively. Jortner [270] gives arguments for a somewhat stronger dependence on R, based on an evaluation of resonance integrals for large aromatic systems. The resulting approximation is

$$T_{ab} = 5.5 \times 10^2 \exp(-1.3\, R) \quad (5)$$

where T_{ab} is in eV, and R is the distance in Å. Because of the exponential dependence of T_{ab} on R, this distance will be close to the edge-to-edge distance defined by Hopfield [268].

---

8 The Gaussian approximation is valid when $(E-\Delta)/\Delta| \ll \mathrm{cotgh}^2\, T_c/2\, kT$ [270], i.e., in general only when $kT \gg T_c$. At low temperatures it leads to an overestimation of the transition probability [270]

The value of T_{ab} can be found from the low-temperature limit of the transfer rate, $T_c/2$ is given by the temperature of the transition to the temperature-activated domain and Δ follows from the activation energy. If W_{ab} is not temperature activated at high temperatures, which exceed the characteristic vibrational temperature, then one might presume that $E_a - E_b \sim \Delta$ and T_{ab} can be directly evaluted.

The tunneling matrix element is closely linked to the magnetic exchange coupling J between two paramgnetic species. It can be shown [271, 272] that J is related to T_{ab}' (which is not necessarily equal to T_{ab} between uncharged species) by

$$\frac{|J|}{2} = \frac{|T_{ab}'|^2}{(\delta E - \Delta)} \tag{6}$$

with $\delta E = E_a - E_b$. Substituting this relation in Eq. (3) and taking into account that only antiparallel spins can recombine, one obtains

$$W_{ab} = \frac{|J|}{2\hbar}\left(\frac{\pi \delta E^{\ddagger}}{kT}\right)^{\frac{1}{2}} e^{-\frac{\delta E^{\ddagger}}{kT}} \tag{7}$$

where $\delta E^{\ddagger} = (\delta E - \Delta)^2/4\Delta$ is the activation energy. Equation (7) contains only two parameters, J and $\delta E^{\ddagger}$, that can be determined experimentally.

If $\delta E = \Delta$, the transfer rate is maximally fast. At room temperature, $kT = 2.6 \cdot 10^{-2}$ eV, and slight variations in the relative amplitude of δE and Δ (each of which will range between, say, 0.05 and 1 eV) may change W_{ab} by many orders of magnitude. This dependence of W_{ab} on $\delta E/\Delta$ presents a way of understanding the great difference in reaction rates between the primary charge separation, $P^*I \rightarrow P^+I^-$, which takes place (in the bacterial photosystem) in about 3 ps (Sect. 2.1.5) and subsequent reactions which are much slower. Let us consider in Fig. 4.2 the schematic diagram of the energy levels of the primary reactions, based on redox potentiometry. The extreme rate of charge separation, and its observed independence[9] of temperature in the range 300–4.2 K [273], suggests that $\delta E = \Delta$. The subsequent reaction, the reduction of X (the ubiquinone-iron complex), is slower by a factor of 60 to 80 and also independent of temperature in the same range. Presumably, a larger distance and/or larger Δ accounts for the slower rate (see below). The back reaction $P^+I^- \rightarrow P\,I$ must be slow, because it is wasteful for the quantum efficiency. In principle, for this reaction $T_{ba} \neq T_{ab}$ because the molecular orbitals involved are different for P and P^*. However, if the assumption is made that T_{ba} is comparable to T_{ab} and that Δ is not much different from that of the charge separation, then the much higher value of δE makes this back reaction much slower (by a factor of more than 10^6 !).

The back reaction $P^+X^- \rightarrow P\,X$ is between nonadjacent reactants, making T_{ba} small. The reaction rate is about $(30\ \text{ms})^{-1}$ between 1.5 and 100 K [98]. However, above 100 K, the reaction rate becomes *slower* with increasing temperature [R17; 328–330]. This increase cannot be explained by the $T^{-\frac{1}{2}}$ preexponential factor for the case that $\delta E = \Delta$. Apparently, thermal expansion effects are important for this reaction.*

9 Note, that if $\delta E = \Delta$, then $W_{ab} \propto T^{-\frac{1}{2}}$. This weak dependence on T may be easily obscured by temperature effects on δE, R, etc

* See *Notes Added in Proof*, p. 324

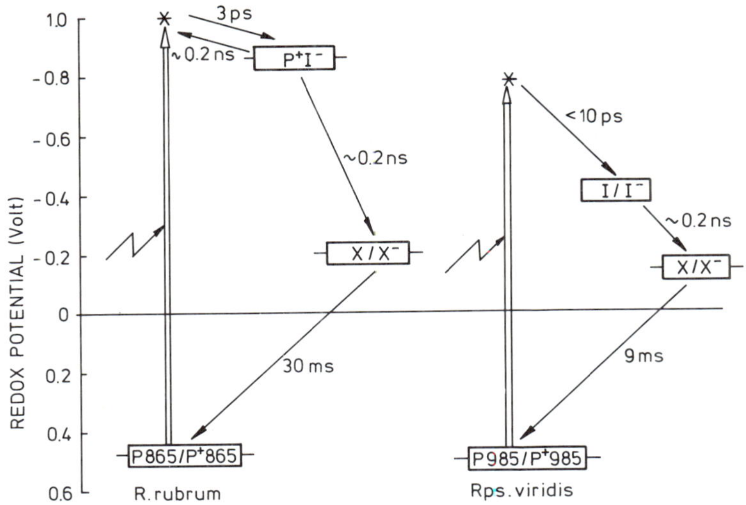

Fig. 4.2. Energy level diagram for primary electron transfer in *R. rubrum* (*left*) and *Rps. viridis* (*right*). The actual position of the P^+I^- level in the latter bacterium has not been established. It might lie above the I/I^- level which was determined by equilibrium redox potentiometry (ref. [122])

What about the back reaction to the excited state of P, $P^+I^- \rightarrow P^*\ I$? Here $T_{ab} = T_{ba}$, so that at room temperature this back reaction will be slower than the forward reaction by the Boltzmann factor $\exp - \delta E/kT \cdot \delta E$ has been estimated to be about 0.35 eV [274], which would give a factor of $4.1 \cdot 10^5$. It seems, however, that this value of δE is overestimated. From the temperature dependence of the fluoressence of chromatophores in the state $P\ I\ X^-$, it was concluded that the back reaction to the excited state has an activation energy of about 0.06 eV [307][10]. This makes this reaction slower than the forward reaction by a factor of only 0.1. However, even this relatively high rate need not be detrimental to the quantum efficiency of charge separation, because after return to the excited state, charge separation can take place anew directly, or the excitation may travel to the surrounding antenna pigments with high probability of renewed trapping. Although the reaction $P^+I^-X \rightarrow P^+I\ X^-$ has a rate comparable to the back reaction to P^*, calculations show that this back reaction lowers the quantum efficiency by at most a few percent [307].

4.2 Configuration of Primary Reactants

It is tempting to apply ET theory to observed reaction rates in photosynthesis in the hope of obtaining structural information, as the distance between reactants, etc. The

10 Previous estimates [74, 275] based on measurements of (delayed) fluorescence of $\delta E \sim 0.12$ eV must be viewed with caution, as the detailed spin statistics within the radical pair P^+I^- were not properly taken into account

key part is the evaluation of T_{ab}. Using the high temperature approximation given by Hopfield [268], or the equivalent expression (5) of Jortner [270], the activation energy can be evaluated from the temperature dependence of the reaction rate at temperatures $T \gg T_c$, and an estimate of Δ is sufficient to obtain T_{ab}. It is, however, not easy to determine whether the high-temperature limit is operative around room temperature or that one is then still in the intermediate temperature regime. Use of the general equation (3a) given by Jortner [270] circumvents this difficulty, but even when the full temperature dependence of W_{ab} is available for a fit, equivalent fits are obtained for values of T_{ab} that differ by two orders of magnitude [B8, Chap. 3]. Thus, one might despair at the practicality of applying existing ET theory, were it not that even large variations in T_{ab} translate into relatively minor variations in R_{ab} because of the exponential relations (4) and (5). In this section, we will therefore use simple arguments based on Hopfied's treatment to extract some approximate figures related to the geometry of the reaction center, with all this bearing in mind the global nature of these calculations.

Let us first focus on the charge-separation reaction. This reaction is only weakly dependent on temperature in the range of 5–300 K [273]. From the high-temperature expression (3) it follows that then $\delta E = \Delta$. This gives the desirable result of maximally fast ET. Taking $\delta E = 0.35$ eV [274, 276] T_{ab} becomes 0.003 eV for $W_{ab} = 2.3 \times 10^{-11}$ s^{-1} [64]. This gives $R(P\ I) \sim 5$ Å [Eq. (4) with $N_a = 36$ and $N_b = 18$] or 9 Å [Eq. (5)]. If $\delta E = \Delta = 0.07$ eV [307] instead of 0.35 eV, T_{ab} is smaller by a factor of 2.2, making R larger by 0.6 to 1.1 Å.

Although the above values of R appear to be reasonable, the application of Eq. (3) to the primary reaction is open to serious doubt. By the application of Eq. (7), taking the properties of the function 'x exp -x' into consideration, one may show that for a given value of $|J|$, the magnetic exchange interaction between P^+ and I^- , the maximal rate of charge separation is given by

$$(k_c)_{max} = 6.7 \cdot 10^6 \ |J| \ s^{-1} \tag{8}$$

with $|J|$ in Gauss. It follows that for $k_c \sim 2.3 \cdot 10^{11}\ s^{-1}$ the minimal value of $|J|$ is $3.4 \cdot 10^4$ G. This is in serious disagreement with estimated values of the exchange interaction between P^+ and I^-averaged over the lifetime of the radical pair. It was recently shown that in prereduced iron-depleted RC transfer of photoinduced electron spin polarization occurs between $I^-$and X^-[331]. From the extent of the polarization of X^-as a function of light intensity a value for the spin polarization of $I^-$was derived, which led to an exchange interaction of 2–5 G [332]. This value agrees well with an estimate of $J(P^*I^-)$ based on the magnetic field dependence of the yield of the triplet state arising from the back reaction $P^+I^-X^- \rightarrow\ P^T I X^-$, which gave $J \leqslant 10$ G [131–133]. The much lower average value of J compared with that expected from the fast rate of charge separation has been taken to indicate that an intermediate electron acceptor exists between P and I, which has a strong interaction with P and a weak interaction with I [332]. It has been shown that the time-averaged value of the exchange interaction between P^+ and I^-may then be much less than calculated from T_{ab} [Eq. (8)] [333]. In support of this view, fast transients on a 30 ps time scale have been observed with picosecond laser absorption spectroscopy, which have been interpreted as being due to an intermediate between P and I [71, 79, 83] (see Sect. 2.1.5). It is, however, uncertain whether the use of the high-temperature approximation [Eq. (3)] in deriving Eq. (8) is justified. The insensitivity of the primary reaction to temperature may not be caused by δE and Δ being approximately equal, but may indicate that even at room rempera-ture the reaction is still in the low-temperature limit, or even that the reaction is not strictly nonadiabatic. In view of the above, it will be clear that the calculated distance between P^+ and I^- has as yet to be taken with a grain of salt.

The dark reaction $I^-X \rightarrow I\,X^-$, with $I \equiv$ Bph and $X \equiv$ ubiquinone, proceeds with a rate of $3\text{–}4 \cdot 10^9\ s^{-1}$ (Sect. 2.1.5).* This slower rate compared with charge separation is readily explained by a somewhat larger δE, or by a larger distance than R(P I). More precise information was obtained by Okamura et al. [271], who exploited Eq. (6) by measuring the paramagnetic exchange interaction between I^- and X^- and the tempera-ture dependence of the rate of the reaction $P^+I^-X^- \rightarrow P^+I\,X^=$ in RC of *Rps. sphaero-ides* R–26, in which the primary acceptor ubiquinone had been extracted and substi-teted by the naphtaquinone vitamin K_1 (phylloquinone). They found that $\delta E^{\ddagger} =$ $= 0.67 \pm 0.03$ eV and J = 60 G. Using this value of J and the measured reaction rate, $W_{ab} = 0.25\ s^{-1}$, a value of $\delta E^{\ddagger} = 0.61$ eV resulted, in reasonable agreement with the experimental value. For ubiquinone-containing centers $\delta E^{\ddagger}$ was found to be 0.42 ± 0.03 eV. Assuming that $\delta E \sim 0.1$ eV, values of Δ result of 2.9 and 1.7 eV for the two preparations, so that T_{ab}' is 10^{-3} and $1.3 \cdot 10^{-4}$ eV, respectively. This in turn yields R = 7.5 (10) Å for vitamin K_1, and R = 10 (12) Å for ubiquinone-containing RC [values between brackets for Eq. (4)].

In *C. vinosum,* T_{ab}' was estimated from J by Dutton [277] using the J-values for the couple I^-/X^- determined in chromatophores (J = 80–100 G) and RC (J = 60 G) [75, 76, 123]. Again, the resulting distances are of the order of 10 Å.

The back reaction $P^+I\,X^- \rightarrow P\,I\,X$ is slow and shows an anomalous temperature dependence [328–330]. This makes it impossible to fit with one of the ET expressions. We may nevertheless estimate T_{ba} from the low-temperature behavior [98] by substi-tuting $\delta E = \Delta = 0.7$ eV, which yields $T_{ab} \sim 10^{-8}$ to 10^{-10} eV and R(P X) > 15–19 Å. The absence of exchange coupling between P^+ and X^- as monitored by ESR (J < 1 G) likewise indicates that R(P X) > 19 A. As also no dipolar coupling is observed (the broadening of P^+ arising from the presence of X^- is smaller than 1 G), R(P X) is prob-ably larger than 25 Å.

The calculation of R(P X) and R(I X) is of necessity approximate. Notably, Eqs. (4) and (5) do not take into account directionality in wave-function overlap which in reality may be rate determining. Nevertheless, the values of R appear to be realistic, and they do agree closely for the different methods of evaluating T_{ab}, which is already quite remarkable. Notably, they are much shorter than distances previously calculated using free-electron tunneling theory [98, 120, 273]. This is an important feature, be-cause if exchange interaction would extend over much larger distances, short circuits in the electron transport chain would become a serious danger [274, 276]

We will now attempt to construct a model of the RC, using the above data and the structural information discussed in the preceding section. In this, we will assume that all data are equally valid for all species of photosynthetic bacteria discussed so far, i.e., we are going to make a composite picture of a reaction center. From the measurement of the electrochromic shift of carotenoid absorption bands in *Rps. sphaeroides* induced by membrane potentials, it is fairly certain that P860 lies close to the center of the

* See *Notes Added in Proof,* p. 324

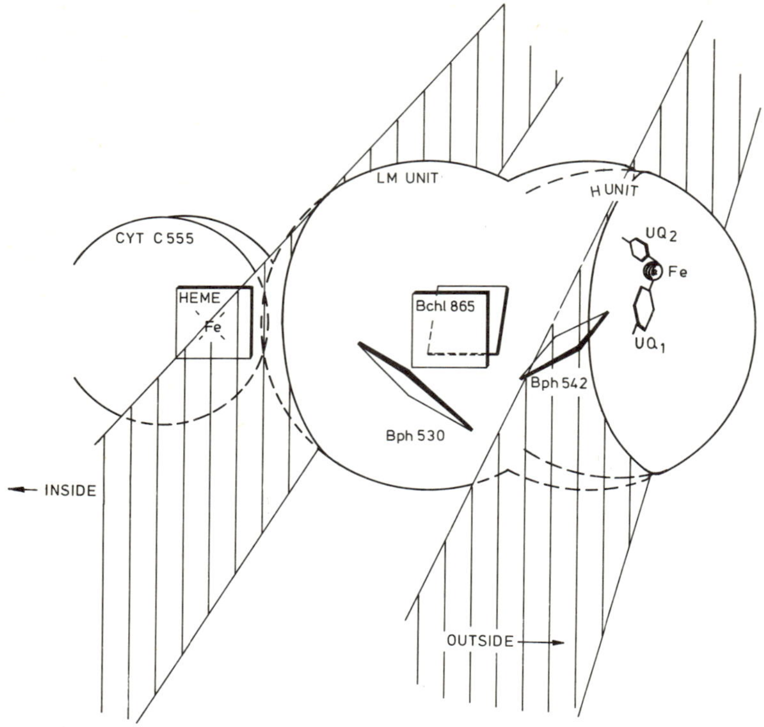

Fig. 4.3. Impression of the arrangement of the reaction center complex in the chromatophore membrane of a (generalized) photosynthetic bacterium. Linear dimensions across the membrane are drawn on scale. For further details see text

membrane [279, 280]. If we take the part of the membrane of low dielectric constant to be 40 Å thick, then P860 lies about 16 Å from the inside boundary of the chromatophore membrane.

We now have to fit the primary dimer, one Bph and one UQ in sequence, such that they span the membrane. The secondary ubiquinone must be located close to the outside of the vesicle, because at this side proton uptake takes place. In Fig. 4.3 a schematic view of a possible configuration is drawn. The plane of the dimer of P860 is roughly perpendicular, its Q_y axis parallel to the membrane [56]. The macrocycles are roughly parallel, the individual Q_y axes are inclined by 25–35° [58, 297] and predominantly colinear. Bph 542, that functions as intermediary acceptor, is drawn to conform to the analysis of Rafferty and Clayton [293], with two possible orientations with respect to Q_y of P860 given. Only one possible orientation of the accessory Bph 530 is drawn. The distances between P, I, and X are taken to be the minimal distances as evaluated from electron transfer rates. Would the distance between P and I be larger (12 Å instead of 5 Å), then it is difficult to place the I = Bph 542 to the right of the dimer. In that case, it could be in front of, or behind, P860. The position of Bph530 is undetermined, but its orientation with respect to Bph542 and Q_y of P860 must conform to Fig. 2.10. It may be relatively far from these two pigments, as it does not take part in electron transport (see also Fig. 2.11).

The primary quinone acceptor lies 7.5 to 10 Å from Bph542. Its orientation has been studied by Hales and Das Gupta [334] who compared the orientation dependence of the ESR signal of UQ, in iron-depleted RC in the state P I X^- which were oriented in a lecithin multilayer, with the orientation dependence of the bacterial triplet, P^TI X^-. They suggested that the plane of the primary quinone is perpendicular to the plane bisecting the planes of the two Bchl molecules that make up P.* With photo-affinity labeling it was shown that UQ_1 is located on or within 5 Å to the H sub-unit of the RC [335]. The Fe atom is taken to bridge the primary and the secondary quinone, as it appears to be involved in electron transfer from UQ_1 to UQ_2 [336].* UQ_2 is pointing to the inside of the membrane as it is in turn oxidized by an oxidant located in the membrane (probably a bound quinone that forms part of a quinoprotein localized between UQ_2 and a cyt *b* molecule [337, 338].

The reaction center protein is drawn after the recent evidence from X-ray work on an RC-lecithin multilayer system, that it is 60 Å long and spans the membrane [281]. The LM unit contains all the primary reactants it is highly hydrophobic [282] and not, or to a minor extent, exposed to the surfaces of the chromatophore membrane, as demonstrated by immunological techniques [85, 283, 299], treatment with proteolytic enzymes [339] and labeling with iodine (RC of *R. rubrum,* [340]). The H unit is more polar than the LM unit, it is exposed to the outer surface of the chromatophore membrane and probably located in front of or behind the LM unit [282, 283, 299]. The iron atom is lost[11] when the RC protein is split into LM and H units [284], in agreement with its location close to the outer surface.*

$P860^+$ is reduced by one, and in some bacteria by two, *c*-type cytochrome(s). The distance between the heme of the cytochromes and P860 must be fairly large (>25 Å), as there is no magnetic interaction between them [285]. In *c*-type cytochromes, one heme edge is exposed to the solvent, so that if this edge is pressed against the RC protein, the distances between the charges on cyt *c* and $P860^+$ is at least 20 Å. The rather large distance is not unreasonable in view of the comparatively slow rate of electron transfer. By chemical cross-linking experiments it was found that in *Rps. sphaeroides*, cyt c_2 is located close to the L and to the M subunit {R (surface to suface) < 9 Å, [299]}. The distance to the H subunit must be (appreciably) larger. The orientation of the heme of cyt c_{555} with respect to the membrane in *C. vinosum* has been determined by ESR spectroscopy [285] to be perpendicular to the membrane plane. Surprisingly, the heme of the low potential cyt c_{553}, which is the preferred electron donor at 4.2 K [286], was found to be parallel to the membrane, increasing the distance for electron transfer by at least 4 Å.

The orientation of the two associated B800 molecules is only partially known. One Q_y transition is parallel to the membrane, the other makes an angle of 50° with the plane of the membrane [297]. For clarity they are omitted in Fig. 4.3, but there seems to be enough room to accomodate two more tetrapyrrole rings.

It must be stressed, that the view presented in Fig. 4.3 is not unique. The relative position of P860 and the Bph's can be greatly varied within the limits imposed by the dichroism data [58, 289, 290, 292, 293, 297]. The angle between the planes of the

---

11 It has recently proven possible to isolate the LM unit while retaining the iron ion [367]

* See *Notes Added in Proof*, p. 324, 325

P860 molecules and the positions of Bph542 and the B800 pigment depend on the viewpoint which of the ΔA bands are engaged in exciton interaction [25, 45, 56, 58, 289, 290, 293, 297], and the orientation of the UQ_2 is unknown. Nevertheless, Fig. 4.3 may present a reasonable description of the gross makeup of the RC and hopefully it will provoke further thought and research.

Acknowledgments. The many discussion with members of the Leiden Biophysics Department are gratefully acknowledged. I am indebted to Drs. J. Amesz, L.N.M. Duysens, and R. van Grondelle for critically reading the manuscript and to Mrs. Tineke Veldhuyzen for her untiring help in its preparation.

Bibliography*

Recent Books

B1. Bioenergetics of Photosynthesis, Govindjee (ed.). New York: Academic Press 1975
B2. Encyclopedia of Plant Physiology, New Series, Vol. 5: Photosynthesis I, Trebst, A., Avron, M. (eds.). Berlin-Heidelber-New York: Springer 1977
B3. Primary Processes of Photosynthesis, Barber, J. (ed.). Amsterdam: Elsevier/North-Holland Biomedical Press 1977
B4. The Photosynthetic Bacteria, Clayton, R.K., Sistrom, W.R. (eds.). New York: Plenum Press 1978
B5. Tunneling in Biological Systems, Chance, B., DeVault, D., Frauenfelder, H., Marcus, R.A., Sutin, N., and Schrieffer, J.R. (eds.). New York: Academic Press 1979
B6. Reynolds, W.L., Lumry, R.W.: Mechanisms of Electron Transfer: New York: Ronald Press 1966
B7. Ulstrup, J.: Charge Transfer Processes in Condensed Media. Berlin-Heidelberg-New York: Springer 1979
B8. Photosynthesis in Relation to Model Systems, Barber, J. (ed.). Amsterdam: Elsevier/North-Holland Biomedical Press 1979

Review Articles

R1. Clayton, R.K.: Primary processes in bacterial photosynthesis. Annu. Rev. Biophys. Bioeng. **2**, 131–156 (1973)
R2. Ke, B.: The primary electron acceptor of photosystem I. Biochim. Biophys. Acta **301**, 1–33 (1973)
R3. Ke, B.: Photosynthetic reaction centers and primary photochemical reactions. Photochem. Photobiol. **20**, 542–546 (1974)
R4. Parson, W.W., Cogdell, R.J.: The primary photochemical reaction of bacterial photosynthesis. Biochim. Biophys. Acta **416**, 105–149 (1975)
R5. Loach, P.A.: Chemical properties of the phototrap in bacterial photosynthesis. In: Progress in Bioorganic Chemistry, Kaiser, E.T. (ed.), Vol. 4, pp. 89–192. New York: Wiley 1976
R6. Loach, P.A.: Primary photochemistry in photosynthesis Photochem. Photobiol. **26**, 87–94 (1977)
R7. Blankenship, R.E. Parson, W.W.: The photochemical electron transfer reactions of photosynthetic bacteria and plants. Annu. Rev. Biochem. **47**, 635–653 (1978)
R8. Campillo, A.J., Shapiro, S.L.: Picosecond relaxation measurements in biology. In: Topics in Applied Physics, Vol. 18, Ultrashort Light Pulses, Shapiro, S.L. (ed.), pp. 317–373. Berlin-Heidelber-New York: Springer 1978
R9. Holten, D., Windsor, M.W.: Picosecond flash photolysis in biology and biophysics. Annu. Rev. Biophys. Bioeng. **7**, 189–227 (1978)

---

* See also *Added Bibliography*, p. 325

R10. Katz, J.J., Norris, J.R., Shipman, L.L., Thurnauer, M.C., Wasielewski, M.R.: Chlorophyll function in the photosynthetic reaction center. Annu. Rev. Biophys. Bioeng. 7, 393–434 (1978)
R11. Ke, B.: The primary electron acceptors in green-plant photosystem I and photosynthetic bacteria. In: Current Topics in Bioenergetics, Sanadi, D.R., Vernon, L.P. (eds.), Vol. 7, pp. 75–138. New York: Academic Press 1978
R12. Knaff, D.B., Malkin, R.: The primary reaction of chloroplast photosystem II. In: Current Topics in Bioenergetics, Sanadi, D.R., Vernon, L.P. (eds.), Vol. 7, pp. 139–172. New York: Academic Press 1978
R13. Levanon, H., Norris, J.R.: The photoexcited triplet state and photosynthesis. Chem. Rev. **78**, 85–198 (1978)
R14. Malkin, R., Bearden, A.J.: Membrane-bound iron-sulfur centers in photosynthetic systems. Biochim. Biophys. Acta **505**, 147–181 (1978)
R15. Olson, J.M., Thornber, J.P.: Photosynthetic reaction centers. In: Membrane Proteins in Energy Transduction, Capaldi, R.A. (ed.), pp. 279–340. New York: Dekker 1979
R16. Seibert, M.: Picosecond events and their measurement. In: Current Topics in Bioenergetics, Sanadi, D.R., Vernon, L.P. (eds.), Vol. 7, pp. 39–73. New York: Academic Press 1978
R17. Hoff, A.J.: Applications of ESR in photosynthesis. Physics Reports **54**, 75–200 (1979)
R18. Dutton, P.L., Prince, R.C., Tiede, D.M.: The reaction center of photosynthetic bacteria. Photochem. Photobiol. **28**, 929–939 (1978)

References*

1. Duysens, L.N.M.: Transfer of Excitation Energy in Photosynthesis. Utrecht, Thesis 1952
2. Niel, C.B. van: Cold Spring Harbour Symp. Quant. Biol. **3**, 138–149 (1935)
3. Duysens, L.N.M.: Brookhaven Symp. Biol. **11**, 10–23 (1958)
4. Ross, R.T., Calvin, M.: Biophys. J. **7**, 595–614 (1967)
5. Knox, R.S.: Biophys. J. **9**, 1351–1362 (1969)
6. Parson, W.W.: Photochem. Photobiol. **28**, 389–393
7. Gromet-Elhanan, Z.: Encyclopedia of Plant Physiology, New Series, Volume 5, Photosynthesis I, Trebst, A., Avron, M. (eds.), pp. 637–662. Berlin-Heidelberg-New York: Springer 1977
8. Duysens, L.N.M.: Nature **173**, 692–693 (1954)
9. Duysens, L.N.M., Huiskamp, W.J., Vos, J.J., Van der Hart, J.M.: Biochim. Biophys. Acta **19**, 188–190 (1956)
10. Goedheer, J.C.: Brookhaven Symp. Biol. **11**, 325–331 (1958)
11. Fuhrop, J.H., Mauzerall, D.: J. Am. Chem. Soc. **91**, 4174–4181 (1969)
12. Loach, P.A., Bambara, R.A., Ryan, F.J.: Photochem. Photobiol. **13**, 247–257 (1971)
13. Loach, P.A.: Biochemistry **5**, 595–600 (1966)
14. Katz, J.J., Norris, J.R., Shipman, L.L., Thurnauer, M.C., Wasielewski, M.R.: Annu. Rev. Biophys. Bioeng. **7**, 393–434
15. Cotton, T.M., Loach, P.A., Katz, J.J., Ballschmiter, K.: Photochem. Photobiol. **27**, 735–749 (1978)
16. Reed, D.W., Clayton, R.K.: Biochem. Biophys. Res. Commun. **30**, 471–475 (1968)
17. Feher, G.: Photochem. Photobiol. **14**, 373–387 (1971)
18. Olson, J.M., Thornber, J.P.: Membrane Proteins in Energy Transduction, Capaldi, R.A. (ed.), pp. 279–340). New York: Dekker 1979
19. Loach, P.A., Walsh, K.: Biochemistry **8**, 1908–1913 (1969)
20. Straley, S.C., Parson, W.W., Mauzerall, D.C., Clayton, R.K.: Biochim. Biophys. Acta **305**, 597–609 (1973)
21. Otten, H.A.: Photochem. Photobiol. **14**, 589–596 (1971)
22. Dutton, P.L., Prince, R.C., Tiede, D.M.: Photochem. Photobiol. **28**, 939–949 (1978)
23. Jennings, J.V., Evans, M.C.W.: FEBS Lett. **75**, 33–36 (1977)
24. Prince, R.C., Olson, J.M.: Biochim. Biophys. Acta **423**, 357–362 (1976)
25. Shuvalov, V.A., Krakhmaleva, I.N., Klimov, V.V.: Biochim. Biophys. Acta **449**, 597–601 (1976)

---

* See also *Added References*, pp. 325, 326

26. Trosper, T.L., Benson, D.L., Thornber, J.P.: Biochim. Biophys. Acta **460**, 318–330 (1977)
27. Clayton, R.K., Clayton, B.J.: Biochim. Biophys. Acta **501**, 478–487 (1978)
28. Clayton, R.K.: Photochem. Photobiol. **5**, 669–677 (1966)
29. Straley, S.C., Clayton, R.K.: Biochim. Biophys. Acta **292**, 685–691 (1973)
30. Beugeling, T., Slooten, L., Barelds-van de Beek, P.G.M.M.: Biochim. Biophys. Acta **283**, 328–333 (1972)
31. Davydov, A.S.: Zhur. Eksptl. Theoret. Fiz. **18**, 210–218 (1948)
32. Kasha, M.: Rev. Mod. Phys. **31**, 162–169 (1959)
33. Kasha, M.: Rad. Res. **20**, 55–71 (1963)
34. Kasha, M., Rawls, H.R., El-Bayoumi, M.A.: Pure Appl. Chem. **11**, 371–392 (1965)
35. Tinoco, I.: J. Chem. Phys. **33**, 1332–1338 (1960); **34**, 1067 (1961)
36. Tinoco, I.: Rad. Res. **20**, 133–139 (1963)
37. Gouterman, M., and Wagniere, G.H.: J. Mol. Spectr. **11**, 118–127 (1963)
38. Weiss, C.: J. Mol. Spectr. **44**, 37–80 (1972)
39. Houssier, C., Sauer, K.: J. Am. Chem. Soc. **92**, 779–791 (1970)
40. McHugh, A., Gouterman, M., Weiss, C.: Theor. Chim. Acta **24**, 346–370 (1972)
41. Dratz, E.A., Schultz, A.J., Sauer, K.: Brookhaven Symp. Biol. **19**, 303–318 (1966)
42. Sauer, K., Dratz, E.A., Coyne, L.: Proc. Natl. Acad. Sci. USA **61**, 17–24 (1968)
43. Philipson, K.D., Sauer, K.: Biochemistry **12**, 535–539 (1973)
44. Reed, D.W., Ke, B.: J. Biol. Chem. **248**, 3041–3045 (1973)
45. Shuvalov, V.A., Asadov, A.A., Krakhmaleva, I.N.: FEBS Lett. **76**, 240–245 (1977)
46. Jablonski, A.: Z. Physik **96**, 236–246 (1935)
47. Albrecht, A.C.: J. Mol. Spectr. **6**, 84–108 (1961)
48. Geacintov, N.E., van Nostrand, F., Becker, J.F.: Proceedings of the 2nd International Congress of Photosynthesis Research, Forti, G., Avron, M., Melandri, A. (eds.), Vol. 1, pp. 283–290. The Hague: Junk 1972
49. Geacintov, N.E., van Nostrand, F., Becker, J.F.: Biochim. Biophys. Acta **267**, 65–79 (1972)
50. Gagliano, A.G., Geacintov, N.E., Breton, J.: Biochim. Biophys. Acta **461**, 460–474 (1977)
51. Rafferty, C.N., Clayton, R.K.: Biochim. Biophys. Acta **502**, 51–60 (1978)
52. Morita, S., Miyazaki, T.: Biochim. Biophys. Acta **245**, 151–159 (1971)
53. Oelze, J., Drews, G.: Biochim. Biophys. Acta **265**, 209–239 (1972)
54. Breton, J.: Biochem. Biophys. Res. Commun. **59**, 1011–1017 (1974)
55. Penna, F.J., Reed, D.W., Ke, B.: Proceedings of the 3rd International Congress of Photosynthesis, Avron, M. (ed.), Vol. 1, pp. 421–425. Amsterdam: Elsevier 1974
56. Vermeglio, A., Clayton, R.K.: Biochim. Biophys. Acta **449**, 510–515 (1976)
57. Mar, T., Gingras, G.: Biochim. Biophys. Acta **440**, 609–621 (1976)
58. Vermeglio, A., Breton, J., Paillotin, G., Cogdell, R.: Biochim. Biophys. Acta **501**, 514–530 (1978)
59. Netzel, T.L., Struve, W.S., Rentzepis, P.M.: Annu. Rev. Phys. Chem. **24**, 473–491 (1973)
60. Campillo, A.J., Shapiro, S.L.: Topics in Applied Physics Vol. 18, Ultra Short Light Pulses, Shapiro, S.L. (ed.), pp. 318–373. Berlin-Heidelberg-New York: Springer 1977
61. Holten, D., Windsor, M.W.: Annu. Rev. Biophys. Bioeng. **7**, 189–227 (1978)
62. Seibert, M.: Current Topics in Bioenergetics, Sanadi, D.R., Vernon, L.P. (eds.), Vol. 7, Photosynthesis, pp. 39–73. New York: Academic Press 1978
63. Netzel, T.L., Rentzepis, P.M., Leight, J.S.: Science **182**, 238–241 (1973)
64. Shuvalov, V.A., Parson, W.W.: Proceedings 5th International Congress of Photosynthesis, Akoyunoglou, G. (ed.), Vol. III, pp. 949–957. Philadelphia: Balaban International Science Service, 1981
65. Dutton, P.L., Kaufmann, K.J., Chance, B., Rentzepis, P.M.: FEBS Lett. **60**, 275–280 (1975)
66. Parson, W.W., Clayton, R.K., Cogdell, R.J.: Biochim. Biophys. Acta **387**, 265–278 (1975)
67. Parson, W.W., Monger, T.G.: Brookhaven Symp. Biol. **28**, 195–212 (1976)
68. Slooten, L.: Biochim. Biophys. Acta **256**, 452–466 (1972)
69. Zankel, K.L., Reed, D.W., Clayton, R.K.: Proc. Natl. Acad. Sci. USA **61**, 1243–1249 (1968)
70. Kaufmann, K.J., Dutton, P.L., Netzel, T.L., Leigh, J.S., Rentzepis, P.M.: Science **188**, 1301–1304 (1975)

71. Rockley, M.G., Windsor, M.W., Cogdell, R.J., Parson, W.W.: Proc. Natl. Acad. Sci. USA **72**, 2251–2255 (1975)
72. Fajer, J., Brune, D.C., Davis, M.S., Forman, A., Spaulding, L.D.: Proc. Natl. Acad. Sci. USA **72**, 4956–4960
73. Clayton, R.K., Yamamoto, T.: Biophys. J. **16**, 222a, abstr. F-PM-D6 (1976)
74. Shuvalov, V.A., Klimov, V.V.: Biochim. Biophys. Acta **440**, 587–599 (1976)
75. Tiede, D.M., Prince, R.C., Reed, G.H., Dutton, P.L.: FEBS Lett. **65**, 301–304 (1976)
76. Tiede, D.M., Prince, R.C., Dutton, P.L.: Biochim. Biophys. Acta **449**, 447–467 (1976)
77. Van Grondelle, R., Romijn, J.C., Holmes, N.G.: FEBS Lett. **72**, 187–192 (1976)
78. Netzel, T.L., Rentzepis, P.M., Tiede, D.M., Prince, R.C., Dutton, P.L.: Biochim. Biophy. Acta **460**, 467–479 (1977)
79. Holten, D., Windsor, M.W., Parson, W.W., Thornber, J.P.: Biochim. Biophys. Acta **501**, 112–126 (1978)
80. Fajer, J., Davis, M.S., Brune, D.C., Spaulding, L.D., Borg, D.C., Forman, A.: Brookhaven Symp. Biol. **28**, 74–103 (1977)
81. Fajer, J., Davis, M.S., Brune, D.C., Forman, A., Thornber, J.P.: J. Am. Chem. Soc. **100**, 1918–1920 (1978)
82. Davis, M.S., Forman, A., Hanson, L.K., Thornber, J.P., Fajer, J.: J. Phys. Chem. **83**, 3325–3332 (1979)
83. Shuvalov, V.A., Klevanik, A.V., Sharkov, A.V., Matreetz, J.A., Kryukov, P.G.: FEBS Lett. **91**, 135–139 (1978)
84. Ke, B.: Current Topics in Bioenergetics, Sanadi, D.R., Vernon, L.P. (eds.), Vol. 7, pp. 75–138. New York: Academic Press 1978
85. Feher, G., Okamura, M.Y., in: The Photosynthetic Bacteria, Clayton, R.K., Sistrom, W.R. (eds.), pp. 349–386. New York: Plenum Press 1978
86. Blankenship, R.E., Parson, W.W.: Annu. Rev. Biochem. **47**, 635–653 (1978)
87. Hoff, A.J.: Physics Reports **54**, 75–200 (1979)
88. Commoner, B., Heise, J.J., Townsend, J.: Proc. Natl. Acad. Sci. USA **42**, 710–718 (1956)
89. Commoner, B., Kohl, D., Townsend, J.: Proc. Natl. Acad. Sci. USA **50**, 638–641 (1963)
90. McElroy, J.D., Feher, G., Mauzerall, D.C.: Biochim. Biophys. Acta **172**, 180–183 (1969)
91. Bolton, J.R., Clayton, R.K., Reed, D.W.: Photochem. Photobiol. **9**, 209–218 (1969)
92. Kohl, D.M., Townsend, J., Commoner, B., Crespi, H.L., Dougherty, R.C., Katz, J.J.: Nature **206**, 1105–1110 (1965)
93. McElroy, J.D., Feher, G., Mauzerall, D.C.: Biochim. Biophys. Acta **267**, 363–374 (1972)
94. Druyan, M.E., Norris, J.R., Katz, J.J.: J. Am. Chem. Soc. **95**, 1682–1683 (1973)
95. Thurnauer, M.C., Bowman, M.K., Cope, B.T. and Norris, J.R.: J. Am. Chem. Soc. **100**, 1965–1966 (1978)
96. Schleyer, H.: Biochim. Biophys. Acta **153**, 427–447 (1968)
97. Corker, G.A., Henkin, B.M., Sharpe, S.A.: Photochem. Photobiol. **29**, 141–146 (1979)
98. McElroy, J.D., Mauzerall, D.C., Feher, G.: Biochim. Biophys. Acta **333**, 261–277 (1974)
99. Ruby, R.H., Kuntz, I.D., Calvin, M.: Proc. Natl. Acad. USA **51**, 515–520 (1964)
100. Clayton, R.K., Sistrom, W.R.: Photochem. Photobiol. **5**, 661–668 (1966)
101. Clayton, R.K.: Annu. Rev. Biophys. Bioeng. **2**, 131–156 (1973)
102. Parson, W.W., Cogdell, R.J.: Biochim. Biophys. Acta **416**, 105–149 (1975)
103. Loach, P.A.: Progress in Bioorganic Chemistry, Kaiser, E.T., Kezdy, F.J. (eds.), Vol. 4, pp. 89–192. New York: Wiley-Interscience 1976
104. Ke, B.: Photochem. Photobiol. **20**, 542–546 (1974)
105. Loach, P.A.: Photochem. Photobiol. **26**, 87–94
106. Sistrom, W.R., Clayton, R.K.: Biochim. Biophys. Acta **88**, 61–73 (1964)
107. Norris, J.R., Uphaus, R.A., Crespi, H.L., Katz, J.J.: Proc. Natl. Acad. Sci. USA **68**, 625–628 (1971)
108. Kip, A.P., Kittel, C., Levy, R.A., Portis, A.M.: Phys. Rev. **91**, 1066–1071 (1953)
109. Katz, J.J., Norris, J.R., Shipman, L.: Brookhaven Symp. Biol. **28**, 16–55 (1976)
110. Seely, B.R., in: Primary Processes of Photosynthesis, Barber, J. (ed.), pp. 1–53. Amsterdam: Elsevier/North-Holland Biomedical Press 1977
111. Feher, G., Hoff, A.J., McElroy, J.D., Isaacson, R.A.: Biophys. J. **13**, 61a (1973)

112. Norris, J.R., Druyan, M.E., Katz, J.J.: J. Am. Chem. Soc. **95**, 1680–1682 (1973)
113. Norris, J.R., Scheer, H., Druyan, M.E., Katz, J.J.: Proc. Natl. Acad. Sci. USA **71**, 4897–4900 (1974)
114. Feher, G., Hoff, A.J., Isaacson, R.A., Ackerson, L.C.: Ann. New York Acad. Sci. **244**, 239–259 (1975)
115. Norris, J.R., Scheer, H., Katz, J.J.: Ann. New York Acad. Sci. **244**, 260–280 (1975)
116. Feher, G.: EPR with Applications to Selected Problems in Biology. New York: Gordon and Breach 1970
117. Abragam, A., Bleaney, B.: Electron Paramagnetic Resonance of Transition Ions. Oxford: Clarendon Press 1970
118. Dorio, M.M., Freed, J.H.: Multiple Electron Resonance Spectroscopy. New York: Plenum Press 1978
119. Hoff, A.J., Möbius, K.: Proc. Natl. Acad. Sci. USA **75**, 2296–2300 (1978)
120. DeVault, D., Chance, B.: Biophys. J. **6**, 825–847 (1966)
121. Dutton, PlL., Leigh, J.S., Seibert, M.: Biochem. Biophys. Res. Commun. **46**, 406–413 (1971)
122. Prince, R.C., Leigh, J.S., Dutton, P.L.: Biochim. Biophys. Acta **440**, 622–636 (1976)
123. Prince, R.C., Tiede, D.M., Thornber, J.P., Dutton, P.L.: Biochim. Biophys. Acta **462**, 467–490 (1977)
124. Prince, R.C.: Biochim. Biophys. Acta **501**, 195–207 (1978)
125. Feher, G., Isaacson, R.A., Okamura, M.Y.: Biophys. J. **17**, 149a, abstr. TH-AM-F12 (1977)
126. Okamura, M.Y., Isaacson, R.A., Feher, G.: Biophys. J. **17**, 149a, abstr. TH-AM-F11 (1977)
127. Levanon, H., Norris, J.R.: Chem. Rev. **78**, 185–198 (1978)
128. Leigh, J.S., Dutton, P.L.: Biochim. Biophys. Acta **357**, 67–77 (1974)
129. Thurnauer, M.C., Katz, J.J., Norris, J.R.: Proc. Natl. Acad. Sci. USA **72**, 3270–3274 (1975)
130. Gast, P., Hoff, A.J.: FEBS Lett. **85**, 183–188 (1978)
131. Blankenship, R.E., Schaafsma, T.J., Parson, W.W.: Biochim. Biophys. Acta **461**, 297–305 (1977)
132. Hoff, A.J., Rademaker, H., Van Grondelle, R., Duysens, L.N.M.: Biochim. Biophys. Acta **460**, 547–554 (1977)
133. Werner, H.–J., Schulten, K., Weller, A.: Biochim. Biophys. Acta **502**, 255–268
134. Hoff, A.J., Rademaker, H., in: Chemically Induced Magnetic Polarization, Muus, L.T., Atkins, P.W., McLauchlan, K.A., Pedersen, J.B. (eds.), Chap. 23. Dordrecht: Reidel 1977
135. Rademaker, H., Hoff, A.J., Duysens, L.N.M.: Biochim. Biophys. Acta **546**, 248–255 (1979)
136. Gorter de Vries, H., Hoff, A.J.: Chem. Phys. Lett. **55**, 395–398 (1978)
137. Clarke, R.H., Connors, R.E., Norris, J.R., Thurnauer, M.C.: J. Am. Chem. Soc. **97**, 7178–7179 (1975)
138. Hoff, A.J.: Biochim. Biophys. Acta **440**, 765–771
139. Clarke, R.H., Connors, R.E., Frank, H.A., Hoch, J.C.: Chem. Phys. Lett. **45**, 523–528 (1977)
140. Hoff, A.J., Gorter de Vries, H.: Biochim. Biophys. Acta **503**, 94–106 (1978)
141. Hägele, W.U.: Thesis, University of Stuttgart (1977)
142. Hägele, W.U., Schmid, D., Wolf, H.C.: Z. Naturforsch. **33a**, 94–97 (1978)
143. Clarke, R.H., Connors, R.E.: Chem. Phys. Lett. **42**, 69–72 (1976)
144. Thornber, J.P.: Annu. Rev. Plant Physiol. **26**, 127–158 (1975)
145. Thornber, J.P., Alberte, R.S., Hunter, F.A., Shiozawa, J.A., Kan, K.–S.: Brookhaven Symp. Biol. **28**, 132–148 (1976)
146. Jacobi, G., in: Encyclopedia of Plant Physiology, New Series, Vol. 5, Photosynthesis I, Trebst, A., Avron, M. (eds.), pp. 543–562. Berlin-Heidelberg-New York: Springer 1977
147. Wessels, J.S.C.: Ibidem, pp. 563–573 (1977)
148. Thornber, J.P., Alberte, R.S.: Ibidem, pp. 574–582 (1977)
149. Boardman, N.K.: Meth. Enzymol. **23**, 268–276 (1971)
150. Ke, B., Sugahara, K., Shaw, E.R.: Biochim. Biophys. Acta **408**, 12–25 (1975)
151. Vernon, L.P., Shaw, E.R.: Meth. Enzymol. **23**, 277–289 (1971)
152. Sane, P.V., Goodchild, D.J., Parker, R.B.: Biochim. Biophys. Acta **216**, 162–178 (1970)
153. Malkin, R.: Arch. Biochem. Biophys. **169**, 77–83 (1975)
154. Shiozawa, J.A., Alberte, R.S., Thornber, J.P.: Arch. Biochem. Biophys. **165**, 388–397 (1974)

155. Malkin, R., Bearden, A.J., Hunter, F.A., Alberte, R.S., Thornber, J.P.: Biochim. Biophys. Acta **430**, 389–394 (1976)
156. Bengis, C., Nelson, N.: J. Biol. Chem **250**, 2783–2788 (1975)
157. Wessels, J.S.C., Van Alphen-van Waveren, O., Voorn, G.: Biochim. Biophys. Acta **292**, 741–752 (1973)
158. Åkerlund, H.-E., Anderson, B., Albertson, P.-Å.: Biochim. Biophys. Acta **449**, 525–535 (1976)
159. Vernon, L.P., Shaw, E.R., Ogawa, T., Raveed, D.: Photochem. Photobiol. **14**, 343–357 (1971)
160. Vernon, L.P., Klein, S.M.: Ann. New York Acad. Sci. USA **244**, 281–296 (1975)
161. Alberte, R.S., Thornber, J.P.: FEBS Lett. **91**, 126–130 (1978)
162. Kok, B.: Biochim. Biophys. Acta **22**, 399–401 (1956)
163. Kok, B.: Acta Bot. Neerl. **6**, 316–336 (1957)
164. Borg, D.C., Fajer, J., Felton, R.H., Dolphin, D.: Proc. Natl. Acad. Sci. USA **67**, 813–820 (1970)
165. Lozier, R.H., Butler, W.L.: Biochim. Biophys. Acta **333**, 465–480 (1974)
166. Shuvalov, V.A., Klimov, V.V., Krasnovsky, A.A.: Mol. Biol. **10**, 326–337 (1976)
167. Ke, B., Shuvalov, V.A., Dolan, E., in: Frontiers of Biological Energetics, Dutton, P.L., Leigh, J.S., Scarpa, A. (eds.), Vol. 1, pp. 234–240. New York: Academic Press 1978
168. Hoch, G.E., in: Encyclopedia of Plant Physiology, New Series, Vol. 5, Photosynthesis I, Trebst, A., Avron, M. (eds.), pp. 136–148. Berlin-Heidelberg-New York: Springer 1977
169. Philipson, K.D., Sato, V.L., Sauer, K.: Biochemistry **11**, 4591–4595 (1972)
170. Van Gorkom, H.J., Tamminga, J.J., Haveman, J., Van der Linden, I.K.: Biochim. Biophys. Acta **347**, 417–438 (1974)
171. Visser, J.W.M.: Thesis, University of Leiden (1975)
172. Hiyama, T., Ke, B.: Arch. Biochem. Biophys. **147**, 99–108 (1971)
173. Demeter, S., Ke, B.: Biochim. Biophys. Acta **462**, 770–774 (1978)
174. Witt, K., Wolff, C.: Z. Naturforsch. **25b**, 387–388 (1970)
175. Haehnel, W.: Biochim. Biophys. Acta **305**, 618–631 (1973)
176. Kok, B.: Biochim. Biophys. Acta **48**, 527–533 (1961)
177. Calvin, M., Androes, G.M.: Science **138**, 867–873 (1962)
178. Knaff, D.B., Malkin, R.: Arch. Biochem. Biophys. **159**, 555–562 (1973)
179. Evans, M.C.W., Sihra, C.K., Slabas, A.R.: Biochem. J. **162**, 75–85 (1977)
180. Williams-Smith, D.L., Heathcote, P., Sihra, C.K., Evans, M.C.W.: Biochem. J. **170**, 365–371 (1978)
181. Heathcote, P., Williams-Smith, D.L., Sihra, C.K., Evans, M.C.W.: Biochim. Biophys. Acta **503**, 333–342 (1978)
182. Seely, G.R., Jensen, R.G.: Spectrochim. Acta **21**, 1835–1845 (1965)
183. Hochstrasser, R.M., Kasha, M.: Photochem. Photobiol. **3**, 317–331 (1964)
184. Shipman, L.L., Norris, J.R., Katz, J.J.: J. Phys. Chem. **80**, 877–882 (1976)
185. Shipman, L.L., Cotton, T.M., Norris, J.R., Katz, J.J.: J. Am. Chem. Soc. **98**, 8222–8230 (1976)
186. Shipman, L.L., Katz, J.J.: J. Phys. Chem. **81**, 577–581 (1977)
187. Shipman, L.L.: J. Phys. Chem. **81**, 2180–2184 (1977)
188. Wasielewski, M.R., Svec, W.A., Cope, B.T.: J. Am. Chem. Soc. **100**, 1961–1962 (1978)
189. Lutz, M., in: Lasers in Physical Chemistry, Joussot-dubien, J. (ed.), pp. 451–463. Amsterdam: Elsevier 1975
190. Lutz, M.: Biochim. Biophys. Acta **460**, 408–430 (1977)
191. Lutz, M., Brown, J.S., Remy, R., in: Proceedings of the CIBA Foundation Symposium, Vol. 61, Porter, G. (ed.), pp. 105–125. Amsterdam: Elsevier 1979
192. Fenna, R.E., Matthews, B.W., Olson, J.M., Shaw, E.K.: J. Mol. Biol. **84**, 231–240 (1974)
193. Fenna, R.E., Matthews, B.W.: Nature **258**, 573–577 (1975)
194. Döring, G., Renger, G., Vater, J., Witt, H.T.: Z. Naturforsch. **24b**, 1139–1143 (1969)
195. Döring, G., Witt, H.T., in: Proceedings of the 2nd International Congress of Photosynthesis, Forti, G., Avron, M., Melandri, A. (eds.), pp. 39–45. The Hague: Junk 1972
196. Gläser, M., Wolff, C., Buchwald, H.–E., Witt, H.T.: FEBS Lett. **42**, 81–85 (1974)

197. Van Best, J.A., Mathis, P.: Biochim. Biophys. Acta **503**, 178–188 (1978)
198. Floyd, R.A., Chance, B., DeVault, D.: Biochim. Biophys. Acta **226**, 103–112 (1971)
199. Mathis, P., Vermeglio, A.: Biochim. Biophys. Acta **396**, 371–381 (1975)
200. Van Gorkom, H.J., Pulles, M.P.J., Wessels, J.S.C.: Biochim. Biophys. Acta **408**, 311–339 (1975)
201. Amesz, J., Van Gorkom, H.J.: Annu. Rev. Plant Physiol. **29**, 47–66 (1978)
202. Sauer, K., Lindsay Smith, J.R., Schultz, A.J.: J. Am. Chem. Soc. **88**, 2681–2688 (1966)
203. Fong, F.K., Koester, V.J.: Biochim. Biophys. Acta **423**, 52–64 (1976)
204. Boxer, S.C., Closs, G.L.: J. Am. Chem. Soc. **98**, 5406–5408 (1976)
205. Wasielewski, M.R., Studier, M.H., Katz, J.J.: Proc. Natl. Acad. Sci. USA **73**, 4282–4286 (1976)
206. Stiehl, H.H., Witt, H.T.: Z. Naturforsch. **24b**, 1588–1598 (1969)
207. Van Gorkom, H.J.: Biochim. Biophys. Acta **347**, 439–442 (1974)
208. Klimov, V.V., Klevanik, A.V., Shuvalov, V.A., Krasnovsky, A.A.: FEBS Lett. **82**, 183–186 (1977)
209. Fong, F.K., Hoff, A.J., Brinkman, F.A.: J. Am. Chem. Soc. **100**, 619–621 (1978)
210a. Fetterman, L.M., Galloway, L., Winograd, N., Fong, F.K.: J. Am. Chem. Soc. **99**, 653–655 (1977)
210b. Fong, F.K., Koester, V.J., Galloway, L.: J. Am. Chem. Soc. **99**, 2372–2375 (1977)
210c. Fong, F.K., Wassam, W.A.: J. Am. Chem. Soc. **99**, 2375–2376 (1977)
211a. Fong, F.K., Polles, J.S., Galloway, L., Fruge, D.R.: J. Am. Chem. Soc. **99**, 5802–5804 (1977)
211b. Fong, F.K., Galloway, L.: J. Am. Chem. Soc. **100**, 3594–3596 (1978)
211c. Galloway, L., Fruge, D.R., Fong, F.K.: Adv. Chem. Ser. **173**, 210–224 (1978)
212. Gregory, R.P.F., in: Primary Processes of Photosynthesis, Barber, J. (ed.), pp. 465–492. Amsterdam: Elsevier/North-Holland Biomedical Press 1977
213. Breton, J., Roux, E.: Biochem. Biophys. Res. Commun. **45**, 557–563 (1971)
214. Breton, J., Michel-Villaz, M., Paillotin, G.: Biochim. Biophys. Acta **314**, 42–56 (1973)
215. Paillotin, G., Breton, J.: Biophys. J. **18**, 63–79 (1977)
216. Breton, J., Becker, J.F., Geacintov, N.E.: Biochem. Biophys. Res. Commun. **54**, 1403–1407 (1973)
217. Geacintov, N.E., Van Nostrand, F., Becker, J.F.: Biochim. Biophys. Acta **347**, 443–463 (1974)
218. Breton, J., Roux, E.R., Whitmarsh, J.: Biochem. Biophys. Res. Commun. **64**, 1274–1277 (1975)
219. Vermeglio, A., Breton, J., Mathis, P.: J. Supramol. Struct. **5**, 109–117 (1976)
220. Breton, J.: Biochim. Biophys. Acta **459**, 66–75 (1977)
221. Junge, W., Eckhof, A.: FEBS Lett. **36**, 207–212 (1973)
222. Junge, W., Eckhof, A.: Biochim. Biophys. Acta **357**, 103–117 (1974)
223. Junge, W., Schaffernicht, H., Nelson, B.: Biochim. Biophys. Acta **462**, 73–85 (1977)
224. Junge, W., Schaffernicht, H., in: Proceedings of the 4th International Congress of Photosynthesis, Hall, D.O., Coombs, J., Goodwin, T.W. (eds.), pp. 21–32. London: The Biochemical Society 1977
225. Junge, W., Schaffernicht, H., in: Proc. CIBA Foundation Symposium, Vol. 61, Porter, G. (ed.), pp. 127–146. Amsterdam: Elsevier 1978
226. Fong, F.K.: Proc. Natl. Acad. Sci. USA **71**, 3692–3695 (1974)
227. Shipman, L., Cotton, T.M., Norris, J.R., Katz, J.J.: Proc. Natl. Acad. Sci. USA **73**, 1791–1794 (1976)
228. Mathis, P., Breton, J., Vermeglio, A., Yates, M.: FEBS Lett. **63**, 171–173 (1976)
229. Commoner, B., Townsend, J., Pake, G.E.: Nature **174**, 689–691 (1954)
230. Commoner, B., Heise, J.J., Lippincott, B.B., Norberg, R.E., Passonneau, J.V., Townsend, J.: Science **126**, 57–63 (1957)
231. Sogo, P.B., Pon, N.P., Calvin, M.: Proc. Natl. Acad. Sci. USA **43**, 387–393 (1957)
232. Commoner, B., in: Light and Life, McElroy, W.D., Glass, B. (eds.), pp. 356–377. Baltimore: Johns Hopkins Press 1961
233. Warden, J.T., Bolton, J.R.: J. Am. Chem. Soc. **94**, 4352–4353 (1972)
234. Warden, J.T., Bolton, J.R.: J. Am. Chem. Soc. **95**, 6435–6436 (1973)

235. Baker, R.A., Weaver, E.C.: Photochem. Photobiol. **18**, 237–244 (1973)
236. Nishi, N., Hoff, A.J., Schmidt, J., Van der Waals, J.H.: Chem. Phys. Lett. **58**, 164–170 (1978)
237. Blankenship, R.E., McGuire, A., Sauer, K.: Proc. Natl. Acad. Sci. USA **72**, 4943–4947 (1975)
238. McIntosh, A.R., Bolton, J.R.: Nature **263**, 443–445 (1976)
239. Dismukes, G.C., McGuire, A., Blankenship, R.E., Sauer, K.: Biophys. J. **21**, 239–256 (1978)
240. Warden, J.T., Bolton, J.R.: Photochem. Photobiol. **20**, 251–262 (1974)
241. Norris, J.R., Uphaus, R.A., Katz, J.J.: Biochim. Biophys. Acta **275**, 162–178 (1972)
242. Malkin, R., Bearden, A.J.: Biochim. Biophys. Acta **505**, 147–181 (1978)
243. Evans, M.C.W., in: Primary Processes of Photosynthesis, Barber, J. (ed.), pp. 433–464. Amsterdam: Elsevier 1977
244. Warden, J.T., Mohanty, P., Bolton, J.R.: Biochem. Biophys. Res. Commun. **59**, 872–878 (1974)
245. Ke, B., Dolan, E., Sugahara, K., Hawkridge, F.M., Demeter, S., Shaw, E.R., in: Photosynthetic Organelles; special issue Plant Cell Physiol. no. 3 187–199 (1977)
246. Evans, E.H., Cammack, R., Evans, M.C.W.: Biochem. Biophys. Res. Commun. **68**, 1212–1218 (1976)
247. McIntosh, A.R., Chu, M., Bolton, J.R.: Biochim. Biophys. Acta **376**, 308–314 (1975)
248. Evans, M.C.W., Cammack, R.: Biochem. Biophys. Res. Commun. **63**, 187–193 (1975)
249. Evans, M.C.W., Sihra, C.K., Bolton, J.R., Cammack, R.: Nature **256**, 668–670 (1975)
250. McIntosh, A.R., Bolton, J.R.: Biochim. Biophys. Acta **430**, 555–559 (1976)
251. Sauer, K., Mathis, P., Acker, S., Van Best, J.A.: Biochim. Biophys. Acta **503**, 120–134 (1978)
252. Ke, B.: Biochim. Biophys. Acta **301**, 1–33 (1973)
253. Friesner, R., Dismukes, G.C., Sauer, K.: Biophys. J. **25**, 277–294 (1979)
254. Fujita, I., Davis, M.S., Fajer, J.: J. Am. Chem. Soc. **100**, 6280–6282 (1978)
255. Van Best, J.A., Duysens, L.N.M.: Biochim. Biophys. Acta **459**, 187–206 (1977)
256. Knaff, D.B., Arnon, D.I.: Proc. Natl. Acad. Sci. USA **63**, 963–969 (1969)
257. Malkin, R., Bearden, A.J.: Proc. Natl. Acad. Sci. USA **70**, 294–297 (1973)
258. Bearden, A.J., Malkin, R.: Biochim. Biophys. Acta **325**, 266–274 (1973)
259. Visser, J.W.M., Rijgersberg, C.P., Gast, P.: Biochim. Biophys. Acta **460**, 36–46 (1977)
260. Malkin, R., Bearden, A.J.: Biochim. Biophys. Acta **396**, 250–259 (1975)
261. Knaff, D.B., Malkin, R., in: Current Topics in Bioenergetics, Sanadi, D.R., Vernon, L.P. (eds.), Vol. 7, pp. 139–172. New York: Academic Press 1978
262. Uphaus, R.A., Norris, J.R., Katz, J.J.: Biochem. Biophys. Res. Commun. **61**, 1057–1063 (1974)
263. Hoff, A.J., Van der Waals, J.H.: Biochim. Biophys. Acta **425**, 615–620 (1976)
264. Hoff, A.J., Govindjee, Romijn, J.C.; FEBS Lett. **73**, 191–196 (1977)
265. Van der Bent, S.J., Schaafsma, T.J., Goedheer, J.C.: Biochem. Biophys. Res. Commun. **71**, 1147–1152 (1976)
266. Shuvalov, V.A., Dolan, E., Ke, B.: Proc. Natl. Acad. Sci. USA **76**, 770–773 (1979)
267. Shuvalov, V.A., Ke, B., Dolan, E.: FEBS Lett. **100**, 5–8 (1979)
268. Hopfield, J.: Proc. Natl. Acad. Sci. USA **71**, 3640–3644 (1974)
269. Potasek, M.J., Hopfield, J.J.: Proc. Natl. Acad. Sci. USA **74**, 229–233 (1977)
270. Jortner, J.: J. Chem. Phys. **64**, 4860–4867 (1976)
271. Okamura, M.Y., Isaacson, R.A., Feher, G.: Biochim. Biophys. Acta **547**, 394–417 (1979)
272. Okamura, M.Y., Fredkin, D.R., Isaacson, R.A., Feher, G., in: Tunneling in Biological Systems, Chance, B., DeVault, D., Frauenfelder, H., Marcus, R.A., Sutin, N., Schrieffer, J.R. (eds.), pp. 729–743. New York: Academic Press 1979
273. Peters, K., Avouris, Ph., Rentzepis, P.M.: Biophys. J. **23**, 207–217 (1978)
274. Hopfield, J., in: Tunneling in Biological Systems, Chance, B., DeVault, D., Frauenfelder, H., Marcus, R.A., Sutin, N., Schrieffer, J.R. (eds.), pp. 417–432. New York: Academic Press 1979
275. Van Grondelle, R., Holmes, N.G., Rademaker, H., Duysens, L.N.M.: Biochim. Biophys. Acta **503**, 10–25 (1978)
276. Hopfield, J.J., in: Electrical Phenomena at the Biological Level, Roux, E. (ed.), pp. 471–490. Amsterdam: Elsevier 1977

277. Dutton, P.L., Leigh, J.S., Prince, R.C., Tiede, D.M., in: Tunneling in Biological Systems, Chance, B., DeVault, D., Frauenfelder, H., Marcus, R.A., Sutin, N., Schrieffer, J.R. (eds.), pp. 319–354. New York: Academic Press 1979
278. DeVault, D., Parker, J.H., Chance, B.: Nature **215**, 642–644 (1967)
279. Jackson, J.B., Dutton, P.L.: Biochim. Biophys. Acta **325**, 102–113 (1973)
280. Takamiya, K., Dutton, P.L.: FEBS Lett. **80**, 279–284 (1977)
281. Pachence, J.M., Dutton, P.L., Torriani, I., Blasie, J.K.: Biochim. Biophys. Acta **548**, 348–373 (1979)
282. Steiner, L.A., Okamura, M.Y., Lopes, A.D., Moskowitz, E., Feher, G.: Biochemistry **13**, 1403–1410 (1974)
283. Steiner, L.A., Lopes, A.D., Okamura, M.Y., Ackerson, L.C., Feher, G.: Fed. Proc. **33**, 1461 (1974)
284. Okamura, M.Y., Steiner, L.A., Feher, G.: Biochemistry **13**, 1394–1403 (1974)
285. Tiede, D.M., Leigh, J.S., Dutton, P.L.: Biochim. Biophys. Acta **503**, 524–544 (1978)
286. Dutton, P.L., Prince, R.C., in: The Photosynthetic Bacteria, Clayton, R.K., Sistrom, W.R. (eds.), pp. 525–610. New York: Plenum Press 1978
287. Reed, D.W.: J. Biol. Chem. **244**, 4936–4941 (1969)
288. Sauer, K., in: Bioenergetics of Photosynthesis, Govindjee (ed.), pp. 115–181. New York: Academic Press 1975
289. Shuvalov, V.A., Asadov, A.A.: Biochim. Biophys. Acta **545**, 296–308 (1979)
290. Rafferty, C.N., Clayton, R.K.: Biochim. Biophys. Acta **545**, 106–121 (1979)
291. Lutz, M., Kleo, J.: Biochim. Biophys. Acta **546**, 365–369 (1979)
292. Clayton, R.K., Rafferty, C.N., Vermeglio, A.: Biochim. Biophys. Acta **546**, 58–68 (1979)
293. Rafferty, C.N., Clayton, R.K.: Biochim. Biophys. Acta **546**, 189–206 (1979)
294. Fraser, R.D.B.: J. Chem. Phys. **21**, 1511–1515 (1953)
295. Cherry, R.J., Hsu, K., Chapman, J.D.: Biochim. Biophys. Acta **267**, 512–522 (1972)
296. Hoff, A.J.: Photochem. Photobiol. **19**, 51–57; **20**, 471 (1974)
297. Paillotin, G., Vermeglio, A., Breton, J.: Biochim. Biophys. Acta **545**, 249–264 (1979)
298. Frank, H.A., McLean, M.B., Sauer, K.: Proc. Natl. Acad. Sci. USA **76**, 5124–5128 (1979)
299. Rosen, D.: Thesis, University of California, La Jolla (1979)
300. Kuznetzov, A.M., Søndergård, N.C., Ulstrup, J: Chem. Phys. **29**, 383–390 (1978)
301. Vadeboncoeur, C., Noël, H., Poirier, L., Cloutier, Y., Gingras, G.: Biochemistry **18**, 4301–4308 (1979)
302. Vadeboncoeur, G., Mamet-Bratley, M., Gingras, G.: Biochemistry **18**, 4308–4314 (1979)
303. Clayton, B.J., Clayton, R.K.: Biochim. Biophys. Acta **501**, 470–477 (1978)
304. Abdourakhmanov, I.A., Ganago, A.O., Erokhin, Yu.E., Solov'ev, A.A., Chugunov, V.A.: Biochim. Biophys. Acta **546**, 183–186 (1979)
305. Ganago, A.O., Erokhin, Yu.E., Solov'ev, A.A.: Stud. Biophys. **77**, 5–12 (1979)
306. Rademaker, H., Hoff, A.J., Van Grondelle, R., Duysens, L.N.M.: Biochim. Biophys. Acta **592**, 240–257 (1980)
307. Rademaker, H., Hoff, A.J.: Biophys. J. **34**, 325–344 (1980)
308. Voznyak, V.M., Elfimov, E.I., Proskuryakov, I.I.: Doklady Akad. Nauk. **242**, 1200–1203 (1978)
309. Voznyak, V.M., Elfimov, E.I., Suskovatitzina, V.K.: Biochim. Biophys. Acta **592**, 235–239 (1980)
310. Kooyman, R.P.H., Schaafsma, T.J.: J. Mol. Struct. **60**, 373–380 (1980)
311. Schaffernicht, H., Junge, W.: Biochim. Biophys. Acta, in press (1981)
312. Warshel, A.: J. Am. Chem. Soc. **101**, 744–746 (1979)
313. Chow, H.C., Serlin, R., Strouse, C.E.: J. Am. Chem. Soc. **97**, 7230–7237 (1975)
314. Sonneveld, A., Rademaker, H., Duysens, L.N.M.: Biochim. Biophys. Acta **548**, 536–551 (1978)
315. Ke, B., Dolan, A.: Biochim. Biophys. Acta **590**, 401–406 (1980)
316. Den Haan, G.A., Warden, J.T., Duysens, L.N.M.: Biochim. Biophys. Acta **325**, 120–125 (1973)
317. Davis, M.S., Forman, A., Fajer, J.: Proc. Natl. Acad. Sci. USA **76**, 4170–4174 (1979)

318. Baltimore, B.G., Malkin, R.: FEBS Lett. **110**, 50–52 (1980)
319. Mathis, P., Sauer, K., Remy, R.: FEBS Lett. **88**, 275–278 (1978)
320. Baltimore, B.G., Malkin, R.: FEBS Lett. **110**, 50–52 (1980)
321. Heathcote, P., Evans, M.C.W.: FEBS Lett. **111**, 381–385 (1980)
322. Klimov, V.V., Allakhverdiev, S.I., Demeter, S., Krasnovsky, A.A.: Dokl. Akad. Nauk. **249**, 227–230 (1980)
323. Malkin, R.: FEBS Lett. **87**, 329–333 (1978)
324. Malkin, R., Barber, J.: Arch. Biochem. Biophys. **193**, 169–178 (1979)
325. Knaff, D.B.: FEBS Lett. **60**, 331–335 (1975)
326. Klimov, V.V., Allakhverdiev, S.I., Krasnovsky, A.A.: Dokl. Akad. Nauk. **249**, 485–488 (1980)
327. Klimov, V.V., Dolan, E., Ke, B.: FEBS Lett. **112**, 97–100 (1980)
328. Hsi, E.S.P., Bolton, J.R.: Biochim. Biophys. Acta **317**, 126–133 (1974)
329. Loach, P.A., Kung, M., Hales, B.J.: Ann. New York Acad. Sci. **244**, 297–319 (1975)
330. Romijn, J.C., Amesz, J.: Biochim. Biophys. Acta **423**, 164–173 (1976)
331. Gast, P., Hoff, A.J.: Biochim. Biophys Acta **548**, 520–535 (1979)
332. Hoff, A.J., Gast, P.: J. Phys. Chem. **83**, 3355–3358 (1979)
333. Haberkorn, R., Michel-Beyerle, M.E., Marcus, R.A.: Proc. Natl. Acad. Sci. USA **76**, 4185–4188 (1979)
334. Hales, B.J., Das Gupta, A.: Biochim. Biophys. Acta **548**, 276–286 (1979)
335. Marinetti, T.D., Okamura, M.Y., Feher, G.: Biochemistry **18**, 3126–3133 (1979)
336. Blankenship, R.E., Parson, W.W.: Biochim. Biophys. Acta **545**, 429–444 (1979)
337. Nishi, N., Kataoka, M., Soe, G., Kakuno, T., Uki, T., Yamasita, J., Horio, T.: J. Biochem. **86**, 1211–1224 (1979)
338. Van den Berg, W.H., Prince, R.C., Bashford, C.L., Takamiya, K., Bonner, W.D., Dutton, P.L.: J. Biol. Chem. **254**, 8594–8604 (1979)
339. Hall, R.L., Doorley, P.F., Niederman, R.A.: Photochem. Photobiol. **28**, 273–276 (1978)
340. Zürrer, H., Snozzi, M., Hanselmann, K., Bachofen, R.: Biochim. Biophys. Acta **460**, 273–279 (1977)
341. Rafferty, C.N., Bolt, J., Sauer, K., Clayton, R.K.: Proc. Natl. Acad. Sci. USA **76**, 4429–4432 (1979)

Chapter 5
Triplet State and Chlorophylls

Haim Levanon [1] and J. R. Norris [2]

Abbreviations

$\|0\rangle$, $\|\pm 1\rangle$	Triplet spin wave function at high magnetic fields
AP	Antenna pigments
A_i (i = x, y, z)	Population rate constant intersystem crossing rate) to a triplet state at zero external magnetic field
Bchl	Bacteriochlorophyll
Bph	Bacteriopheophytin
CIDEP	Chemically induced dynamic electron polarization
CIDNP	Chemically induced dynanic nuclear polarization
Chl	Chlorophyll
ENDOR	Electron nuclear double resonance
EPR	Electron paramagnetic resonance
H_2TPP	Tetraphenylporphyrin
ISC	Intersystem crossing
k_i (i=x, y, z)	Depopulation rate constant from a triplet spin substate to the ground singlet at zero magnetic field
MgTBP	Magnesium tetrabenzoporphine
MgTPP	Magnesium tetraphenylporphyrin
MPS	Magnetophotoselection
ODMR	Optical detection of magnetic resonance
OEPR	Optical electron paramagnetic resonance
OMR	Optical magnetic resonance
ONMR	Optical nuclear magnetic resonance
$P\gamma$	A term indicating the wavelength of the absorption spectrum of a pigment, employed most frequently in photosynthesis
P^F	A term indicating the fast transient which is involved with the charge separation of the special pair
P^R	A term indicating a transient that originates from P^F, this intermediate has a longer lifetime than P^F
PS	Photosynthesis
RC	Reaction Center
RP	Radical Pair
S	Spin operator
SO	Spin orbit
SLR	Spin-lattice relaxation; the rate for this process is designated as W
TM	Triplet mechanism
T_0, $T_{\pm 1}$	Spin substates of the first excited triplet state at high magnetic field ($g\beta H \gg D$)
$\|T_i\rangle$(i = x, y, z)	Pure spin wave functions at zero external magnetic field
T_i (i = x, y, z)	Spin substates of the first excited triplet state at zero external magnetic field; in some cases, when it is necessary, we express this state explicitly as T_i (i = x, y, z)
TPC	Tetraphenylchlorine
ZFS	Zero field splitting; it can be expressed either in terms of the triplet energy levels X, Y, Z or the proper combination D and E

---

1 Department of Physical Chemistry, The Hebrew University, Jerusalem, Israel, and and Radiation Laboratory, University of Notre Dame, Notre Dame, IN 46556, USA
2 Chemistry Division, Argonne National Laboratory, Argonne, IL 60439, USA

1 Introduction

The basic mechanism of green plant photosynthesis (PS) has been recognized to require two photosystems, photosystem I (PSI) and photosystem II (PSII), which are functionally connected by an intersystem electron transport system [1]. This formulation, in which water serves as an electron donor and $NADP^+$ as an essential electron acceptor for the combined photosystems, represents the most generally accepted description for photosynthesis of green plant. The formulation representing this mechanism is expressed by the so-called Z scheme mechanism of photosynthesis [1] (Fig. 1). The antenna pigments of PSI are mainly Chl *a*, carotenoids and a few accessory molecules. The antenna pigments of PSII contain both Chl *a* and Chl *b*. The RCs of both photosystems contain Chl *a*.

The overall reactions describing the splitting of water are represented by [2, 3]

$$2\,A + 2\,H_2O \xrightarrow[\text{PSII}]{h\nu_{II}} 2AH_2 + O_2$$

$$2AH_2 \xrightarrow[\text{PSI}]{h\nu_{I}} 2A + H_2 .$$

In general, bacterial PS can be described with appropriate modifications to green plant PSI [4]. This certainly appears true when regarding the primary photoact with respect to structure and function. In green plant PS, $E_3 - E_1 = h\nu_{II}$ (900 mV) and $E_4 - E_2 = h\nu_{II}$ (800 mV). In photosynthetic bacteria $E_4' - E_2' = h\nu$ (800 mV).

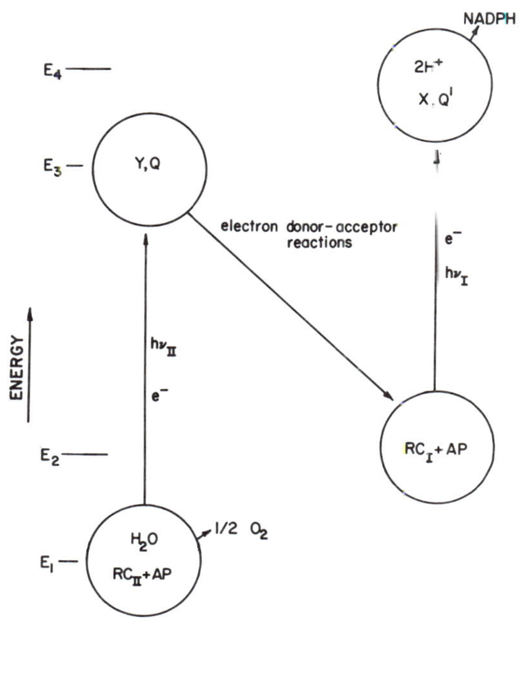

Fig. 1. Light-driven reactions in photosynthesis. Green plant photosynthesis is described by system I reactions, $h\nu_I$ (reduction chemistry) and system II reactions, $h\nu_{II}$, (oxidation chemistry). Bacterial photosynthesis is described by a single system analogous in some ways to the system II. *RC*, reaction center; *AP*, antenna pigments; *O*, quinones; *Y* and *X*, electron acceptor intermediates

The photophysics and photochemistry of PSII are apparently more complex and less understood than either PSI or bacterial PS. The availability of isolated bacterial photoreaction centers resulted in extensive work on the early light-induced primary chemical events occurring in bacteria. Moreover, antenna free RC preparations of green plants are not available and consequently much more is known about bacterial PS. Thus, in many cases, mechanisms established only for bacteria are applied without direct evidence to green plants PS. We, therefore, order our knowledge on photosynthesis as follows:

$$\text{PS–Bacteria} > \text{PSI} > \text{PSII}.$$

Thus, in the survey we will primarily emphasize bacterial and PSI photosynthesis. It is generally accepted that bacterial PS in its natural "unblocked"[3] mode does not involve the photoexcited tripled state as a precursor, whereas in green plant PS this question has not yet been fully settled. Nevertheless, this spectroscopic state undoubtedly serves as a probe for structure and mechanism which are related to the primary act.

The fundamental physical process is the absorption of light quanta over a span of wavelengths including the entire visible spectrum followed by an efficient, fast light-induced chemical reaction [5–9]. Magnetic resonance and optical spectroscopy has established that the primary process results in an electron-hole separation within the constituents of the RC [10–36]. Picosecond optical spectroscopy studies, in photosynthetic bacteria, proved that indeed an electron transfer occurs in less than 6 ps [37–40]. Very recent sub-picosecond measurements suggest an initial electron transfer reaction occurring in less than 3 ps [41], and in green plants less then 10 ps [42]. This charge separation, which is formed via an electronically photoexcited state, provides the potential energy for the "dark" chemistry which follows the primary act. A full understanding of the early stages should be considered as the key for usefully applying the concept, most frequently used, known as solar energy conversion.

In bacterial PS, the light induced photophysical and photochemical changes are schematically described in Fig. 2. Step 1 in Fig. 2 represents the photoexcitation within the singlet manifold; included are the energy transfer via the light harvesting antenna pigments. The process described by step 2 is the charge separation between two Bchl molecules and one Bph molecule (time scale within 6 ps). A secondary electron transfer, designated as step 3 , occurring to a quinone-iron complex in a time scale of ~ 200 ps. Recent experimental evidence suggests the existence of two types of acceptors, a quinone and quinone-iron complex [43–53]. Since the iron interaction may oscillate between the two quinone molecules, we write the secondary acceptor as [Q–Fe–Q].

Much evidence exist showing that the charge separation gives rise to a radical pair (RP) as indicated by step 2 in Fig. 2 [47–58]. This evidence includes extensive EPR studies of the photoexcited triplet state of in-vitro and in-vivo preparations [54]. More recently, two relatively new experimental approaches have been applied to bacterial and green plants PS, namely chemically induced dynamic electron polarization CIDEP

3 The term "unblocked" will be explained in subsequent sections

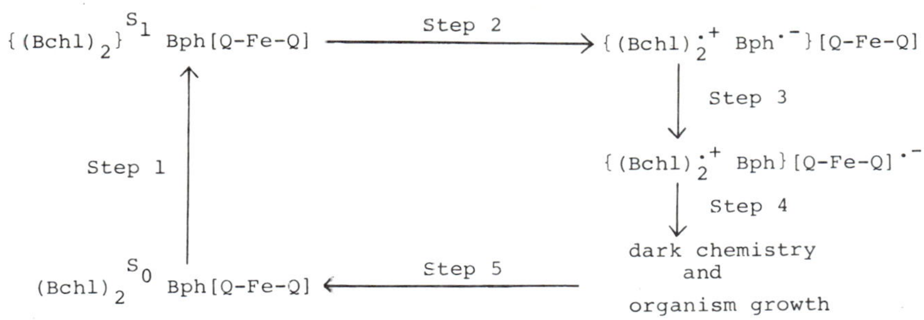

Fig. 2. Reaction scheme for bacterial photosynthesis

[59–60] and external magnetic field dependence studies of the triplet products [56–57]. Both techniques are related to the general theory of chemically induced magnetic polarization [61]. It is obvious therefore that even if the photoexcited triplet state does not participate in the main part of PS, an understanding of the triplet state is important for investigating such things as the RP mechanism and the structure of the constituents participating in the primary PS.

In this review the discussion of the triplet state revolves around the primary events of PS. Several other excellent reviews have not emphasized the triplet state [62–70]. Recent reviews on triplet states in photosynthesis have not discussed in much detail [71] the CIDEP effects observed in photosynthetic systems. The question of whether a triplet state can be detected in photosynthetic constituents is no longer contemporary, for now the triplet is a well-documented state in photosynthetic organisms. Rather, the possible significance of the triplet state to the energetics, mechanisms, and structures of the primary events of photosynthesis is discussed in this chapter.

2 Optical-Magnetic Resonance Spectroscopy

2.1 Triplet Detection. Zero Field Experiments

Let us consider a typical organic photoexcited triplet state which is formed via intersystem crossing (ISC) from the singlet manifold [72–73]. The removal of degeneracy in the triplet state in the absence of an external magnetic field is mainly due to the dipolar interaction between the two unparied electrons

$$\mathscr{H}_d = S \cdot D \cdot S \tag{1}$$

where S is the spin operator and D is the traceless dipolar operator which transforms like a second-rank tensor [74]. Expansion of Eq. (1) gives rise to different representations of the dipolar Hamiltonian. In terms of a coordinate system which diagonalizes the dipolar tensor the Hamiltonian is given by

$$\mathcal{H}_d = -(XS_x^2 + YS_y^2 + ZS_x^2) \qquad (2)$$

where X, Y, and Z are the expectation values of the dipolar Hamiltonian along the principal axes for the spatial wave functions. Since $X+Y+Z = O$, Eq. (2) can be replaced by another common representation

$$\mathcal{H}_d = D[S_z^2 - \tfrac{1}{3} S^2] + E[S_x^2 - S_y^2] \qquad (3)$$

where $|D|$ and $|E|$ are the zero field splitting (ZFS) parameters, which are expressed in terms of X, Y and Z.

$$|D| = -\tfrac{3}{2}|Z| \qquad (4)$$
$$|E| = \tfrac{1}{2}|(Y - X)|.$$

The first excited triplet state, T_1, in most cases is comprised of three nondegenerate spin sublevels. Thus, in the absence of an external magnetic field, the zero field wave functions are $|T_{1i}\rangle$, i = x, y, z.

The introduction of an external magnetic field will cause an admixture between the spin wave functions such that the spin energy levels are field dependent. The Hamiltonian must now include a Zeeman term such that

$$\mathcal{H} = \beta H\hat{g}S + \mathcal{H}_d \qquad (5)$$

where g behaves like a second-rank tensor, and in many cases the average diagonal value almost coincides with the free electron g_e factor, 2.0023. In high magnetic field, the triplet field wave functions are designated as $|T_0\rangle$, $|T_{+1}\rangle$ and $|T_{-1}\rangle$.

In most aromatics, the photoexcited triplet state is reached via an ISC mechanism. The singlet state carries no spin angular momentum, whereas the triplet state does. Thus, when a singlet-triplet transition occurs, spin angular momentum of the system necessarily changes. Such a change requires a special mechanism. In most situations the mechanism for ISC involves spin-orbit coupling giving rise to nonvanishing matrix elements of the type between singlets and triplets namely, $\langle S_j|\mathcal{H}_{so}|T_k\rangle$. The spin-orbit Hamiltonian is proportional to the scalar product $L \cdot S$, where L is the orbital angular momentum operator. In systems of a single symmetry, the coupling between the first excited singlet and first excited triplet state is given by

$$\langle S_1|\mathcal{H}_{so}|T^1_{sp}\, T_i\rangle \equiv \langle S_1|\ \ _{so}|T_{1i}\rangle \qquad (6)$$

for i = x, y, z, where, T^1_{sp} is the spatial orbital part of the triplet state, T_i is the triplet spin wave functions and S_1 is the first excited singlet state. From symmetry considerations the nonvanishing matrix elements must obey the general relation

$$\Gamma(\text{triplet orbital}) \times \Gamma(\text{triplet spin}) = \Gamma(\text{singlet orbital}) \qquad (7)$$

where Γ is the irreducible representation of the corresponding point group. Thus, in general, ISC is highly selective with respect to the three spin sublevels, predominantly populating only a single level or at most a pair of levels in significant excess to the remaining levels or level. The ISC rates will thus be proportional to the square of matrix elements of the type [75]

$$|\langle S_1 | \mathscr{H}_{so} | T^1_{sp} T_i \rangle|^2 \alpha \text{ populating rates,} \qquad (8)$$

and

$$|\langle S_o | \mathscr{H}_{so} | T^1_{sp} T_i \rangle|^2 \alpha \text{ depopulating rates.}$$

The photoexcited triplet state can be investigated by employing optical or optical-magnetic resonance spectroscopy. Optical absorption spectroscopy is employed because of high time resolution and high sensitivity. Time domain flash and laser photolysis are the most common techniques used to monitor photochemical reactions. Recently, time resolution of a few tenths of ps has become possible using advanced laser spectroscopy. The primary advantages of optical absorption spectroscopy are high time resolution and high sensitivity. However, optical absorption spectroscopy is incapable of providing for a detailed analysis with regard to the appropriate transitions within the triplet manifold, i.e., the spin transitions. For that reason optical-magnetic resonance (OMR) spectroscopy [76–84] is considered a complementary technique providing additional and independent data. OMR spectroscopy can be further divided into zero field experiments which are familiar as optical detection of magnetic resonance (ODMR) [76–81] and optical-electron paramagnetic or optical-nuclear magnetic resonance spectroscopy (OEPR) [82] and (ONMR) [83–84], respectively.

In ODMR, an optical parameter of the system under study, for example phosphorescence, is monitored as a function of microwave frequency and/or power. In porphyrins and chlorophylls, where phosphorescence does not exist or is very low, the optical parameter, which is normally monitored, is the fluorescence which provides indirect information about the photoexcited triplet state. Experiments are also performed on triplet absorption transitions. In all these types of experiments double resonance conditions are necessary, namely one of optical energy ($\geq\sim 10^4$ cm^{-1}) and one of microwave energy (0.1 to 0.01 cm^{-1}).

Since the technique of ODMR depends on significant deviation from Boltzman populations within the three spin sublevels, as produced by the highly selective ISC rates, it is preferable if these deviations are preserved. The communication between the triplet sublevels is controlled by spin lattice relaxation (SLR) times which are highly temperature dependent. At temperatures between 1 and 2K the SLR times of the system in general exceed the triplet life time. Thus, ODMR requires very low temperatures (typically 2K), which may be considered as the main limitation of this technique [85]. In terms of kinetic response, ODMR has the time resolution and sensitivity of optical spectroscopy, although in general the perturbing microwave pulses are in the ns region and thus limit the overall time resolution.

2.2 The Triplet State and the EPR Experiment

Electron paramagnetic resonance (EPR) spectroscopy covers transition energies of the order 0.3–1 cm^{-1} compared to $10^4 - 5 \cdot 10^7$ cm^{-1} in the optical methods described above. Variation in the EPR line widths reflects dynamic processes which may be studied over a wide range of times, i. e., $10^{-4} - 10^{-10}$ s. Those two properties make the EPR method very powerful in the unique determination of molecular structure and many inter- and intramolecular dynamic processes.

Figure 3 describes the magnetic field dependence of the spin energy level for three orientations of the magnetic field relative to the molecular axes. The directions shown in Fig. 3 are the so-called canonical orientations.

It is the magnitude of |D| and |E| which, in part, determines whether one should be able to apply EPR spectroscopy to detect the triplet state. Thus, in general, EPR triplet detection requires that the ZFS parameters |D| and |E| should be smaller than the external magnetic field. In this case, for a particular canonical orientation, three transitions are possible: two at the so-called $\Delta M_S = \pm 1$ (high field region) and one at the so-called $\Delta M_S = \pm 2$ (low field region).

The possibility to record EPR spectra (with a reasonable S/N ratio) of randomly oriented triplets [86–87] is of importance for compounds that cannot be grown into

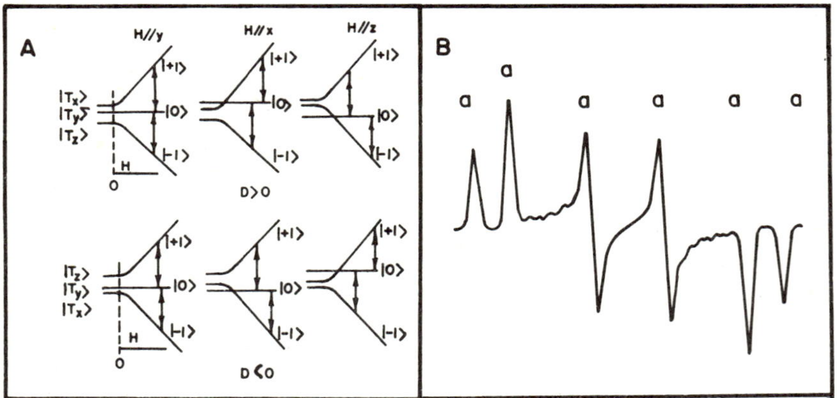

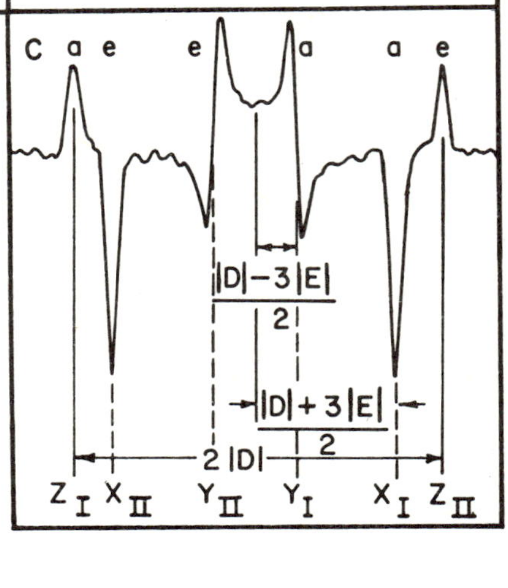

Fig. 3. **A** Magnetic field dependence of the spin energy levels for the three canonical orientations, *X, Y, Z*. The $|0\rangle \longleftrightarrow |-1\rangle$ and $|0\rangle \longleftrightarrow |+1\rangle$ transitions for each canonical orientation will be defined as Ii and IIi (i = x, y, z), respectively. **B** First derivative of a randomly oriented triplet, $|D| > 3|E|$. The line shape and peak intensities are typical of a triplet state in thermal equilibrium; i. e., all signal intensities are in the absorption mode indicated by *a*. **C** First derivative of a randomly oriented polarized triplet, $|D| > 3|E|$. The line shape and peak intensities are given for a triplet in a nonthermal equilibrium; i. e., part of the intensities are in emission, *e*, and part in enchanced absorption, *a*. *X, Y,* and *Z* are the position of the magnetic field associated with the canonical orientation

single crystals [88]. This is the case in many porphyrins and chlorophyll molecules in vitro and certainly in most in-vivo preparations.

Figure 3 B demonstrates schematically a conventional $\Delta M_s = \pm 1$ EPR spectra of randomly oriented triplets. The line shape represents the first derivative of the susceptibility X'' with respect to the external magnetic field. The line shape is typical of molecules with a rhombic contribution such that $|D| > 3|E| \neq 0$. Also, Fig. 3 C illustrates how the ZFS parameters can be evaluated from this type of spectrum. For molecules of axial symmetry (x and y principal axes are indistinguishable) $|E|$ vanishes. This results in the coalescence of S_{px} and S_{py} peak intensities in the EPR spectrum.

The absolute signs of D and E depend on the choice of a molecular axis system. In conventional EPR detection the sign determination from the experimental high temperature triplet spectrum is not possible. However, the application of the magnetophotoselection (MPS) method enables one to determine the sign of the zero field splittings if the directions of the optical transition moments in the molecular frame are known [89–90]. This method can be applied to randomly oriented triplets whose EPR spectra show distinct peaks at the canonical orientations. These unique peaks are magnetically selected. On the other hand, the populating transitions $S_1 \leftarrow S_0$ are polarized according to the type of transition (e.g., $\pi^* \leftarrow \pi$ or $\pi^* \leftarrow n$). By using plane-polarized light one can select the electric field transitions, and subsequently the peak intensities in the canonical orientation will be affected accordingly. Thus, if the polarizations of the optical transitions with respect to the molecular framework are known, one can study the magnetic transitions or vice versa.

As in ODMR, the EPR technique enables one to elucidate details concerning the dynamics associated with the triplet manifold. In the high field approximation and for $|D| \ll g\beta H$ the ISC triplet decay rate constants are given by

$$k_{\pm 1} = \tfrac{1}{2}(k_j + k_k) \qquad (9)$$
$$k_0 = k_i \text{ for } i \,||\, H,\ i = x, y, z.$$

These high field relations (9) are expressed in terms of zero field rate constants. The same type relations hold for the population rates A_0 and $A_{\pm 1}$. From Eq. (9) it follows that the three ISC rates are not necessarily the same. This is, in fact, the case in most compounds and if the SLR rates are not fast enough to cause thermalization, one should observe two types of EPR transitions between each pair of levels. One transition will be enhanced absorption (a) and the other an emission (e). This will also be reflected by the kinetic curves which will exhibit anomalous patterns in the early stage of triplet formation. This phenomenon has gained the general nomenclature of electron spin polarization (ESP) [75, 91]. The mechanism of spin-orbit-ISC (SO-ISC) of Eq. (8) for selectively populating and depopulating the triplet sublevels has been treated extensively [91–97]. The actual ESP of Fig. 3 C is exceptional and will be treated in Sect. 4.1.3.

It is for this reason that high time resolution is required for a proper analysis of the kinetic data. Recent modification of EPR detection either by employing direct detection combined with pulsed laser excitation or the application of EPR spin echo have reduced the time resolution of the EPR experiment to ~ 100 ns [98] and ~ 6 ns [99], respectively. The extreme importance of high time resolution will be demonstrated in the subsequent sections.

2.3 The Triplet Yield vs Magnetic Field

The employment of ns optical spectroscopy to hydrocarbons-amines systems clearly demonstrated the phenomenon of magnetic field dependence (modulation) on the recombination yields of radical ion pairs which were generated via photoelectron transfer reactions [100]. In general, the triplet yield at high magnetic fields is lower than triplet yield at low magnetic fields. Thus, to account for these observations it was necessary to invoke the radical-pair-ISC (RP-ISC) combined with the general theory of chemically induced magnetic polarization. At zero magnetic field all three RP spin components T_x, T_y, T_z are equally populated, whereas at high magnetic fields the $T_{\pm 1}$ components are field dependent and are apart in energy from the singlet component, T_0. If indeed the RP-ISC mechanism is dominant, T_0 will be overpopulated at any molecular orientation with respect to magnetic field. Therefore, at high magnetic fields the triplet yield should decrease [56–57].

The relative magnetic field effect R defined as

$$R = 1 - \phi_T(H)\,\phi_T(H=O) \tag{10}$$

where ϕ_T is the triplet yield depends on the particular system studied [100–107]. The important implications of these observations described above to primary PS are clear so that these types of studies should be pursued in both bacterial and green plants PS. It is expected that the triplet yield will depend on the rate parameters and also on the exchange interaction between the unpaired electron spin on the donor and primary acceptors radical pair. (See also Sect. 4.2.3.)

2.4 Triplet Photochemistry. The CIDEP Method

2.4.1 What is CIDEP?

Chemically induced dynamic electron polarization [61], CIDEP, involves the study of doublet-state free radicals which are formed in the course of certain chemical reactions, e.g., a photoreaction. These doublet-state free radicals are unusual in that their associated chemistry leads to non-Boltzmann electron spin populations, e.e., the radicals exhibit ESP. Thus, as in the polarized triplet spectrum, the EPR doublet spectrum will show anomalous line shapes and kinetics. Typical polarized EPR spectra include magnetic resonance transitions that exhibit enhanced absorption, emission or perhaps just no measurable resonance. Thus, the idea behind CIDEP is to deduce the nature of the chemical act which is responsible for the polarized EPR spectra.

In general, the two following basic CIDEP mechanisms exist: (1) the triplet mechanism (TM) [108–111], and (2) the radical pair mechanism (RPM) [112–113].

2.4.2 Triplet Precursor vs Triplet Mechanism

When radical pairs are involved in chemical reactions the possibility of CIDEP exists. One example may be two random doublet-state radicals reacting to form ultimately a diamagnetic product. Another typical system is the abstraction of the hydrogen atom of an alcohol by triplet excited state quinone to form two double-state radicals. In the process of photosynthesis the photoinduced electron transfer reaction involves several radical pairs such as

$$(\mathrm{Bchl}_2)^{\cdot +}\ \mathrm{Pheo}^{\cdot -}\mathrm{Q\ Fe\ Q}$$

$$(\mathrm{Bchl}_2)^{\cdot +}\ \mathrm{Pheo\ Q}^{\cdot -}\mathrm{FeQ}$$

and thus ESP may occur via mechanism associated with CIDEP.

If the radical pair chemistry involves a triplet precursor as in the above quinone example, then the triplet mechanism (TM) may be in operation. The TM has a characteristic ESP pattern (Table 1) and will be explained below. At this point, we emphasize that although the radical pair chemistry may involve a triplet precursor, the ESP pattern of the observed radicals need not be characteristic of the TM. Thus, the ESP pattern alone does not establish whether or not a radical pair reaction has triplet precursor or singlet precursor.

Most frequently the TM mechanism, when observed, produces systems in which both radicals are in emission. Other possibilities are also listed in Table 1. Furthermore, at times the extent of TM polarization can be relatively large. However, even large polarization can be lost via a fast SLR time in the triplet state such that resulting pair of radicals is essentially born with a normal Boltzman spin population, i.e., with no ESP. Again, we point out that the resulting EPR spectrum can not reveal the presence of a triplet precursor.

We now introduce an additional mechanism, the radical pair RP mechanism. In order to do so we assume that the triplet state is the precursor, but that only Boltzman populations are created in the initial radical pair, $A^{*T} + BH \rightarrow [AH\cdot + B\cdot]_{RP}$. The initial radical pair with Boltzman population is in what is known as a correlated triplet radical pair since the immediate precursor involves a triplet state. In this example the TM is not in operation even though the triplet state is a precursor: At this point the radical pair mechanism is to be invoked. The most common example of the RP mechanism involves ST_0 (singlet-triplet) mixing. This mechanism always produces both absorption and emission. This is in contrast to the TM mechanism which produces only enhanced absorption or emission, but not both. Again, ESP developed by the RP mechanism need not be observed since fast chemistry or SLR times may be important. In practice, events too fast to be observed by the EPR spectrometer can go undetected.

In summary, radical pair derived from triplet state precursors, as sometimes has been suggested for PS, can exhibit a variety of ESP patterns. Such ESP can be characteristic of the TM, of the RP mechanism, of a combination of each, or of none. Since these spin polarization mechanisms produce a variety of ESP patterns, caution is advised in concluding the singlet or triplet nature of the chemical reaction. In Table 1 are listed most of the possible chemical reaction mechanisms along with the corresponding ESP mechanism. For the sake of completeness we have also included the random radical pair mechanism in which two random radicals encounter each other and eventually react to form the diamagnetic product. During the encounter period ST mixing occurs to produce ESP in the unreacted species. This is a mechanism of liquid solution and is not very likely to be important in the study of PS by CIDEP.

Table 1. Typical ESP patterns for the triplet mechanism (TM) and radical pair mechanism (RPM)

A. Triplet mechanism (TM): Triplet precursor to produce a radical pair R_1R_2 e.g.: $R_1{}^{*T} + R_2H \rightarrow R_1H\cdot + R_2\cdot$

$D>0$[a]	$D<0$[a]	$D>0$[a]	$D<0$[a]
$N_x+N_y>2N_z$	$N_x+N_y<2N_z$	$N_x+N_y<2N_z$	$N_x+N_y>2N_z$
R_1 (e)[b], R_2 (e)		R_1 (a)[c], R_2 (a)	
R_1 (B)[d], R_2 (c)		R_1 (B), R_2 (a)	
R_1 (e), R_2 (B)		R_1 (a), R_2 (B)	
R_1 (B), R_2 (B)		R_1 (B), R_2 (B)	

B. Radical pair mechanism (RPM) ST ± mixing: singlet or triplet precursor[e] $R_1{}^{*S}$ or $R_1{}^{*T}$, respectively, e.g., $\overset{*}{R}_1 + R_2 \rightarrow R_1^{\cdot-} + R_2^{\cdot+}$; random pair recombination, singlet or triplet sink[f] $R_1{}^{*S}$ or $R_1{}^{*T}$, respectively, e.g., $R_1^{\cdot-} + R_2^{\cdot+} \rightarrow R_1{}^* + R_2 \rightarrow R_1 + R_2$.

$J>0$	$J<0$	$J<0$	$J>0$
Singlet or triplet or precursor[e]	Random pair singlet or triplet sink[f]	Singlet or triplet precursor[e]	Random pair singlet or triplet sink[f]
R_1 (e), R_2 (e)		R_1 (a), R_2 (a)	
R_1 (B), R_2 (e)		R_1 (B), R_2 (a)	
R_1 (e), R_2 (B)		R_1 (a), R_2 (B)	
R_1 (B), R_2 (B)		R_1 (B), R_2 (B)	

C. Radical pair mechanism (RPM) ST_0 mixing (see B for chemical examples).

$J>0$ $g_1>g_2$[g]	$J<0$ $g_1>g_2$[g]	$J<0$ $g_1>g_2$[g]	J>) $g_1>g_2$[g]
Random pair triplet sink[f] of singlet precursor[c]	Triplet precursor[e] or random pair singlet sink[f]	Random pair triplet sink[f] or singlet precursor[e]	Triplet precursor[e] or random pair singlet sink[f]
R_1 (e), R_2 (a)		R_1 (a), R_2 (e)	
R_1 (B), R_2 (a)		R_1 (B), R_2 (e)	
R_1 (e), R_2 (B)		R_1 (a), R_2 (B)	
R_1 (B), R_2 (B)		R_1 (B), R_2 (B)	

D. Radical pair mechanism (RPM) ST_0 mixing (see B for chemical examples).

$J<0$ hf[h]	$J>0$ hf[h]	$J>0$ hf[h]	$J<0$ hf[h]
Random pair triplet sink[f] or singlet precursor	Triplet precursor[e] or random pair singlet sink[f]	Random pair triplet sink[f] or singlet precursor[e]	Triplet precursor[e] or random pair singlet sink[f]
R_1 (a, e) [i]R_2 (a, e)		R_1 (e, a), R_2 (e, a)	
R_1 (B), R_2 (a, e)		R_1 (B), R_2 (e, a)	
R_1 (a, e), R_2 (B)		R_1 (e, a), R_2 (B)	
R_1 (B), R_2 (B)		R_1 (B), R_2 (B)	

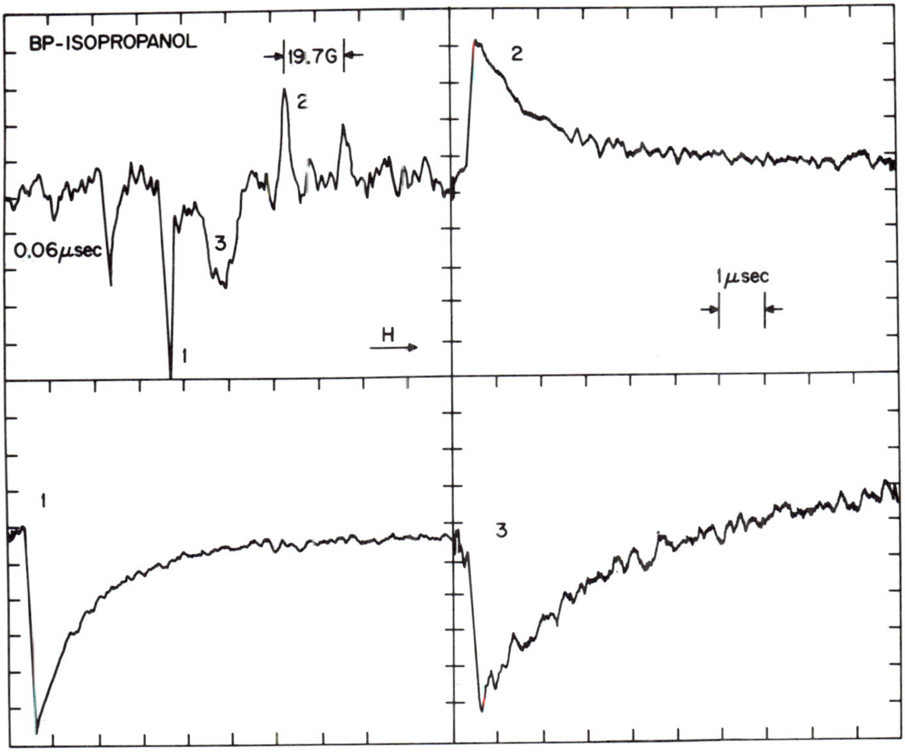

Fig. 4. Electron spin echo studies of the benzophenone-isoproponal light induced hydrogen abstraction reaction. Electron spin echo field spectrum taken 0.06 µs after a 10 ns laser pulse. Kinetic traces are referenced to the field spectrum and represent 40 ns time resolution due to the electron spin echo technique

We demonstrate a typical example of CIDEP in Fig. 4. Triplet excited state benzophenone (BP) abstracts a hydrogen atom from isoproponal to form a ketyl radical RH·, and the isopropoxide radical R_2 . Both radical exhibit ESP and in the context of this discussion have the same triplet precursor, namely triplet benzophenone. However, the ketyl radical is mostly in emission and thus its ESP is dominated by the TM. In contrast, the isopropoxide radical exhibits ESP primarily characteristic of the hyperfine-RP mechanism since the low field spectral components are in emission, whereas the corresponding high field components are in enhanced absorption.

◄ [a] $|D|>3|E|$ where 2D represents approximately the largest splittings (along the Z canonical orientation in the EPR triplet spectrum; [b] e = emission; [c] a = enhanced absorption; [d] Boltzman population; [e] initially correlated pairs; [f] initially uncorrelated pairs; [g] ST_0 mixing via a $\Delta g = g_1 - g_2$ mechanism; [h] ST_0 mixing via the hyperfine mechanism. Note that this is the only ESP mechanism that is not transferable in sequential electron transfer reaction; [i] (a, e) = low field signal component in enhanced absorption, high counterpart in emission

2.4.3 The Triplet Mechanism

In most cases solid-state studies of the photoexcited triplet reflect photophysics as discussed previously. There are, however, photochemical processes which take place in the photosynthetic pathway and involve the participation of a photoexcited state, singlet or triplet.

Traditionally, photoexcited triplet states have been important in liquid solution photochemistry, especially the triplet state of the chlorophylls. At the same time "solid-state-like" photochemistry, such as may be involved in photosynthesis, may also invoke excited triplet states of chlorophylls. Thus, the direct tool of studying the triplet state in liquid solution has been by optical spectroscopy, since direct EPR spectroscopy is not suitable for detection of photoexcited triplets in the liquid phase because of line broadening [87]. However, EPR and NMR can be most useful in studying solution photochemistry which may or may not derive from triplet precursors. This involves the recently developed magnetic resonance techniques (CIDEP), and the nuclear spin analog (CIDNP). In this review we shall be focusing mainly on CIDEP because of recent studies of primary photosynthesis that may have direct connections with chlorophyll triplet states.

As mentioned earlier in this survey, ISC to triplet substates of aromatic molecules or chlorophylls is a selective process. In other words, different ISC rates occur to the three zero magnetic field spin sublevels, T_x, T_y and T_z. For our purposes $N_x > N_y > N_z$ or $N_y > N_x > N_z$, where N_i represents the zero field population of T_i. Selective ISC occurs to the high magnetic field substates, T_0, $T_{\pm 1}$, and the high field populations are given by (cf. Sect. 2.2).

$$N_{\pm 1} \cong (N_j + N_k)/2 \quad \text{and} \quad N_0 = N_i \text{ for } H//i. \tag{11}$$

To a first approximation when $|g\beta H| > |D|$, then $T_{\pm 1}$ and T_{-1} are equally populated but differently from T_0. However, a more exact solution shows that the T_{-1} and T_{+1} levels are in general never equal in populations except either when $|E| = |D| = 0$, or in a magnetic field of infinite strength. In practice, the reaction of a triplet leading to an electron transfer (abstract or donation) involves all three sublevels, T_0, $T_{\pm 1}$. The selectivity in ISC rates and the sign (and size) of the ZFS parameter D are sufficient to deduce the polarization patterns of the radicals formed in the photochemical process.

If the ZFS parameter D is positive, i.e., the order of the triplet sublevels are as in Fig. 3a, T_x and T_y will mix into $T_{\pm 1}$ more than T_z[4]. It is obvious therefore that if $N_x > N_y, N_z$, T_{+1} will be overpopulated relative to T_{-1} state is represented by a wave function $|\alpha\alpha\rangle$, and since the two electrons of the triplet molecule C^{T*} are involved in the chemical reaction with the singlet scavengers, one would observe emission from the doublet products C· and S·, i.e., $C^{T*} + S^S \rightarrow C\cdot + S\cdot$. Likewise, enhanced absorption would arise from the chemical reaction involving T_{-1}, i.e., $N_z > N_x, N_y$. Finally,

4 This statement is correct, although the coefficients in the linear combination of T_x, T_y, T_z vary as the molecule rotates

no EPR polarization for the doublets would arise from the T_0 level since this sublevel contains an equal amount of α and β spins. The polarization patterns of the doublet radicals in liquid solution which are formed via TM can be predicted by knowing the populations N_i (i = x, y, z) and the ZFS parameter D from the expression [109]

$$N_{1/2} - N_{-1/2} \sim \frac{4}{15} \frac{D}{H} (N_x + N_y - 2 N_z). \tag{12}$$

We have summarized these triplet rules for CIDEP in Table 1A.

2.4.4 The Radical Pair Mechanism: $ST_{\pm 1}$ Mixing

The correct description of a radical pair requires mixtures of RP singlet and RP triplet eigenfunctions with each radical contributing one electron spin to these eigenfunctions. The energy separation between the singlet level and the average of the three triplet levels is given by the exchange interaction parameter, 2J. The sign of J determines whether the singlet lies lowest ($J > 0$) or the triplet lies lowest ($J > 0$). The energy gap in the triplet manifold is determined by the external magnetic field through the Zeeman energy $g\beta H$ and the ZFS parameters, D and E. Most frequently, $g\beta H > D$ such that the energy differences between triplet sublevels is controlled primarily by the external magnetic field. Now if $2J \simeq 0$, then the RP singlet level is closest to the RP triplet sublevel T_0. We shall cover ST_0 radical pair mixing mechanisms in the next section. If, however, 2J is close to the Zeeman energy, $g\beta H$, then either ST_{+1} mixing occurs or ST_{-1} mixing occurs, depending on the sign of J. In other words, if the singlet level is lower in energy than the average of the triplet levels by an amount approximately equal to the Zeeman energy, ST_{-1} mixing occurs and chemistry may occur using the T_{-1} sublevel ($J < O$). For example, a triplet radical pair in the T_{-1} sublevel could dimerize to form a covalent bond where the product is a singlet. If such were the case, then the β spins of the unreacted doublets observed by EPR would be depleted and emission would be observed for those radicals which had not yet undergone chemistry In ST_{+1} mixing ($J > 0$) the average triplet energy lies lower by the amount of the Zeeman energy, and thus the singlet level is near T_{+1} in the remaining radicals, resulting in absorption for the unreacted doublets.

The above examples demonstrate the case in which "depletion chemistry" occurs via radical recombination to produce singlet or triplet state (singlet or triplet sink), which further relax to their diamagnetic ground states. It is obvious that the encounter radicals in those cases can be initially uncorrelated and the mechanism will still function. Now, instead of "depletion chemistry" we mention the case of "production chemistry" in which radical pairs are formed through singlet or triplet precursors and polarization develops via initially correlated pairs in the cage before the radicals diffuse apart. The CIDEP rules for $ST_{\pm 1}$ mixing are summarized in Table 1B. When ST_{+1} mixing occurs in systems with the radical pair mechanism, the net process involves the "flipping" of one of the electrons. The "flipping" of an electron requires a magnetic field and in these cases an internal magnetic field is required. These internal magnetic fields are provided via hyperfine interactions, g value differences, dipolar fields, etc.

This mechanism is easily understood since one imagines a complete spin flip. This is in sharp contrast to the spin phasing mechanism of ST_0 mixing of the next section.

2.4.5 The Radical Pair Mechanism: ST_0 Mixing

By far the most difficult CIDEP mechanism to understand physically is that involving ST_0 mixing. In the ST_0 case no electron is "flipped" in the process of ISC. Instead, the electron spin phases are correlated and uncorrelated via the exchange interaction in combination with a local magnetic field [113, 114]. First of all, ST_0 mixing involves a pair of electrons, one α and one β. In a radical pair electron 1 is essentially on one molecule (R_1) and electron 2 is essentially on another molecule (R_2). If both radicals have exactly the same local magnetic fields, i.e., identical symmetries with the same hyperfine, with the same set of nuclear spin eigenvalues, and the same g values, then the two electrons, one α and one β, precess at exactly the same rate and always maintain the same phase angles. Thus, if the two spins add together such that the net spin is zero (see Fig. 5), then the radical pair is singlet remains singlet forever. Likewise, the phase relationship may be such that the combined spin has one unit of spin angular momentum normal to the external magnetic field to form a pure T_0 state as shown in Fig. 5. Again, as long as the local magnetic fields of electron 1 and electron 2 are identical, the system remains in T_0. However, if the magnetic field difference causes electron 1 to precess faster or slower than electron 2, then the radical pair oscillates between singlet and triplet at a rate determined by the magnitude of this local field difference. For example, suppose that both radicals of the radical pair are chemically identical, each electron interacting with a single dominant proton, proton 1 of radical 1 or proton 2 of radical 2. In this case four types of radical pairs exist, namely

I	$R_1\ (\alpha_e\ \alpha_N)$	$R_2\ (\beta_e\ \alpha_N)$
II	$R_1\ (\alpha_e\ \beta_N)$	$R_2\ (\beta_e\ \alpha_N)$
III	$R_1\ (\alpha_e\ \alpha_N)$	$R_2\ (\beta_e\ \beta_N)$
IV	$R_1\ (\alpha_e\ \beta_N)$	$R_2\ (\beta_e\ \beta_N)$

(The case of β_e on R_1 and α_e on R_2 is not necessary for this discussion.)

In case I electron 1 and electron 2 experience the same magnetic field; the same holds for case IV. Singlet or triplet phase coherency is maintained in cases I and IV and thus no RP-ISC occurs. However, in cases II and III electron 1 sees a hyperfine field of the opposite sign as electron 2 and thus singlet-triplet phase angle varies in time. The II and the III systems oscillate between singlet and triplet radical pair descriptions as shown in Fig. 5. The rate of this oscillations is in this example determined by the hyperfine coupling constant. In general, not only hyperfine interactions contribute to the local magnetic environment, but other sources of local fields exist such as g-value differences. Later we will, however, lump all local field differences into a single set of parameters A_i where i can include all possible combinations (such as I, II, III, and IV) in the above example. This local field difference, in conjunction with the chemistry, is sufficient to provide a simple explanation of the CIDNP that leads to nuclear polarization in diamagnetic products [114]. However, it

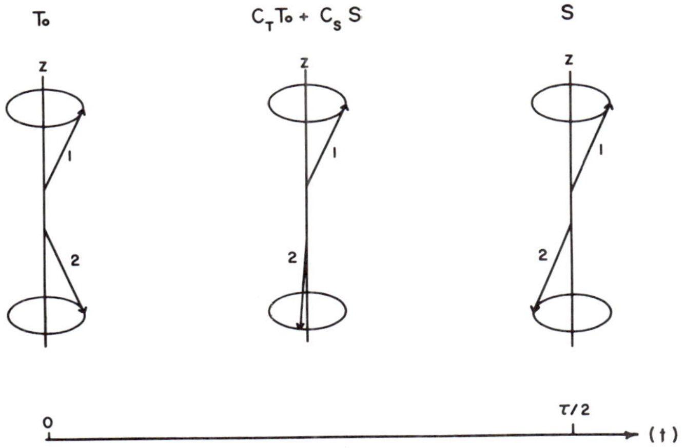

Fig. 5. Diagramatic representation of radical pair intersystem crossing involving ST_0 mixing as a function of time

totally inadequate for explaining ST_0 CIDEP mechanisms. Not only is the explanation inadequate, but just ISC via radical pair ST_0 mixing does not produce any CIDEP [112]. In addition to ST_0 mixing the exchange interaction represented in the exchange parameter, $2\vec{J}$, is also necessary [112]. The effect of the exchange term is not easily explained in a simple physical picture. The best physical type description so far developed is the ad hoc precessing vector model due to Monchick and Adrian [113]. This vector model is extremely useful for the experimentalist who intends to employ the CIDEP technique, especially since the ST_0 radical pair mechanism of CIDEP has been the most prevalent to date.

In the vector model (Fig. 6) the sample magnetization (the parameter to be measured by EPR) is associated with a vector $\vec{\rho}$ (t) where t takes into account the time dependence of the sample (see Fig. 6). When $\vec{\rho}$ is along +z, then the sample is pure singlet and no magnetization exists. When $\vec{\rho}$ (t) is along −z, then the sample is pure triplet and again the sample magnetization, although one exists in the radical pair, is unmeasurable in the separated radicals. According to this vector model, only when $\vec{p}$ (t) has a projection along the x direction does EPR activity exists and ESP can be observed in the radical product $R_1\cdot$ or R_2.

So far we have only defined the necessary properties of $\vec{\rho}$ (t). The usefulness of the vector diagram lies in its ability to describe the time dependence of $\vec{\rho}$ as a function of local field difference which we will call $\vec{A}_i$ and the exchange interaction $\vec{J}$. The actual time dependence of $\vec{\rho}$ (t) is described by precession of $\vec{\rho}$ (t) about another vector which Monchick and Adrian calls $\vec{\Omega}$, where $\vec{\Omega} = 2\,\vec{J} + \vec{A}_i$. As shown in the inset of Fig. 6, $-2\,\vec{J}$ lies along the minus z axis and $-\vec{A}_i$ lies along the minus x axis. Take the case of singlet $\vec{\rho}$ (0) such that $\vec{\rho}$ (0) lies along the plus z axis. Also, let $\vec{J} = 0$ with a finite $\vec{A}_i$ values. Referring to Fig. 6 the vector $\vec{\Omega}$ coincides with and lies along the −x direction. Thus, $\vec{\rho}$ (t) will precess about the x axis which means that it oscillates between +z (pure singlet) and −z (pure triplet). When $\vec{\rho}$ (t) lies in the

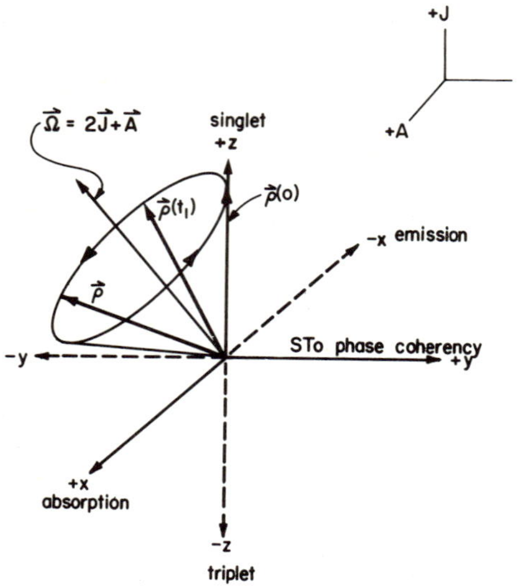

Fig. 6. Vector diagram for CIDEP effects due to the ST_0 radical pair mechnism. The exchange interaction is represent by a vector $\vec{J}$ which lies along the *Z-axis*. The ST_0 mixing term is represented by a vector $\vec{A}$ which lies along the *X-axis*. $\vec{A}$ can include both hyperfine and/or Δg effects. The time evolution of the system is represented by a vector $\vec{\rho}\,(r)$ which precesses about $\vec{\Omega} = 2\vec{J} + \vec{A}$. The singlet, triplet, absorptive and emissive character associated with $\vec{\rho}\,(t)$ can be deduced from this vector diagram

xy plane the radical pair is exactly half singlet and half triplet. Thus rotation about $\vec{\Omega} = -\vec{A}_i$ is consistent with the above explanation for oscillating ST_0 radical pair ISC. In this description, CIDEP occurs when $\vec{\rho}$, has a projection along ±x. Notice that no CIDEP is produced without exchange since $\vec{\rho}$ (t) still has no projection along the plus or minus x axis. However, now let us take the case where $2\,\vec{J} > \vec{A}_i$ and points almost along z, but $\vec{\Omega}$ is still tipped significantly toward x in the xz plane. In this case $\vec{\rho}$ (t) (where $\vec{\rho}$ (0) = pure singlet) precesses as shown in Fig. 6, producing a small admixture of triplet into the singlet radical pair, but more importantly a significant amount of CIDEP polarization.

In practice, the usual explanation of liquid solution is the following: Suppose we form a radical pair by the following reaction:

$$R_1{}^{*S} + R_2 \longrightarrow R_1{}^{\cdot -} + R_2{}^{\cdot +}.$$

This means that the radical pair is initially prepared in a singlet state ($\vec{\rho}$ (0) is along z). Since the radicals are close together in order for the chemical reaction to occur, than $2\,|\,\vec{J}\,| > |\,\vec{A}_i|$[5]' and the vector $\vec{\rho}$ (t) does not appreciably change (refer to Fig. 6). Now the radicals diffuse apart and $\vec{J}$ essentially vanishes. In this step $\vec{\Omega}$ is along x such that $\vec{\rho}$ (t) rotates in the zy plane toward the y axis. At this point RP-ISC has occured but still no CIDEP polarization exists. (We have assumed that the initially formed radical pair diffused apart so rapidly that $\vec{\rho}$ (t) is unaffected by the $\vec{J} \gg |\vec{A}|$ that existed around t = 0.) However, if the same two radicals reencounter each other i. e., diffuse together again, then J becomes large again and now $\vec{\Omega} \cong -2\,\vec{J}$ such that $\vec{\rho}$ (t) now precesses essentially about z. In other words, the $\vec{\rho}$ (t) component along y

5 J is larger at short distances of separation between R_1 and R_2 and vanishes at larger distances of separation

quickly precesses toward x (Fig. 6) producing large CIDEP polarizations (cf. Table 1 C and D).

Thus, this vector diagram [113] can be used to predict a veriety of ST_0 radical pair situations. In the solid-state type environment of photosynthesis this description is one of the most useful tools available for guiding the experimentalist.

3 Triplet State Studies of Model Chlorophyll Compounds

The numerous investigations of model porphyrins [115] and chlorophylls in vitro serve as a basis for interpreting the magnetic resonance studies of intact photosynthetic cells or biological preparations such as reaction centers, chromataphores, chloroplasts, etc. The fundamental gross feature for interpreting the spectroscopy of the chlorophylls is the basic skeleton, porphyrin-like ring system. Thus, many model compounds which consist of free base and substituted porphyrins as well as various types of chlorophylls in vitro have been investigated using the techniques previously described. Table 2 summarizes the optical, magnetic, and kinetic parameters of a variety of model compounds and systems. We note that the property of weak phosphorescence for many porphyrins and most chlorophylls has restricted ODMR studies to the fluorescence or triplet-triplet absorption detection techniques.

Table 2 illustrates the similarity of the ZFS parameters in a particular series of compounds. The relatively small values for the ZFS parameters result from extensive delocalization over the large macrocycle skeleton common to this type compound. The signs of the ZFS parameters D and E, i. e., the ordering of the ZFS energy levels X, Y, and Z, are important for molecular structures and mechanisms which are involved in the primary events of photosynthesis. The well-established optical properties of the in vitro chlorophylls [124] have allowed the magnetophotoselection (MPS) method a means of ZFS sign determination for many chlorophylls (Table 2) [112]. Although in vivo ZFS sign determination is not the major goal of the MPS method applied to photosynthetic systems, the application of the MPS method to in vivo chlorophylls is of structural interest and appears promising for the study of antenna as well as reaction center chlorophyll [125].

All the model compounds listed in Table 2 are capable of exhibiting an electron spin polarized triplet state (ESP) since the triplet sublevels (T_x, T_y, T_z) are coupled unequally to the singlet manifold. Typically only one spin component, T_x, T_y, or T_z, is the most active with respect to population and depopulation of the triplet state, resulting in significant ESP even at relatively high temperatures.

In these compounds, containing a large essentially planar macrocycle, the most commonly used rectangular coordinate system in the molecular framework defines the z component as perpendicular to the macrocycle plane. Earlier, it was assumed that the dipolar splitting is maximum along this z principal axis, i. e., T_z lies lowest in energy. Recently this was confirmed in several chlorophylls using MPS techniques [122]. Thus, Table 2 shows that in general the active kinetic spin component is quantized in a direction in the plane of the macrocycle, namely the T_{xy} plane. However, in the zinc porphyrins, the out-of-plane component, T_z, is the most active and has

Table 2. Kinetic rate parameters (in units of s^{-1}) and ZFS parameters (in units of cm^{-1}) for some in vitro compounds

Molecule	Solvent	T, K	k_x[a]	k_y[a]	k_z[a]	A_x[a]	A_y[a]	A_z[a]	$\|D\|$[a,b]	$\|E\|$[a,b]	Method and Ref.
Chl a	Pyr: Tol	~5							+0.0273	−0.0040	EPR 122
	n-Octane	85	800	1200	360	0.9	1.0	0.2	0.0262	0.0022	EPR 119
	n-Octane	2	661	1255	241	0.3	1.0	1.0	0.0280	0.0038	ODMR 118
	EtOH	95	710	2710	310	0.38	1.0	0.15			EPR 118
Chl b	Pyr:Tol	~5							+0.0293	−0.0052	EPR 122
	n-Octane	85	320	550	40	1.0	0.6	0.1	0.0286	0.0037	EPR 119
	n-Octane	2	268	570	34	0.3	1.0	0	0.0320	0.0041	ODMR 118
	EtOH	95	310	850	65	0.4	1.0	0.1			EPR 118
Chl c_1		~5							0.0269	0.0055	EPR 116
Chl c_2		~5							0.0276	0.0050	EPR 116
Ph a	Pyr:Tol	~5							+0.0342	−0.0033	EPR 122
	MeTHF	103	1040	1300	820	0.7	1	0.6	0.0341	0.0033	EPR 118
Ph b	Pyr:Tol	~5							+0.0347	−0.0038	EPR 122
	MeTHF	103	590	870	420	0.6	1	0.5	0.0358	0.0046	EPR 118
Bchl a	Pyr:Tol	~5							+0.0224	−0.0053	EPR 122
Bchl a	THF	2	2287	3321	661				0.0238	0.0069	ODMR 120
Bchl	Pyr:Tol	~5							+0.0212	+0.0055	EPR 122
Bph a	Pyr:Tol	~5							+0.0259	+0.0046	EPR 122
Bph b	Pyr:Tol	~5							+0.0249	+0.0050	EPR 122
H_2TPP	n-Octane	80	300	600	150	0.31	0.54	0.14			EPR 92
H_2P	n-Octane	4.2	75	230	6	0.31	0.68	0.01	0.0435	0.0063	ODMR 76, 123
TPC	Tol:EtOH (5:1)	77	400	700	240	0.56	1.0	0.40	0.0364	0.0063	EPR 121
	n-Octane	105	500	900	250	0.6	1.0	0.36	0.0364	0.0063	EPR 121
MgTPP	toluene	140	Av decay rate 21[c]						0.0294[e]	0.0076[e]	EPR 133
	ethanol	140	Av decay rate 15[c]						0.0288[e]	0.0028[e]	EPR 133
MgTBP	Tol:Pyr	5	Av decay rate 4.3[c]						0.0334	0.0065	EPR 118

ZnP	n-Octane	1.2	4.9	5.9	9.0	0.03	>0.04	0.93	0.035	0.009[d]	ODMR	117
ZnTPP	n-Octane	140	Av decay rate ~20[c]						0.0301[e]	0.0079[e]	EPR	133
	Ethanol	140	Av decay rate ~20[c]						0.0302[e]	0.0100[e]	EPR	133
ZnTPC	toluene	140							0.0315[e]	0.0069[e]	EPR	133
ZnChl a	n-Octane	2	346	330	660				0.0306	0.0042	ODMR	119
Znchl b	n-Octane	2	122	250	622	0.3	0.7	1.0	0.0328	0.0032	ODMR	119

[a] Coordinate systems are chosen such that x, y are in the molecular plane an z is out plane. The order of X, Y, Z energy levels are X>Y>Z. For experimental errors see appropriate references. [b] The sign in front of the ZFS parameter indicates that it has been determined uniquely. Those values without a sign indicate the absolute value. [c] These decay rate correspond to the total triplet lifetime. [d] At 77K in EPA glass; the same authors report $|E| \approx 0$. [e] These values for the ZFS parameters are attributed to the monomeric species of the metalloporphyrins as derived from the line shape analysis [133]

been interpreted qualitatively, using approaches developed by Metz et al. for aromatic molecules [126]. Clarke et al. [95] demonstrated that in planar aromatic compounds, characterized by $\pi\pi^*$ triplet, the in-plane components are the most active. According to this view, insertion of the light magnesium atom into the macrocycle does not appreciably affect the $\pi\pi^*$ character of the observed triplet state. On the other hand, the heavier zinc moiety, via the d orbitals, is capable of spin-orbit coupling with the π macrocycle system, resulting in the z-axis as dominant in ISC [119]. Clarke's work suggests that the magnesium is of minor direct importance in affecting the triplet dynamics of these in-vitro models. Comparison of the triplet state dynamics of the TPC vs Chl led Nissani et al. [121] to conclusions similar to those of Clarke et al.

Investigations by ordinary optical techniques on the average triplet lifetimes of normal and fully deuterated Chl a in dry and wet pyridine led Bowman to propose solvent effects on ISC [127]. According to Bowman the ISC mechanism in the Chl molecule is achieved via a static distortion in which the central magnesium is forced out of the molecular plane by solvent interactions. In other words, the Mg-solvent coordination indirectly affects the ISC nature of the triplet. In pheophytin molecules a dynamic model, in which ISC is achieved by out-of-plane N-H vibrations, was invoked. The ESP patterns in the model compounds appear in accord with these mechanisms of spin-orbit ISC coupling which generate the selective ISC rates.

Apart from ESP effects described above, valuable information regarding the mechanism of energy transfer within porphyrins and chlorophylls can be achieved from triplet EPR line shape analysis. It has been known for some time that randomly oriented photoexcited triplets of magnesium and zinc porphyrins reveal anomalous EPR line shapes which do not reflect an axial symmetry [128–130]. Since to a first approximation the point group symmetry of ZnTPP (or MgTPP) in its ground state is D_{4h} [131–132], the disagreement between the experimental and calculated spectra is apparent. On the other hand, ODMR performed at ≤ 4.2K on ZnTPP, demonstrated unambiguously the existence of a rhombic contribution to the dipolar spin Hamiltonian giving rise to a nonvanshing $|E|$ value [131].

In a study aimed at interpreting the triplet EPR spectra of metalloporphyrins, the randomly oriented triplet EPR spectra, of MgTPP, ZnTPP at different temperatures and solvents, were analyzed [133]. For the transition from one type of spectrum, observed at low temperature, T ~100K, into another type of spectrum observed at ~180K, an activation energy of the order of 1000 cm^{-1} was calculated. This value for an activation energy is an order of magnitude larger than that expected from a dynamic Jahn-Teller mechanism.

The triplet spectral line shape was found to be most sensitive over a narrow range of temperatures around the freezing point of the solvent, where the soft properties of the solid matrix are still maintained. These observations were interpreted in terms of an equilibrium process between two forms: $P^* (D_{4h}) \rightleftarrows D^* (D_4)$. The $P^* (D_{4h})$ species are characterized by their axial symmetry, D_{4h}, and the $D^* (D_4)$ species are excimers which are composed of weakly interacting monomeric subunits.

At high temperatures, the main species which contribute to the EPR spectrum are of D_{4h} symmetry. On Lowering the temperature, ligation to the central metal takes place which results in removal of the in-plane symmetry to give nonaxial monomeric species of D_{2h} symmetry. In this treatment, it is proposed that around the freezing point of the solvent, where diffusional motion is allowed, the triplet monomers are interacting to form excimers. The structure of these excimers is of parallel configuration in which the out of plane principal Z-axes are common for both interacting nonaxial species, and the XY principal axes are mutually rotated by $\pi/2$. Therefore, an overall D_4 symmetry will result for these excimers. The coupling between the two constituents is via intramolecular triplet exciton hopping should affect the EPR spectrum. Therefore, the overall triplet line shape variations are due to a combination of chemical dynamics together with energy transfer mechanism.

The special chemical properties of the triplet state that may be operative have prompted a different experimental and theoretical approach in studies of intact photosynthetic preparations. For example, since the primary stage in photosynthesis involves a charge separation process, obviously photoreduction and photooxidation processes occur in the special reaction center. The important question, whether this photochemistry involves an excited singlet or an excited triplet as a precursor, is frequently asked. Studies have been reported on the photoreduction of porphyrins and porphyrin-like molecules [124–136]. Gouterman et al. have investigated the electron transfer reactions involving Bph and p-benzoquinone [135, 136]. Bph is of interest since it is probably a primary electron acceptor in bacterial photosynthesis. Likewise, the quinone is a known in vivo primary acceptor. Gouterman et al. demonstrated that the excited triplet of Bph reacts with the quinones to form separate radicals $Bph.^{+}$ and $Q.^{-}$ Thus, they proposed a photomechanism requiring a triplet charge transfer complex $(Bph^{\cdot +}\ Q^{\cdot -})^{T}$. These results may be relevant to interpreting photoelectron transfer processes of the in vivo reaction center.

More recent work has demonstrated in vitro photoinduced electron transfer from a synthetic version of chlorophyll special pair [137–140] to a pheophytin-like molecule [141]. Since the electron transfer occurs in less than 6 ps, triplet state participation was ruled out [141]. This latter case appears to duplicate the primary events in photosynthesis and supports the view that the triplets are not involved in the main pathway of photosystem I or in bacterial photosynthesis.

4 In-Vivo Chlorophyll Triplets

4.1 Introduction

The major impetus for recent investigations of triplet states in photosynthetic organisms is due mostly to the work of Dutton, Leigh, and co-workers. In pioneering observations Leigh and Dutton recorded an intense, highly polarized EPR triplet spectrum in photosynthetic bacterial [142–146]. They invoked the chlorophyll special pair to explain that the ZFS parameters were very small. More precisely, they interpreted the ESP signal in terms of a triplet radical pair within the Bchl special pair, i. e., $Bchl^{\cdot +}\ Bchl^{\cdot -}$. In addition, they noted that the ESP pattern could be explained by an excess spin population in the T_0 sublevel (the middle triplet level at high magnetic field) for all canonical orientations. Also of great importance, they demonstrated that the triplet state observed by EPR was the result of blocking the primary electron transport pathway, presumably at the iron-quinone complex. Later work suggested that the ESP pattern was most unusual and resulted from a radical pair intermediate consisting of oxidized special pair and a primary electron acceptor (probably Bph), i. e., $[(Bchl)_2^{\cdot+} Bph^{\cdot-}]$. In contrast to bacteria, the triplet state previously observed in algae is not closely related to the reaction center [147]. Furthermore, the algae exhibited monomeric ZFS parameters and a normal SO-ISC ESP pattern. In summary, these early investigations demonstrated that the triplet state probably originated in the Bchl special pair and was structurally and mechanistically important.

4.2 Bacterial Photosynthesis

4.2.1 The Triplet State in Bacterial Photosynthesis

An intense triplet signal is observed when the quinone species is reduced prior to the photoexcitation. Otherwise much weaker triplet EPR signals are observed. Complete removal of the quinone in reaction centers of R26 mutants is equivalent to the reduction of quinone because, again, intense triplet EPR signals are observed [148]. In both cases the electron flow is stopped at the preprimary electron acceptor (i. e., probably Bph). Since R26 RC contains no carotenoids either Bchl or Bph is the origin of the triplet state signal. Actually, the RC contains six molecules that are possibly involved in the triplet observations. Two molecules of Bchl absorb at 800 nm (P800), two at 870 nm (P870-special pair), and two Bph molecules at 760 nm. Direct measurement of the singlet-triplet transition has never been reported for either Bchl or Bph. In comparison, only very weak $T_1 \rightarrow S_0$ phosphorescence has been observed for Chl *a*, Chl *b*, Ph *a*, and Ph *b* [149–150]. The triplet-singlet energy gap measured in these type chlorophylls is found to be summarized by the following empirical relation.

$$\Delta E_{T_1 \rightarrow S_0} \cong K \Delta E_{S_1 \leftarrow S_0} . \tag{13}$$

These determinations place K at approximately 0.7 [151a]. Extension to bacteria implies that the triplet state of Bchl special pair lies below that of Bph by about 100 cm^{-1}(or by 170 cm^{-1} if one employs $\Delta E_{S_1 \leftarrow S_0}$ for the monomer Bchl). These observations are in agreement with Gouterman and Holten [136] and also with in-vitro experiments by Boxer and Closs [151b] using the optical-NMR line-broadening technique. Parson et al. [152] reported direct observation on photosynthetic bacterial reaction centers in which flash photolysis induced a transient absorption attributable to a triplet-triplet transition. This transient absorption with a lifetime of microseconds was similar to the triplet-triplet absorption in monomeric Bchl (or Bph). It is important that as long as this triplet-triplet transition was observed, P870 was bleached. In contrast, the P800 absorption is changed only slightly, probably in the form of a slight shift. A small shift in P800 is expected if the nearby special pair P870 is "bleached" via triplet formation. This requires that $S_1 \leftarrow S_0$ in P800 is slightly coupled to the corresponding transition in P870. In addition, such an interpretation is supported by the quantitative optical changes that take place when P870 is oxidized to form the cation radical [153]. Again, P800 in the RC is only slightly shifted. In the triplet case, Bph also exhibits small changes in its optical absorption spectrum. Thus, the most likely origin of the observed triplet is primarily the $(Bchl)_2$ special pair. Neverthless, none of these experiments eliminate in a rigorous manner the participation to some extent of Bph or P800 in the triplet state. For example, a charge transfer complex between P870 and a nearby Bph may be involved. Such a case would require that the triplet reside mostly only in the P870, but is contaminated by a slight amount of $[Bchl_2{}^{\cdot +}\ Bph^{\cdot -}]$ triplet radical pair. Nevertheless, we take the approach of explaining the triplet state ZFS of bacteria as residing only in P870 Bchl special pair.

4.2.2 Zero Field Splitting Parameters

ZFS parameters have been measured for several photosynthetic bacteria. The $|D|$ and $|E|$ values obtained for all bacteria are noticeably smaller than those of the corresponding monomeric chlorophylls species listed in Table 3. The reduced $|D|$ value indicates that the observed triplet state is delocalized over more than a single molecule of Bchl, as in the Bchl P870 special pair. The $E \neq 0$ signifies deviations from axial symmetry requiring that E vanishes.

The mechanisms for reduction of $|D|$ are important and are discussed below. Solvent or environmental effects on $|D|$ are generally small and, therefore, are essentially ruled out here. The remaining mechanisms are the following: (1) formation of a triplet radical pair (RP) such as $[Bchl^{\cdot +}\ Bchl^{\cdot -}]$ or $[(Bchl)_2{}^{\cdot +}\ Bph^{\cdot -}]$, (2) triplet delocalization in an aggregate as in, for example, $Bchl^{T_1}\ Bchl^{S_0} \rightleftarrows Bchl^{S_0}\ Bchl^{T_1}$, and (3) a combination of (1) and (2). In the pure triplet RP, i. e., the (1) mechanism, one expects to observe a triplet EPR spectrum with characteristically small $|D|$ and $|E|$ values. Furthermore, D must be less than zero. The samll zero field parameters are easily explained. Because one unpaired triplet electron is on another molecule, e. g., $Bchl^{\cdot -}$, the dipole-dipole interaction responsible for the ZFS should be smaller than in monomeric Bchl, where the two unpaired triplet electrons are always confined within a single macrocycle framework. In these compounds the ZFS is almost exclusively due to the dipole-dipole magnetic interaction between the triplet electrons. One expects therefore, because of the inverse cube effect with distance of the dipole-dipole interaction, that the ZFS of $[Bchl^{\cdot +}\ Bchl^{\cdot -}]$ would be very much smaller than the ZFS of monomeric Bchl triplets. Since the observed reduction is not very large, the *pure* RP mechanism is not considered as likely in these bacteria of Table 3. Thus, we are left with mechanism (2) in order to account for the fairly small reduction of the ZFS parameters as observed. Actually both mechanisms, i. e., mechanism (3) may operate simultaneously, but in the case of bacteria, mechanism (2) seems the dominant factor in the reduction of $|D|$. However, the above arguments are not rigorous in the elimination of a dominant RP mechanism. For example, a RP in which the conjugated planes are stacked closely above each other with a very short electron-electron distance between almost all sites may exhibit quite large $|D|$ and $|E|$ values. In other words, such an arrangement, in principle, can have ZFS parame-

Table 3. Kinetic rate parameters (in units of B^{-1}) and ZFS parameters (in units of cm^{-1}) for some in vivo preparations

Species	T, K	k_x[a]	k_y[a]	k_z[a]	A_x[a]	A_y[a]	A_z[a]	$\|D\|$[a, b]	$\|E\|$[a, b]	Method and Ref.	
				In vivo bacteria							
^{1}H-*R. rubrum*	4.8–9							+0.0185[c]	0.0033	EPR	122
^{2}H-*R. rubrum*	4.8–9							0.0185	0.0034	EPR	122
^{1}H-*Rps. spheroides*	4–9							0.0182	0.0035	EPR	122
^{2}H-*Rps. spheroides*	4–9							0.0183	0.0032	EPR	122
^{1}H-*Rps. palustris*	4–9							0.00182	0.0035	EPR	122
^{2}H-*Rps. palustris*	4–9							0.0184	0.0031	EPR	122
^{1}H-*Rps. gelatinosa*	4–9							0.0184	0.0028	EPR	122
^{1}H-*Rps. spheroides*	4–9							0.0183	0.0031	EPR	154
^{1}H-*Rps. spheroides* Ga	4–9							0.0185	0.0031	EPR	154
^{1}H-*R. rubrum* G9	4–9							0.0185	0.0031	EPR	154
^{1}H-Chromatium D	4–9							0.0178	0.0033	EPR	154
^{1}H-*Rps. spheroides* Ga	4–9							0.0185	0.0031	EPR	154
^{1}H-*Rps. gelatinosa*	4–9							0.0186	0.0027	EPR	154
^{1}H-*Rps. spheroides* R-26	4–9							0.0186	0.0031	EPR	154
R. spheroides (wild type)	1.8	9300	8500	2100	0.437	0.449	0.094	0.01872	0.00312	ODMR	155
R. spheroides R-26	1.8	9000	8000	1400	0.484	0.445	0.071	0.01878	0.00322	ODMR	155
R. rubrum	1.8	9000	8000	1400	0.488	0.437	0.0075	0.0190	0.0034	ODMR	155
R. rubrum	1.8	2105	2885	1335				0.0190	0.0034	ODMR	156,157
R. spheroides R-26	1.8	2660	3183	1596				0.0188	0.0031	ODMR	157
R. spheroides (2.4.1)	1.8	2674	3033	1600				0.0189	0.0032	ODMR	157
Chromatium vinosum strain D	1.8							0.0181	0.0034	ODMR	157
				In vivo green plants							
Anacystis nidulans											
A	4.2							0.0283	0.0038	ODMR	159
B	4.2							0.0348	0.0021	ODMR	161
C	4.2							0.0311	0.0038	ODMR	159

D	4.2							0.0366	0.0020	ODMR	159
Porphyridium cruentum	4.2							0.0283	0.0037	ODMR	159
Chlorella A	4.2							0.0288	0.0038	ODMR	159
Chlorella C	4.2							0.0311	0.0038	ODMR	159
Chlamydonomas reinhardi	35	850	1300	320	0.7	1:0	0.2	0.0280	0.0032	EPR	66
		500	560	60	0.6		0.1				
Spinach chloroplasts	4.5							0.0284	0.0039	EPR	57

[a] Coordinate systems are chosen such that x, y are in the molecular plane and z is out plane. The order of X, Y, Z energy levels are X > Y > Z. For experimental errors see appropriate references.

[b] The sign in front of the ZFS parameter indicates that it has been determined uniquely. Those values without a sign indicate the absolute value.

[c] For sign determination see [161] and [162]

ters as large as or larger than a monomer. Thus, how can this be ruled out in a convincing manner? The distinguishing feature of RP triplets is that $D < 0$ as mentioned above, whereas monomeric chlorophylls exhibit $D > 0$. The $D < 0$ for a RP is expected since an end-to-end [→→] dipolar interaction must dominate the stacked model in order to achieve the large $|D|$ and the sign of this interaction is negative. The explanation is based on studies of triplets in a high magnetic field. Of course a side-by-side [↑↑] also occurs but is much weaker [160]. In the planar monomers the weaker [↑↑] dipolar interaction dominates by symmetry considerations and the sign of the interaction is opposite the [→→] dipolar interaction case, i. e., $D > 0$. In R. Rubrum, Norris and Thurnauer [161–162] have determined that $D > 0$ and thus a dominant or pure RP mechanism must not be in operation in the reduction of the $|D|$ value. However, this does not mean that the triplet observed by EPR is not preceded by a transient radical pair such as $[(Bchl)_2{}^{\cdot +}\ Bchl^{\cdot -}]$. On the contrary, it is generally agreed that such a transient radical pair is likely to be responsible for the ESP observed in the triplet EPR spectra obtained from bacteria. Furthermore, we emphasize that mechanism (3) can be in operation with the restriction that (2) is dominant.

Mechanism (2) allows for increased delocalization of the pair of triplet electrons via migration and thereby may reduce the ZFS. In distinction with mechanism (1), both electrons are confined to the same molecule as far as the dipole-dipole interaction is concerned. Thus, the electron-electron distances are not significantly reduced. How then does one explain a reduction in ZFS parameters $|D|$ and $|E|$? The magnitude and sign of these parameters also depend on "magnetic" symmetry. As already mentioned, symmetry associated with planar delocalization explains that $D > 0$. By similar arguments, if the aggregate structure moves to higher magnetic symmetry, then the ZFS becomes smaller. Suppose that x and y are in the plane of the monomer macrocycle such that z is normal to that plane and that the canonical orientation along z gives the largest dipolar splitting. Then, in an aggregate of size two, as long as the z-axes are parallel or antiparallel, D will not decrease in mechanism (2). If the two z-axes are parallel or antiparallel and one molecule has been rotated by 90°, then $|E|$ vanished even though $|D|$ remains the same. Of course, the "migration" rate or the "delocalization" must exceed at least by an order of magnitude the $|E|$ value of the monomers. In either example, the dimer has magnetic axial symmetry and x and y of the aggregate are indistinguishable magnetically (neglecting additional factors such as hyperfine interactions). In this latter case, a reduction in $|E|$ has been produced by an increase in symmetry of the dipole-dipole interaction via aggregation. Similarly, $|D|$ can be decreased by mechanism (2) when the z-axes are not parallel.

In fact, in the special pair with all canonical orientations of the two molecules lined up (0 to 180° rotation about any canonical axis in one member of the dimer) no ZFS reduction occurs for mechanism (2). Only a measurement of the hyperfine interaction can detect these special cases and then only if the "hopping" rate is greater than the inverse of the hyperfine interaction. Unfortunately, hyperfine interactions in triplets of random order are difficult to measure. A new method is the electron spin echo envelope modulation technique which has been used to observe interactions in doublet state radicals. Extension of this technique to randomly

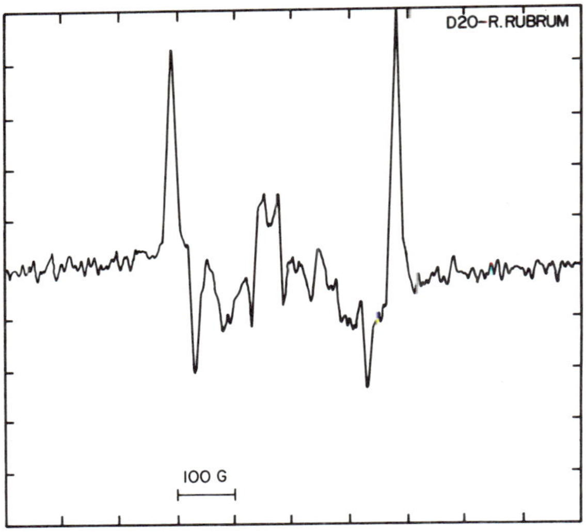

Fig. 7. An electron spin echo triplet spectrum ($\Delta M = \pm 1$) during steady state illumination of ^{2}H-R˙ Rubrum taken at 4.5 K. Echo intensity resulted from a 90–180° pulse sequence (pulse separation ~.5 μs). The first derivative presentation was obtained by computer differentiation

oriented triplets appears to be promising since the steady-state electron spin echo spectrum has been observed in ^{2}H R˙ Rubrum (Fig. 7).

4.2.3 Electron Spin Polarization in Triplets

In photosynthetic bacteria, the ESP pattern is a (Iz) e (IIx) e (IIy) a (Iy) a (Ix) e (IIz) [54, 116, 142]. This pattern is shown in Fig. 3 C. In the in-vitro case, a change in the polarization direction is observed in transitions Ii or IIi (i = x, y, z), whereas in the in-vivo case no change in the polarization direction is noticed. For the case of D > O the ESP can be explained if the center level T_0 is overpopulated relative to the upper and lower levels $T_{\pm 1}$.

Let us assume that the three populations of the spin sublevels at zero magnetic field are N_i, i = x, y, z. The nine corresponding populations at high magnetic field, neglecting SLR processes, are given to first order by Eq. (11) in Sect. 2.4.3 above. Eq. (11) is based on the usual spin-orbit-ISC described previously. The ESP pattern, as observed in bacteria, requires T_0 to be overpopulated for all canonical orientations, namely

$$N_0 = N_i > N_{T_\pm} \; i \,|\, |H \; i = x, y, z. \tag{14}$$

It is easy to show that Eq. (14) combined with Eq (11) leads to the inequality which is clearly impossible.

$$\sum_i N_i > \tfrac{1}{2} \left(2 \sum_i N_i\right), \; i = x, y, z. \tag{15}$$

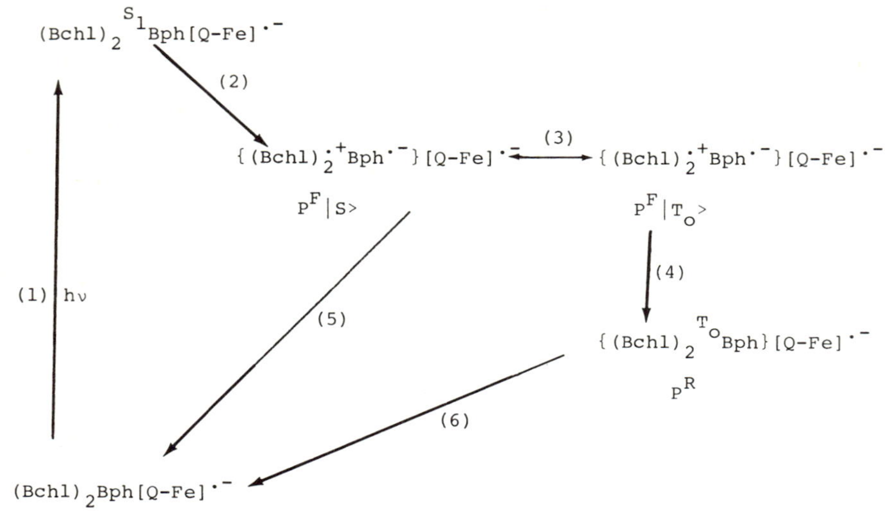

Fig. 8. Radical pair explanation of the triplet polarization patterns observed in photosynthetic bacteria; $[Q\text{-}Fe]^{\cdot -}$ is the reduced iron-quinone complex

Thus, ordinary ISC will never produce overpopulation of T_0 at all high-field canonical orientations. Variations of the ordinary ISC that includes SLR still are not capable of accounting for the observed high-field EPR spectra.

The simplest explanation for overpopulation of only T_0 for all high-field canonical orientations is via the so-called ST_0RP-ISR mechanism. This mechanism was discussed extensively above in describing the phenonmena of chemically induced dynamic nuclear or electron polarization of CIDNP or CIDEP, respectively. It is easy to account for the experimental observations in in-vivo bacterial preparations in terms of the ST_0 RP-ISC mechanism. Figure 8 represents schematically the photoelectron transfer which takes place immediately after the primary photoabsorption.

Prior to process 2, the electrons are correlated as a singlet state and thus the initial electron transfer produces a singlet state. Because of the large distance between the elctrons in the radical pair, the correlation begins to change owing to different Larmor frequencies of the electron spins on $(Bchl)_2^{\cdot +}$ and $Bph^{\cdot -}$. Hence, this dephasing process causes the spin angular momentum of the electron pair to change back and forth from $M_{s=o} = 0$ to $M_{s=1} = 0, T_0$. The difference in magnetic environment may be provided by hyperfine interaction, g-factor differences, or interaction of either spin with other paramagnetic species in the RC. Conversion of singlet radical pair to the $T_{\pm 1}$ radical pair is also possible, but as long as the electrons are sufficiently separated, this type of mixing is negligible. Once the triplet character is developed, a back electron transfer produces an ordinary triplet (process 4). Processes 5 and 6 result in the ground-state singlet.

We summarize the mechanism that seems to fit most observations about the triplet state in bacterial PS in Fig. 9. Notice that observation of the triplet is possible only in the blocked mechanism (B) where the acceptor Q-Fe has been reduced (or

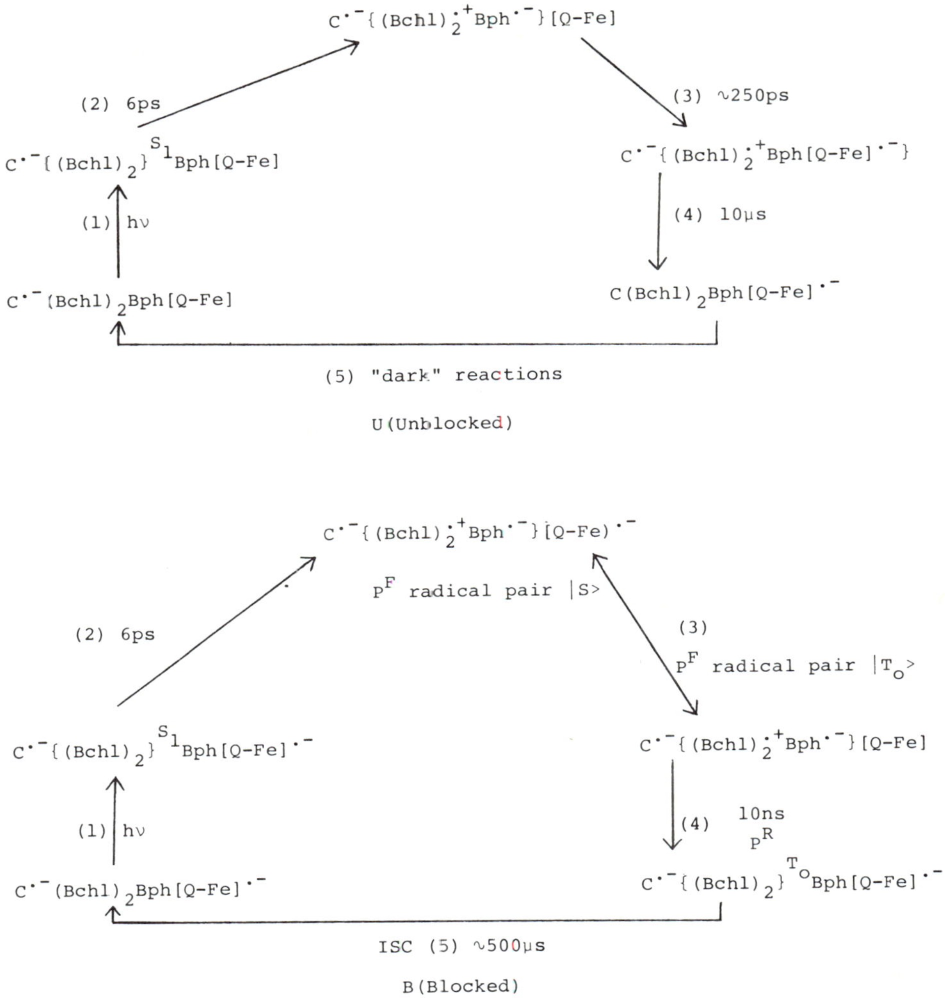

Fig. 9. "Blocked" (*B*) and "unblocked" (*U*) reactions in the bacterial photosynthetic reaction center. A singlet state which is indicated without an index, i.e., *S*, is due to a radical pair singlet configuration. The term *C* corresponds to cyctochrome

removed). The observed EPR triplet is basically an ordinary delocalized triplet which is produced by process 4, and decays by the usual ISC mechanism. The unusual feature of this state is in its birth which is via a RP-ISC mechanism. This triplet state is commonly referred to as P^R [152]. Notice that the P^F state, which is a triplet radical pair, is not observed owing to its very short lifetime on the EPR time scale. When the system is unblocked (U), the ~250 ps for process 3 does not allow the conversion of a significant amount of singlet radical pair into triplet radical pair. Instead, a new radical pair is formed via process 3. However, if the state produced via process 3 were to back-react directly, not enough energy is present to reach the $(Bchl)_2$ in the T_0 state and instead $(Bchl)_2$ in the ground state is formed. Back-reaction of this state is generally avoided in the natural course of events by a fast μs reaction of reduced cytochrome with the cation special pair. Schemes U and B represent the most frequently encountered explanation of the primary events in bacteria and the role of the triplet state. The reaction times of U and B represent either optical data or magnetic resonance lifetimes. The radical pair state as shown in Fig. 9 employs $(Bchl)_2$ and Bph [163]. The magnetic resonance data can directly detect only the P^R state, and thus invokes a radical pair precursor via indirect means. Any radical pair with sufficiently fast conversion into T_0 is sufficient to explain the triplet ESP pattern of bacteria.

The most direct evidence on the chemical nature of the radical pair state P^F is from fast time-resolved optical studies. The optical transient associated with this state agrees best with $[(Bchl)_2^{\cdot +} Bph^{\cdot -}]$. The spectral changes occur within ~6 ps and last for ~250 ps in an unblocked system [37. 38] (Fig. 9 U), and ~10–30 ns in a blocked system [164] (Fig. 9 B). The observed spectral differences include bleaching of P_{870} with a corresponding increase at 1250 nm where $P_{870}^{\cdot +}$ absorbs. Also, bleaching occurs near 540 nm where Bph absorbs. Thus, the state P^F has tentatively been identified by some authors as a $[(Bchl)_2^{\cdot +} Bph^{\cdot -}]$ radical pair [37–40, 54, 163].

There is some experimental evidence that prior to the reduction of Bph an intermediate process in which Bchl molecules are reduced occurs in the reaction center [165]. On the other hand, magnetic resonance experiments cannot determine unambiguously whether Bchl or Bph are the primary reduced donors, leaving this problem as an open question.

In the mechanism of Fig. 9 (U and B) the triplet state is not on the essential photosynthetic pathway. The radical pair state P^F of the primary charge separation occurs from a singlet state. Moreover, in most of the in-vivo experiments, the corresponding reactions indicate the involvement of the photoexcited singlet in electron reactions. Fong, however, has proposed a charge transfer in the Bchl special pair that requires a triplet state on the essential pathway [166–173]. We present Fong's scheme in Fig. 10. This scheme requires that the charge-transfer state of the special pair is initially in a triplet state which is formed by the usual ISC mechanism. Thus in Fong's model the RP state, that gives rise to the observed triplet when quinone is blocked, starts out in a triplet state where all three spin sublevels $T_{\pm 1}$ and T_0 must be populated [174]. At this point, this mechanism would not be able to explain the ESP pattern observed in bacteria. Additional ESP will develop in the radical pair when

$$\{(Bchl)_2^{\cdot +}Bph^{\cdot -}\}^{T_0} \xrightarrow[RP]{ISC} \{(Bchl)_2^{\cdot +}Bph^{\cdot -}\}^{S} \rightarrow \{(Bchl)_2^{S_0} Bph\} \quad (16)$$

In this case T_0 of the radical pair P^F would be depleted, thus giving a polarization pattern (in the P^R state) e (Iz) a (IIx) a (IIy) e (Iy) e (Ix) a (IIz) which is exactly opposite to the observed a (Iz) e (IIx) e (IIy) a (Iy) a (Ix) e (IIz). The above mechanisms are compared schematically by Eqs. (17) and (18):

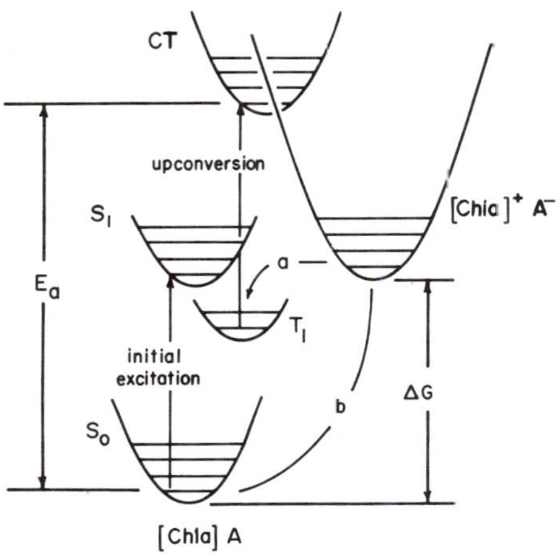

Fig. 10. Fong's model involving the triplet precursor prior to the charge separation

$$|S_0> \to |S_1> \to |S>_P^F \xrightarrow{\text{RP-ISC}} |T_0>P^F$$
$$\to |T_0>P^R \xrightarrow{\text{SO-ISC}} |S_0> \quad (17)$$

$$|S_0> \to |S_1 \xrightarrow{\text{SO-ISC}} |T_1> \to |^T CT>$$

$$\{|T_0>|T_\pm>\}P^F \to \{|T_0>|T_\pm|>\}P^R \quad (18)$$
$$\downarrow \text{RP-ISC} \qquad\qquad \downarrow \text{SO-ISC}$$
$$|S>P^F \longrightarrow |S_0>$$

The former process is carried out via a singlet precursor, and the latter process is carried out via a triplet precursor.

Regardless of the nature of P^R, i. e., its exact identity, the observed ESP pattern of bacteria is unexplained by the Fong mechanism. We emphasize that although this appears to be a problem for the Fong mechanism in bacteria, the situation in green plants may be different.

Supporting evidence for the RP-ISC mechanism in bacterial PS comes from recent studies on the triplet and fluorescence yields as a function of the external magnetic field. In general, the triplet yield has been found to decrease upon increasing the external magnetic field [56, 57, 175]. This is expected in terms of the general formulation of the RP mechanism in CIDNP and CIDEP. At zero magnetic field all three triplet components – T_x, T_y, T_z – are almost equally mixed with the photoexcited singlet S_1, whereas in an external magnetic field the singlet and triplet are mixed most efficiently via $S_1 T_0$. Evidently, in the absence of a magnetic field the triplet yield should increase as compared with the case where an external field is present. In a quantitative treatment of the field effect, the various electron transfer reaction rates (Fig. 8) together with the exchange interactions between the radical pair and the acceptor, e. g., the [Q-Fe] complex and also the hyperfine fields, should be taken into account.

4.3 Green Plant Photosynthesis

4.3.1 Photoexcited Triplet State Detection in Green Plants

With regard to ESP observed in the photoexcited triplet state in green plants preparation, the published experimental reports are scarce [111, 142, 147, 176, 177]. Although ESP has been observed in green plants, there is no conclusive evidence that the polarization effects in the triplet state originate in the RC's. On the other hand, the recent noticeable improvements in EPR time resolution [97, 98] enabled the successful application of fast optical EPR spectroscopy to the studies of spin polarization in the double states of the initial radical pairs.

Green plant photosynthesis appears much more complicated than bacterial photosynthesis. Thus, it is not surprising that the study of the triplet state in green plant systems is incomplete and is being extensively pursued. Several groups, including the pioneering work of Dutton and Leigh, reported the observation of triplets in green organism [121, 142, 147, 176, 177]. All these studies have failed to establish the significance of the triplet state in green plant photochemistry. In agreement, however, is the common feature that in all these studies the ZFS parameters D and E are very close to those reported for monomeric species of in-vitro chlorophyll *a* and *b*. Also, in most instances the observed triplet signal is quite small. Moreover, the polarization (ESP) patterns in the triplet EPR spectrum are consistent with the normal ISC common to the monomeric chlorophylls. Thus, it would appear that previous observations of the triplet state in green plants cannot at this time be interpreted in terms of reaction center photochemistry. Certainly, the question arises as to what is the biological origin of these triplets.

More recently, Frank et al. have observed the triplet polarization pattern consistent with radical pair intersystem crossing in spinach chloroplasts. This is the most direct evidence yet for a triplet state signal which arises from excitation of the reaction center [178].

4.3.2 CIDEP Studies of Photosynthesis

The role of the triplet state in photosynthesis is most difficult to establish by direct experiments. For triplet state investigations, green plant PS is more difficult to study than bacterial PS. In green plants, as yet, it has been difficult to prepare the antenna free reaction centers which is so helpful in ps optical studies. Thus, indirect observations regarding the triplet state are required at present. A promising approach was first introduced to the field of photosynthesis when Blankenship et al. [179] applied CIDEP techniques to green plants. Experimentally, the CIDEP method required EPR observations in the μs time domain. Currently, the three following methods of EPR are capable of providing spectral data at sufficiently short times: (1) Standard EPR using field modulation which has a time resolution of from 400 to 1 μsec, (2) Direct detection (e. g., no field modulation) with time resolution of ~100 nsec, and (3) Spin echo EPR with time resolution of ~6 nsec. All three techniques appear to observe polarized doublet EPR spectra shortly after narrow excitation pulses. Although each method applied to green plant PS observes polarized spectra of different features, currently the general consensus is that the polarization is the result of RP-ST mixing. Recently Dismukes et al. [180] and Friesner et al. [181] reported on experiments using $1MH_z$ field modulation, where the chloroplasts flow rate through the EPR sample cavity was changed (flowing or no flow). The system with flow exhibited both emission and absorption, but the sample with no flow exhibited only emission. These authors [179, 182] argued against the TM originally invoked to explain the polarization because of the mixed emission and absorption character observed in the samples with flow [180, 181]. Presumably, the flow rate orders the components in the photosynthetic membrane, producing primarily a Δg effect.

Since the basic observations of Sauer and his collaborators [179] this work has been confirmed by other workers [100, 183–185] and these experimental observations have induced the appearance of several theoretical treatments specific to the CIDEP process in photosynthesis [181–186].

Likewise, Hoff and Rademaker [100] observed polarized EPR signals in bacteria and again have the interpretation which is consistent with a radical pair mechanism and excited singlet chlorophyll as precursor. In this case, ps optical data rule out the participation of triplet state chlorophyll on the main stream of photosynthesis. Thus by analogy with bacteria, Friesner et al. [181] have dismissed the possibility of triplet chlorophyll in green plant photosynthesis. Thus, the question of the role of triplet chlorophyll is not yet rigorously resolved by these direct EPR experiments in green plants.

In addition, Friesner et al. [181] as well as other workers have interpreted their results in terms of an electron transport scheme which requires an initial transient chlorophyll-like acceptor [187]. Such an interpretation was invoked primarily because of the change polarization patterns, that occured when flowing the sample required a sequence of electron acceptors, $[(\text{chl chl})_{sp}^{\cdot +} A_1^{\cdot -} X]$, where A_1 is the transient intermediate electron acceptor. This explanation has been questioned by Warden [188] since the change in polarization pattern, as a function of flowing or not flowing the sample, is not present in experiments in which magnetic field modulation is absent. Since the purpose of this review is to explore the role of triplet chlorophyll, we shall not pursue further the question of the electron transport scheme. Instead, we have emphasized that the CIDEP technique is a relatively new and important method for studying the chemistry of excited triplet states of in vivo and in vitro chlorophyll. We also point out that the CIDEP method must be used with caution when establishing the role of the triplet state of chlorophyll in photosynthesis.

In conjunction with the role of the triplet state, CIDEP theory predicts that the TM produces both radicals with the same sign of polarization, and the net polarization of the mechanism is not zero. In contrast, the RP mechanism produces a net polarization of zero or, in other words, enhanced absorption and emission occur in such a way as to overal balance out. Therefore, in principle, the proof of a TM mechanism can depend on establishing the overall polarisation. In practice, this requires observation of both radicals that are involved in the reaction. Most of the time only one radical is observed, and to determine explicitly which mechanism is in operation is difficult. At least two obvious reasons for the inability to observe both radicals can be stated. Either the spin-lattice relaxation (SLR) time of one radical is so short that the polarization is quickly lost, or one of the polarized radical reacts too fast for observation. Both of these mechanisms are likely in photosynthesis. For example, the long lived $Chl^{\cdot +}$ is easily observed. The acceptor molecule, on the other hand, is not so easily observed as it has a much shorter lifetime. And since it is likely to be closer to the paramagnetic iron, it is expected to have a shorter SLR time. In order to overcome the above handicaps, the time resolution of the EPR spectrometer should be improved significantly. Indeed, recent reports on fast EPR detection have been shown to be extremely useful in its application to primary photosynthesis [98, 183].

5 Summary

The act of PS necessarily must involve singlet-excited chlorophyll. However, most mechanisms for the primary act of PS do not invoke the excited triplet states of chlorophyll implicitly or explicitly. Rigorously, radical pair states must be classified in terms of singlets and triplets. RP states are surely formed in the photoinduced electron flow system of PS. These RP states appear to arise from singlet-excited Chl states since such a mechanism explains subsequent triplet states of the special pair in bacteria as well as the CIDEP observed in green plants.

If the RP were derived from an initially excited chlorophyll involving triplets, the triplet state magnetic resonance spectroscopy of bacterial photosynthesis would be difficult to explain. Actually, we have argued in this chapter that the triplet state P^R observed in bacteria occurs after backward electron flow. In this mechanism the triplet state in bacterial PS is not on the main pathway. Also, the RP mechanism for production of P^R is consistent with optical density changes of P^R monitored as a function of magnetic field. In addition, chemical evidence has been presented that argues for production of triplet states P^R from a RP-ISC mechanism. When the electron flow close to the primary donor special pair is chemically blocked, presumably by reduction of Bph, the triplet P^R is no longer observed. Since the initial electron transfer is prevented and concomitantly the initial RP state never forms, no subsequent highly polarized triplet state should be observable according to the RP-ISC mechanism. Although in this view much evidence exists in support of the triplet state P^R as resulting from an undesirable backward flow of electrons, much structural and mechanistic information remains to be obtained from studies of the triplet state observed in bacteria.

In green plants a less definitive description of the role of the triplet state in the primary act of photosynthesis exists. In most studies, the observed triplet states have been extremely dilute, no clear relationship with the primary act has been established, or if the triplet signals observed were strong, the evidence supported the signals as arising from nonreaction center chlorophyll. In our view, the most promising technique for resolving the role of chl triplets in green plant PS is via CIDEP as discussed above.

In conclusion, we note that this chapter has not discussed the possible mechanism of selective ISC via triplet-triplet fusion or singlet-singlet fission $T_1 + T_1 \rightarrow$ $\rightarrow S_0 + S_1$ or $S_0 + S_1 \rightarrow T_1 + T_1$, respectively. Such mechanisms can possibly explain some of the data for bacterial systems. However, these fusion and fission mechanisms occur if the sum of two triplet state energies is approximately equal to one excited singlet state energy. Since, except for one report which place K = 0.5 in the relation $\Delta ES_1 \rightarrow S_0 = K\Delta ES_1 \rightarrow S_0$, most studies report a significantly larger value of K = ~0.7, we have not considered the multiple state mechanism. We suggest, therefore, that more direct measurements of the triplet state energy relative to the singlet manifold are of importance in general as well as in particular for settling this question.

Acknowledgment. The research described herein was supported by the Office of Basic Energy Sciences of the Department of Energy. This is Document No. SR–47 from the Notre Dame Radiation Laboratory.

References and Notes

1. Govindjee, E.: Bioenergetics of Photosynthesis. New York: Academic Press 1975
2. Van Niel, C.B.: Photosynthesis of bacteria. In: Contributions to Marine Biology, p. 158. Palo Alto, CA: Stanford University Press 1930
3. Van Niel, C.B.: Bacteriol. Rev. **8**, 10 (1944)
4. For a recent review see Clayton, R.K., in: The Photosynthetic Bacteria, Clayton, R.K., Sistrom, W.R (eds.), New York: Plenum Press 1977
5. Goedheer, J.C.: Spectral and redox properties of bacteriochlorophyll in its natural state, Biochim. Biophys. Acta **38**, 389–399 (1960)
6. Kuntz, I.D., Loach, P.A., Calvin, M.: Absorption changes in bacterial chromatophores. Biophys. J. **4**, 227–249 (1964)
7. Doring, G., Witt, H.T.: Proc. 3rd Int. Congr. Photosyn. Res. 1971, 39 (1972)
8. Reed, D.W., Ke, B.: Spectral properties of reaction center preparations from *rhodopseudomonas spheroides.* J. Biol. Chem. **248**, 3041–3045 (1973)
9. Wraight, C.A., Clayton, R.K.: The absolute quantum efficiency of Bchl photooxidation in Rx. centres of *Rhodopseudomonas spheroides*. Biochim. Biophys. Acta **333**, 246–260 (1973)
10. Commoner, B., Heise, J.J., Townsend, J.: Light-induced paramagnetism in chloroplasts. Proc. Natl. Acad. Sci. USA **42**, 710–718 (1956)
11. Sogo, P., Jost, M., Calvon, M.: Free radical production in photosynthesizing systems. Radiat. Res. Suppl. I, 511–518 (1959)
12. Loach, P.A., Androes, G.M., Maksim, A.F., Calvin, M.: Variation in electron paramagnetic resonance signals of photosynthetic systems with the redox level of their environment. Photochem. Photobiol. **2**, 443–454 (1963)
13. Kohl, D.H., Townsend, J., Commoner, B., Crespi, H.L., Dougherty, R.C., Katz, J.J.: Effect of isotopic substitution on electron spin resonance signals in photosynthetic organisms. Nature **206**, 1105–1110 (1965)
14. Loach, P.A., Sekura, D.L.: Comparison of decay kinetics of photo-produced absorbance, EPR and luminescence changes in chromatophores of *Rhodospirillum rubrum.* Photochem. Photobiol. **6**, 381–393 (1967)

15. Loach, P.A., Sekura, D.C.: Primary photochemistry and electron transport in *Rhodospirillum rubrum.* Biochemistry 7, 2642–2649 (1968)
16. Mcelroy, J.D., Feher, G., Mauzerall, D.C.: On the nature of the free radical formed during the primary process of bacterial photosynthesis. B.B.A. **172**, 180–183 (1969)
17. Fuhrhop, J.H., Mauzerall, D.C.: The one-electron oxidation of metalloporphyrins. J. Am. Chem. Soc. **91**, 4174–4181 (1969)
18. Bolton, J.R., Clayton, R.K., Reed, D.W.: An identification of the radical giving rise to the light-induced electron spin resonance signal in photosynthetic bacteria. Photochem. Photobiol. **9**, 209–218 (1969)
19. Fajer, J., Borg, D.C., Forman, A., Dolphin, D., Felton, R.H.: π-cation radicals and cations of metalloporphyrins. J. Am. Chem. Soc. **92**, 3451–3459 (1970)
20. Borg, D.C., Fajer, J., Felton, R.H., Dolphin, D.: The π-cation radical of chlorophyll a. Proc. Natl. Acad. Sci. USA **67**, 813–820 (1970)
21. Norris, J.R., Uphasus, R.A., Crespi, H.L., Katz, J.J.: Electron spin resonance of chlorophyll and the origin of signal I in photosynthesis. Proc. Natl. Acad. Sci. USA **68**, 625–628 (1971)
22. McElroy, J.D., Feher, G., Mauzerall, D.C.: Characterization of primary reactants in bacterial photosynthesis I. comparison of the light-induced EPR signal (g = 2.0026) with that of a bacteriochlorophyll radical. Biochem. Biophys. Acta **267**, 363–374 (1972)
23. Norris, J.R., Uphaus, R.A., Katz, J.J.: Electron spin resonance in ^{13}C-labeled algae. Biochim. Biophys. Acta **275**, 616–168 (1972)
24. Warden, J.T., Jr., Bolton, J.R.: Simultaneous quantitative comparison of the optical changes at 700 nm (P700) and electron spin resonance signals in system I of green plant photosynthesis. J. Am. Chem. Soc. **95**, 6435–6436 (1973)
25. Druyan, M.E., Norris, J.R., Katz, J.J.: Electron spin resonance of (^{25}Mg) chlorophyll a. J. Am. Chem. Soc. **95**, 1682–1683 (1973)
26. Katz, J.J., Norris, J.R.: Chlorophyll and light energy transduction in photosynthesis. In: Current Topics in Bioenergetics, Sanadi, D.R., Packer, L. (eds.), Vol. 5, pp. 41–75. New York: Academic Press 1973
27. McElroy, J.D., Mauzerall, D.C., Feher, G.: Characterization of primary reactants in bacterial photosynthesis II. Kinetic studies of the light-induced EPR signal (g = 2.006) and the optical absorbance changes at cryogenic temperatures. Biochim. Biophys. Acta **333**, 261–277 (1974)
28. Fajer, J., Borg, D.C., Forman, A., Felton, R.H., Dolphin, D., Vegh, L.: The cation radicals of free base and zinc bacteriochlorin, bacteriochlorophyll and bacteriopheophytin. Proc. Natl. Acad. Sci. USA **71**, 994–998 (1974)
29. Norris, J.R., Druyan, M.E., Katz, J.J.: Electron nuclear double resonance of bacteriochlorophyll free radical in vitro and in vivo. J. Am. Chem. Soc. **95**, 1680–1682 (1973)
30. Feher, G., Hoff, A.J., Isaacson, R.A., McElroy, J.D.: Abstr. Biophys. Soc. **61a**, 17 (1973)
31. Norris, J.R., Scheer, H., Druyan, M.E., Katz, J.J.: "An electron-nuclear double resonance (ENDOR) study of the special pair model for photo-reactive chlorophyll in photosynthesis. Proc. Natl. Acad. Sci. USA **71**, 4897–4900 (1974)
32. Norris, J.R., Scheer, H., Katz, J.J.: Models for antenna and reaction center chlorophylls. Ann. N.Y. Acad. Sci. **244**, 260–280 (1975)
33. Feher, G., Hoff, A.J., Isaacson, R.A., Ackerson, L.C.: ENDOR experiments on chlorophyll and bacteriochlorophyll in vitro and in the photosynthetic unit. Ann. N.Y. Acad. Sci. **244**, 239–259 (1975)
34. Scheer, H., Katz, J.J., Norris, J.R.: Proton-electron hyperfine coupling constants of the chlorophyll *a* cation radical by ENDOR spectroscopy. J. Am. Chem. Soc. **99**, 1372–1381 (1977)
35. Katz, J.J., Dougherty, R.C., Crespi, H.L., Strain, H.H.: Nuclear magnetic resonance studies of plant biosynthesis. A bacteriochlorophyll isotope mirror experiments. J. Am. Chem. Soc. **88**, 2856–2857 (1966)
36. Dougherty, R.C., Crespi, H.L., Strain, H.H., Katz, J.J.: Nuclear magnetic resonance studies of plant biosynthesis bacteriochlorophyll. J. Am. Chem. Soc. **88**, 2854–2855 (1966)
37. Rockley, M.G., Windsor, M.W., Cogdell, R.J., Parson, W.W.: Picosecond detection of an intermediate in the photochemical reaction of bacterial photosynthesis. Proc. Natl. Acad. Sci. USA **72**, 2251–225 (1975)

38. Kaufman, K.J., Dutton, P.L., Netzel, T.L., Leigh, J.S., Rentzepis, P.M.: Picosecond kinetics of events leading to reaction center bacteriochlorophyll oxidation. Science **188**, 1301–1304 (1975)
39. Fajer, J., Brune, D.C., David, M.A., Forman, A., Spaulding, L.D.: Primary separation in bacterial photosynthesis: Oxidized chlorophylls and reduced pheophytin. Proc. Natl. Acad. Sci. USA **72**, 4956–4960 (1975)
40. Dutton, P.L., Kaufmann, K.J., Chance, B., Rentzepis, P.M.: Picosecond kinetics of the 1250 nm band of the *Rhodopseudomonas sphaeroides* reaction center: The nature of the primary photochemical intermediary state. FEBS Lett. **60**, 275–280 (1975)
41. Parson, W.W.: 6th Int. Biophysics Congr. Kyoto, Japan (1978)
42. Fenton, J., Pelling, M., Govindjee, Kaufmann, K.: Chlorophyll *a* cation formation in photochemical system I of green plant photosynthesis occurs within ten picoseconds. FEBS Lett. **100**, 1–4 (1979)
43. Loach, P.A., Hall, R.L.: The question of the primary electron acceptor in bacterial photosynthesis. Proc. Natl. Acad. Sci. USA **69**, 786–790 (1972)
44. Feher, G., Okamura, M.Y., McElroy, J.D.: Identification of an electron acceptor in reaction centers of *Rhodopseudomonas sphaeroides* by EPR. Biochem. Biophys. Acta **267**, 222–226 (1972)
45. Vermeglio, A.: Secondary electron transfer in reaction centers of *Rhodopseudomonos sphaeroides.* Out-of-phase periodicity of two for the formation of ubisemiquinone and fully reduced uleiquinone. Biochem. Biophys. Acta **459**, 516–524 (1977)
46. Feher, G., Isaacson, R.A., McElroy, J.D., Ackerson, L.C., Okamura, M.Y.: On the question of primary acceptor in bacterial photosynthesis: Manganese substituting for iron in reaction centers of *Rhodopseudomonas sphaeroides*, R-26 strain. Biochim. Biophys. Acta **368**, 135–139 (1974)
47. Debrunner, P.C., Schultz, C.E., Feher, G., Okamura, M.Y.: Biophys. J. **15**, 226a, Abstr. TH-PM-L12 (1975)
48. Bolton, J.R., Cost, K.: Flash photolysis – ESR: A kinetic study of endogenous light induced free radicals in reaction center preparations from *Rhodopseudomonas sphaeroides.* Photochem Photobiol. **18**, 417–421 (1973)
49. Prince, R.C., Thronber, J.P.: A novel EPR signal associated with the "primary" electron in isolated photochemical reaction centers of *Rhodospirillum rubrum.* FEBS Lett. **81**, 233–237 (1977)
50. Wraight, C.A.: FEBS Lett. **93**, 283–288 (1978)
51. Barouch, Y., Clayton, R.K.: Biochem. Biophys. Acta **462**, 785–788 (1978)
52. Vermeglio, A., Clayton, R.K.: Kinetics of electron transfer between the primary and secondary electron acceptor in reaction centers from *Rhodopseudomonas sphaeroides.* Biochem. Biophys. Acta **461**, 159–165 (1977)
53. Wraight, C.A.: Electron acceptors of photosynthetic bacterial reaction centers. Direct observation of oscillatory behavior suggesting two closely equivalent ubiquinones. Biochem. Biophys. Acta **459**, 525–531 (1977)
54. Thurnauer, M.C., Katz, J.J., Norris, J.R.: The triplet state in bacterial photosynthesis: Possible mechanisms of the primary photoact. Proc. Natl. Acad. Sci. USA **72**, 3270–3274 (1975)
55. Clayton, R.K., Yamamoto, T.: Photochemical quantum efficiency and absorption spectra of reaction centers from *Rhodopseudomonas sphaeroides* at low temperature. Photochem. Photobiol. **24**, 67–75 (1976)
56. Blankenship, R.E., Schaafsma, T.J., Parson, W.W.: Magnetic field effects on radical pair intermediates in bacterial photosynthesis. Biochem. Biophys. Acta **461**, 297–305 (1977)
57. Hoff, A.J., Rademaker, H., Van Grondelle, R., Duysens, L.N.M.: On the magnetic field dependence of the yield of the triplet state. Biochim. Biophys. Acta **460**,
58. Netzel, T.L., Rentzepis, P.M., Tiede, D.M., Prince, R.C., Dutton, P.L.: Effect of reduction of the reaction center intermediate upon the picosecond oxidation reaction of bacteriochlorophyll dimer in *Chromatium vinosurn* and *Rhodopseudomonas vinidis.* Biochim. Biophys. Acta **460**, 467–479 (1977)
59. Sauer, K.H., Blankenship, R.E., Dismukes, G.C., McGuire, A.: Biophys. J. Abstr. F-AM-F2 (1977)

60. Dismukes, C., Friesner, R., Sauer, K.: Biophys. J. Abstr. F-AM-F3 (1977)
61. For general reviews see (a) Lepley, A.R., Closs, G.L. (eds.): Chemically Induced Magnetic Polarization. New York: Wiley 1973. (b) Muus, L.T., Atkins, P.W., McLauchlan, K.A., Pedersen, J.B. (eds.): Chemically Induced Magnetic Polarization. Boston: Reidel 1977
62. Van der Meulen, D.L., Govindjee: Is there a triplet state in photosynthesis? J. Sci. Ind. Res. **32**, 62–69 (1973)
63. Bolton, J.R., Warden, J.T.:, in: Creation and Detection of the Excited State, Vol. 2, Ware, W.R., (ed.), p. 63. New York: Dekker 1974
64. Parson, W.W., Cogdell, R.J.: The primary photochemical reaction of bacterial photosynthesis. Biochim. Biophys. Acta **416**, 105–149 (1975)
65. Loach, P.A., in: Progress in Bioorganic Chemistry, Vol. 4, Kaiser, E.T. (ed.), p. 90. New York: Wiley 1976 (and references therin)
66. Norris, J.R.: Triplet state and photosynthesis. Photochem. Photobiol. **23**, 449–450 (1976)
67. Loach, P.A.: Primary photochemistry in photosynthesis. Photochem. Photobiol. **26**, 87–94 (1977)
68. Sauer, K.: Photosynthetic membranes. Acc. Chem. Res. **11**, 257–264 (1978)
69. Sauer, K.: Photosynthesis – the light reactions. Ann. Rev. Phys. Chem. **30**, 155–178 (1979)
70. Hoff, A.J.: Applications of ESR in photosynthesis. Phys. Reports **54**, 75–200
71. Levanon, H., Norris, J.R.: The photoexcited triplet state and photosynthesis. Chem. Rev. **78**, 185–198 (1978)
72. For a general textbook, see Birks, J.B.: Photophysics of Aromatic Molecules. New York: Wiley-Interscience 1970
73. For a general textbook and references, see McGlynn, S.D., Azumi, J., Kinoshita, M.: Molecular Spectroscopy of the Triplet State. Englewood Cliffs, NJ: Prentice-Hall 1969
74. For a complete analysis of the triplet spin-Hamiltonian, see Hutchison, C.A., Jr., in: The Triplet State, Zahlan, A.B. (ed.), p. 63. New York: Cambridge University Press 1967
75. Hausser, K.H., Wolf, H.C.: Optical spin polarization in molecular crystals. Adv. Mag. Res. **8**, 85–121 (1976)
76. Van Dorp, W.H., Schaafsma, T.J., Soma, M., Van der Waals, J.H.: Investigation of the lowest triplet state of free base porphin by microwave induced changes in its fluorescence. Chem. Phys. Lett. **21**, 221–225 (1973)
77. Kwiram, A.L.: Optical detection of paramagnetic resonance in phosphorescent triplet state. Chem. Phys. Lett. **1**, 272–275 (1967)
78. Schmidt, J., Hasselmann, I.A.M., de Groot, M.S., Van der Waals, J.H.: Optical detection of electron resonance transitions in phosphorescent quinoxaline. Chem. Phys. Lett. **1**, 434–436 (1967)
79. Sharnoff, M.: ESR-produced modulation of triplet phosphrescence. J. Chem. Phys. **46**, 3263–3204 (1967)
80. Clarke, R.H., Hayes, J.M.: Microwave induced triplet-triplet absorption in organic molecules. J. Chem.Phys. **59**, 3113–3118 (1973)
81. El-Sayed, M.A.: Double resonance and the properties of the lowest excited triplet state of organic molecules. Annu. Rev. Phys. Chem. **26**, 235–258 (1975)
82. Levanon, H., in: Multiple Electronic Resonance, Dorio, M., Freed, J.H. (eds.), Chap. 13 and references therein. New York: Plenum 1979
83. Boxer, S.G., Closs, G.L.: NMR of photoexcited triplet states. I. The measurement of the rate of degenerate singlet-triplet exchange for anthracene in solution. J. Am. Chem. Soc. **97**, 3268–3270 (1975)
84. Cocivery, M.: Triplet energy transfer in solution studied by NMR spectroscopy. Chem. Phys. Lett. **2**, 529–532 (1968)
85. In contrast to the OMR method where the measuring temperatures is very critical ($<$4.2K), ordinary triplet EPR spectra can be detected over a wide range of low temperature (for EPR at room temperature few examples have been also reported; see, for example, Frankevich, E.L. et al.: Magnetic resonance of short-lived triplet exciton pairs detected by fluorescence modulation at room temperature. Chem. Phys. Lett. **47**. 304–308 (1977); Kim, S.S., Weissman, S.I.: Detection of transient EPR. J. Mag. Res. **24**, 167–169 (1976)

86. For a discussion and references on EPR detection of randomly oriented triplets, see ref. [73].
87. Weissmen, S.I.: On detection of triplet molecules in solution by ESR; J. Chem. Phys. **29**, 1189–1190 (1958)
88. For an exception see Gonclaves, A.M.P., Burgner, R.P.: EPR of the triplet state of porphyrins oriented in phtalic acid single crystals. J. Chem. Phys. **61**, 2975–2978 (1974)
89. El-Sayed, M.A., Siegel, S.: Method of magnetophotoselection of the lowest excited triplet state of aromatic molecules. J. Chem. Phys. **44**, 1416–1423 (1966)
90. Siegel, S., Judeikis, H.S.: A magnetophotoselection study of the polarization of the absorption bands of some structurally related hydrocarbons and heterocyclic molecules. J. Phys. Chem. **70**, 2205–221 (1966). A different method for ZFS sign determination has been pointed out by El-Sayed. This method utilizing the analysis of the ESP patterns combined with the determination of the sign of the g factor
91. Levanon, H., Weissman, S.I.: Spin polarization in the photoexcitation of the triplet state of pherazine in a rigid glass. Isr. J. Chem. **10**, 1–5 (1972)
92. Levanon, H., Vega, S.: Analysis of the transient EPR signals in the photoexcited triplet state. Application to porphyrin molecules. J. Chem. Phys. **61**, 2265–2274 (1974)
93. Schwoerer, M., Sixl, S.: Optische spin-polarisation in Triplet Zustand von Naphthalin. Z. Naturforsch. **24A**, 952–967 (1969)
94. Kleibeuker, J.F., Schaafsma, T.J.: Spin plarization in the lowest triplet state of chlorophyll. Cehm. Phys. Lett. **29**, 116–122 (1974)
95. Winscom, C.J.: Analysis of spin polarization transients in periodically excited photoexcited triplet states. Z. Naturforsch. **309**, 571–502 (1975)
96. Van der Bent, S.J., de Jager, A., Schaafsma, T.J.: Rev. Sci. Instrum. **47**, 117–121 (1976)
97. Felix, C.C., Kim, S.S., Weissman, S.I.: Temperature dependence of kinetics of triplet formation in phenazine analysis by analog computer. Chem. Phys. Lett. **48**, 29–30 (1977)
98. Trifunac, A.D., Thrunauer, M.C., Norris, J.R.: Submicrosecond time-resolved EPR in laser photolysis. Chem. Phys. Lett. **57**, 471–473 (1978)
99. Trifunac, A.D., Norris, J.R.: Nanosecond time-resolved EPR spectroscopy. EPR time profile via electron spin echo. CIDEP. Chem. Phys. Lett. **59**, 140–142 (1978)
100. Schulten, K., Staerk, H., Weller, A., Werner, H.–J., Nickle, B.: Z. Phys. Chem. N.F. **101**, 371–390 (1976)
101. Hoff, A.J., Rademaker, H.: Light-induced magnetic polarization in photosynthesis. In: Chemically Induced Magnetic Polarization, Muus, L.T. et al. (eds.), pp. 399–404. Boston: Reidel 1977
102. Haberkorn, R.: Theory of magnetic field modulation of radical recombination reactions I. Chem. Phys. **19**, 165–179 (1976)
103. Haberkorn, R.: Density matrix description of spin selective radical pair reactions. Mol. Phys. **32**, 1491–1493 (1976)
104. Werner, H.–J., Schulten, Z., Schluten, K.: Theory of the magnetic field modulated geminate recombination of radical ion-pairs in polar solvents: Application to the pyrene- N,N-demethylaniline system. J. Chem. Phys. **67**, 646–663 (1977)
105. Brocklehurst, B.: Yields of excited states from geminate recombination of hydrocarbon radical ions. Chem. Phys. Lett. **28**, 357–360 (1974)
106. Brocklehurst, B.: Chem. Phys. Lett. **29**, 635–635 (1974)
107. Brocklehurst, B.: J. Chem. Soc. Faraday Trans. II, **72**, 1869–1884 (1976)
108. Wong, S.K., Hutchison, D.A., Wan, J.K.S.: Chemically induced dynamic electron polarization. II. A general theory for radicals produced ions photochemical reactions of excited triplet carbonyl compounds. J. Chem. Phys. **58**, 985–989 (1973)
109. Atkins, P.W., Evans, G.T,: Electron spin polarization in a rotating triplet. Mol. Phys. **27**, 1633–1644 (1974)
110. Wan, J.K.S., Elliott, A.J.: Chemically induced dynamic magnetic polarization in photochemistry. Acc. Chem. Res. **10**, 161–166 (1977)
111. Dobbs, A.J.: Experimental observations of chemically induced dynamic electron polarization. Mol. Phys. **30**, 1073–1084 (1975)

112. Adrain, F.J.: Role of diffusion-controlled reaction in chemically induced nuclear spin polarization. J. Chem. Phys. **53**, 3374–3375 (1979); and: II. General theory and comparison with experiment. J. Chem. Phys. **54**, 3912–3917 (1971); Theory of anomalous ESR spectra of free radicals in solution role of diffusion-controlled separation and reencounter of radical pairs. J. Chem. Phys. **54**, 3918–3923 (1971)
113. Monchick, L., Adrain, F.J.: On the theory of chemically induced electron polarization (CIDEP); vector model and an asymptotic solution. J. Chem. Phys. **68**, 4376–4383 (1978)
114. Closs, G.L.: Chemically induced dynamic nuclear polarization. A tool in radiation chemistry. Special lectures. Proc. 23rd Int. Congr. Pure Appl. Chem. Spec. Lett. **4**, 19– (1971)
115. For recent textbook on porphyrins, see Smith, K.M. (ed.): Porphyrins and Metalloporphyrins. Amsterdam: Elsevier 1975
116. Norris, J.R., Uphaus, R.A., Katz, J.J.: ESR of triplet states of chlorophylls a, b, c_1, c_2 and bacteriochlorophyll a. Application of ZFS and electron spin polarization to photosynthesis. Chem. Phys. Lett. **31**, 157–161 (1975)
117. Chan, I.Y., Van Dorp, W.G., Schaafsma, T.J., Van der Waals, J.H.: The lowest triplet state of Zn porphin I. Modulation of its phosphorescence by microwaves. Mol. Phys. **22**, 741–751 (1971) and the lowest triplet state of Zn porphin II. Investigation of its dynamics by microwave induced delayed phosphorescence. Mol. Phys. **22**, 753–760 (1971); also, the reader is referred to in W. G. Van Dorp, Thesis, Leiden (1975)
118. Kleibeuker, J.F., Platenkamp, R.J., Schaafsma, R.J.: Optically induced electron spin polarization in the triplet state of chlorophyll and its model compounds. Chem. Phys. Lett. **41**, 557–561 (1976); also, the reader is referred to J.F. Kleibeuker, Thesis, Wageningen (1977)
119. Clarke, R.H., Connors, R.E., Schaafsma, T.J., Kleibeuker, J.K., Platenkamp, R.J.: The triplet state of chlorophylls. J. Am. Chem. Soc. **98**, 3674–3677 (1976)
120. Clarke, R.H., Conners, R.E., Frank, H.A., Hoch, J.C.: Investigation of the structure of the reaction center in photosynthetic systems by optical detection of triplet-state magnetic resonance. Chem. Phys. Lett. **45**, 523–528 (1977)
121. Nissani, E., Scherz, A., Levanon, H.: The photoexcited triplet state of tetraphenyl chlorin, magnesium tetraphenyl porphyrin and whole cells of *Chlamydomonas Reinhardi.* A lieght modulation – EPR study. Photochem. Photobiol. **25**, 93–101 (1977)
122. Thurnauer, M.C., Norris, J.R.: The ordering of the zero field triplet spin sublevels in the chlorophylls. A magnetophotoselection study. Chem. Phys. Lett. **47**, 100–105 (1977)
123. Van Dorp, W.G., Schoemaker, W.H., Soma, M., Van der Waals, J.H.: Mol. Phys. **30**, 1701 (1975). However, these authors show that a straight extrapolation may result in erroneous results because of the nonlinear dependence at low light intensities
124. Weiss, C. Jr.: The Pi electron structure and absorption spectra of chlorophyll in solution. J. Mol. Spectrosc. **44**, 37–80 (1972)
125. Thurnauer, M.C., Norris, J.R.: Magnetophotoselection applied to the triplet state observed by EPR in photosynthetic bacteria. Biochim. Biophys. Res. Commun. **73**, 503–506 (1976)
126. Metz, F., Friedrich, S., Hohlneicher, G.: What is the leading mechanism for the nonradiative decay of the lowest triplet state of aromatic hydrocarbons. Chem. Phys. Lett. **6**, 353–358 (1972)
127. Bowman, M.K.: Intersystem crossing in photosynthetic pigments. Chem. Phys. Lett. **48**, 17–21 (1977)
128. Lhoste, J.M., Helen, C., Ptak, M., in: The Triplet State, Zahlan, A.B. (ed.), p. 487. New York: Cambridge University Press 1967
129. Gribova, Z.P., Kayushin, L.P.: ESR of the excited triplet state of biologically important molecules. Usp. Klim. **41**, 287–320 (1972)
130. Hoffman, B.M.: Triplet state electron paramagnetic resonance studies of zinc porphyrins and zinc substituted hemoglobins and myoglobins. J. Am. Chem. Soc. **97**, 1688–1694 (1975)
131. Langhoff, S.R., Davidson, E.R., Goutermann, M., Leenstra, W.R., Kwiram, A.L.: Zero field splitting of the triplet state of porphyrins. J. Chem. Phys. **62**, 169–176 (1975)
132. Scheidt, W.R., Kastner, M.E., Hatano, K.: Steriochemistry of the toluene solvates of $\alpha, \beta, \gamma, \delta$-tetraphenylporphinatozinc (II). Inorg. Chem. **17**, 706–710 (1978)

133. Scherz, A., Levanon, H.: International Symposium on Magnetic Resonance in Chemistry, Biology and Physics in honor of Professor S.I. Weissman, Argonne Chicago June 1979; and Scherz, A., Levanon, H.: The photoexcited triplete state and chemical dynamics in frozen solutions of metalloporphyrins. An optical EPR study J. Phys. Chem. **84**, 324–336 (1980)
134. Harel, Y., Manassen, J., Levanon, H.: Photoreduction of phorphyrins to chlorins by tertiary amines in the visible spectral range. Optical and EPR studies. Photochem. Photobiol. **23**, 337–341 (1976)
135. Holten, D., Gouterman, M., Parson, W.W., Windsor, M.W., Rockley, M.G.: Electron transfer from photoexcited singlet and triplet bacteriopheophylin. Photochem. Photobiol. **23**, 415–413 (1976)
136. Gouterman, M., Holten, D.: Electron transfer from photoexcited singlet and triplet bacteriopheophytin. II, Theoretical. Photochem. Photobiol. **25**, 85–92 (1977)
137. Boxer, S.G., Closs, G.L.: A covalently bound dimeric derivative of pyrochlorophyllide a. A possible model for reaction center chlorophyll. J. Am. Soc. **98**, 5406–5408 (1976)
138. Anton, A.J., Kwong, J., Loach, P.A.: Synthesis of covalently linked porphyrin dimers and trimers. J. Heterocycl. Chem. **13**, 717–725 (1976)
139. Wasielewski, M.R., Studier, M.H., Katz, J.J.: Covalently linked chlorophyll a dimers. A biomimetic model of special pair chlorophyll. Proc. Natl. Acad. Sci. USA **73**, 4282–4286 (1976)
140. Wasielewski, M.R., Smith, U.H., Cope, B.T., Katz, J.J.: A synthetic biomimetic model of special pair bacheriochlorophyll a. J. Am. Chem. Soc. **99**, 4172–4173 (1977)
141. Pellin, M.J., Kaufmann, K.J., Wasielewski, M.R.: In vitro duplication of the primary light induced charge separation in purple photosynthetic bacteria. Nature **278**, 54–55 (1979)
142. Dutton, D.L., Leigh, J.S., Seibert, M.: Primary process in photosynthesis: In situ ESR studies on the light-induced oxidized and triplet state of reaction center bacteriochlorophyll. Biochim. Biophys. Res. Commun. **46**, 406–413 (1972)
143. Leigh, J.S., Dutton, D.L.: The primary electron acceptor in photosynthesis. Biochim. Biophys. Res. Commun. **46**, 414–421 (1972)
144. Leigh, J.S., Dutton, D.L.: Reaction center bacteriochlorophyll triplet state: Redox potential dependence and kinetics. Biochim. Biophys. Acta **357**, 67–77 (1974)
145. Dutton, D.L., Leigh, J.S., Reed, D.W.: Primary events in photosynthetic reaction center from *Rhodopseudomonas sphaeroides*, strain R26, triplet and oxidized states of bacteriochlorophyll and the identification of the primary electron acceptor. Biochim. Biophys. Acta **292**, 654–664 (1973)
146. Optical evidence for triplet states in bacteria was also provided by Seibert, M., Devault, D.: Photosynthetic reaction center transients P435 and P434 in chromatium vinosum strain D. Biochim. Biophys. Acta **253**, 396–411 (1971)
147. Uphaus, R.A., Norris, J.R., Katz, J.J.: Triplet state in photosynthesis. Biochim. Biophys. Res. Commun. **61**, 1057–1063 (1974)
148. Okamura, M.Y., Isaacson, R.A., Feher, G.: The primary acceptor in bacterial photosynthesis: The obligatory role of ubiquinone in photoactive reaction centers of *Rhodopseudomonas sphaeroides*. Proc. Natl. Acad. Sci. USA **72**, 3491–3495 (1975)
149. Krasnovskii, A.A. Jr., Lebedev, N.N., Litvin, F.F.: Spectral characteristics of phosphorescence of chlorophylls and pheophytins A and B. Dokl. Akad. Nauk. SSSR **216**, 1406–1409 (1974)
150. Mau, A.W.H., Puza, M.: Phosphorescence of chlorophylls. Photochem. Photobiol. **25**, 601–603 (1977)
151. Connolly, J.S., Gorman, D.S., Seely, G.R.: Ann. N.Y. Acad. Sci. **206**, 649–669 (1973)
a. However, a value of $K \simeq 0$ exists. b. Boxer, S., Closs, G.L., private communication.
152. Parson, W.W., Clayton, R.K., Cogdell, R.J.: Excited states of photosynthetic reaction centers at low redox potentials. Biochim. Biophys. Acta **387**, 265–278 (1975)
153. Wraight, C.A., Leigh, J.S., Dutton, D.L., Clayton, R.K.: The triplet state of reaction center bacteriochlorophyll: Determination of a relative yield. Biochem. Biophys. Acta **333**, 401–408 (1974)
154. Prince, R.C., Leigh, J.S., Dutton, D.L.: Thermodynamic properties of the reaction center of *Rhodopseudomonas viridis*. In vivo measurements of the reaction center bacteriochlorophyll-primary acceptor intermediate electron carrier. Biochim. Biophys. Acta **440**, 622–636 (1976)

155. Hoff, A.J.: Kinetics of population and depopulation of the components of the photoinduced triplet state of the photosynthetic bacteria *Rhodosperillum rubrum, Rhodopseudomonas sphaeroides* (wild type) and its mutant R-26 as measured by ESR in zero field. Biochim. Biophys. Acta **440**, 765–771 (1976)
156. Clarke, R.H., Connors, J.R., Norris, J.R., Thurnauer, M.C.: Optically detected zero-field magnetic resonance studies of the photoexcited triplet state of the photosynthetic bacterium *Rhodospirillum rubrum.* J. Am. Chem. Soc. **97**, 7178–7179 (1975)
157. Clarke, R.H., Connors, R.E.: Optically detected zero-field triplet state magnetic resonance in photosynthetic bacteria. Chem. Phys. Lett. **42**, 69–72 (1976)
158. Clarke, R.H., Hofeldt, R.H.: Optically detected zero-filed magnetic resonance studies of the photoexcited triplet state of chlorophyll a and b. J. Chem. Phys. **61**, 4582–4587 (1974)
159. Van der Bent, S.J., Schaafsma, T.J., Goedheer, J.C.: Detection of triplet states in algae by zero field resonance. Biophys. Biochem. Res. Commun. **71**, 1147–1152 (1976)
160. At a constant separation because of the $1-3\cos^2\theta$ factor, the end-to-end interaction is twice as strong and opposite in sign to the side-by-side interaction.
161. Norris, J.R., Thurnauser, M.C.. in: Proceeding 3rd International Seminar on Energy Transfer in Condensed Matter. Prague, Czechoslavakia, June 1976. Fiala, J. (ed.), Univerzita Karlova Praha, p. 12, 1978
162. Norris, J.R., Thurnauer, M.C.: Biophys. J. **16**, 224a (1976), Abstract F-PM-Dll
163. We use the notation of $[(Bchl)_2{}^{\cdot +}Bph^{\cdot -}]$ pair for generalization. The reader should be aware of the fact that the identity of the anion half is not completely settled.
164. Parson, W.W.: Flash-induced absorbance changes in *Rhodospirillum rubrum* chromatophores. Biochim. Biophys. Acta **131**, 154– (1967)
165. Shuvalov, V.A., Klevanik, A.V., Sharkov, A.V., Matveetz, Ju.A., Krukov, D.G.: Picosecond detection of Bchl-800 as an intermediate electron carrier between selectively excited P_{870} and bacteriopheophytin in *Rhodospirillum rubrum* reaction centers. FEBS Lett. **91**, 135–139 (1978)
166. Fong, F.K.: Energy upconversion theory of the primary photochemical reaction in plant photosynthesis. J. Theor. Biol. **46**, 160, 407–420 (1974)
167. Fong, F.K.: Molecular symmetry and the interaction in photosynthetic primary events. Appl. Phys. **6**, 151–166 (1975)
168. Fong, F.K.: Energy upconversion and the minimum quantum requirement in photosynthesis. J. Am. Chem. Soc. **98**, 7840–7843 (1976)
169. Fong, F.K.: Molecular basis for the photosynthesis primary process. Proc. Natl. Acad. Sci. USA **71**, 3692–3695 (1974)
170. Fong, F.K., Koester, V.J.: Ester and keto carbonyl linkages in chlorophyll a pyrochlorophyll a and protochlorophyll a. J. Am. Chem. Soc. **97**, 6888–6890 (1975)
171. Fong, F.K.: Bonding interactions in anhydrons and hydrated chlorophyll a. J. Am. Chem. Soc. **97**, 6890–6892 (1975)
172. Winograd, N., Shepard, A., Karweik, D.H., Koester, V.J., Fong, F.K.: X-ray photoelectron spectroscopic studies of the thermal stability of chlorophylla monohydrate. J. Am. Chem. Soc. **98**, 2369–2370 (1976)
173. Fong, F.K., Winograd, N.: In vitro conversion after the primary light reaction in photosynthesis. Reversible photogalvanic effects of chlorophyll-quinhydron half cell reaction. J. Chem. Soc. **98**, 2287–2289 (1976)
174. Fong interprets the observed triplet as significantly invelved with P800 on grounds of optical changes of P800 during the lifetime of P^R.
175. Clarke, R.H., Connors, R.E., Keegan, J.: Magnetic field effect on the low temperature triplet state population of an organic molecule. J. Chem. Phys. **66**, 358–359 (1977)
176. Hoff, A.J., Govindjee, Romjin, J.C.: Electron spin resonance in zero magnetic field of triplet state of chloroplasts and subchloroplasts particles. FEBS Lett. **73**, 191–196 (1977)
177. Hoff, A.J., Van der Waals, J.H.: Zero field resonance and spin alignment of the triplet state of chloroplasts at 2˚K. Biochim. Biophys. Acta **423**, 615–620 (1976)
178. Frank, H., McLean, M., Sauer, K.: Triplet state in photosystem I of spinach chloroplast and subchloroplst particles. Proc. Natl. Acad. Sci. USA **76**, 5124–5128 (1979)

179. Blankenship, R., McGuire, A., Sauer, K.: Chemically induced dynamic electron polarization in chloroplasts at room temperature: Evidence for triplet state participation in photosynthesis. Proc. Natl. Acad. Sci. USA **72**, 4943–4947 (1975)
180. Dismukes, G.C., McGuire, A., Blankenship, R., Sauer, K.: Electron spin polarization in photosynthesis and the mechanism of electron transfer in photosystem I. Experimental observations. Biophys. J. **21**, 239–256 (1978)
181. Friesner, R., Dismukes, G.C., Sauer, K.: Development of Electron Spin Polarization in Photosynthetic Electron Transfer by the Radical Pair Mechanism. Biophys. J. **25**, 277–294 (1979)
182. McIntosh, A.R., Bolton, J.R.: Triplet state involvement in primary photochemistry of photosynthetic photosystem II. Nature (London) **263**, 443–445 (1976)
183. Norris, J.R., Thurnauer, M.C., Bowman, M.K., Trifunac, A.D.: Electron spin echo spectroscopy and photosynthesis. Front. Biol. Energ. **1**, 581–591 (1978)
184. Thrnauer, M.C., Bowman, M.K., Norris, J.R.: Time-resolved electron spin echo spectroscopy applied to the study of photosynthesis. FEBS Lett. **100**, 309–312 (1979)
185. Adrianowycz, O., Kinnally, K.W., Warden, J.T.: Biophys. J. **25**, M-AM-P61, 52a (1979)
186. Pedersen, J.B.: Determination of the primary reactions of photosynthesis from transient ESR signals. FEBS Lett. **97**, 305–310 (1979)
187. Sauer, K., Mathis, P., Acker, S., Van Best, J.A.: Electron acceptors associated with P-700 in triton solubilized photosystem I particles from spinach chloroplasts. Biophys. Acta **503**, 120–134 (1978)
188. Warden, J.T.: Private communication.

Chapter 6
The Chlorophyll Triplet State and the Structure of Chlorophyll Aggregates

Richard H. Clarke[1]

1 Introduction

In the series of photophysical events that are initiated by the absorption of light in green plants the chlorophyll pigment molecules play a central role. The chlorophylls, present in all organisms that carry out photosynthesis with the evolution of molecular oxygen, have long been recognized as functionally formulated for efficient participation in the mechanisms of radiant energy capture and conversion in green plant systems [1]. Chlorophyll is particularly well designed for the formation of molecular aggregates, groupings of pigment molecules arranged to carry out specific functions within the plant. It is generally considered that chlorophyll in the plant is organized into photosynthetic units consisting of about 300 pigment molecules [1]. The bulk of the chlorophyll in such a unit appears to be arranged in a photochemically inert network, which functions primarily to gather light and is, therefore, referred to as antenna chlorophyll [1]. The incident light energy collected by the antenna chlorophyll system is then conducted to a smaller collection of interacting chlorophylls constituting the photoreaction centers where the energy transduction of photosynthesis occurs [1]. It is this functional duality, the utilization of the same molecule for two energetically distinct functions within the photosynthetic unit, that points up the importance of the aggregated structures of chlorophyll. The distinction between antenna and reaction center chlorophyll is one of molecular arrangement, rather than chemical composition. If one is to understand the mechanisms of energy capture and conversion in green plant systems, it is necessary to have a complete understanding of these structural units involving chlorophyll aggregates, since it is the structure of these chlorophyll complexes in vivo that controls the initial photophysical processes of photosynthesis [1, 2].

Magnetic resonance spectroscopy in the photoexcited triplet state, a field pioneered by Hutchinson and Mangum [3, 4], has become an increasingly valuable method of investigating structural and dynamical features of chlorophyll systems. The photoexcited triplet state EPR spectrum of a chlorophyll molecule was first observed by Rikhireva et al. in 1963, who reported the $\Delta m = 2$ transition in chlorophyll *b* [5], and by Lhoste in 1968, who determined the chlorophyll *b* zero-field splitting parameters [6]. But the recent successful observations of triplet state EPR in photosynthetic bac-

1 Department of Chemistry, Boston University, Boston, MA 02215 USA

teria by Leigh, Dutton and coworkers [7-10] has revived interest in the triplet state in photosynthesis and in the investigation of triplet state properties of chlorophyll as an isolated molecule, in aggregated units and in its naturally occuring environments [11-17].

The purpose of this article is to explore the application of a specific type of excited state EPR, zero-field optically detected magnetic resonance (ODMR), to the investigation of the photoexcited triplet state of the chlorophyll molecule and its aggregate systems in vitro and in vivo. ODMR offers unique advantages in studying the triplet state properties of chlorophyll systems by combining optical spectroscopy and magnetic resonance techniques [18, 19]. The overall goal of our work in this field has been to provide a detailed description of the triplet state properties of the chlorophyll molecule (and its derivatives) and to utilize this molecular information to elucidate structural aspects of chlorophyll in aggregated form. Such information is expected to be valuable in formulating molecular mechanisms of photosynthesis which might then allow artificial simulation of the photosynthetic apparatus [1, 20].

In the development of this article we shall first provide the necessary background to understand the zero-field ODMR experiment as it is applied to chlorophylls. Then the results of ODMR on the chlorophyll molecule will be presented and discussed in terms of the electronic structure and dynamical features of the chlorophyll triplet state. Finally, we shall explore the utilization of the chlorophyll triplet, through ODMR spectroscopy, as a structural probe into the geometrical features of chlorophyll aggregates in vitro and in vivo. It should be noted at the outset that a significant amount of important research on the chlorophyll triplet state has been accomplished, using a variety of EPR techniques other than zero-field ODMR [21]; nor is this article meánt to be a thorough review of the progress of chlorophyll ODMR to date. Rather, we seek mainly to give an overview of the application of ODMR techniques in studying photosynthetic pigments, to show the approach used, some of the progress obtained to date (with apologies to many of my colleagues working in the field whose work may have been omitted) and to provide a perspective on the contributions of ODMR spectroscopy by photosynthesis which may be anticipated in the future.

2 Optically Detected Magnetic Resonance in the Triplet State

The observation of optically detected magnetic resonance transitions at zero-field among the three spin sublevels in the photoexcited triplet state of organic molecules by monitoring either optical absorbance or emission has led to considerable activity in the application of these techniques to a wide variety of molecular problems, including those in the area of biological systems [13-15, 22, 23]. Since the original demonstrations that the phosphorescence intensity of organic molecules at low temperatures could be modulated by a microwave field in resonance with the magnetic sublevels of an emitting triplet state [24], an analogous microwave-induced effect has been found for triplet-triplet absorbance [25], fluorescence [19, 26, 27], and singlet-singlet absorbance [28] for nonphosphorescing molecules such as chlorophyll. Basically, all of these last intensity changes arise from a microwave-induced coupling of isolated triplet spin sublevels with different rates of intersystem crossing, leading to an adjustment in the steady-state triplet population and a subsequent intensity change in the monitored optical spectrum. Such ODMR experiments provide two important pieces of molecular information – the energetics of the triplet state sublevels in the form of the microwave transition frequencies, and the dynamics of the sublevels in the form of the time dependence of their response to the microwave field. The former measures the zero-field splitting parameters, D, E, associated with the distribution of electrons in the triplet state, while the latter monitors the rates of intersystem crossing into and out of the individual triplet spin sublevels. Both properties are important for an understanding of electronic structure and mechanisms of triplet energy conversion in chlorophyll, as well as for providing information about the surrounding environmental and perturbing interactions affecting the pigment system [12, 13, 17, 21].

The microwave-induced changes in absorption or emission intensity for a nonphosphorescing molecule may be understood qualitatively by consideration of the example given schematically in Fig. 1. Since intersystem crossing from the lowest singlet state S_1 of a molecule occurs with different rates to each of the magnetic sublevels of T_1 [19, 29], and in general with different rates from the magnetic sublevels back to the ground state S_1, a molecular situation could be imagined in which population of T_1 was predominantly to one spin sublevel and depopulation was from a second, as illustrated in Fig. 1. The result of continuous optical pumping of such a system at temperatures sufficiently low to prevent spin-lattice relaxation within T_1 (usually $<4°$ K) will be to build up a large population in T_1 at the expense of the ground state. The example illustrates a kind of triplet state bottleneck in which population is driven by photoexcitation and intersystem crossing into an isolated triplet spin sublevel which has a very long lifetime (low probability of intersystem crossing back to the ground state). When microwaves are applied at a frequency corresponding to the energy difference between the populated T_1 sublevel and the sublevel with a large probability for conversion back to S_0, the triplet bottleneck is released, the saturating microwave field acting as an infinitely strong spin-lattice relaxation mechanism between the two previously isolated sublevels, the overall triplet state population is reduced and the ground state (and the subsequently repopulated S_1 level) population increases. Thus, if the triplet state population is monitored by following the intensity of ground state absorption (S_0 population), fluorescence (S_1 population), or triplet-triplet absorption (T_1 population) at low temperatures, a microwave-induced population change in T_1 may be observed in the optical spectra whenever the frequency of the applied microwave field is in resonance with a zero-field transition of T_1 [19, 29]. In principle, of course, such microwave-induced T_1 population changes may be monitored by observing the intensity of *any* of the electronic transitions detectable for a molecular system, since all the energy levels are dynamically coupled [29]. In practice, ease of detection or convenience of observation of a particular region of the optical absorption or emission spectrum will dictate which is most efficient for an ODMR experiment [29]. Because of the strong fluorescence characteristic of chlorophyll systems, most of the ODMR experiments on chlorophylls have utilized fluorescence detection [13-15, 19]. However, triplet absorption detection [30] may prove useful if nonfluorescent chlorophyll aggregate species are encountered in vivo [1].

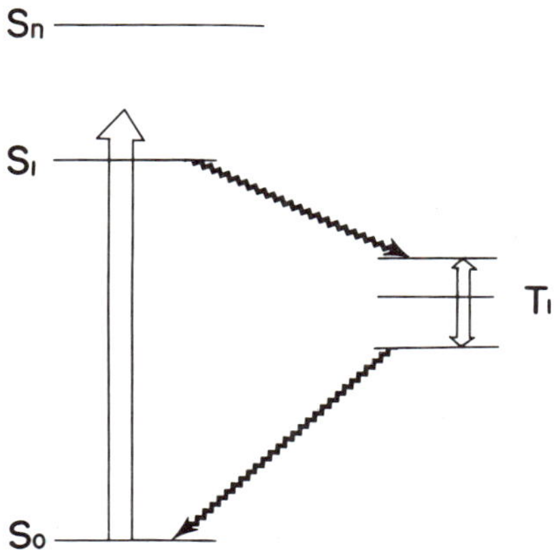

Fig. 1. Optical pumping ($S_0 \rightarrow S_1$), intersystem crossing ($S_1 \rightsquigarrow T_1$, $T_1 \rightsquigarrow S_0$) and microwave saturation within the triplet state of T_1 of chlorophyll

When the zero-field ODMR transition frequencies have been determined, the intersystem crossing rate constants for each of the triplet spin sublevels may also be obtained through time-dependent ODMR measurements. The description of time-dependent population energy levels induced by light can be very complex, since the radiation field is interacting with a collection of (nonstationary) energy levels, all of which are dynamically coupled. The relative efficiency of the relaxation processes among the electronic and vibrational levels of the molecule will allow some simplifying approximations, but, in general, the problem remains complex.

A generalized description of the time dependence of population changes among the spin sublevels of the lowest triplet state has been worked out and provides a practical foundation for the analysis of rates of intersystem crossing in chlorophyll systems from time-dependent ODMR data [29]. We consider a nonphosphorescing system with a lowest triplet state T_1 which is undergoing population buildup and decay upon (indirect) photoexcitation, as described in Fig. 2. We have omitted from our description any processes involving singlet states higher than the first, S_1, and any processes between T_n and the singlet manifold. For most chlorophyll triplet states our description will be sufficient, but an extension of the description in cases were $T_n \rightarrow S_1$ processes may be important has been analyzed for chlorophyll by van der Bent and Schaafsma [31]. We assume that the triplet spin sublevels are isolated (no spin-lattice relaxation).

The time-dependent change of population in each magnetic sublevel i of T_1 is given by

$$\frac{dT_1^i}{dt} = k_2^i S_1 - k_3^i T_1^i - k_4^i T_1^i + k_5^i T_n^i \qquad (1)$$

where i = 1,m,n, the principal axis designation at zero-field of spin sublevels of T_1 and T_n. The principal axis for T_n is taken to be the same as that for T_1 purely for convenience; inclusion of the general case of different principal axis systems for T_1 and T does not affect the triplet-triplet absorption description and adds nothing new to our dynamics description, as has been shown previously [19].

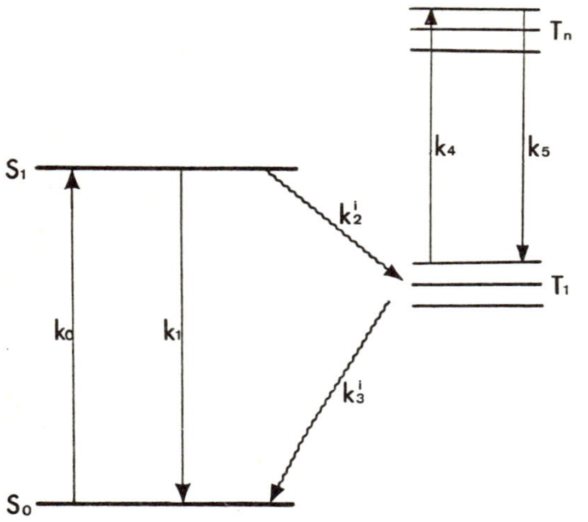

Fig. 2. Rate constants used to describe the ODMR experiment

The population of the lowest singlet excited state is given by

$$\frac{dS_1}{dt} = k_0 S_0 - (k_1 + k_2^\ell + k_2^m + k_2^n)S_1 \equiv k_0 S_0 - (k_1 + k_2)S_1 \tag{2}$$

and for the higher triplet state T_n

$$\frac{dT_n^i}{dt} = k_4^i T_1^i - k_5^i T_n^i \,. \tag{3}$$

Since processes which occur within a spin manifold are normally much more rapid than those occurring between singlet and triplet manifolds, we can assume steady state population for S_1 and T_n, when viewed on the time scale associated with the population changes taking place in T_1 [19, 29].

Consider the effect of introducing a saturating microwave field in resonance with magnetic sublevels T_1^ℓ and T_1^m. Assuming that the microwaves equalize (saturate) the populations in T_1^ℓ, T_1^m, the kinetic description of the population in the two spin sublevels will proceed as if T_1^ℓ and T_1^m were a single, combined sublevel with dynamics given by

$$\frac{d(T_1^\ell + T_1^m)}{dt} = (k_2^\ell + k_2^m)S_1 - 1/2(k_3 + k_3^m)(T_1^\ell + T_1^m)$$
$$- 1/2(k_4^\ell + k_4^m)(T_1^\ell + T_1^m) + 1/2(k_5^\ell + k_5^m)(T_n^\ell + T_n^m)\,. \tag{4}$$

Since the T_1^i, S_1, and S_0 are coupled through Eqs. (1) and (2), population changes induced by microwaves in T_1^ℓ, and T_1^m will affect S_0, S_1. As S_0, S_1 change, it is clear that T_1^n will also change upon microwave saturation of T_1^ℓ, T_1^m, as the dynamically coupled levels adjust population within the singlet and triplet manifold. Therefore, we must solve Eqs. (1) and (4) simulatenously, Eq. (1) for T_1^n and Eq. (4) for T_1^ℓ, T_1^m, to determine the *overall* time-dependent triplet population change induced by the saturating microwave field in T_1. By substituting the steady-state relationships for S_1 and T_n into Eqs. (1) and (4), we obtain a pair of coupled dynamical equations for the triplet spin sublevels of T_1:

$$\frac{dT_1^{\ell m}}{dt} = -(Kk'k_2^{\ell m} + k_3^{\ell m}/2)T_1^{\ell m} - Kk'k_2^{\ell m}T_1^n + NKk_2^{\ell m} \quad (5)$$

$$\frac{dT_1^n}{dt} = -Kk'k_2^n T_1^{\ell m} - (Kk'k_2^n + k_3^n)T_1^n + NKk_2^n \quad (6)$$

where $K = k_0/(k_0 + k_1 + k_2)$, $k' = (k_4 + k_5)/k_4$ and where we have adopted the shorthand notation $T_1^{\ell m} = T_1^\ell + T_1^m$, $k_2^{\ell m} = k_2^\ell + k_2^m$, $k_3^{\ell m} = k_3^\ell + k_3^m$. The complete solution to these coupled equations has been detailed [29]; we present only the results of interest to chlorophyll triplets.

We are interested in the change with time of the *total* population in T_1 upon application of microwaves in an ODMR experiment. The total triplet population is given by (assuming small K)

$$T_1(t) = T_1^{\ell m}(t) + T_1^n(t) = -NK\frac{(k_2^\ell/k_3^\ell - k_2^m/k_3^m)(k_3^\ell - k_3^m)}{k_3^\ell + k_3^m} e - 1/2k_3^{\ell m}t \quad (7)$$

$$+ NK[\frac{k_2^n}{k_3^n} + \frac{k_2^{\ell m}}{1/2k_3^{\ell m}}]$$

where terms in higher powers of K have been neglected.

The result obtained in Eq. (7) shows that in the limit of small K:

1. the total triplet population varies in time as a single exponential when the microwaves are turned on;
2. the triplet population reaches a steady state value which is independent of the rates k_4, k_5 involving the higher triplet state T_n;
3. the total triplet population at any time after turning on the microwaves is independent of k', the triplet absorption factor.

When the microwave field is turned off in the presence of the exciting light, the triplet spin sublevels return to their equilibrium populations with the time dependence for each given by Eq. (1). Replacing S_1, T_n in Eq. (1) by their steady-state values, a set of three coupled equations is obtained:

$$\frac{dT_1^\ell}{dt} = -(Kk'k_2^\ell + k_3^\ell)T^\ell - Kk'k_2^\ell T_1^m - Kk'k_2^\ell T_1^n + NKk_2^\ell \quad (8)$$

$$\frac{dT_1^m}{dt} = -Kk'k_2^m T_1^\ell - (Kk'k_2^m + k_3^m)T_1^m - Kk'k_2^m T_1^m + NKk_2^m \quad (9)$$

$$\frac{dT_1^n}{dt} = -Kk'k_2^n T_1^\ell - Kk'k_2^n T_1^m - (Kk'k_2^n + k_3^n)T_1^n + NKk_2^n \qquad (10)$$

which is, essentially, the same result obtained in Eq. (6). Equations (8-10) describe the population variation with time in the three spin sublevels – each level contributes as it varies to the change in singlet population, and, indirectly, to the subsequent population adjustment in the other triplet spin sublevels.

We shall only outline the solution here [29]. To simplify what will be a very complex solution to the above equations, we shall assume $dT_1^n/dt = 0$. That is, we adopt the viewpoint that the small change in T_1^n on returning to its equilibrium value produces a negligible indirect effect on the populations of T_1^ℓ, T_1^m as the sublevels come to equilibrium after removal of the saturating microwave field. The justification of this approximation comes from the fact that again we shall be interested in a solution for small values of K (low exciting light power); in this case we have shown in previous discussion [29] that $dT_1^n/dt = 0$ is a good approximation, since under these conditions T_1^n shows a small variation upon microwave saturation. Thus, we again have two coupled equations to solve. These equations give the solution in the limit of small K (low exciting light, k_0):

$$T_1(t) = T_1^\ell + T_1^m + T_1^n = NK \frac{(k_2^m/k_3^m - k_2^\ell/k_3^\ell)}{k_3^\ell + k_3^m} [k_3^m e - k_3^\ell t - k_3^\ell e - k_3^m t]$$

$$+ NK[k_2^\ell/k_3^\ell + k_2^m/k_3^m + k_2^n/k_3^n]. \qquad (11)$$

Upon removal of the microwave field, the overall triplet population returns to its equilibrium value with a time dependence determined by the difference in two exponential terms, each of which reflects a first-order decay of one of the spin sublevels released from microwave saturation [19, 29]. It should be noted that the preexponential factors in Eq. (11) essentially multiply each first-order decay by the depopulation rate constant k_3^m or k_3^ℓ. The exponential terms representing the first-order decay of T_1^ℓ, T_1^m are each weighted by the value of the depopulating rate constant for other level. Thus, the longer-lived exponential is multiplied by the larger depopulating rate constant and is expected to dominate the time-dependent variation of T_1 when the microwaves are turned off, unless k_3^ℓ, k_3^m are of comparable value (k_3^ℓ, k_3^m cannot be exactly the same or no microwave signal will be detected) in which case a nonexponential decay characteristic of the difference of two exponential terms will be observed.

The population changes described in the previous section can, in general, be observed by monitoring the intensity of any convenient optical transition, for either absorption or emission spectra. Since the microwave-induced population changes within T_1 determine the population adjustments in S_0 and S_1, the description of the *overall* microwave-induced intensity changes is the same for absorption detected [25, 28] or emission detected [19, 26] ODMR. However, for chlorophyll fluorescence detection often has a resolution advantage in that a single band can be detected in a structured spectrum [19] and emission is usually more readily observed optically than triplet absorption. However, in cases where the triplet absorption extinction coefficient is very large, triplet absorption detection may produce stronger ODMR signals than does fluorescence detection from the same sample (most notable example is the benz(α) pyrene ODMR spectrum [32]). In the absence of any emission, triplet absorption detection is preferred over $S_0 \rightarrow S_1$ detection [28], if observable, since it is a direct probe of the T_1 population and will in these cases result in stronger ODMR spectra.

Once the method of ODMR has been selected, the determination of intersystem crossing rates proceeds in the same manner for all. In principle, Eqs. (7) and (11) provide a method for determining the $T_1 \rightarrow S_0$ rate constants for the three individual spin sublevels of T_1. One monitors the microwave-induced intensity change in emission or absorption as a function of time after turning the microwave field on and off by square wave modulation of the microwave source. With three-zero-field transitions to be monitored, these expressions provide six measurements for the determination of k_3^ℓ, k_3^m, k_3^n – three when the microwaves are turned on, and three when the microwave field is removed [29]. Relative populating rates may be obtained from intensity ratios among the ODMR peak as described previously [19, 25].

In actual experiments it is usually the case that only two of the microwave transitions provide appreciable ODMR signals, the intensity is determined by the difference in steady-state population among the spin sublevels $(k_3^\ell k_3^m \mp k_3^\ell k_2^m)$ [19, 29]. The third transition frequently involves two spin sublevels with steady-state populations very similar and the ODMR transition is, therefore, correspondingly difficult to detect. With two transitions the rates must be determined from the microwaves-on and two microwaves-off measurements. From Eq. (7), when the microwave field is applied, the intensity change in time is expected to follow a single exponential, providing the value of $1/2(k_3^\ell + k_3^m)$ for each transition. The third measurement, required to determine the k_3^ℓ, k_3^m, k_3^n individually must then come from analysis of the intensity changes when the field is turned off. From Eq. (11) the time-dependent intensity changes when the microwaves are removed will be the resultant of two exponential terms. If there is a large difference between rates among the sublevels, one can wait for the decay of the faster level and analyze the decay of the persisting (slower) level. Consistency among the results obtained from the two microwaves-off curves will provide a check on the analysis. If the rates of decay are within a factor two or three of one another, one can attempt to analyze the microwaves-off dynamics for two exponentials. As an alternative (or as a check) a measurement of the *overall* triplet lifetime $\Upsilon(T_1) = 3/(k_3^\ell + k_3^m + k_3^n)$ can be utilized, along with the two single-exponential curves giving $1/2(k_3^\ell + k_3^m)$ from the intensity change with time when the microwaves are turned on. The overall lifetime may be obtained in several ways. If observable, the lifetime may be obtained directly from a 77°K measurement of the decay of the triplet-triplet absorption. One may also obtain Υ from a 77°K measurement of the fluorescence intensity vs time after turning on the exciting light as the triplet population comes to steady-state equilibrium [33]. Also, by utilizing two microwave sources, in a double-resonance experiment saturating two zero-field transitions simultaneously (for the change in intensity vs time when all three zero-field levels are connected) the kinetics results become simplified to a single exponential, with rate constant (at small K) equal to $1/3(k_3^\ell + k_3^n + k_3^m)$.

It is important to emphasize that all the results reduce to simple expressions only in the limit of K small relative to the $T_1 \rightarrow S_0$ intersystem crossing rates. In general, our equation predicts that the ODMR signals will have a complicated dependence on optical power, k_0. The pumping power factor in these equations will have the effect of providing nonexponential intensity changes and observed rates *larger* than the intrinsic molecular rates. Thus, microwave-induced intensity changes must be measured as a function of incident power for each ODMR transition to determine the actual rate constants $k_3^{\ell}, k_3^{n}, k_3^{m}$. In practice, this provides no serious difficulty in most cases, since K is already a number less than one, and results in the limit of neglibigle K can be achieved by systematic power variation and appropriate signal averaging of the square-wave modulated intensities changes [19, 28] until no further change is observed in the rate constants with power reduction. Also, each ODMR transition will have a different K dependence, as can be seen from the complete solutions [29].

It should be noted that an alternative approach to determining triplet state dynamics by ODMR methods has been proposed by van Dorp et al., [27]. This method measured the optical response to a pulsed microwave signal and has been utilized successfully in the study of triplet state intersystem crossing rates in porphyrins [27].

3 Application of ODMR to the Chlorophyll Triplet State

The application of zero-field ODMR spectroscopy to the triplet state of the chlorophyll molecule is of interest from several different points of view. First, from a purely molecular perspective, the excited state electronic structure and intersystem crossing mechanisms active in a large, conjugated aromatic ring system such as is characteristic of the chlorophylls is of value for understanding nonradiative excited state energy conversion in multiring organic molecules. Second, in relation to its function in photosynthesis, a characterization of the triplet state properties and the factors which influence them in the isolated molecule is necessary for utilization of the chlorophyll triplet as a magnetic probe into more complex chlorophyll containing systems [17]. In this section we shall attempt to review what is known about the triplet state of chlorophyll and its derivatives from zero-field ODMR measurements.

Typical chlorophyll fluorescence spectra are shown in Figs. 3 and 4. The features of the fluorescence spectra may vary with solvent and concentration [1, 19], but in general fluorescence features provide a reliable, reproducible optical monitor of the characteristics of the chlorophyll system of interest [1, 14, 17, 19]; phosphorescence is usually very weak and in a spectroscopic region (in the infrared) difficult to detect [34]. Thus, fluorescence detection is the usual method of obtaining ODMR spectra of the chlorophyll triplet state.

Some typical ODMR results for chlorophylls at 2 K are shown in Figs. 5 and 6. Line widths of chlorophyll ODMR spectra are generally sharp. At zero field hyperfine interactions between the triplet electrons and nuclei around the ring are negligible [35], and the ODMR line shapes are primarily due to inhomogeneous broadening from interactions of the excited molecules with varying local solvent environments (sites) within the frozen sample [19, 36, 37]. Often the specific fluorescence detection wavelength will influence the structure observed in a chlorophyll ODMR spectrum, if multiple sites are resolvable in the fluorescence spectrum [37]. ODMR transitions (at zero field)

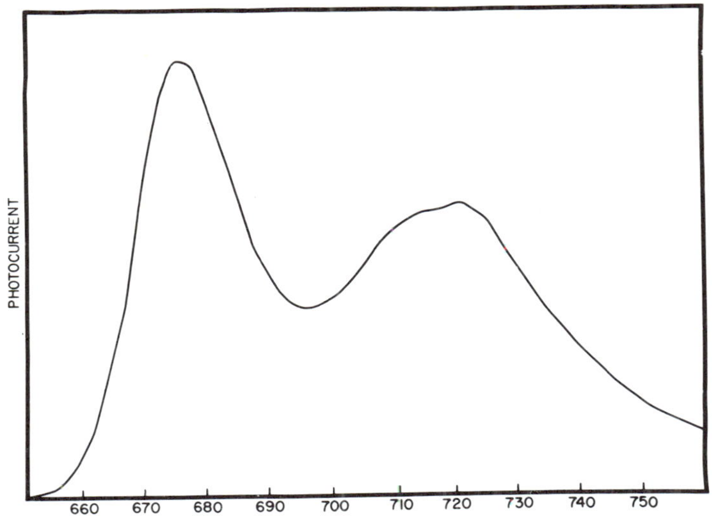

Fig. 3. Fluorescence spectrum of chlorophyll *a* in hydrocarbon solution containing water at 77 K. Wavelength scale in nanometers

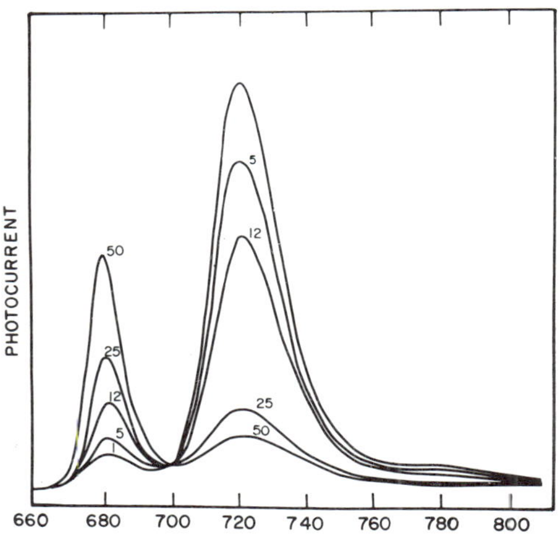

Fig. 4. Fluorescence spectrum of the covalently bound dimeric derivative of zinc pyrochlorophyllide in toluene at 77 K as a function of added percent methanol. Wavelength scale is nanometers

have also been observed for chlorophylls using triplet absorption detection [30, 37] and singlet (ground state) absorption detection [28], but these alternative methods offer no present advantage over fluorescence detection for the investigation of the chlorophyll triplet state in the molecule.

Although the ODMR lines shapes may be of interest in providing information about the range of solvent interactions experienced by the photoexcited chlorophyll molecule in solution, the most valuable triplet state features obtained by ODMR are the highly precise (accuracy typically to better than 1%) measurements of the zero-field transition frequencies, from which the triplet state zero-field splitting parameters may be calculated, and the individual spin sublevel intersystem crossing rate constants. We shall concentrate on these two properties in the remaining discussion.

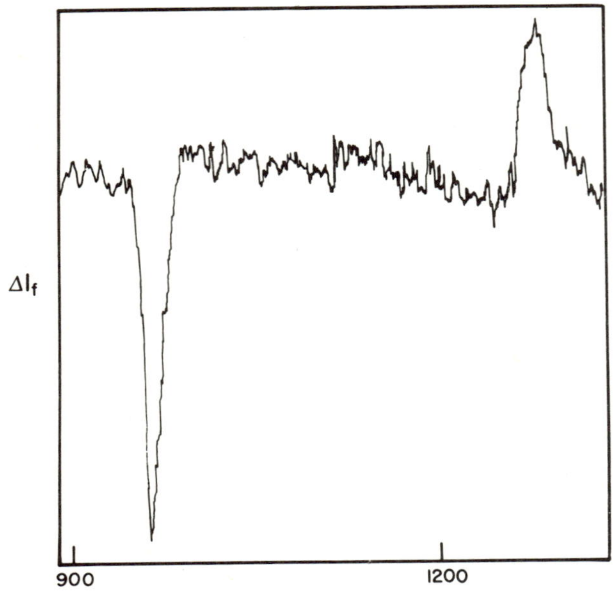

Fig. 5. Fluorescence detected ODMR transitions of pheophytin *b* in n-octane at 2 K. Fluorescence monitored at 667 nm, using 457 nm laser excitation. Frequency scale in megahertz

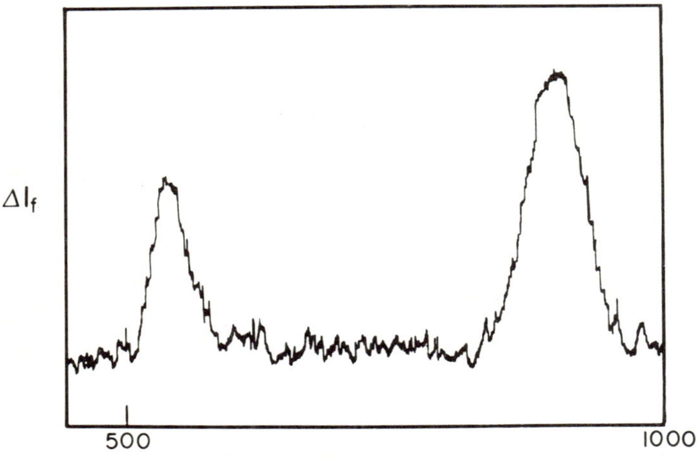

Fig. 6. Fluorescence detected ODMR transitions of bacteriochlorophyll *a* in tetrahydrofuran at 2 K. Fluorescence monitored at 778 nm, using 600 nm dye laser excitation. Frequency scale in megahertz

3.1 Chlorophyll Triplet State Zero-Field Splittings

The triplet state zero-field splitting parameters, D, E, the values of which represent the spin-dipolar interaction energy between the unpaired electrons in the excited state configuration for the triplet states, are obtained by fitting to the well-known spin Hamiltonian, H_S, of the form [4]

$$H_S = D(S_z^2 - 1/3S^2) + E(S_x^2 - S_y^2) \qquad (12)$$

/D/ and /E/ are immediately obtainable from a zero-field ODMR experiment, since the ODMR transition will occur at frequencies, /D/ + /E/, /D/ – /E/ and 2/E/, as shown in Fig. 7. The absolute signs of D, E are usually not experimentally fixed at zero field, but require high field measurements for their determination. Such an experiment has been done by Thurnauer and Norris for chlorophyll [38].

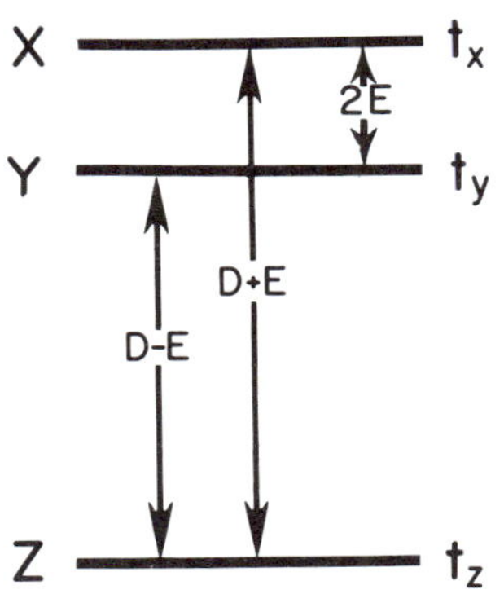

Fig. 7. The designation of zero-field triplet state ODMR transitions

The parameter D is a measure of the extent of delocalization of the unpaired electrons in the triplet state; E is related to the symmetry of the ring system [21]. More explicitly, the zero-field splitting parameter /D/ is defined as:

$$/D/ = 3/4g^2\beta^2 \; \frac{\Upsilon_{12}^{\;2} - 3Z_{12}^{\;2}}{\Upsilon_{12}^{\;5}} \qquad (13)$$

where g is the electronic g factor, β is the electronic Bohr magneton, Υ_{12} is the distance between the two unpaired electrons and the integral is over the orbital part of the triplet state wavefunction. The /D/ value essentially represents a measure of the extent of π-electron interaction over the chlorophyll ring system, proportional to $1/\Upsilon^3$. Changes in electronic distribution within the chlorophyll triplet state, by ring substitution metal center replacement, solvent ligation, or aggregation, all will produce a corresponding change in the /D/ value and a characteristic shift in ODMR frequency. Although all chlorophylls will have roughly the same /D/ value because of the large aromatic ring system characteristic of chlorophyll [21], small changes in electronic structure can be measured in a zero-field ODMR experiment, because of the inherently high resolution of the zero-field measurement.

Zero-field splitting parameters for several chlorophyll systems are presented in Table 1. Trends in these data have been discussed in several previous publications [12,

Table 1. Triplet state zero-field splitting parameters of chlorophylls (cm^{-1})

Molecule	|D|	|E|
Chlorophyll *a*	0.0280 ± 0.0003	0.0038 ± 0.0003
Zn chlorophyll *a*	0.0306 ± 0.0003	0.0042 ± 0.0003
Bacteriochlorophyll *a*	0.0238 ± 0.0005	0.0069 ± 0.0003
Pheophytin *B*	0.0368 ± 0.0008	0.0049 ± 0.0004
Chlorophyll *b*	0.0320 ± 0.0010	0.0041 ± 0.0006
Ca Chlorophyll *b*	0.0396 ± 0.0008	0.0039 ± 0.0003
Zn chlorophyll *b*	0.0328 ± 0.0003	0.0032 ± 0.0003
Cd chlorophyll *b*	0.0326 ± 0.0005	0.0036 ± 0.0003

21, 39]. A few features are worth noting. Metal substitution for magnesium in the chlorophyll ring has the effect of raising the D value observed in the triplet state; the implication is that metal substitution reduces the extent of electron delocalization, most probably in the region around the pyrrole nitrogens [40]. The largest zero-field splitting is observed for calcium-substituted chlorophyll *b*; since the large (ionic radius 0.99Å) positively charged calcium ion is most likely positioned out of the molecular plane [39], it will exert a strong attractive influence on the π-electron system in the vicinity of the central nitrogen atoms, increasing the value of D. Both pheophytin *a* and pheophytin *b* also exhibit larger D values than thecorresponding magnesium-substituted chlorophylls; again these differences appear to be reflecting restricted delocalization in the vicinity of the nitrogens when the central protons are bound to two pyrrole rings. The smallest observed D value is found for the triplet state of bacteriochlorophyll *a*; in this molecule the path of conjugation is extended by a carbonyl ring substituent along an axis drawn through the cyclopentanone carbonyl and the metal center, producing a possibility of greater delocalization than chlorophyll *a* and a subsequently smaller D value [34]. Note that the zero-field splitting parameters show a consistent trend when one compares molecules with the chlorophyll *a* structure to those of the chlorophyll *b* structure. In each case the molecule derived from the chlorophyll *b* structure has a larger value than its counterpart with the chlorophyll *a* structure. Since the substituent change in the chlorin ring involves the presence of a CHO unit in chlorophyll *b* instead of the CH_3 group found in chlorophyll *a*, the electron-withdrawing nature of the aldehyde substituent on the chlorophyll ring must decrease the extent of triplet delocalization around the ring system. This effect produces a somewhat greater electron spin density in the vicinity of the CHO unit, resulting in an increased spin-spin interaction in molecules with the chlorophyll *b* structure [40].

3.2 Chlorophyll $T_1 \rightarrow S_0$ Intersystem Crossing Rates

With the zero-field ODMR transitions for chlorophylls determined, time-dependent ODMR experiments, as described in Sect. 2 are utilized to obtain the individual triplet spin sublevel decay rates; Table 2 gives values for these decay rates for a series of chlorophylls. No phosphorescence has been observed for the metal-substituted chlorophylls [39, 40]; very weak phosphorescence emission has been reported for

Table 2. Rate constants for triplet state deactivation in chlorophylls (sec^{-1})

Molecule	Solvent	k_x	k_y	k_z
Chlorophyll *a*	n-octane	661 ± 89	1255 ± 91	241 ± 15
Zn Chlorophyll *a*	n-octane	346 ± 50	330 ± 32	660 ± 70
Bacteriochlorophyll *a*	THF	2287 ± 280	3321 ± 572	661 ± 74
Pheophytin *b*	n-octane	311 ± 44	1000 ± 24	54 ± 5
Chlorophyll *b*	n-octane	268 ± 34	570 ± 54	34 ± 4
Ca chlorophyll *b*	n-octane	207 ± 24	480 ± 38	32 ± 6
Zn chlorophyll *b*	n-octane	122 ± 13	250 ± 50	622 ± 47
Cd chlorophyll *b*	MTHF	370 ± 27	462 ± 37	722 ±50

chlorophyll *a*, chlorophyll *b*, pheophytin *a*, and pheophytin *b* [34, 41]. Therefore, the decay rates listed in Table 2 are expected to be dominated by *nonradiative* processes from the photoexcited triplet to the ground state.

Perhaps the most interesting feature of the rate data presented in Table 2 is the fact that, unlike the zero-field splittings, the triplet intersystem crossing rate constants are highly sensitive to the detailed changes in the structure of the molecule. This sensitivity may prove of significant value in the application of triplet state ODMR to in-vivo chlorophylls.

The most general trend in the chlorophyll intersystem crossing in the table is the dominance of the x and y (the arbitrarily designated in-plane molecular axes) spin sublevels in the decay of all the molecules listed, with the exception of the heavy-metal substituted (zinc and cadmium) chlorophylls. Since this same behavior is characteristic of simpler polycyclic hydrocarbons, these data suggest that the $T_1 \rightarrow S_0$ intersystem crossing of chlorophyll may be described in terms of models advanced for radiationless transitions in the simpler prototype planar aromatic systems [42]. For such systems the rate of depopulation from spin sublevel i of triplet state T_1 into an isoenergetic vibrational level of the ground state, $S_{0,\nu}$ can be written as

$$k_i(T_1{}^i \rightarrow S_{0,\nu}) = (2\pi/\hbar)|\langle T_1{}^i|\Omega|S_{0,\nu}\rangle|^2 \rho(E) \tag{14}$$

where Ω is the level shift operator capable of mixing the states $T_1{}^i$ and $S_{0,\nu}$, and ρ (E) is the density of states. Identification of the mixing operator with the spin-orbit coupling interaction operator and straightforward expansion of $S_{0,\nu}$ and $T_1{}^i$ as pure-spin adiabatic Born-Oppenheimer states leads to a simplification of the rate expression for the planar aromatic hydrocarbons which has been shown [43, 44] to reduce essentially to the following expression for x and y

$$k_i(T_1{}^i \rightarrow S_{0,\nu}) \propto |\delta/\delta Q\langle S_0|H_{so}|T_1{}^i\rangle|^2 (F.C.) \tag{15}$$

a product of Franck-Condon factor (F.C.), which is the same for all spin sublevels i and a spin-vibronic coupling term $\delta\langle S_0|H_{so}|T_1{}^i\rangle|\delta Q$ (Q is the nuclear displacement coordinate and H_{so} is the spin-orbit coupling operator for the molecule), which depends strongly on the spin-orbit coupling mechanisms available to each triplet spin sublevel. Work by Metz et al. has shown that for planar aromatic systems one would expect that in calculating the spin-vibronic coupling term, the spin sublevels corresponding to the in-plane molecular axes (x and y) associated with a $\pi\pi^*$ triplet state derive their intersystem crossing activity from first-order, one-center spin-orbit coupling with singlet $\sigma\pi^*$ and $n\pi^*$ levels in Eq. (15), whereas the z spin sublevel must utilize higher order one-center terms to gain appreciable activity; therefore, the population and decay rates for the z level will always be smaller than for x and y [44]. The dominance of the x and y spin sublevels in the chlorophyll triplet dynamics suggest that spin-vibronic coupling between the chlorophyll $\pi\pi^*$ triplet state and $\sigma\pi^*$ (and $n\sigma^*$) singlet states derived from nitrogen and oxygen centers around the chlorin ring, is sufficient to account for the overall observed intersystem crossing kinetics without the necessity of invoking any magnesium involvement in the coupling scheme. Thus, the decay of a chlorophyll $\pi\pi^*$ triplet state by nonradiative energy conversion to the ground state follows intersystem crossing mechanisms typical of planar aromatic systems; the efficiency of intersystem crossing for each triplet spin sublevel can be evaluated essentially by analyzing the symmetry-selected one-center spin-orbit coupling mechanisms available to each spin sublevel [40, 42].

In the case of large planar aromatic systems, the evaluation provides the simple prediction that spin sublevels associated with in-plane molecular axes (x, y in our description) are more efficiently depopulated than out-of-plane (z) spin sublevels [12, 39]. In Table 2 it is seen that the rates for pheophytin *b*, chlorophyll *b*, bacteriochlorophyll *a*, and calcium-substituted chlorophyll *b*, all follow well such a prediction. The metal centers in these chlorophylls introduce no new spin-orbit coupling mechanisms over those active in the metal-free pheophytin *b* π system that would affect the dominance of the in-plane spin sublevel activity in $T_1 \rightarrow S_0$ intersystem crossing.

The most interesting features of the isc rates are seen in the trends in the values k_x, k_y through Table 2. These values illustrate dramatically the importance of the one-center spin-orbit coupling centered on the nitrogens in determining the rates of isc in chlorophylls. The principal axes of the spin-dipolar interaction tensor have not been definitely assigned, but there is good evidence, the most convincing that recently obtained from magnetophotoselection EPR experiments by Thurnauer and Norris [38], that the in-plane principal axes x and y pass through the pyrrole nitrogens as indicated in Fig. 8. This axis assignment is also consistent with that found by high-field EPR for the triplet state of the similar π-electron system, free-base porphin [46]. Such an assignment is supported by a comparison of rates in pheophytin *b* and chlorophyll *b*. For a $\pi\pi^*$ triplet state the most important spin-orbit coupling terms for intersystem crossing involve one-center mixing with singlet states derived from $\sigma\pi$ configurations; the lowest energy $\sigma\pi$ states will be the most important, since the intersystem crossing experssions expanded from Eq. (15) will contain an energy denominator term involving the energy difference of the mixing states [39,40]. Among the lowest-lying such states are those involving nonbonding electrons centered on the pyrrole nitrogen atoms, the $^1n\pi^*$ states, as are found in pheophytin *b*. The effect of the metal center introduced into the ring will be to interact with the nonbonding nitrogen electrons, lowering (stabilizing) the energy of

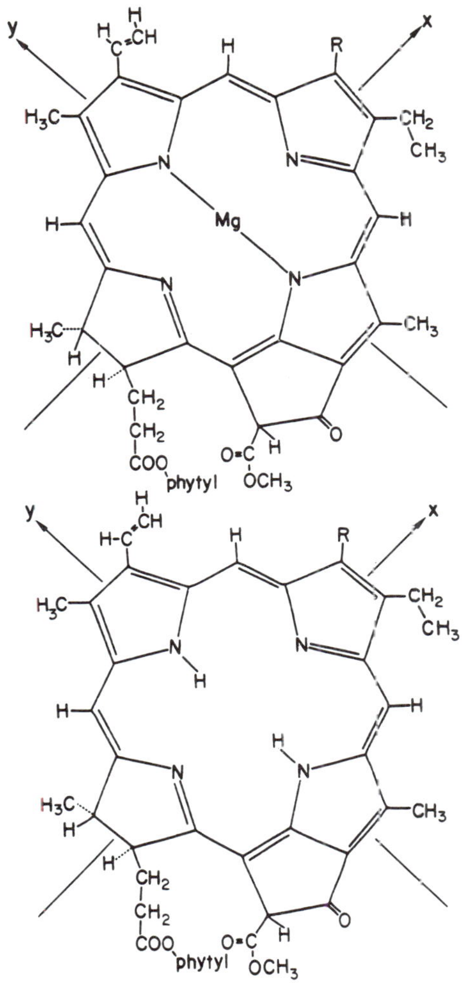

Fig. 8. Molecular structure and principal axis designations used in discussing triplet state ODMR data on chlorophylls. For chlorophyll *a*, *R* is a CH_3 group; for chlorophyll *b*, *R* is a *CHO* unit

such σ orbitals and raising the energy of the $\sigma\pi$ singlet states derived from these orbitals. In the case of pheophytin *b*, if the principal axes are taken as pictured in Fig. 8, it is clear that since nonbonding electrons point along the x-axis, $^1n\pi^*$ states derived from these orbitals will be spin-orbit coupled to the $\pi\pi^*$ triplet state by spin-orbit coupling operators directed along the y-axis [45]; the spin sublevel associated with the x-axis direction mixes with the $\sigma\pi$ states whose σ orbital components point along y, the lowest energy of which is expected to involve the nitrogen centers bonded to hydrogen atoms [46, 47]. Since the y spin sublevel spin-orbit couples with the more energy accessible $^1n\pi^*$ states (smaller energy gap) and the x spin sublevel mixes with higher energy $\sigma\pi$ singlets, we expect for pheophytin *b* that k_y is greater than k_x, as is observed. When a magnesium ion is coordinated, the nonbonding orbital x drops to lower energy and the value of k_y is reduced; k_x remains essentially unchanged, since the Mg-N interaction is not expected to be substantially different than that for the N-H bond. The four Mg-N bonds, however, are not equivalent in chlorophyll since the molecule no longer has the high symmetry of porphyrins. The molecular structure determination of ethyl chlorophyllide confirms this fact in that the bond lengths for the four bonds are considerably different [48]. Since they are not equivalent, the nitrogen σ orbitals involving these bonds and resulting $\sigma\pi$ states are also not equivalent, and this is reflected in the observation that $k_X \neq k_Y$. Thus, on comparing pheophytin *b* with chlorophyll *b* , essentially only the value of k_Y is affected, consistent with one-center spin-orbit coupling contributions to spin sublevels directed along in-plane axes as diagrammed in Fig. 8.

In a similar manner, the other metal-center substituted chlorophylls have rates for k_X, k_Y which can be understood by examining the effect of the metal on the energy of the nitrogen orbital directed toward the metal center and interacting with the metal in the plane of the molecule. As the metal-nitrogen interaction increases, the $\sigma\pi$ states involving these nitrogen electrons shift to higher energy, reducing the efficiency of one-center spin-orbit coupling to these states and lowering the value of the isc rates. On comparing the rates in the chlorophyll *b* series for Chl *b*, CaChl *b*, ZnChl *b*, one sees the rates of k_X, k_Y decreasing, but keeping a constant ratio of $\sim$1 : 2. As one progresses from the smaller Mg ion to the larger Ca ion, the interactions with the central nitrogen orbitals increase, lowering the values of k_X, k_Y. In the case of Zn, one expects even larger interactions than Mg and Ca since the ion is expected to fit better into the ring system (it is the most stable ion in the series with respect to removal [39]) and has additional d orbital interactions with the nitrogens. The increased interactions raise the energy of nitrogen-based $\sigma\pi$ states, and k_X, k_Y are both reduced, again retaining a 1 : 2 ratio, indicating both rates are equally affected by the metal interaction. Finally, when one examines the heaviest metal center in the series, cadmium, the k_X, k_Y values have now increased, presumably due to the fact that the heavy atom effect becomes an important factor in evaluating the spin-orbit coupling terms in the presence of the cadmium.

For the heavy-atom substituted chlorophylls, zinc chlorophyll and cadmium chlorophyll, it is the z (out-of-plane) spin sublevel which dominates the $T_1 \rightarrow S_0$ intersystem crossing, rather than the in-plane spin sublevels characteristic of single planar aromatic molecules. This difference is understood by consideration of the additional spin-orbit coupling activity associated with the d electrons of these metal centers (27) (39) (40). The presence of the d electrons provides a new one-center spin-orbit coupling term mixing the t_Z (out-of-plane) component of T_1 with the low-lying $\pi\pi^*$ singlet states (27) (39) (40), whereas t_X and t_Y are still restricted to mixing with the higher energy $\sigma\pi^*$ singlet levels [39, 40]. Such a situation has also been observed for the triplet state of zinc porphin [27]. Note that cadmium-substituted chlorophyll *b* has an even larger value of k_Z than the zinc-substituted systems [39], due most probably, to additional terms for t_Z involving the 4d electrons on cadmium, as well as to the overall increased spin-orbit mixing from the heavier atom.

An interesting difference in intersystem crossing rate constants is found in comparing the rates of triplet spin sublevel decay of chlorophyll *a* with those of chlorophyll *b*, where only a simple structural difference exists between the two molecules (see Fig. 8). The substitution of an electron-withdrawing aldehyde group in chlorophyll *b* for the methyl substituent in chlorophyll *a* leads to a dramatic lowering of the values of the rate constants for triplet decay (cf. Table 2), preserving, however, the same trends in the values (viz., $k_y > k_x > k_z$). In view of the comparison of metal-free and metal-substituted chlorophylls it is clear that differences in the energies of the mixing singlet $\sigma\pi^*$ states will result in differences in triplet decay constants. The smaller k values for chlorophyll *b* as compared to chlorophyll *a* may be explained by the fact that aldehyde substitution acts to withdraw electron density from the interior of the ring, giving rise to more positive central nitrogen atoms, and raising the energy of the $\sigma\pi^*$ states associated with the metal-nitrogen bonds [12, 40]. Further, reduction in electron density on the chlorophyll *b* nitrogens will also reduce the coefficients of the atomic orbitals on these centers, leading to smaller contributions from these centers to the rate expressions in Eq. (15) which describe chlorophyll *b* triplet state deactivation [40]. Thus, in comparing the triplet spin sublevel rate constants of chlorophyll *a* with those of chlrophyll *b*, upon aldehyde substitution the overall effect of electron density reduction in the pyrrole nitrogens is to raise the energy of the singlet mixing states and reduce the contribution of one-center spin-orbit coupling terms involving the nitrogen centers, effecting a reduction of *all* the triplet spin sublevel decay rates in chlorophyll *b*.

In Table 2 the largest depopulating rate constants (fastest spin sublevel lifetimes) are found for bacteriochlorophyll *a*. The structural changes in going from chlorophyll to bacteriochlorophyll (Fig. 8) lead to large red shifts in its absorption and fluorescence bands [2], which is reasonably expected to be accompanied by a significant lowering of the bacteriochlorophyll triplet state energy. The trend of in-plane (k_x, k_y) spin sublevels dominating the $T_1 \rightarrow S_0$ intersystem crossing in chlorophylls is still observed in bacteriochlorophyll, indicating that the same intersystem crossing mechanisms are active in promoting $T_1 \rightarrow S_0$ decay. Presumably, the faster rate at which depopulation occurs in bacteriochlorophyll compared with chlorophyll *a* and *b* is due to an extremely low-lying triplet state [40]. A rough estimate of the triplet energy of bacteriochlorophyll *a* may be obtained by assuming an inverse proportionality between the average triplet decay rates, $1/3(k_x + k_y + k_z)$ and the energy gap between the triplet and the ground state for the chlorophylls [49]. Based on a triplet energy for chlorophyll *a* of 10,400 cm^{-1} as determined from phosphorescence measurements [34, 41], the triplet energy of bacteriochlorophyll *a* is estimated to be about one-half the chlorophyll *a* value. This result is consistent with calculated values which place the bacteriochlorophyll triplet state at about 5000 cm^{-1} [50, 51].

In a general perspective, although chlorophylls possess considerable complexity in their conjugated ring system and attached substituents, the ODMR rate data show that systematic trends in nonradiative triplet state intersystem crossing processes do exist for these systems. Notably, the data given in Table 2 reveal that in all the systems studied there is one spin sublevel which dominates the triplet decay process. In the chlorophyll systems which contain metal centers with d electrons the most efficient depopulating level is the lowest energy spin sublevel, corresponding to the out-of-plane principal axis; in the other chlorophylls an in-plane (middle energy) spin sublevel is the most active. The analysis of these data suggest that the details of triplet state intersystem crossing can be understood by applying arguments previously advanced for the description of radiationless energy conversion from the triplet to the ground state in simple planar aromatic systems. In attempting to characterize the controlling features active in the triplet state intersystem crossing process for chlorophylls and related systems, the following aspects are of central importance:

1. the energy of the triplet state relative to the ground state;
2. a consideration of the one-center spin-orbit coupling involving the four central nitrogens, including the energies of the mixing states as a function of interactions with the central metal;
3. additional spin-orbit coupling contributions from the central metal (i.e., those involving d electrons).

A consideration of these factors provides a general basis with which to properly evaluate trends in intersystem crossing expected for the entire class of chlorophyll pigment and prototype molecules.

4 Application of Triplet State ODMR to Chlorophyll Aggregate Structure

There has been much speculation over the structural features and interactions among the chlorophyll pigment molecules which lead to the spectral characteristics exhibited by these pigment systems in vivo. Several models have been proposed for the aggregated chlorophylls in photosynthetic systems which have been developed to explain the spectral features in vivo [1, 52-54]; such models are valuable in providing insight into possible mechanisms for the light-induced chemistry occuring the photosynthetic systems [1]. In addition, several groups have studied the properties of the pigment systems in vitro whose physical and chemical features appear to parallel those found in naturally occuring photosynthetic systems [1, 48, 54, 55]. The detailed understanding of such in-vitro systems and their relationship to the models put forth to rationalize the activity of pigment systems in vivo are, naturally, of considerable importance for understanding of the physical and chemical processes occuring in photosynthesis [1].

The chlorophyll aggregate structures considered to be important in photosynthesis have been visualized and modeled around the basic physical features of the chlorophyll molecule. In 1963 Closs et al. showed by NMR (subsequently confirmed by further NMR and IR studies) that the central magnesium atom in chlorophyll is coordinately unsaturated in the chlorophyll ring system and has a strong tendency to pick up an electron donor in one or both of its axial positions [56]. This finding led to speculation concerning the coordination of water in forming chlorophyll aggregates in vitro and in vivo [1]. Water is an important ligand due to its bifunctional nature, since the oxygen atom can be coordinated to the Mg atom of one chlorophyll molecule and the two hydrogen atoms are then available for hydrogen bonding to other chlorophylls at various carbonyl positions around the chlorin ring [1, 56]. Recent work, using both optical spectroscopy [2] and magnetic resonance [57, 58], has shown that the reaction center in photosynthesis systems contains a pair of strongly interacting chlorophyll molecules, the so-called special pair, which are involved in the initial photophysical processes of the reaction center [1]. Models for the geometry of such a reaction center dimer have been proposed by Fong [52] and by Katz and co-workers [1, 54]. All the models considered for the reaction center structures of photosynthetic systems involve water-linked chlorophyll (or bacteriochlorophyll) dimers. Fong and his group first suggested that the chlorophyll dimer in the reaction center is held together by two water molecules coordinated and hydrogen-bonded to provide a plane-parallel dimer structure with C_2 symmetry [52]. In the most recent work Fong postulates that such dimers may be formed either the chlorophyll monohydrate $(Chla{\cdot}H_2O)_2$ or the dihydrate $(Chla{\cdot}2H_2O)_2$, which have been proposed, respectively, to be the P700 and water-splitting reaction centers in plant photosynthesis [53, 59, 60]. The two dimer structures differ in that the C10 ester carbonyl group is used for hydrogen bonding by the Mg-coordinated water molecule of the chlorophyll monohydrate, whereas in the chlorophyll dihydrate the C9 keto carbonyl unit is utilized [52, 53, 59, 60]. Katz and co-workers had originally described a reaction center dimer containing one water molecule linking two plane-parallel chlorophylls (although) this model does follow for freedom of motion away from the plane-parallel configuration) [1]. In their most recent work, however, they also invoke a dimer locked into a plane-parallel arrangement by two water molecules coordinated to the central Mg atoms, but hydrogen-bonded to the chlorin rings by the C9 keto carbonyl group [54]. Further, the Katz group proposes a structure for the antenna system requiring no water linkages but, rather, utilizing self-coordination among chlorophylls, with one chlorophyll acting as a donor via its carbonyl substituents to the Mg atom of another [1]. The carbonyl-Mg self-aggregateion can lead to the formation of large networks, or oligomers, of chlorophylls making up the antenna system [1]. In both groups extensive optical, IR and NMR spectroscopic comparisons of in-vivo chlorophyll systems with a variety of chlorophyll preparations in vitro have been used to support the models proposed [1, 54, 55]. A review article which assesses the spectroscopy of all the model systems for chlorophyll aggregation has recently appeared [61] and is a thorough reference source on the work in this field to date.

With a background in the triplet state of the chlorophyll molecule from the previous section, we explore the use of the photoexcited triplet state as a paramagnetic probe into the structural features of chlorophyll complexes. The usefulness of paramagnetic centers as probes for structural features of atomic and molecular systems has been recognized for virtually as long as magnetic resonance techniques have been in existence [62, 63]. The investigation by EPR of the interaction of the electron's magnetic moment with the atomic or molecular surroundings with which it interacts has been successfully exploited in evaluating geometrical details of the system and its environment for a wide variety of materials, ranging from inorganic solids to complete, biologically active units [63-65]. Such EPR experiments essentially utilize the magnetic interactions within and around the paramagnetic center to probe structural and dynamical aspects of the center and its surroundings [22, 64, 65]. In this section we shall evaluate the utilization of the triplet state of chlorophylls as a particularly convenient type of "spin label" [66], since it can be introduced into the system of interest nondestructively by photoexcitation.

Since the original observation by Leigh and Dutton of triplet state EPR signals in reaction centers of purple photosynthetic bacteria in 1974, it has been recognized that the application of triplet state EPR spectroscopy to the problem of the makeup of the chlorophyll aggregates found in vivo may be a valuable structural tool [8-11]. Most of the effort has been focused on bacterial systems because of their less complex makeup and because their photosynthetic activity proceeds through a single photosystem unit, rather than the dual photosystem structure found in plants and algae [2]. Conventional high-field triplet state EPR experiments on bacteria in vivo have been complemented and extended by triplet state studies employing zero-field ODMR [13, 62, 68]. The first triplet ODMR experiments on an intact photosynthetic system in vivo were done on *R. Rubrum* in our laboratory in 1975 [67]. Following the original ODMR experiments on purple bacteria, additional ODMR results have been obtained for reaction center and whole cell preparations of bacteria [13, 68], and, to a lesser extent, in chloroplast and algae preparations [14, 15].

In the present section we show that the triplet zero-field splitting and intersystem crossing rates can be utilized to provide structural information in chlorophyll aggregates. Specifically, the structure of a chlorophyll dimer such as might exist in the reaction center "special pair" [1], may be inferred from the triplet properties as interpreted within the framework of a simple triplet excition treatment [17, 69]. If the assumptions of the triplet excition approach, as applied successfully in the description of triplet excition states in organic molecular solids [70, 71], are applicable to the chlorophyll aggregate, then both the values of the zero-field splittings and the intersystem crossing rate constants in the chlorophyll dimer are expected to be directly dependent on the relative orientation of the chlorophyll molecules in the pair [17, 72]. The use of a triplet excition treatment to interpret dimer structure in chlorophylls has also been discussed recently by Hagele, Schmid, and Wolf [73] and by Bowman and Norris [74].

The ability to investigate by triplet state EPR the structural aspects of specific chlorophyll aggregate units within a given sample is dependent on the ability to select and isolate from what may be a complex system the unit of interest. This is particularly important when several forms of the chlorophyll aggregate may exist simultaneously, as in a solution containing equilibrium mixtures of monomer, dimer, and higher aggregates, or in vivo, where several chlorophyll aggregate forms are expected in the reaction centers and antenna systems. In such cases optically detected magnetic resonance methods for investigating the chlorophyll triplet state become valuable [19]. ODMR allows an optical screening of the system, monitoring the triplet state EPR spectrum on fluorescence peaks which originate from the unit of interest. And by varying the optical detection wavelength among the fluorescence peaks, structural features from different units within the same sample may be evaluated and compared under identical conditions.

We shall first develop the triplet excition model for the interpretation of triplet ODMR data in terms of aggregate structural features, then evaluate its applicibility to the chlorophyll dimer in vitro. Finally, we shall examine structural features of in-vivo chlorophylls through their triplet ODMR spectra.

4.1 The Triplet Exciton Model

As a first approximation, the triplet states associated with a pair of identical chlorophyll molecules may be conveniently described through the simple exciton formalism, a formalism which has been successfully employed in the description of triplet levels in molecular crystals [70, 71, 75, 76]. If necessary, additional corrections to the exciton state functions may be included [11], depending on the adequacy of prediction of physical observables derived from the simple exciton description, to take into account charge transfer, localized or inequivalent environmental interactions, or other terms important to the representation of a specific chlorophyll pair.

Sternlicht and McConnell [75] first formulated the exciton description for identical interacting organic molecules excited into a triplet state in which the intermolecular interaction between molecules is larger than the zero-field splitting. Following their approach, diagonalization of the matrix for the interacting dimer system in a molecular product basis leads to two sets of (triply degenerate)triplet dimer levels [71, 75]. The spin Hamiltonian for the triplet states of the system can then be represented as the average of the spin Hamiltonians for the isolated molecules [75, 76]. In this limit the intermolecular interactions force quantization of the dimer spin system about a new set of principal axes whose orientations are essentially the average of the relative positions of the two monomer principal axis systems [70, 71, 76]. If the dimer system possesses a symmetry axis, one of the "averaged" directions will necessarily fall along the symmetry axis [71, 75, 76]. For a pair of interacting chlorophyll molecules displaying the relatively small zero-field splitting characteristic of extensively delocalized π-electron systems, we may expect that the chlorophyll dimer is adequately described in the strong spin coupling limit of Sternlicht and McConnell [77]. Diagonalization of the dimer matrix, as described by Hochstrasser and Lin [71], leads to the state functions for the triplet dimer levels of the form:

$$^3\psi_i^0(\pm) = 2^{-1/2}(T_A^0 t_A^i S_B^0 \pm S_A^0 T_B^0 t_B^i) \qquad (16)$$

where the subscripts A, B refer to the two molecules of the pair, i = x*, y*, z* designates a specific dimer spin sublevel in the "averaged" principal axis system, $T_{A,B}^0$ refers to the triplet state from space function in the zeroth vibrational level localized on molecule A or B, $t_{A,B}^i$ is the appropriate spin function for the ith triplet spin sublevel (on molecule A or B) and the $S_{A,B}^0$ are the complete ground state functions, also in their lowest-energy vibrational level. The $T_{A,B}^0$ are most conventiently represented by pure spin adiabatic Born-Oppenheimer wave functions for manipulation in expressions for the intramolecular radiationless transitions of molecule A or B, as described by Siebrand et al. [42, 43, 78]. The triplet spin functions $T_{A,B}^0$ contain directly information on the geometrical features of the dimer, since they can be expressed in the form:

$$t_A^i = \xi_{Ax}^i t_{Ax} + \xi_{Ay}^i t_{Ay} + \xi_{Az}^i t_{Az} \qquad (17)$$

where ξ_A^i are the set of direction cosines which express the averaged principal axis system for the dimer triplet in the molecular-based principal axis system centered on molecule A, with a similar expression for t_B^i. Essentially, the dimer state functions in the stron spin coupling limit are the usual product functions of one molecule excited, the second in its ground electronic state, but with the triplet spin functions now quantized along a new axis system determined by the geometry of the pair [71]. This description is valid independently of any symmetry considerations within the dimer, the coefficients $\xi^{i}_{A,B}$ determining the new averaged dimer axis system being obtained directly from the matrix diagonalization [71].

The three spin sublevels in each of the symmetric $({}^3\psi_i^0(+))$ and antisymmetric $({}^3\psi_i^0(-))$ dimer states are initially degenerate. Inclusion of the spin-spin interaction leads to expression for the dimer zero-field energy, X*, Y*, Z*, identical for both the symmetric and antisymmetric dimer levels, of the form given by several authors [72-74] for each of the three dimer spin sublevels

$$X^* = 1/2(\xi_{Ax}^{x^*2} + \xi_{Bx}^{x^*2})X + 1/2(\xi_{Ay}^{x^*2} + \xi_{By}^{x^*2})Y + 1/2(\xi_{Az}^{x^*2} + \xi_{Bz}^{x^*2})Z \tag{18}$$

$$Y^* = 1/2(\xi_{Ax}^{y^*2} + \xi_{Bx}^{y^*2})X + 1/2(\xi_{Ay}^{y^*2} + \xi_{By}^{y^*2})Y + 1/2(\xi_{Az}^{y^*2} + \xi_{Bz}^{y^*2})Z \tag{19}$$

$$Z^* = 1/2(\xi_{Az}^{z^*2} + \xi_{Bx}^{z^*2})X + 1/2(\xi_{Ay}^{z^*2} + \xi_{By}^{z^*2})Y + 1/2(\xi_{Az}^{z^*2} + \xi_{Bz}^{z^*2})Z \tag{20}$$

where, again the $\xi^{i}_{A,B}$ are the coefficients found in the expressions for the dimer triplet spin functions, the direction cosines for the transformation from monomer to dimer principal axes, and the X, Y, Z are the triplet zero-field energies of the monomer [79]. The applicability of these expressions has been verified by EPR studies of triplet excitons in organic crystals [70, 76, 80, 81], and the expressions have been utilized in the study of the geometry of organic aggregates in solution [82,84].

If the dimer system contains a twofold symmetry axis, as is predicted in the current models for the water-linked chlorophyll dimer, the expressions for the zero-field energy, I, simplify further to those given in previous work [17, 74, 77], viz.,

$$I = \xi_x^{i2}X + \xi_y^{i2}Y + \xi_z^{i2}Z \qquad (i = x^*, y^*, z^*). \tag{21}$$

Since the coefficients of the spin functions in the triplet dimer state expressions written in Eqs. (16) and (17) are determined by the geometry of the complex, it is clear that expressions for the probability of transitions from the spin sublevels of the metastable triplet level to the ground state will also reflect the geometry of the pair. Triplet-to-ground state transitions in chlorophylls are predominantly nonradiative [12]. Expressions for nonradiative transition probabilities from any of the six triplet dimer levels ${}^3\psi_i^0(\pm)$ into an isoenergetic vibronic level of the ground state may be straightforwardly expressed using the state functions of Eq. (16). Within our exciton description the ground state vibronic levels can be formally expressed as

$$1_{\psi}{}^{\upsilon}(\pm) = 2^{-1/2}(S_A^{\upsilon}S_B^0 \pm S_A^0 S_B^{\upsilon}) \tag{22}$$

where the quantum number υ refers to a molecular vibrational state of appropriate energy for intersystem crossing. The rate of intersystem crossing from dimer spin sublevel i is determined by the probability of triplet energy conversion to both the symmetric and antisymmetric ground state vibrational exciton levels, i.e., [72]

$$k^{i}_{isc}(\pm) = k_{isc}[^3\psi^0_i(\pm) \rightarrow {}^1\psi^{\upsilon}(+)] + k_{isc}[^3\psi^0_i(\pm) \rightarrow {}^1\psi^{\upsilon}(-)]$$

$$= 2\pi/\hbar|\langle^3\psi^0_i(\pm)|\Omega|^1\psi^{\upsilon}(+)\rangle|^2\rho(E) + 2\pi/\hbar|\langle^3\psi^0_i(\pm)|\Omega|^1\psi^{\upsilon}(-)\rangle|^2\rho(E) \quad (23)$$

where Ω is an operator which contains all terms, intramolecular and intermolucular, which can promote mixing of triplet and singlet dimer functions. Substitution of dimer state functions (16) and (22) into Eq. (23) leads to:

$$k^{i}_{isc}(\pm) = 2\pi/\hbar|1/2[\langle T^0_A t^i_A|\Omega|S^{\upsilon}_A\rangle \pm \langle T^0_B t^i_B|\Omega|S^{\upsilon}_B\rangle + \langle T^0_A t^i_A S^0_B|\Omega|S^0_A S^{\upsilon}_B\rangle \pm \langle S^0_A T^0_B t^i_B|\Omega|S^{\upsilon}_A S^0_B\rangle]|^2\rho(E)$$

$$+ 2\pi/\hbar\, 1/2[\langle T^0_A t^i_A|\Omega|S^{\upsilon}_A\rangle \pm \langle T^0_B t^i_B|\Omega|S^{\upsilon}_B\rangle - \langle T^0_A t^i_A S^0_B|\Omega|S^0_A S^{\upsilon}_B\rangle \pm \langle S^0_A T^0_B t^i_B|\Omega|S^{\upsilon}_A S^0_B\rangle]|^2\rho(E). \quad (24)$$

In each of the two squared expressions in Eq. (24) the first two terms are essentially monomer intersystem crossing rate expressions centered on either molecule A or B, while the last two terms represent *intermolecular* conversion with triplet state energy on one molecule being converted into vibrational energy on the second.

If one neglects the intermolecular terms, Eq. (24) may be simplified and rewritten in terms of only intramolecular spin sublevel intersystem crossing rate expressions for the monomeric nonradiative conversion of triplet energy centered on either molecule A or B [72-74]. This can be accomplished straightforwardly by expansion of the dimer spin functions $t^i_{A,B}$ in the molecular framework [Eq. (17)] and reducing Eq. (24) to expressions involving the individual monomer spin sublevel intersystem crossing rates k_X, k_Y, k_Z [72-74]. Such a procedure has been worked out in detail for the case in which the dimer is essentially undergoing incoherent energy conversion from the triplet to the ground state [73, 74]. However, since ODMR experiments are carried out at low temperatures (1-4 K), it is also necessary to consider the case in which the energy conversion originates from coherent dimer levels. In this case the random phase approximation is not appropriate and the expansion of $t^i_{A,B}$ in Eq. (2), substituted into Eq. (24), again neglecting the intermolecular terms, leads to the result:

$$k^{i}_{isc}(\pm) = 1/4|(\xi^i_{xA} \pm \xi^i_{xB})k_x^{1/2} + (\xi^i_{yA} \pm \xi^i_{yB})k_y^{1/2} + (\xi^i_{zA} \pm \xi^i_{zB})k_z^{1/2}|^2$$

$$+ 1/4|(\xi^i_{xA} \mp \xi^i_{xB})k_x^{1/2} + (\xi^i_{yA} \mp \xi^i_{yB})k_y^{1/2} + (\xi^i_{zA} \mp \xi^i_{zB})k_z^{1/2}|^2. \quad (25)$$

Again, the result demonstrates that the intersystem crossing rate constants for the individual spin sublevels i of the triplet dimer states are expressible as functions of the dimer geometry through the set of direction cosines $\xi^i_{\ell A,B}$ (ℓ = x, y, z) relating monomer and dimer principal axes and measurable properties of the molecular units making up the dimer [72]. If one assumes a twofold symmetry axis and the random phase approximation, this result reduces immediately to that of Bowman and Norris [74], i.e.,

$$k^i_{isc}(\pm) = \xi^{i2}_x k_x + \xi^{i2}_y k_y + \xi^{i2}_z k_z \qquad (i = x^*, y^*, z^*). \tag{26}$$

A similar result has also been obtained by Hagele, Schmid and Wolf [73].

The full expression [24] for the rate of intersystem crossing for a triplet dimer spin sublevel demonstrates clearly that the interpretation of triplet dimer rate data for chlorophyll pair geometries can be a nontrivial problem. Apart from the question of whether the appropriate dimer dynamics description is coherent or incoherent, the presence of the intermolecular terms in Eq. (24) makes the dimer intersystem crossing expressions less simply interpretable when attempting to utilize these results in the determination of dimer geometries. One might expect that since the necessary intermolecular form of Ω would involve derivatives of the spin-orbit coupling operator centered on one molecule with respect to nuclear motions of the second, such terms would be small. However, there are, potentially, a large number of such terms, especially if one considers the triplet energy to be distributed among the vibrational levels of *both* molecules simultaneously on intersystem crossing, and in some systems where there exist centers of large spin-orbit coupling activity their commulative effect may sum to a nonnegligible result [72].

4.2 Application to the Chlorophyll Dimer In Vitro

It can be seen immediately from the form of Eqs. (18-20) that for a chlorophyll dimer of the type suggested by Fong for the chlorophyll dihydrate [53, 60] and by Shipman et al. for a symmetrical dimer [54] in which the molecular axes are all parallel the zero-field splittings expressions for the dimer reduce to the values of the monomer [69]. All the direction cosines, ξ^i, will have the values of either 1 ($\theta = 0°$) or 0 ($\theta = 90°$) and the Eqs. (18-20) predict that the zero-field ODMR transitions in such a dimer will be at the same frequencies as those for the monomer. Two cases provide particularly appropriate tests of the application of the simple exciton approach to the chlorophyll dimer triplet – the Fong-Koester water-saturated chlorophyll hydrocarbon solution (chlorophyll dihydrate dimer) [55, 60], and the Boxer-Closs pyrochlorophyllide *a* covalently linked dimer [85], both of which have been interpreted as containing a symmetrical dimer with parallel axes.

The presence of chlorophyll aggregate species in each of these systems is assessed through their characteristic fluorescence spectra at 2 K. The fluorescence typical of a low temperature hydrocarbon solution containing chlorophyll and either alcohol or water is shown in Figs. 3 and 4. The system whose fluorescence spectrum is depicted in Fig. 3, a frozen hydrocarbon solution containing chlorophyll *a* and an excess of water, is that proposed as an in-vitro model system for the reaction center pair by Fong and Koester [55]. This solution exhibits an absorbance band at 700 nm (A-700 in the Fong-Koester notation), which they conclude is due to a water-linked dimer of chlorophyll *a*. The solution also exhibits fluorescence bands at 680 nm and 720 nm (L-720 in their notation) at low temperatures [55]. The L-720 peak is identified with the water-linked chlorophyll dimer; the fluorescence band at 680 nm is most probably the fully ligated monomer [55]. This spectral assignments are supported by the recent work of Boxer and Closs, who synthesized a covalently linked dimeric derivative of pyrochlorophyllide *a* which can be folded over into a parallel dimer structure in hydrocarbon solvents by the addition of water [85]. In the presence of water the folded pyrochlorophyllide dimer (and zinc-substituted pyrochlorophyllide dimer) exhibits an absorbance at 700 nm and a long-wavelength fluorescence at 720-730 nm; in its unfolded form the system fluoresces at 680 nm (see Fig. 4). We have utilized fluorescence detected ODMR to measure the triplet state zero-field splittings at 2 K for the monomer (680 nm) and the dimer (720 nm) chlorophyll in the Fong-Koester hydrocarbon solution, and the results are presented in Table 3. For further comparison chlorophyll-ethanol solution ODMR transitions were detected on the fluorescence bands at 680 nm and 730 nm, assumed to be the (fully ligated) monomeric and ethanol-linked dimeric forms of chlorophyll *a*, and a high concentration chlorophyll *a*, ethanol, toluene system, in which the ODMR was first obtained at the fluorescence maximum, at 760 nm, and then, after dilution with toluene, at the usual monomer fluorescence band at 680 nm; their ODMR transition frequencies are provided in Table 3.

Table 3. ODMR transitions of chlorophyll *a* dimer systems. All microwave frequencies reported are accurate to ± MHz

	Detection wavelength (nm)	υ_1 (MHz)	υ_2 (MHz)
Chl *a* (10^{-5} M) and	680	728	947
H_2O (10^{-2} M) in	720	707	925
Methylcyclohexane-pentane (1 : 1)			
Chl *a* (10^{-4} M) in	680	707	1026
ethanol	730	639	874
Chl *a* (10^{-5} M) and	680	765	1035
H_2O (10^{-5} M) in			
toluene			
Chl *a* (10^{-2} M) and	760	698	894
EtOh (10^{-2} M) in			
toluene			
Zn Chl *a*(10^{-5} M) and	680	774	1038
H_2O (10^{-2} M) in	720	724	955
methylcyclohexane-pentane (1 : 1)			
Pyro *a* dimer in	682	722	940
toluene-methanol (3 : 1)	733	701	909
Zn pyro *a* dimer in	675	785	1015
toluene-methanol (3 : 1)	720	739	943

From Table 3, for the Fong-Koester system the two sets of ODMR zero-field transitions, detected on the monomer and dimer fluorescence bands, are indeed within $\sim$20 MHz, suggesting that the dimeric species must be very close to an all-axes parallel (or antiparallel) configuration. The zero-field splittings alone do not determine uniquely all direction cosines in Eqs. (18-20) fixing the orientation of the dimer molecules relative to one another [69], but the range of angles fitting the dimer zero-field energies from Eqs. (18-20) requires that the monomer principal axes in the pair all be within 10-14 degrees of parallel. The small calculated deviation from exactly parallel (calculated from the small difference in monomer and dimer zero-field transitions) is most likely a reflection of the approximations inherent in the triplet exciton approach rather than a real geometrical feature of the dimer [69].

Table 3 also provides the zero-field transitions frequencies for the triplet states of the covalently bound dimeric derivative of pyrochlorophyllide *a* and zinc pyrochlorophyllide *a*, as prepared by Boxer and Closs [85], in frozen toluene solutions at 2 K. In hydrocarbon solutions containing water (or methanol) a structural assignment has been made for these systems, consistent with NMR data, in which the two pyrochlorophyllide ring systems fold over and are locked into a plane-antiparallel dimer structure with two water molecules stabilizing the configuration by coordination to the metal (magnesium or zinc) of one chromatophore, and hydrogen bonding to the keto-carbonyl of the second [85]. The covalently linked dimer systems can be converted reversibly between the folded (dimer) and unfolded (monomer) configurations by varying the amount of hydrogen bonding ligand added to the toluene solution and monitored through the fluorescence spectrum (Fig. 4). Again, as was found for the Fong-Koester solutions, the observed zero-field transitions in the magnesium-containing system are within $\sim$30 MHz in the unfolded (monomer) and folded (dimer) forms. For the zinc-containing covalently linked derivative the ODMR transitions are at higher frequencies than for the magnesium-containing dimer, consistent with the previously noted fact [12] that zinc substitution raises the zero-field splitting of chlorophylls. But, again, both the folded and unfolded forms of the zinc dimer display similar transition frequencies, as is expected for a chlorophyll pair in the configuration in which all principal molecular axes are close to parallel (or antiparallel).

In the cases of the chlorophyll-ethanol and high concentration chlorophyll-toluene solutions differences, comparable in both systems, in zero-field ODMR transition frequencies, on what are expected to be monomer and dimer fluorescence peaks, are larger than observed in the Fong-Koester hydrocarbon solution, although the ODMR frequencies for the aggregates are still within 50-100 MHz of the monomer values. We calculate from the zero-field splittings and Eqs. (18-20) that the normal to the molecular planes are in a range of 30-36° in these two systems, considerably outside the range calculated for the Fong-Koester dimer or the pyrochlorophyllide *a* dimer systems. The greater difference in monomer and dimer zero-field splittings in these two cases suggests the possibility of solvent-induced distortion of the dimer complex in ethanol, due to saturation of hydrogen bonding positions around the chlorophyll rings by the solvent, and the possibility of self-complexing of the chlorophyll molecules in the high concentration chlorophyll solution, resulting in a non-plane prallel dimer geometry, as has been suggested for antenna chlorophyll aggregates by Katz et al. [1].

For the Fong-Koester water-saturated chlorophyll hydrocarbon solution and for the covalently linked pyrochlorophyllide *a* dimer systems the triplet state zero-field splitting parameters have indicated that a symmetrical dimer with all axes close to parallel is a reasonable model. In considering the intersystem crossing rate constants for these systems, first neglecting intermolecular effects, Eq. (25) provides the result that for a dimer in which the molecules are planeparallel and the in-plane axes are aligned parallel (or antiparallel), the dimer spin sublevel intersystem crossing rate constants are equal to the monomer spin sublevel rates. Note that for this choice of geometry, viz., all exes parallel , the coherent result described in Eq. (25) and that calculated from an incoherent model [Eq. (26)] predict the same result – monomer and dimer rate constants the same. Further, both models predict that the dimer rates are the same in both the symmetric ($^3\psi(+)$) and the antisymmetric ($^3\psi(-)$) dimer levels [72-74].

First considering the pigment systems listed in Table 4, one sees that all chlorophyll systems, including pyrochlorophyllide *a* (structurally similar to chlorophyll *a*), provide essentially the same overall lifetime, whether monomeric or dimeric, as expected from the triplet exciton model. Comparison of the results for the Fong-Koester solutions (chlorophyll *a*, H_2O, methylcyclohexane: pentane) detected on the monomer and dimer fluorescence bands shows that the prediction of Eq. (25) for an all-axes parallel dimer are qualitatively verified – the order of the rate constants in both monomer and dimer is $k_y > k_x > k_z$. The near equal value of k_z (z referes to an out-of-plane direction in the molecular axis system) for both monomer and dimer clearly indicates consistency with a planeparallel dimer structure. Further, the in-plane spin sublevel rate constants, k_x, k_y, for the dimer are reasonably close, considering the experimental error associated with the rate measurements, to those for the monomer values measured in several solvent environments. For the pyrochlorophyllide *a* dimer in its folded form the dynamics are again reasonably similar to those for the parent chlorophyll monomer, as its is overall lifetime. For these systems, therefore, considering both the triplet zero-field splittings and intersystem crossing rate constants, the triplet exciton model provides a description of triplet dimer properties consistent with a symmetrical dimer structure as proposed by Fong [60], and by Boxer and Closs [85].

In the case of the zinc-substituted chlorophyll systems, however, a quite different result is obtained, as seen in Table 4. For the zinc-chlorophyll dimer (and zinc-pyrochlorophyllide *a* dimer) the triplet zero-field splittings were observed to be close to what is expected for a symmetrical all-axes-parallel dimer, as was the case for the magnesium-containing chlorophyll dimer systems. Therefore, one expects from the previous discussion on intersystem crossing rates that the zinc-chlorophyll dimer will also display intersystem crossing rate constants similar to those for the monomer. However, in Table 4 one finds that the overall triplet lifetime is increased substantially on aggregation in the zinc-chlorophyll, water, methylcyclohexane-pentane system and in the folded form of the zinc-pyrochlorophyllide *a* dimer. For the individual spin sublevels, as well, the values in the dimer solution are consistently larger than those observed previously for the zinc chlorophyll *a* monomer [12]. For the zinc pyrochlorophyllide *a* covalently linked dimer, the intersystem crossing rates measured in the unfolded configuration agree quite well with those for zinc chlorophyll *a*, an unexpected result, since the small substituent differences between the chlorophyll and pyrochloro-

Table 4. $T_1 \rightarrow S_0$ intersystem crossing rates for the individual spin sublevels of chlorophyll dimer systems (in sec^{-1})

System	Detection wavelength (nm)	k_X	k_Y	k_Z	K_{calc}[a]	K_{calc}[b]
Chl *a* (10^{-5} M) and H_2O (10^{-2} M) in	680	564 ± 100	1330 ± 200	176 ± 50	690	690 ± 30
methylcyclohexane-pentane (1 : 1)	720	732 ± 110	980 ± 170	180 ± 30	630	690 ± 50
Pyro *a* dimer in toluene-methanol (3 : 1)	733	830 ± 100	1070 ± 80	170 ± 20	690	
Pyro *a* molecule in toluene	680	---	---	---	---	720 ± 20
Zn Chl *a* (10^{-5} M) and H_2O (10^{-2} M) in	680	See Table 2				
methylcyclohexane-pentane (1 : 1)	720	525 ± 25	585 ± 25	1220 ± 50	770	800 ± 50
Zn pyro *a* in	675	340 ± 20	380 ± 20	660 ± 40	460	470 ± 50
toluene-methanol (3 : 1)	720	600 ± 30	620 ± 30	1490 ± 60	903	980 ± 50

[a] $K_{calc} = 1/3\ (k_X + k_Y + k_Z)$

[b] K_{obs} obtained from the overall triplet state lifetime at 77K

phyllide ring systems should not substantially alter the intersystem crossing mechanisms [12, 42]. When the dimer is folded, the rates increase for each spin sublevel, all values larger by approximately a factor of two over those found in the unfolded form, In both systems, the zinc chlorophyll *a* solution and the zinc pyrochlorophyllide *a*, the measured triplet lifetime agrees well with the average of the individual spin sublevel rates, indicating reliability in the measurements. We interpret these increases in intersystem crossing system efficiency upon aggregation, when the zero-field splittings remain about the same, as due to the additional intermolecular terms in the rate expression given in Eq. (24), now important due to the presence of the heavy atom metal centers in the dimer system. In the case of magnesium-containing chlorophyll systems the relatively light metal center did not enhance the nonradiative transition probability over that for the monomer; but with zinc atom substitution the intermolecular terms become important and increase the intersystem crossing efficiency in the complex compared with the isolated monomer units. It is important to emphasize that this substantial increase in intersystem crossing rate upon aggregation of the zinc-containing chlorophyll and pyrochlorophyllide *a* is accompanied by no major change in zero-field splitting. These two results seem consistently explained by invoking an additional metal-induced mechanism for triplet energy conversion in the zinc-containing dimer and, further, suggest that Eq. (25) [or Eq. (26)] without intermolecular terms is indeed adequate to explain normal magnesium-containing chlorophyll aggregation dynamics within the triplet exciton model.

The value of ODMR spectroscopy to the investigation of structural features of chlorophyll aggregates is evident in the results on the in-vitro chlorophyll dimer. Using optical selectivity it is possible to measure triplet state properties of monomeric and dimeric chlorophyll systems under identical conditions in the same solution; this situation minimizes difficulties in applying results of the triplet exciton description when comparing monomer and dimer triplet state properties to evaluate geometrical features of the dimer. Further, ODMR provides the possibility of identifying, in terms of their molecular or aggregate makeup, the origin of features in chlorophyll optical spectra from their distinctive triplet state ODMR properties [14, 15, 17].

The results of triplet state ODMR measurements for the chlorophyll dimer systems studied provide a basis for the applicability of the simple triplet exciton model to the study of chlorophyll aggregate structure. Both the zero-field splittings and individual spin sublevel intersystem crossing rate constants for the water-linked chlorophyll dimer and pyrochlorophyllide *a* dimer are reasonably predicted by the exciton relationships, in terms of the monomer properties and the geometry of the dimer [72-74]. In particular, Eqs. (18-20) provide an adequate description of the zero-field splittings, and Eq. (25) gives a reasonable representation of the intersystem crossing rates, suggesting that no new intermolecular terms are introduced into the intersystem crossing mechanism upon dimerization of ordinary chlorophylls. It is not, however, possible to distinguish between a coherent or an incoherent description for the intersystem crossing on the basis of the in-vitro chlorophyll systems. Heavy metal substitution may complicate the simple exciton description of intersystem crossing, but from the present results such effects are not expected to be important for in-vivo chlorophyll.

Finally, the results of the ODMR investigations of solvent-linked chlorophyll dimer systems provide data consistent with the model proposed by Fong for the chlorophyll dihydrate dimer [53, 60], and by Boxer and Closs for the covalently linked dimer [85]. The results of these experiments support the formation of chlorophyll dimers in vitro, in which the solvent linkages provide a plane antiparallel configuration [52, 54, 60, 85].

4.3 Chlorophyll Aggregate Structure In Vivo

The first applications of zero-field triplet state ODMR to in-vivo photosynthetic systems were to the purple bacteria, systems for which the existence of high-field triplet state EPR in chemically reduced bacteria had been reported [7-10]. The triplet state EPR signals for the bacteria have been interpreted as arising from bacteriochlorophyll molecules within the reaction center, observed when the primary electron acceptor has been reduced [2, 10, 11]. If the bacteriochlorophyll within the reaction center exists in the form of a "special pair", the dimeric form expected for reaction center chlorophylls [1, 2, 11], then one expects to be able to relate the triplet state properties observed in vivo to the geometrical features of the reaction center bacteriochlorophyll dimer as has been described in the previous section. Using the triplet state zero-field splittings and intersystem crossing rate expressions from Eqs. (18-20, 25, 26), and the bacteriochlorophyll *a* molecular triplet state properties from Tables 1 and 2, one can determine the relative orientation of the reaction center bacteriochlorophyll pair [17, 72].

The fluorescence-detected zero-field ODMR transitions have been observed for a variety of chemically reduced purple bacteria; a typical ODMR spectrum, that for the bacterium *Rhodopseudomonas Spheroides*, is presented in Fig. 9. The most striking feature of the ODMR spectra observed for the purple bacteria was that all produced very similar sharp, symmetrical and structureless ODMR spectra [13, 17]. Further, there was a strong similarity in $T_1 \rightarrow S_0$ rate constants given in Table 5. The similarity in triplet state zero-field transition frequencies, intensities, and depopulating rates among the bacteria presumably reflects the fact that the reaction centers in each of the bacterial systems have a common structure in terms of reaction center makeup and the relative orientations and interactions of the pigment molecules within the reaction center [13]. The sharp, symmetrical, structureless ODMR bands are consistent with a uniformity in structure among the reaction centers in the cell and with a delocalization of the triplet excitation within the reaction center [13].

The similarity in reaction center structure is particularly evident from the zero-field splitting parameters. D and E values have been reported by EPR for a wide variety of photosynthetic bacteria, for both those containing the bacteriochlorophyll *a* pigment and the less common bacteriochlorophyll *b* [21]. These zero-field splittings provide a first indication of the geometry of the pair through Eqs. (18-20) developed in Sect. 4.1. Using these expressions we have calculated the angle between the planes for the reaction center dimers from the most recent reported zero-field splittings for the photosynthetic bacteria measured to date [21]. Each calculation produced a range of angles, all consistent with the monomer pigment zero-field splitting, the zero-field splitting for the bacterium and the standard deviations reported for the zero-field splitting measurements. The calculations on bacteria containing bacteriochlorophyll *a* were performed using the zero-field splitting calues in Table 5. The bacteriochlorophyll *b* monomer

Table 5. Triplet state zero-field splitting parameters and intersystem rate constants[a,b] for some photosynthetic bacteria

System	ODMR detection wavelength (nm)	υ_{yz} (MHz)	υ_{xz} (MHz)	$/D/(cm^{-1})$	$/E/(cm^{-1})$	$k_x(sec^{-1})$	$k_y(sec^{-1})$	$k_z(sec^{-1})$
R. Rubrum	920	467 ± 5	668 ± 5	0.0189 ± 0.0003	0.0032 ± 0.0002	2105 ± 182	2885 ± 400	1335 ± 58
R. Spheroides (strain 2.4.1)	908	470 ± 3	660 ± 5	0.0188 ± 0.0002	0.0032 ± 0.0001	2675 ± 300	3033 ± 200	1600 ± 180
R. Spheroides (R-26 mutant)	908	469 ± 3	656 ± 3	0.0187 ± 0.0002	0.0031 ± 0.0001	2660 ± 130	3183 ± 246	1595 ± 75
Chromatium Vinosum	932	440 ± 3	645 ± 5	0.0181 ± 0.0002	0.0034 ± 0.0001	--	--	--

[a]Axis designations x, y, z are arbitrarily chosen

[b]Values for intersystem crossing rate constants have also been obtained for the purple bacteria by Hoff [68]. Although the values above agree with Hoff's for k_z, there is substantial disagreement over the values for k_x, k_y. Whether these differences are intrinsic to the specific systems studied or to the conditions of the experiments has yet to be fully explained; however, work by T.J. Schaafsma and S.J. van der Bent (Proc. Third Specialized Colloque Ampere, Dublin, 1977, to be published) indicates that Hoff's values for k_x, k_y can be extrapolated to the values above by appropriate treatment of the dependence on triplet-triplet absorption effects and excitation power of the ODMR rate data

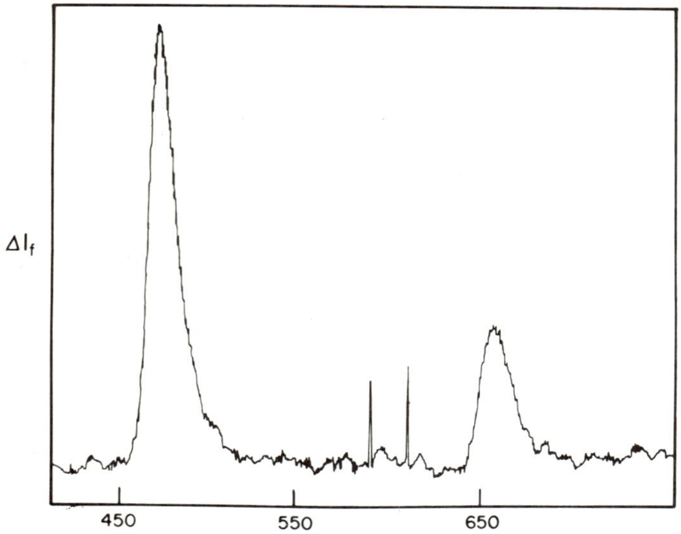

Fig. 9. Fluorescence-detected ODMR spectrum of *R. Spheroides* cells (chemically reduced) at 2 K. Sharp features at 560 and 600 are frequency markers; frequency scale in megahertz

zero-field splittings were the most recent values determined by Thurnauer and Norris. Results of all these calculations are presented in Table 6.

The most striking feature of the table is the uniformity of angles calculated among the bacteria. All give angles around 50° (±10°), even with different pigment molecules, as in the case of *Rps. viridis* which contains bacteriochlorophyll *b*. If the triplet state EPR spectra do arise from the special pair within the reaction center [2, 10, 11], then there appears to be a generally preferred orientation of the planes of the molecules making up the pair for all these bacteria, with the pigment molecules tipped relative to one another by about 50°. Of course, the zero-field splitting calculations give no indication of the degree of difference in relative alignment of the in-plane axes of the two molecules, not enough parameters are provided from the zero-field splittings along to fix all the direction cosines in Eqs. (18-20), but the overall orientation of the pigment planes appears roughly constant in all the varieties of bacteria, supporting the point that the reaction centers seem to have a common structure.

If one includes the intersystem crossing rates for the purple bacteria, listed in Table 5, triplet exciton Eqs. (25, 26) for the triplet sublevel decay give further information on the geometry of the reaction center special pair in photosynthetic bacteria. The results for the in-vitro chlorophyll dimer from the previous section indicate that the intermolecular terms in the rate expressions may be neglected, and it is assumed that this conclusion applies as well to the bacteriochlorophyll pair. We have attempted to fit the data, using both the coherent and incoherent intersystem crossing expressions. Because of the appreciable standard deviation associated with measurements of triplet intersystem crossing rate constants (compared with <1% error in the measurement of zero-field transition frequencies), there is a considerable range of angles which are consistent with the intersystem crossing rate data [72]. The general result of fitting

Table 6. Calculated angle between the molecular planes in the reaction center bacteriochlorophyll dimer of photosynthetic bacteria from the triplet state zero-field splitting

Bacterium		Calculated angle (degrees)
Rps. viridis		
strain NHTS	Cells	51 ± 9
	Chromatophores	50 ± 10
C. vinosum		
strain D	Cells	51 ± 9
	Chromatophores	51 ± 9
Rps. sphaeroides		
strain 2.4.1	Cells	48 ± 8
strain Ga	Cells	46 ± 10
	Chromatophores	48 ± 8
strain R-26	Cells	46 ± 10
	Chromatophores	46 ± 10
	Reaction centers	48 ± 10
Rps. capsulata		
strain St. Louis	Cells	48 ± 8
strain SB 25	Cells	48 ± 8
strain BY 761	Cells	48 ± 8
Rps. gelatinosa		
strain I	Cells	48 ± 8
Rps. palustris		
strain 2.1.6	Cells	50 ± 6
R. rubrum		
strain S 1	Cells	46 ± 10
strain G 9	Cells	48 ± 8

both the triplet zero-field splittings and intersystem crossing rates (including their associated error limits) through the triplet dimer expressions is a bacteriochlorophyll special pair with the molecular planes inclined at an angle of about 50 degrees and the in-plane angles twisted relative to one another by roughly 50 to 90 degrees. Although there is a fair degree of uncertainty in such a calculation, the triplet data do provide a general picture of a rection center special pair in bacteria distorted appreciably away from a simple planeparallel geometry.

In the recent literature there has been some controversy over the values of the triplet intersystem crossing rate constants obtained for the chemically reduced preparations of purple bacteria. In addition to our observations of the rate constant given in Table 5, Hoff has reported values for the same bacteria which were in rough agreement with out values for k_Z, but with considerably larger values for k_X, k_Y [68]. Further, the average values of the individual spin sublevel rate constants obtained by Hoff are in good agreement with optical transient signals (orginating in a state referred to as P^R by Parson et al. [2], whereas our values give an overall average decay rate slower by a factor of 2-3. All other features of the ODMR spectra of the bacteria observed by the two groups are identical [28, 68]. The reasons for the differences in

k_x, k_y remain unclear. A recent discussion for Hoff et al. [90] is not an appropriate resolution to the difference, since it relies on assumptions about microwave sweep rates which are not applicable to the experimental methods used for our measurements (our experiments measure the dynamics of building to steady-state population with a *cw* source, rather than with swept of pulsed microwaves). Whether the differences are due merely to other technical problems, or to intrinsic differences in the systems examined (our measurements were made on freeze-dried cell preparations; Hoff investigated freshly prepared cells) is still not clear. But both ODMR results, on the basis of zero-field splittings alone [69], still provide information on the geometrical features of the reaction center dimer in photosynthetic bacteria through the triplet exciton model.

A comparison of the triplet state results obtained for the chlorophyll complexes in solution and those for reaction center bacteriochlorophyll in photosynthetic bacteria provide an interesting contrast in possible dimer geometry. Triplet state ODMR spectra for several of the in-vitro chlorophyll systems investigated in Sect. 4.2 display zero-field splittings which confirm the general features of models for chlorophyll dimerization by solvent linkage proposed by Fong [60] and by Katz et al. [54], namely dimer structures in which the molecules are held together in a plane-antiparallel configuration. For the case of in-vivo measurements on photosynthetic bacteria reaction centers, however, the zero-field splittings clearly indicate significant deviations from such a planar arrangement in all bacteria investigated.

The contrast between the results of these calculations and those for the in-vitro chlorophyll dimer systems immediately suggests a difference in dimer structure between the in-vivo and in-vitro complexes. Such a result is perhaps not unexpected. In solution, as predicted by Fong [60] and by Katz et al. [54], chlorophyll locks into a dimeric structure with a configuration determined principally by the interaction of solvent molecules coordinating to the metal center of one chlorophyll molecule and hydrogen bonding to a second. The calculated geometry in several of these systems is very close to the predicted parallel orientation (Fong-Koester solution, covalently bound pyrochlorophyllide dimer), with possible deviations from a planeparallel configuration by additional solvent interactions in ethanol solution. In the case of the photosynthetic bacteria, however, the calculations leave little doubt that the dimer structure is considerably removed from a planeparallel orientation. In these systems the dimer orientation will be determined not only by interactions linking the pair, but also by the environmental interactions from surrounding proteins and other nearby pigment molecules, all of which must be considered in arriving at the most stable configuration of the special pair in the reaction center in vivo.

Application of triplet state EPR spectroscopy to green plant systems has been far more limited than for photosynthetic bacteria, due, in part, to the higher complexity of the photosynthetic apparatus involved. Some preliminary reports of triplet state zero-field fluorescence detected ODMR in algae [15, 31] and spinach chloroplast preparations [14] have been published, with interesting results. Hoff and van der Waals have reported ODMR spectra of chloroplast suspensions reduced with dithionite and obtain transition frequencies very close to those observed for the chlorophyll *a* molecule in hydrocarbon solution [14, 16]; they have attributed these transitions as due to a pigment species associated with photosystem II [14]. Van der Bent und Schaafsma have observed these same transitions when monitoring the fluorescence of a series of algae, along with an additional set of ODMR transitions to higher frequencies generated after prolonged excitation, whose origin is as yet unidentified [15].

The models of Fong [52, 53, 60] and Katz et al. [1, 54] for the reaction center chlorophyll dimer structure and the equations developed in Sect. 4.1 provide very specific predictions for the triplet state zero-field splittings and intersystem crossing rate constants to be expected from such dimer structures in vivo. To date no triplet state data have been obtained which can unambiguously be related to reaction center chlorophylls in a green plant system to test the models.

Because of the complexity of the dual photosystem arrangement in green plants [1] and of the fluorescence originating from it, a useful approach may be to study separately photosystem I and photosystem II particles, preparations of which are known [86, 87], simplifying the interpretation of any triplet ODMR signals observed. Such an approach has been utilized by Junge et al. [88] in attempting to evaluate the reaction center geometry of photosystem I by photoinduced linear dichroism. Some limited work has been done on purified subchloroplast preparations. Hoff et al. have examined enriched photosystem I particles by fluorescence-detected ODMR using high microwave power, and find ODMR transitions in the 1000 to 1300 MHz region, frequencies larger than those typically observed for the chlorophyll triplet state [89]. They speculate that these signals may originate from pheophytin triplet states present in their preparations. Our laboratory has also examined subchloroplast particles, which have been chemically reduced with sodium dithionite [91]. We find ODMR signals when detecting fluorescence at 680 nm and 723 nm, regions assigned by Butler [92] to be associated with the light-harvesting chlorophyll-protein complex and photosystem I, respectively. Both peaks provide ODMR transition frequencies lower than those observed at 670 nm, assigned as free chlorophyll pigment solubilized in the fractionation procedures. We believe that the ODMR transitions observed at 680 nm and 723 nm arise from antenna pigment molecules. Careful preparative work still remains for unambiguous spectroscopic examination of reaction center pigment systems.

The identification of triplet state spectra for photosystem I and photosystem II reaction center chlorophyll remains an important future goal for ODMR. Such measurements will provide a critical assessment of the appropriateness of the solvent-linked planeparallel chlorophyll dimer model for reaction center chlorophyll in plant systems.

Acknowledgments. I am grateful to the U.S. Department of Energy, the U.S. Army Research Office and the Petroleum Research Fund, administered by the American Chemical Society for their support of various aspects of this work. I am also deeply grateful to the people in my laboratory, especially P.A. Chiha, R.E. Connors, H.A. Frank, D.R. Hobart, R.H. Hofeldt, and W.R. Leenstra, whose hard work and helpful discussion have contributed substantially to this paper.

References

1. Katz, J.J., Norris, J.R.: Curr. Top. Bioenerg. **5**, 41 (1973)
2. Parson, W.W., Cogdell, R.J.: Biochim. Biophys. Acta **416**, 105 (1975)
3. Hutchinson, C.A., Mangum, B.W.: Chem. Phys. **29**, 952 (1958)

4. Hutchinson, C.A., Mangum, B.W.: J. Chem. Phys. **34**, 908 (1961)
5. Rikhireva, G.T., Gribova, Z.B., Kayushin, L.P., Umrikhina, A.V., Krasnovskii, A.A.: Dokl. Acad. Nauk SSSR, **159**, 196 (1964)
6. Lhoste, J.M.: C.R. Acad. Sci. **226D**, 1059 (1968)
7. Dutton, P.L., Leigh, J.S., Siebert, M.: Biochem. Biophys. Res. Commun. **46**, 406 (1972)
8. Dutton, P.L., Leigh, J.S., Reed, D.W.: Biochim. Biophys. Acta **292**, 654 (1973)
9. Wraight, C.A., Leigh, J.S., Dutton, P.L., Clayton, R.K.: Biochim. Biophys. Acta **333**, 401 (1974)
10. Leigh, J.S., Dutton, P.L.: Biochim. Biophys. Acta **357**, 67 (1974)
11. Thurnauer, M.C., Katz, J.J., Norris, J.R.: Proc. Natl. Acad. Sci. USA **72**, 3270 (1975)
12. Clarke, R.H., Connors, R.E., Schaafsma, T.J., Kleibeuker, J.F., Platenkamp, R.J.: J. Am. Chem. Soc. **98**, 3674 (1976).
13. Clarke, R.H., Connors, R.E.: Chem. Phys. Lett. **42**, 69 (1976)
14. Hoff, A.J., Van der Waals, J.H.: Biochim. Biophys. Acta **423**, 615 (1976)
15. Van der Bent, S.J., Schaafsma, T.J., Goodheer, J.C.: Biochem. Biophys. Res. Commun. **71**, 1147 (1976).
16. Nissani, E., Schertz, A., Levanon, H.: Photochem. Photobiol. **25**, 93 (1977)
17. Clarke, R.H., Connors, R.E., Frank, H.A.: Biochem. Biophys. Res. Commun. **71**, 671 (1976)
18. El Sayed, M.A.: J. Chem. Phys. **54**, 680 (1971)
19. Clarke, R.H., Hofeldt, R.H.: J. Chem. Phys. **61**, 4582 (1974)
20. Fong, F.K., Galloway, L.: J. Am. Chem. Soc. **100**, 3594 (1978)
21. For a thorough recent review article on the chlorophyll triplet state see Levanon, H., Norris, J. R.: Chem. Rev. **78**, 185 (1978)
22. Maki, A.H., Co, T.: Biochemistry **15**, 1229 (1976)
23. Chiha, P.A., Clarke, R.H., Kramer, E.: Chem. Phys. Lett. **50**, 61 (1977)
24. Sharnoff, M.: J. Chem. Phys. **46**, 3263 (1967)
25. Clarke, R.H., Hayes, J.M.: J. Chem. Phys. **59**, 3113 (1973)
26. Van Dorp, W.G., Schaafsma, T.J., Soma, M., Van der Waals, J.H.: Chem. Phys. Lett. **21**, 221 (1973)
27. Van Dorp, W.G., Schoemaker, W.H., Soma, M., Van der Waals, J.H.: Mol. Phys. **30**, 1701 (1975)
28. Clarke, R.H., Connors, R.E.: Chem. Phys. Lett. **33**, 365 (1975)
29. Chiha, P.A., Clarke, R.H.: J. Magn. Reson. **29**, 535 (1978).
30. Clarke, R.H., Hofeldt, R.H.: J. Am. Chem. Soc. **96**, 3005 (1974)
31. Van der Bent, S.J.: Thesis, Agricultural University, Wageningen, 1977
32. Clarke, R.H., Hayes, J.M.: Chem. Phys. Lett. **27**, 556 (1974)
33. Avarmaa, R.: Chem. Phys. Lett. **46**, 279 (1977)
34. Kleibenker, J.F.: Thesis, Agricultural University, Wageningen, 1977
35. Hutchinson, C.A., Nicholas, J.V., Scott, G.W.: J. Chem. Phys. **53**, 1906 (1970),
36. Chan, I.Y., Van Dorp, W.G., Schaafsma, T.J., Van der Waals, J.H.: Mol. Phys. **22**, 753 (1971)
37. Clarke, R.H., Hofeldt, R.H.: Proceedings of the International Conference on Excited States of Biological Molecules. Birks, J.B. (ed.) p. 309. New York: Wiley 1975
38. Thurnauer, M.C., Norris, J.R.: Chem. Phys. Lett. **47**, 100 (1977)
39. Clarke, R.H., Frank, H.A.: Chem. Phys. Lett. **51**, 13 (1977)
40. Frank, H.A.: Ph.D. thesis, Boston University, 1977
41. Krasnovskii, A.A., Lebeder, N.N., Litvin, F.F.: Dokl. Akad. Nauk SSR **216**, 1406 (1974)
42. Clarke, R.H., Frank, H.A.: J. Chem. Phys. **65**, 39 (1976)
43. Lawetz, V., Orlandi, G., Siebrand, W.: J. Chem. Phys. **56**, 4059 (1972)
44. Metz, F., Friedrich, S., Hohlneicher, G.: Chem. Phys. Lett. **16**, 353 (1972)
45. Antheunis, D.A., Schmidt, J., Van der Waals, J.H.: Mol. Phys. **27**, 1521 (1974)
46. Van Dorp, W.G., Soma, M., Kooter, J.A., Van der Waals, J.H.: Mol. Phys. **28**, 1551 (1974)
47. Spangler, D., McKinney, R., Christoffersen, R.E., Maggiora, G.M., Shipman, L.L.: Chem. Phys. Lett. **36**, 427 (1975)
48. Chow, H., Serlin, R., Strouse, C.E.: J. Am. Chem. Soc. **97**, 7230 (1975)
49. Siebrand, W.: J. Chem. Phys. **47**, 2411 (1967)
50. Connolly, J.S., Gorman, D.S., Seely, G.R.: Ann. N.Y. Acad. Sci. **206**, 649 (1973)
51. Weiss, C.: J. Mol. Spectr. **44**, 37 (1972)

52. Fong, F.K.: Appl. Phys. **6**, 151 (1975).
53. Fong, F.K., Koester, V.J.: J. Am. Chem. Soc. **97**, 6888 (1975)
54. Shipman, L.L., Cotton, T.M., Norris, J.R., Katz, J.J.: Proc. Natl. Acad. Sci. USA **73**, 1791 (1976)
55. Fong, F.K., Koester, V.J.: Biochim. Biophys. Acta **423**, 52 (1976)
56. Closs, G.L., Katz, J.J., Pennington, F.C., Thomas, M.R., Strain, H.H.: J. Am. Chem. Soc. **85**, 3809 (1963)
57. Norris, J.R., Druyan, M.E., Katz, J.J.: J. Am. Chem. Soc. **96**, 1680 (1973)
58. Norris, J.R., Uphans, R.A., Crespi, H.L., Katz, J.J.: Proc. Natl. Akad. Sci. USA **68**, 625 (1976)
59. Fong, F.K.: J. Am. Chem. Soc. **98**, 7840 (1976)
60. Fong, F.K., Koester, V.J., Galloway, L.: J. Am. Chem. Soc. **99**, 2372 (1977)
61. Brown, J.S.: Photochem. Photobiol. **26**, 319 (1977)
62. Jones, M.T., Phillips, W.D.: Annu. Rev. Phys. Chem. **17**, 323 (1966)
63. Abragam, A., Bleaney, B.: Electron Paramagnetic Resonance of Transition Ions. New York: Oxford University Press 1970
64. Bolton, J.R., Borg, D., Schwartz, H.: Biological Applications of Electron Spin Resonance Spectroscopy. New York: Wiley-Interscience 1972
65. Maki, A.H., Zuchlich, J.A.: Top. Curr. Chem. **54**, 115 (1975)
66. McConnell, H.M., McFarland, B.G.: Quart. Rev. Biophys. **3**, 91 (1970).
67. Clarke, R.H., Connors, R.E., Norris, J.R., Thurnauer, M.C.: J. Am. Chem. Soc. **97**, 7178 (1975)
68. Hoff, A.J.: Biochim. Biophys. Acta **440**, 765 (1976).
69. Clarke, R.H., Hobart, D.R.: FEBS Lett. **82**, 155 (1977)
70. Haarer, D., Schmid, D., Wolf, H.C.: Phys. Stat. Sol. **23**, 633 (1967)
71. Hochstrasser, R.M., Lin, T.S.: J. Chem. Phys. **49**, 4929 (1968)
72. Clarke, R.H., Hobart, D.R., Leenstra, W.R.: J. Am. Chem. Soc.
73. Hagele, W., Schmid, D., Wolf, H.C.: Z. Naturforsch. **33a**, 94 (1978)
74. Bowman, M.K., Norris, J.R.: Chem. Phys. Lett. **54**, 45 (1978)
75. Sternlicht, H., McConnell, H.M.: J. Chem. Phys. **35**, 1793 (1961)
76. Haarer, D., Wolf, H.C.: Mol. Cryst. **10**, 359 (1970)
77. Clarke, R.H., Connors, R.E., Frank, H.A., Hoch, J.C.: Chem. Phys. Lett. **45**, 523 (1977).
78. Henry, B.R., Siebrand, W.: J. Chem. Phys. **54**, 1072 (1971)
79. The zero-field splitting energies X, Y, Z are related to the previously defined D and E parameters by D = 3/2 Z, E = 1/2(Y-X)
80. Schwoerer, M., Wolf, H.C.: Mol. Cryst. **3**, 177 (1967)
81. Yarmus, L., Rosenthal, J., Chopp, M.: Chem. Phys. Lett. **16**, 477 (1967)
82. Schmidt, H.: Z. Naturforsch. **26a**, 763 (1971).
83. Schmidt, H., Zellhofer, R.: Z. Physik. Chem. N. F. **91**, 204 (1974)
84. Schweitzer, D., Hauser, K.M., Kirrstetter, R.G.H., Staub, H.A.: A. Naturforsch. **21a**, 1189 (1976)
85. Boxer, S.G., Closs, G.L.: J. Am. Chem. Soc. **98**, 5406 (1976)
86. Bengis, C., Nelson, N.: J. Biol. Chem. **250**, 2783 (1975)
87. Wessels, J.S., Van Alphen-van Waveren, O., Voorn, G.: Biochim. Biophys. Acta **292**, 741 (1973)
88. Junge, W., Schaffernicht, H., Nelson, N.: Biochim. Biophys. Acta **462**, 73 (1977)
89. Hoff, A.J., Govindjee, Romijn, J.C.: FEBS Lett. **73**, 191 (1977)
90. Hoff, A.J., De Vries, H.G.: Biochim. Biophys. Acta **503**, 94 (1978)
91. Clarke, R.H., Jagannathan, S.P., Leenstra, W.R.: In: Lasers in Photomedicine and Photobiology, Pratesi, R. (ed.). Berlin-Heidelberg-New York: Springer 1980
92. Butler, W.L., Kitajima, M.: Proceedings of the Third International Conference of Photosynthesis, Auron, M. (ed.), pp. 13-15. Amsterdam: Elsevier 1975.

Chapter 7
Synthetic Approaches to Photoreaction Center Structure and Function

Michael R. Wasielewski[1]

1 Introduction

An ever-expanding body of evidence supports the concept that photosynthesis both in green plants and in photosynthetic bacteria depends on the cooperative interaction of large numbers of chlorophyll molecules [1-4]. The vast majority of chlorophyll molecules in photosynthetic organisms function as a light-harvesting or antenna system. Incident photons striking these antenna chlorophylls produce an electronic excitation that propagates along the antenna array. A small number of chlorophylls in these organisms reside in a specialized environment proximate to electron donors and acceptors. The entire specialized assembly is called a photoreaction center. The chlorophylls in the photoreaction center trap the electronic excitation produced in the antenna, leading to rapid oxidation of the photoreaction center chlorophyll. The charge separation thus produced drives a series of dark electron transfer reactions, ultimately producing the chemical potential gradient necessary to carry out the organism's metabolism.

The detailed nature of this initial charge separation will be the primary concern of this chapter. The photoreaction center chlorophylls P865 of the purple photosynthetic bacterium *Rhodopseudomonas sphaeroides* [5] and P700 of photosystem I in green plants [6] have been best characterized thus far. Trapping of the electronic excitation in the particular antenna system associated with each of these phototraps results in one electron oxidation of the trap to yield $P865^+$ and $P700^+$. These oxidations are accompanied by the observed formation of a Gaussian EPR signal which is in each case line-narrowed by about a factor of $1/\sqrt{2}$ from that of the corresponding signal produced via one electron oxidation of bacteriochlorophyll *a* (BChl *a*) or chlorophyll *a* (Chl *a*) in vitro [7, 8]. This result is consistent with the delocalization of spin over two chlorophyll molecules in each phototrap. An elegant determination of the hyperfine splittings of the radical cations present in $P865^+$ by ENDOR spectroscopy provides strong evidence for the chlorophyll special pair nature of the primary photochemical electron donor in P865 [7–10].

In what might be called the traditional approch to studies of reaction center structure and function a great deal of success has been achieved in separating the

---

1 Chemistry Division, Argonne National Laboratory, Argonne, IL 60439, USA

antenna pigment system from the attendant reaction center proteins. The greatest success in this respect is the isolation of reaction centers from purple photosynthetic bacteria completely free of the antenna system [11-13]. The availability of these reaction centers has not only made possible the identification of the molecules responsible for the primary photochemistry, but has also enabled a rather detailed analysis of the electron transfer kinetics in these systems to be accomplished.

On the other hand, a similar separation of the antenna system from P700 in green plants has proven more difficult. The best preparations to date yield particles having 20-40 antenna Chl-*a* molecules per reaction center Chl *a* [14]. The added complexity of this system has precluded direct observation of many reaction center characteristics.

The voluminous work that has appeared concerning P700 and P865 has yielded many suggestions as to the molecular geometry and distances involved in the electron transfer processes taking place in the reaction center. As yet these questions remain for the most part unresolved. The high efficiency with which photoinduced charge separation occurs in the reaction center undoubtedly depends strongly on these geometric factors.

For the past few years a handful of workers, ourselves included, have undertaken what can be termed a new "synthetic" approach to the mechanism of photosynthetic charge separation. In this context the word synthetic is used in its broadest sense. Our goal has been to prepare in the laboratory well-defined chlorophyll structures which model the known physical parameters of the photosynthetic reaction center well enough that these models may be used to make predictions concerning the natural photoreaction center. These models may be used to test mechanistic ideas and hopefully advance some of these derived concepts into the realm of the probable. An understanding of the in-vivo chemistry via an in-vitro model leads naturally to the laboratory duplication of the natural energy conversion chemistry.

We have developed criteria for examining the utility of a given model system that form the basis of our own work and are used to examine other work performed in this area. For example, the most important feature of any photoreaction center model is that it be structurally well defined. They key to understanding photosynthetic charge separation is to know the relative placement of the electron donors and acceptors with respect to one another. Once a plausible structure for a model has been established, the electronic states responsible for the ensuing charge transfer chemistry can be fully characterized.

The first attempts to model reaction center structures utilized porphyrins rather than chlorophylls. There remains today a strong tendency in the literature to make this substitution. The primary reason for the use of porphyrins is one of apparent convenience in chemical synthesis. Unfortunately, as we shall see below, the use of porphyrin models diminishes the information content of the model as it applies to reaction center chemistry. A summary of the porphyrin work appears in the following section.

2 Porphyrin Models of Photoreaction Center Chlorophylls

The study of covalently linked porphyrin dimers has given rise to two basic types of model. In one class of models the porphyrins are joined by a single carbon chain usually via diester or diamide linkages. The second class of models introduced more recently combines two or more such linkages to yield bis(porphyrin) cyclophanes. As might be expected, the structural rigidity of the second class of molecules yields models of more well-defined geometry.

Schwarz et al. [15] carried out the first study of energy transfer in systems containing two covalently linked porphyrins. These workers prepared a series of molecules in which a zinc porphyrin was linked covalently to either a copper or cobalt porphyrin by amide linkages containing either a p-phenylene or ethylene group (Fig. 1). In rigid solutions in 2-methyltetrahydrofuran at 77K intramolecular energy transfer occurs between the linked zinc porphyrins as manifested by a shortening of the zinc porphyrin triplet state lifetime from 46 to 34 ms in the ethylene-bridged dimer. In a cobalt-zinc substituted ethylene-bridged dimer triplet energy transfer from the zinc chromophore was more difficult, as shown by the complete quenching of the zinc porphyrin phosphorescence. These workers found no evidence for triplet-triplet energy transfer in the p-phenylene-bridged bis-porphyrins. More importantly, singlet-singlet energy transfer was not observed in any of the above systems. It is known that the chlorophyll singlet state is primarily responsible both for energy transfer interactions in the antenna system [16] and for the photochemistry of the reaction center [17]. The paucity of interannular effects observed in these systems may be attributable, as the authors note, to the inability of these linked porphyrin macrocycles to assume a stacked conformation in which energy transfer should be more efficient.

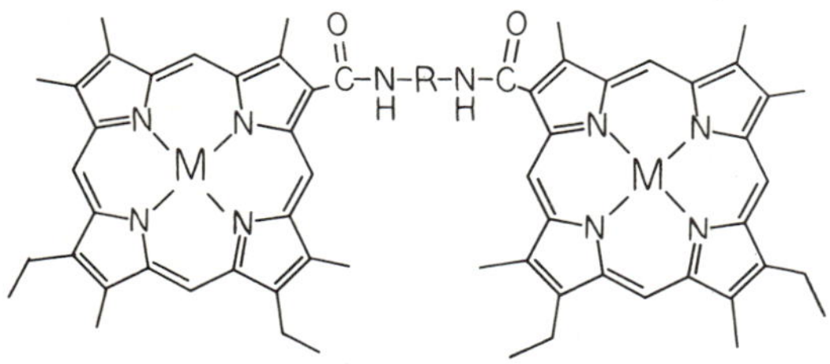

Fig. 1. Porphyrin dimers of Schwarz et al. (ref. [15]) R = $-CH_2CH_2-$, p-C_6H_4

Anton et al. [18] have also prepared a series of porphyrin dimers and trimers for use in energy transfer studies. Some of these studies have recently appeared [19]. Three covalently linked porphyrin hybrid dimers were synthesized, each containing a metallo-tetraarylporphyrin [Zn(II), Cu(II), and Ni(II)] and a free-base tetraarylporphyrin (Fig. 2). Transfer of singlet excitation energy from the metalloporphyrin center to the free base was characterized by fluorescence studies. The zinc hybrid dimer displayed 85% intramolecular transfer of energy from the excited singlet state of the Zn(II)-containing chromophore to the free base. No evidence for such transfer of excited singlet state energy was found in the Ni(II) or Cu(II) analogs. However, fluorescence from the free-base chromophore readily occurred when it was directly excited. Thus,

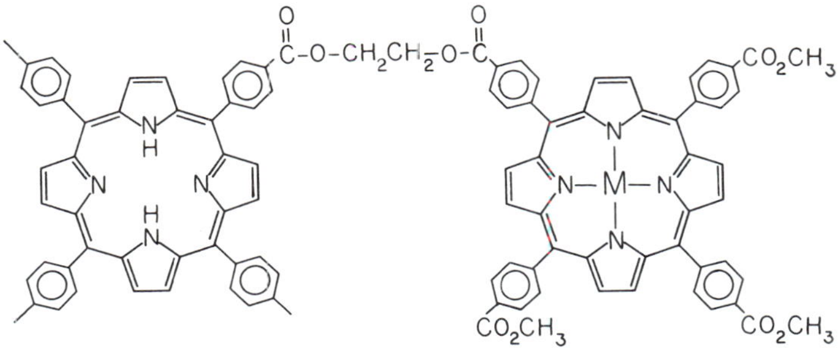

Fig. 2. Porphyrin dimers of Anton et al. (ref. [18]) M = Zn(II), Cu(II), Ni(II)

whatever the process is which competes with fluorescence for utilization of the excited singlet state energy in the Cu(II) and Ni(II) complexes, it occurs appreciably faster than singlet energy transfer to the free-base porphyrin.

Measurements of fluorescence lifetimes for the zinc monomer porphyrin (3 ns), the free-base monomer porphyrin (20 ns), and the hybrid dimer (10 ns), indicate that the hybrid dimer has the same quantum yield of fluorescence (0.2) as the free-base porphyrin. The authors conclude that the covalent attachment of another fluorescent porphyrin center effectively doubled the antenna size without decreasing the quantum yield even though the fluorescence quantum yield of the Zn(II)-containing monomer is substantially less (0.03) than that of the free-base porphyrin.

Little has prepared still another type of covalently linked porphyrin dimer in which the porphyrin macrocycles are joined via a diether linkage (Fig. 3) [20]. Examination of the visible absorption spectra of vanadyl and copper(II)-substituted dimers showed no interaction between the two chromophores save a splitting of the Soret band. The number of states responsible for the electronic transitions comprising the Soret band is sufficiently large for a clearcut interpretation of this splitting being impossible. Electron spin resonance observation of these paramagnetic dimers did not reveal any significant interaction between the macrocycles.

Paine et al. [21, 22] have prepared several dimeric porphyrins linked by hydrocarbon chains (Fig. 4).

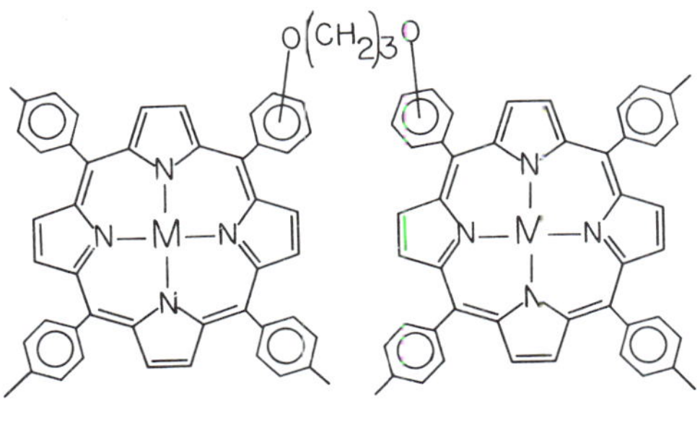

Fig. 3. Porphyrin dimers of Little (ref. [20]) M = VO, Cu(II)

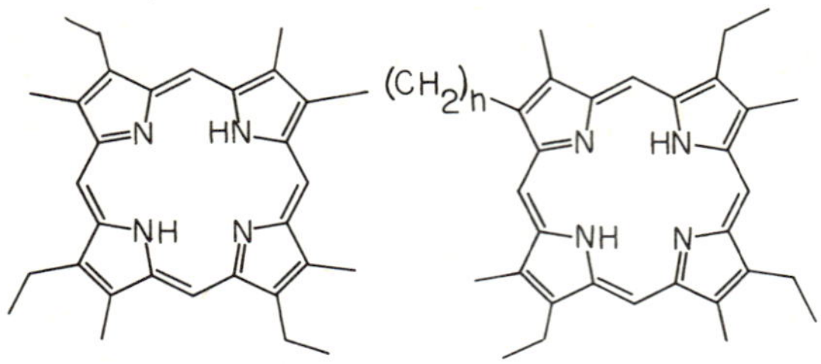

Fig. 4. Porphyrin dimer of Paine et al. (refs. [21, 22])

A unique feature of these dimers is that they were prepared by building the porphyrin macrocycles around linked pyrroles. The authors claim that the hydrocarbon link provides a solubility advantage for these models over those linked with esters or diamides. As might be expected the excitonic interaction between the two macrocycles increases as n decreases. For the compounds in which n = 0 and n = 1 the Soret band is broadened and red shifted. In addition, the low energy absorption bands of the macrocycles are slightly red shifted.

The interaction between the macrocycles was also monitored with [^{13}C] NMR spectroscopy. Once again, when n = 0,1 the [^{13}C] spectra are nondegenerate, indicative of an interaction between the macrocycles. No additional spectral data were provided in this preliminary report.

As stated above, the singly linked covalent porphyrin dimers are limited even in the type of energy transfer data they can provide. Stacking of the chromophores and control of the relative orientation of the transition moment axes can be achieved only in rigid models. Moreover, these porphyrin dimers do not display the photoredox properties characteristic of the chlorophyll special pairs in P865 and P700.

Pursuing bichromophoric porphyrin models one step further several investigators have prepared bis(porphyrin) cyclophanes. The first examples of this type of molecule to appear were described by Ogoshi et al. [23] and Collmann et al. [24]. Ogoshi's pair of isomeric double porphyrins are depicted in Fig. 5. Each covalent link is seven atoms in length. The authors did not observe any electronic interaction between the chromophores, as revealed by the visible absorption spectra of the molecules. Molecular models were used to estimate a 7 A interplanar distance. In light of the work that immediately followed that of Ogoshi, the reported absence of an interaction between the macrocycles seems a bit puzzling.

Utilizing compounds of the type shown in Fig. 6 Collmann et al. [24] found that porphyrins held rigidly in stacked conformations with the Q_x and Q_y transition moments of each macrocycle aligned showed distinct blue shifts of 15-17 nm in their Soret bands. Further, these workers found that spin exchange between paramagnetic metals in the Co(II) and Cu(II) complexes was rapid on the EPR time scale. Thus, these rigid double porphyrins experience some interannular interaction. Since the preparation of these models was primarily for the study of nitrogenase and oxidase activity, information concerning the magnesium- or zinc-substituted derivatives was not presented.

Kagan et al. [25] have reported another bis(porphyrin) cyclophane based on tetraphenylporphine derivatives (Fig. 7), in which the two macrocycles are linked at all four methine positions by p, p′-2-phenoxyethoxycarbonylphenyl chains. Both the electronic absorption and fluorescence emission spectra of the bis(porphyrin) cyclo-

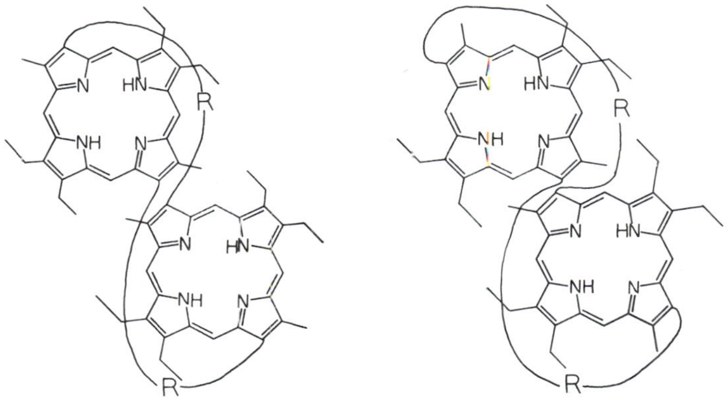

Fig. 5. Bis(porphyrin) cyclophanes of Ogoshi et al. (ref. [23])

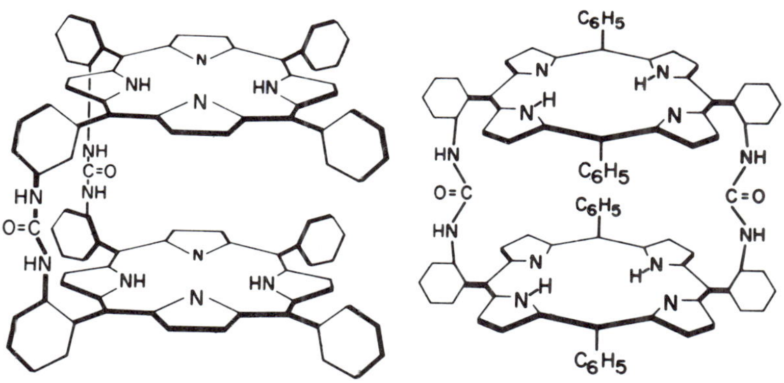

Fig. 6. Bis(porphyrin) cyclophane of Collmann et al. (ref. [24])

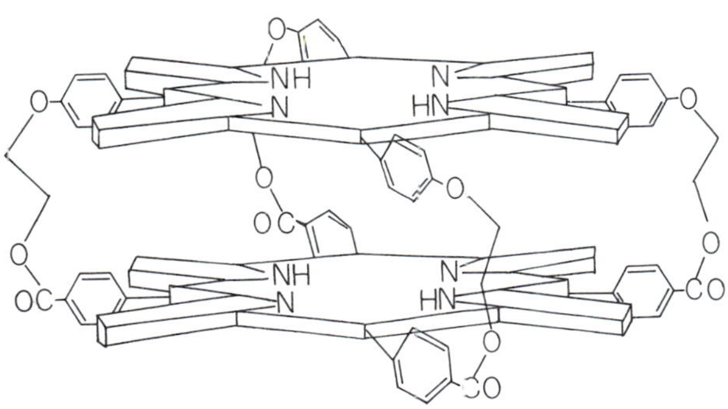

Fig. 7. Bis(porphyrin) cyclophane of Kagan et al. (ref. [25])

phane showed evidence of interaction between the component porphyrins. The Soret band was significantly broadened (40% increased half width) but was unshifted, while the visible absorptions were both broadened and red shifted (500 cm^{-1} for the red-most band). The fluorescence intensity was also red shifted (660 cm^{-1}) and broadened, but the quantum yield of fluorescence was almost unchanged from that of the monomeric components. The corresponding zinc chelate showed an unmoved but broadened Soret band (60% increased half width) and shift in the red-most band of 445 cm^{-1}. Its fluorescence emission was greatly broadened and quenched by 50%. A purely excitonic interaction would have shifted the Soret band to the blue and would have had minimal effect on the visible bands. The authors concluded that the spectra of both the free base and the zinc chelate can be best explained as due to "weak incipient charge transfer interactions."

The most recent report of interannular interactions in bis(porphyrin) cyclophanes is that of Chang [26]. Chang synthesized three bis(porphyrin) cyclophanes possesssing sufficiently long alkyl chains to facilitate solubility of the compounds in a variety of media (Fig. 8). Moreover, these model systems systematically vary the interplanar distance. The interplanar distance between the macrocycles was determined from the dipolar interaction of two Cu(II) paramagnetic centers (Table 1). As the interplanar distance decreased, the Soret band blue shifted, while the visible absorption bands broadened and underwent small red shifts. These observations contrast with the Kagan model in which the Soret band is broadened but unshifted in wavelength. However, like the Kagan model Chang's models possess red-shifted fluorescence maxima. The fluorescence maxima became increasingly red shifted and the fluorescence quantum yield dropped off dramatically as the interplanar distance was reduced. In Kagan's free-base cyclophane no diminution in fluorescence quantum yield between monomer and dimer was observed.

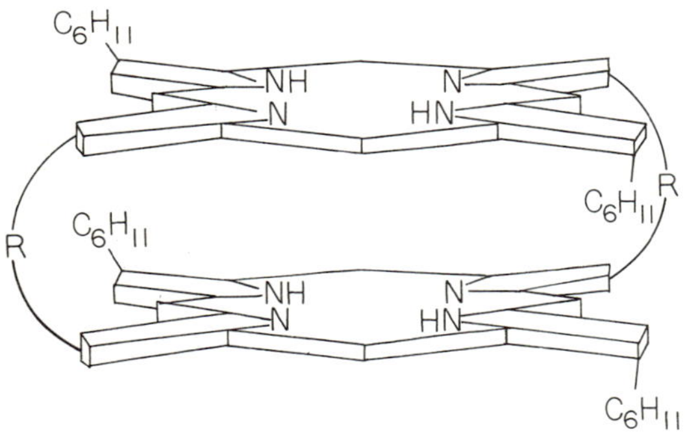

Fig. 8. Bis(porphyrin) cyclophane of Chang (ref. [26])

In addition to the free-base bis(porphyrin) cyclophanes, Chang et al. examined several di-metallated complexes including Cu-Cu Zn-Zn, and Mg-Mg cyclophanes. In each case they observed a blue-shifted Soret maximum. Solvent polarity and temperature had little effect on the absorption or emission spectra. The authors admit that this result is quite different from that expected of a model for P700. For example, the Mg-Mg complex in dry toluene at 23° gave absorption maxima at α 581, β 543,

Table 1. Selected data pertaining to Chang's bis(porphyrin) cyclophanes

Compound	Soret band[a]	Nmr N-H $CDCl_3$	Fluorescence[b] Q(0,0)	ϕf^c	Cu-Cu Zero-field splitting[d]	Interplanar separation
Monomeric methyl diester reference	398 nm (ϵ = 169.9 mM)	- 3.8δ	619 nm	0.094	-	-
Cyclophane with 7 atom links	383 nm (191)	- 6.2δ	628 nm	0.035	0.011 cm^{-1}	6.4Å
Cyclophane with 6 atom links	381 nm (209)	- 6.6δ	630 nm	0.021	0.0205 cm^{-1}	5.4Å
Cyclophane with 5 atom links	373 nm (201)	- 8.5δ	630 nm	0.007	0.0415 cm^{-1}	4.2Å

[a] Absorption spectra recorded in dichloromethane at 23°
[b] 77K in toluene glass
[c] Fluorescence quantum yields assuming yield for etioporphyrin I = 0.09
[d] Epr spectra were recorded at 77K in dichloromethane/-toluene

S 402 nm; while in aqueous dimethylformamide these maxima occurred at α 583, β 546, S 403 at 23° and α 581, β 545, S 406 at 77 K.

In one sense these results are not surprising for a model which by design is fairly rigid and at the same time possesses functional groups conjugated to the macrocyclic π electronic system that can be significantly perturbed by solvent interactions. In addition, the effect of chelation of a fifth ligand to Mg incorporated into porphyrins results in much less perturbation of the porphyrin electronic states than does a similar interaction in chlorophylls [27, 28].

Although the porphyrin dimer models have shown that interannular interactions are a strong function of the conformation of one macrocycle relative to the other, they remain limited in their ability to model photoreaction center chlorophyll function in photosynthesis. This limitation arises from the absence of the unique combination of donor and acceptor functionality in each chlorophyll molecule. These coordinative donor-acceptor properties depend primarily on the nucleophilic character of the ring V keto C = 0 function in the chlorophylls and the electrophilic nature of the central magnesium atom due to its coordination unsaturation [29]. Moreover, since the chlorophylls are dihydro- and tetrahydroporphyrins, the electronic transition moment of the longest wavelength absorption band is strongly polarized along the ring I-ring III axis and is greatly affected by polarization of the keto C = 0 group at C-9 [30]. Thus, the energies of the electronic states of the chlorophylls differ substantially from those of the porphyrins [27, 28]. Moreover, these properties and how they determine the spectral characteristics and functioning of specialized pairs of chlorophyll molecules in the photoreaction center cannot be adequately tested by simple porphyrin models.

3 Noncovalent Chlorophyll Special Pair Models

A great deal of effort has been directed toward the characterization of chlorophyll coordination interactions both with various ligands and with other chlorophylls. A major portion of this work has been summarized [29]. As a preliminary to a discussion of the preparation of synthetic photoreaction center models based on chlorophyll, several salient features of this chemistry should be briefly touched on.

As is mentioned in the discussion above, all chlorophylls exhibit ambident coordination properties. The electrophilic central magnesium atom possesses one or two axial ligands depending on the nucleophilicity of the ligand. In addition, all chlorophylls contain both ester and keto carbonyl functions, which are capable of acting as ligands in their own right. For example, in Chl *a* evidence from both IR[31] and NMR[32] spectra in addition to theoretical calculations [30] support the fact that the keto carbonyl function at C-9 is a much stronger ligand than either of the ester carbonyls. In BChl *a*, the keto group at C-9 and the acetyl group at C-2 appear to have comparable donor strengths. More recently, both IR and NMR evidence for Chl *a* in which the C-9 ketone has been chemically replaced by two hydrogen atoms show that the donor ability of the ester carbonyl is approximately 1000 times less than that of the C-9 ketone based on equilibria for self-aggregation [33].

In the absence of extraneous nucleophiles chlorophylls can self-aggregate by C-9 keto-Mg interactions. Moreover, in "soft" non-nucleophilic solvents, such as benzene or CCl_4, this interaction generates dimers over a wide range of concentrations. The introduction of nucleophilic ligands that are also capable of hydrogen bonding into dry chlorophyll solutions in these soft solvents leads to a different class of aggregation phenomena which bear more directly on photoreaction center modeling problems.[34].

In this context two sets of experiments demand specific comment. Taking their lead from the observations of Brody and Brody concerning the formation of a 700 nm absorbing Chl *a* species in ethanol, Shipman et al. [35] carried out a series of experiments involving the Chl-*a*-toluene-ethanol system. The visible absorption spectrum of a 0.1 M solution of Chl *a* in dry toluene with a 1.5 molar ratio of ethanol: Chl *a* as a function of temperature is shown in Fig. 9. As the temperature is lowered, there is a shift from the predominance of a species that absorbs near 670 nm to one that absorbs near 700 nm. The carbonyl stretching region of the infrared spectrum of a similar Chl-*a* solution as a function of temperature is shown in Fig. 10. As the temperature is lowered, there is very little change in the area of the ester carbonyl absorption and the ester peak shifts slightly from 1737 cm^{-1} to 1733 cm^{-1}. On the other hand, the transformation of the 670 nm absorbing species to the 700 nm absorbing species is accompanied by a large decrease in the area of the free C-9 keto carbonyl absorption. A new peak at 1658 cm^{-1} also appears in the low-temperature spectrum and this peak is assigned to the bound C-9 keto carbonyl. These workers also found that irradiation of a sample containing 1 mM Chl *a*, 1.5 mM ethanol, and 100 mM tetranitromethane as an electron acceptor at 96 K with 700 nm light, results in the production of a 7.5 G Gaussian EPR line. This signal is quite similar to the line-narrowed signal attributed to $P700^+$ in vivo and suggests that each spin is delocalized over an average of two macrocycles in this preparation.

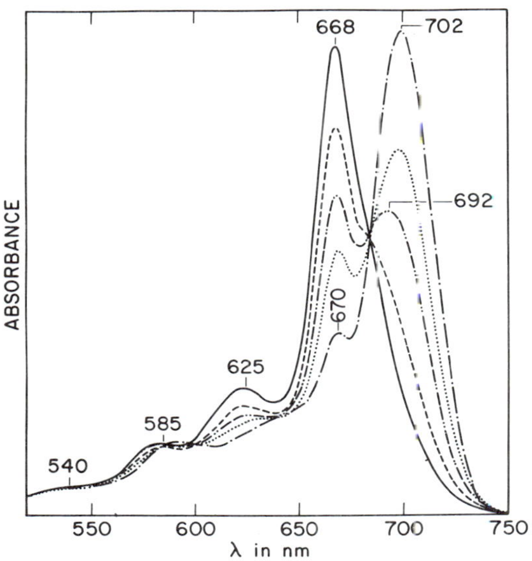

Fig. 9. Visible absorption spectrum of 0.094 M chlorophyll *a*, 0.14 M ethanol in toluene at: ——— 298 K, – – – – 273 K, – – . . – – 247 K, ... 224 K, – . – 175 K. The absorbance scale covers the range from 0 to 1.0

These results led Shipman et al. to propose that the two Chl-*a* molecules in this preparation are jointed at low temperatures by coordinating the magnesium atom of one Chl *a* to the oxygen atom of the ethanol hydroxyl function, which in turn hydrogen bonds to the C-9 keto carbonyl group of the second Chl *a*. Two such interactions result in a structure possessing C_2 symmetry with an approximate 8.9 Å Mg-Mg distance and a 3.6 Å interplanar separation between the macrocycles based on molecular models. This model is illustrated in Fig. 11.

It is important to note that the very same model was proposed simultaneously by Boxer and Closs [36] and the Argonne group. Fortunately, the evidence presented in support of this model by both groups is to a large extent complementary. The work of Boxer and Closs regarding a 700 nm absorbing dimeric species joined by the coordination scheme in depicted Fig. 12 is based on proton NMR observations of covalently linked PChl-*a* dimers and is especially suggestive of this structure. Further discussion of this work is presented below.

The results presented in the previous discussion may be generalized to include other nucleophilic groups capable of hydrogen bonding such as NH and SH. Of particular importance is the role of water in this scheme. Fong and Koester presented a series of experiments in which Chl *a* possessing a single water molecule coordinated to its

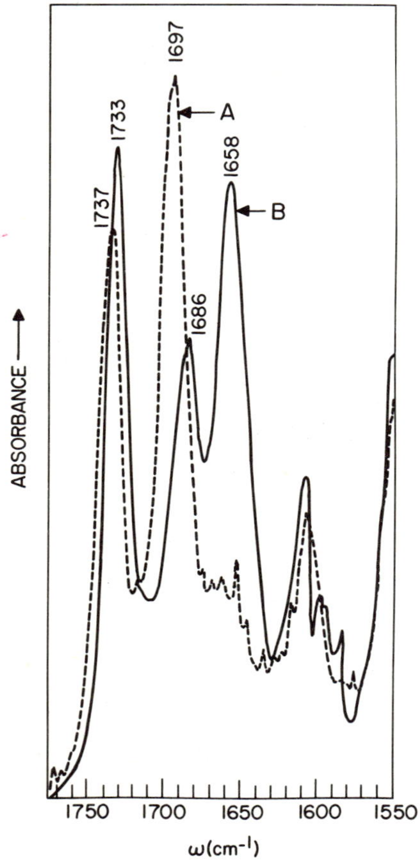

Fig. 10. Difference infrared absorption spectra of 0.064 M chlorophyll *a*, 0.096 M ethanol in toluene against toluene at 298 K (spectrum *A*) and 183 K (spectrum *B*)

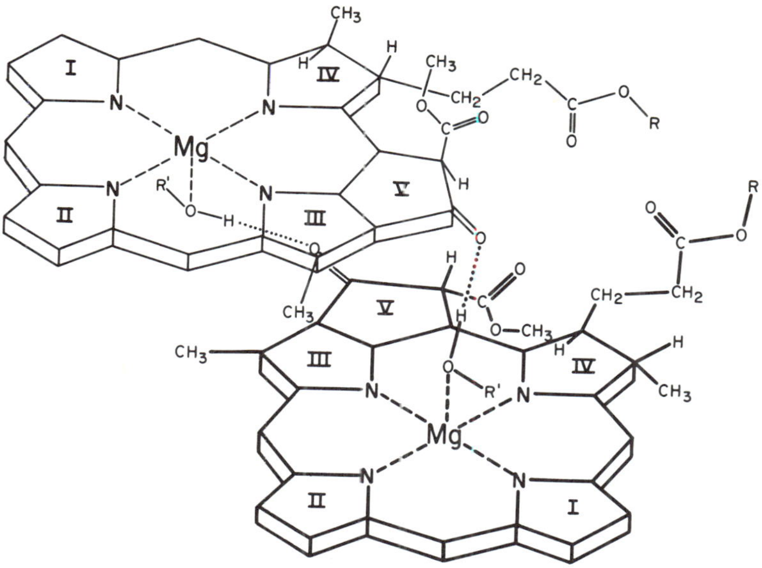

Fig. 11. Schematic representation of a proposed structure of special pair chlorophyll *a* (ref. [35]). For clarity the groups attached to rings I and II are not shown. R = phytyl and R' = H, ethyl, or protein side chain

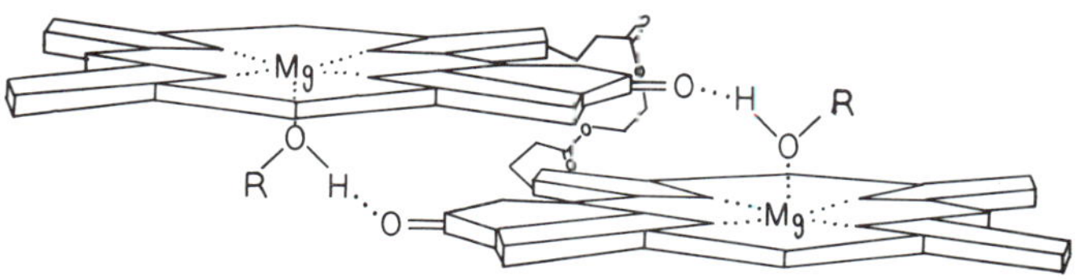

Fig. 12. C_2 symmetric structure proposed for folded bis(pyrochlorophyllide *a*) ethylene glycol diester

magnesium atom was dissolved in a 1 : 1 mixture of methylcyclohexane and pentane and cooled [37]. Lowering the temperature of these solutions again results in formation of a species that absorbs near 700 nm. Deconvolution of the resulting low-temperature absorption spectra and analysis of the resulting components under an exciton formalism led these workers to conclude that the 700 nm absorbing species they obtain does not have the same structure as that obtained by Boxer and Closs [36] and Shipman et al. [35]. Fong and Koester interpret these results in terms of a structure in which the C-10 carboxymethoxy carbonyl group is the coordinating donor group rather than the C-9 keto group (Fig. 13). It is the view of this reviewer that this interpretation conflicts with a large body of evidence which strongly indicates that the C-9 keto carbonyl is the principal coordinating donor ligand in these systems. A lack of corroborative IR or NMR data for the Chl a-H_2O system does not aid in establishing their interpretation.

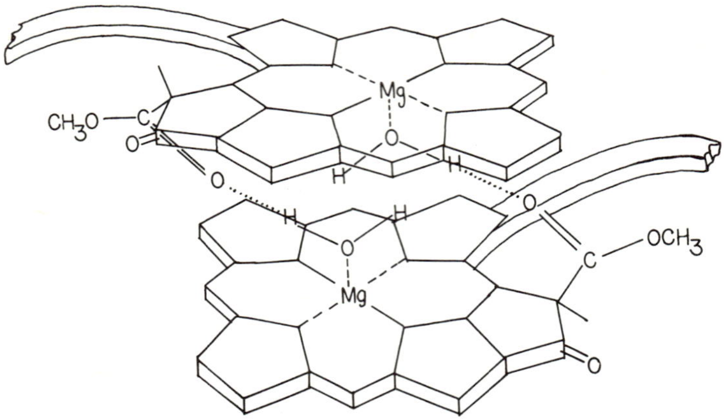

Fig. 13. Proposed structure of special pair chlorophyll *a* based on the experiments of Fong and Koester (ref. [37])

These brief comments serve to point out the difficulties encountered in structural interpretation inherent in chlorophyll coordination chemistry. All these Chl-*a* systems were assembled from two monomers by cooling to near the freezing point of the solvent or in organic glasses near 77 K. The ΔG of association of these in-vitro P700 models is dominated by the entropy term as indicated by the need for either high Chl-*a* concentrations, low temperatures, or both. The formation of these low-temperature species depends on Chl-*a* concentration, the ratio of Chl *a* to nucleophile, the solvent, the rate of cooling, etc. There is reason to suppose that mixtures of various species of poorly defined structure are present in even the best of these preparations. Visible absorption spectra in themselves provide little basis for a structural assignment. The use of corroborative structural probes such as IR and NMR is an absolute necessity in this work.

The relative ease with which all chlorophylls form aggregates with widely varying electronic spectra provides strong impetus for the employment of a wide variety of spectroscopic evidence in assigning specific structures to models. The uncertainties in composition and structure and the experimental problems and restrictions imposed by working at low temperatures in organic glasses has limited the information that can be derived from such model systems and has stimulated efforts to produce in-vitro models of P700 and P865 that are better defined structurally and that can exist under more physiological conditions.

4 Preparation of Singly Linked Covalent Chlorophyll Dimers

The synthesis of covalently linked dimers of Chl *a* and BChl *a* requires reaction conditions that are significantly milder than those usually employed in the preparation of porphyrin dimers. The β keto ester group in ring V of both Chl *a* and BChl *a* is particularly susceptible to strong acids and bases, nucleophiles, and oxidative reagents including prolonged exposure to light and air [38]. In addition, the 3 and 4 positions of BChl *a* are subject to oxidation yielding 2-acetylchlorophyll *a* [38]. Finally, excessive heating of chlorophylls, particularly in basic solvents such as pyridine, results in decarbomethoxylation of C-10 yielding the corresponding pyrochlorophylls [38].

Noting these facts, Boxer and Closs chose the more stable pyrochlorophyll *a* (PChl *a*) as the starting material in their synthesis of a model for P700 photoreaction center dimers [36]. They transesterified the propionic acid side chain of pyropheophytin *a* with ethylene glycol to yield the glycol monoester. This monoester was coupled with a molecule of pyropheophorbide *a* using N,N-carbonyldiimidazole in THF at 70° in a sealed tube. Extensive chromatography followed by remetallation of the macrocycles by the method of Baum et al. [39] led to bis(pyrochlorophyllide *a*) ethylene glycol diester.

The synthetic route described above is much too harsh to utilize in the preparation of the corresponding diesters of Chl *a* and BChl *a*. Ideally, the required chemistry should be performed at room temperature utilizing neutral or at best mildly basic reaction conditions. In addition, the reactions must be fairly rapid and selective to minimize side product formation especially in the dimer formation step. After examining a wide variety of mild esterification methods, Wasielewski et al. [40, 41] developed synthetic schemes that resulted not only in the preparation of covalently bound dimeric species of both Chl *a* and BChl *a*, but have general utility in the preparation of PChl-*a* dimer and many other chlorophyll derivatives requiring high yield esterification at the propionic acid side chain.

A key reagent in effecting this chemistry is benzotriazole-1-methanesulfonate [42]. Using this reagent, the propionic acid side chain of most chlorophylls may be rapidly esterified with primary alcohols in excellent yield at room temperature over a period of 30 min to 1 h in a weakly basic reaction mixture. In a typical example BChl *a* obtained from *R. sphaeroides* was treated with 80% aqueous trifluoroacetic acid to obtain bacteriopheophorbide *a* (BPhide *a*), which was esterified at 25°C with ethylene glycol using benzotriazole-1-methanesulfonate and triethylamine in dry THF. The glycol monoester was coupled with an equivalent of BPhide *a* using the same esterification method, but substituting 4-dimethylaminopyridine as the base and methylene chloride as the solvent.

Reinsertion of the magnesium atoms [43] results in the structure illustrated in Scheme 1. At each stage of the synthesis the product is separated from the starting materials and one or two minor side products by preparative scale high-pressure liquid chromatography. Each bis(chlorophyllide) ethylene glycol diester is prepared in an analogous fashion and is obtained in 30%-50% overall yield based on the starting chlorophyll. Complete analytical profiles of each compound (mass spectrum, [^{1}H] NMR, IR, and elemental analysis) are consistent with the assigned structures.

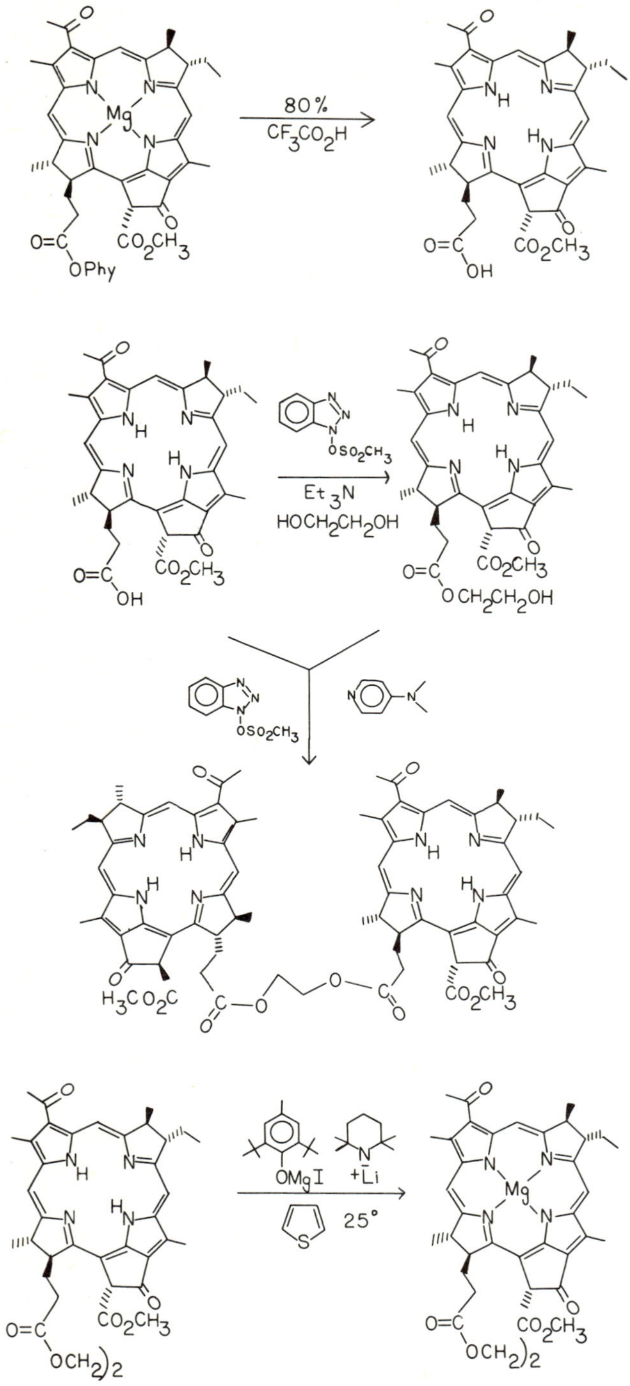
80%
CF_3CO_2H
O=C
OPhy
CO_2CH_3
OH
OSO_2CH_3
Et_3N
$HOCH_2CH_2OH$
OCH_2CH_2OH
H_3CO_2C
OMgI
Li
25°
$OCH_2)_2$

Scheme I

5 Solvent-dependent Structure of Singly Linked Covalent Chlorophyll Dimers

An accurate assessment of the utility of the covalently linked dimeric chlorophylls as models for photoreaction center chlorophylls depends strongly on a secure knowledge of their solution structures. The structure of each covalently linked dimer under differing solvation conditions is based on proton NMR spectral evidence. The [^{1}H] NMR spectrum of each compound dissolved in nucleophilic solvents, i.e., those solvents which provide the magnesium atoms with a fifth or sixth ligand, e.g., acetone, tetrahydrofuran, pyridine, is identical with that of the corresponding monomer under the same conditions except for an additional singlet at $\sim\delta 4.0$ due to the protons of the glycol linkage. This result is illustrated in Fig. 14 for the BChl-*a* covalent dimer and presented in Table 2 for the PChl-*a*, Chl-*a*, and BChl-*a* covalent dimers. At the same time, the electronic transition spectrum of each dimer dissolved in nucleophilic solvents is identical to that of its corresponding monomer, indicating that there is little or no electronic interaction between the macrocycles (see below).

Table 2. Proton magnetic resonance data for covalently linked chlorophyll dimers

	Chemical shift, δ (ppm)					
Position	Open PChl a_2	Folded PChl a_2	Open Chl a_2	Folded Chl a_2	Open BChl a_2	Folded BChl a_2
1	3.34	3.31	3.27	3.25	3.16	3.02
2a	8.11	8.06	8.05	8.04		
2b	6.07	6.10	6.10	6.11	2.79	2.67
3					3.95	3.95
3a	3.16	3.20	3.21	3.23	1.54	1.47
4					3.95	3.95
4a	3.60	3.92	3.65	3.95	3.32	3.50
4b	1.63	1.12	1.60	1.10	0.62	1.20
5	3.56	1.76	3.55	1.70	3.40	1.20
7					3.95	3.95
7a	1.8-2.4	1.8-2.4	1.8-2.4	1.8-2.4	1.8-2.4	1.8-2.4
7b	1.8-2.4	1.8-2.4	1.8-2.4	1.8-2.4	1.8-2.4	1.8-2.4
$-OCH_2CH_2O-$	4.15	4.16	4.10	4.10	4.00	4.31
8					3.95	3.95
8a	1.65	1.50	1.47	1.30	1.28	1.33
10	5.06;5.23	4.98;4.26	6.50	5.30	6.30	6.06
10b			3.60	3.50	3.40	3.02
α	9.66	9.55	9.62	9.50	9.59	9.13
β	9.83	9.72	9.82	9.70	8.36	8.18
δ	8.47	8.35	8.43	8.35	8.22	7.95

Dissolving each dimer in a dry non-nucleophilic solvent, such as benzene, toluene, or carbon tetrachloride, results in broad, ill-defined proton NMR spectra as illustrated in Fig. 14. This change is accompanied in each case by a shift in the longest wavelength

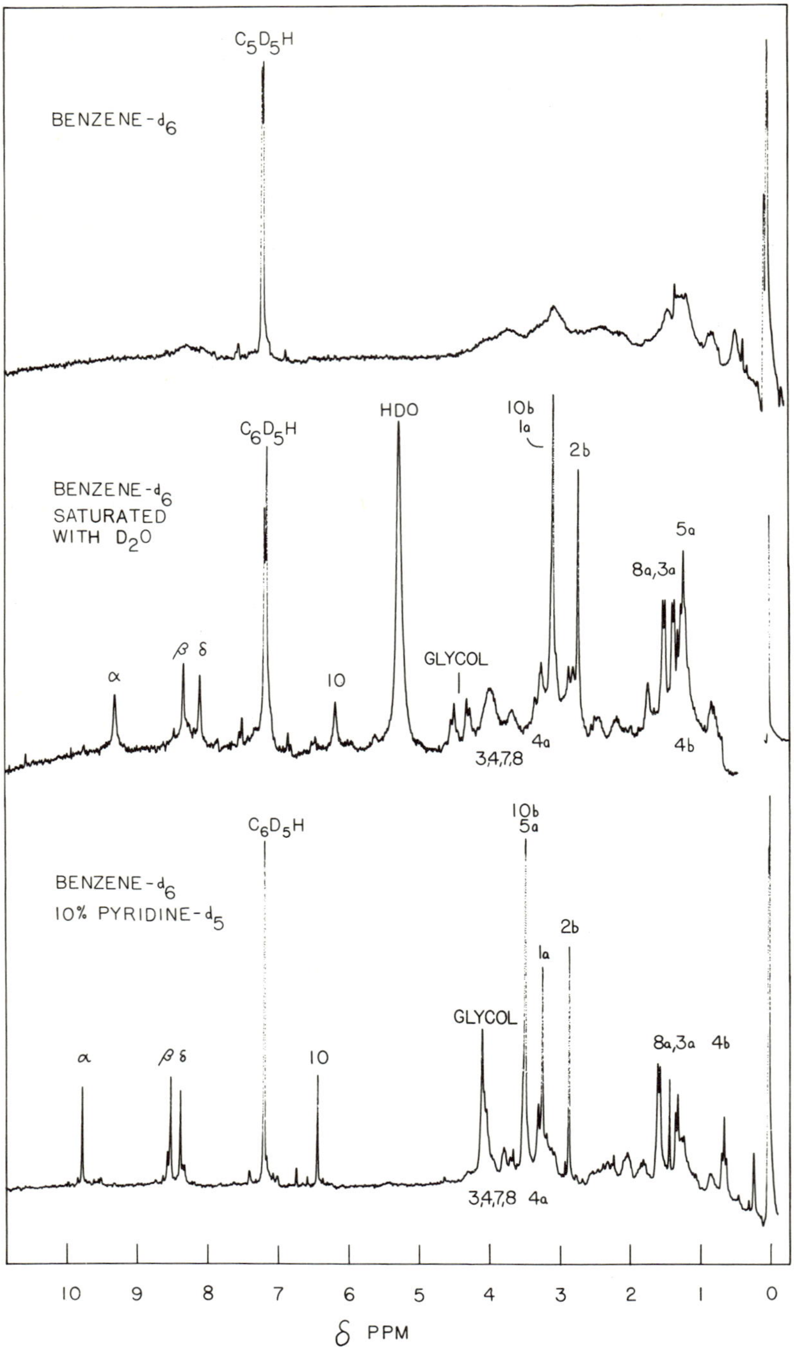

Fig. 14. Proton NMR spectra at 220 MHz of the covalently linked BChl-*a* dimer in three different solvents

absorbance in the electronic transition spectrum to even longer wavelength, 663 → 667 nm for PChl-*a* and Chl-*a* dimers, and 775 → 780 for BChl-*a* dimer. These features are illustrated in Figs. 15-17, respectively.

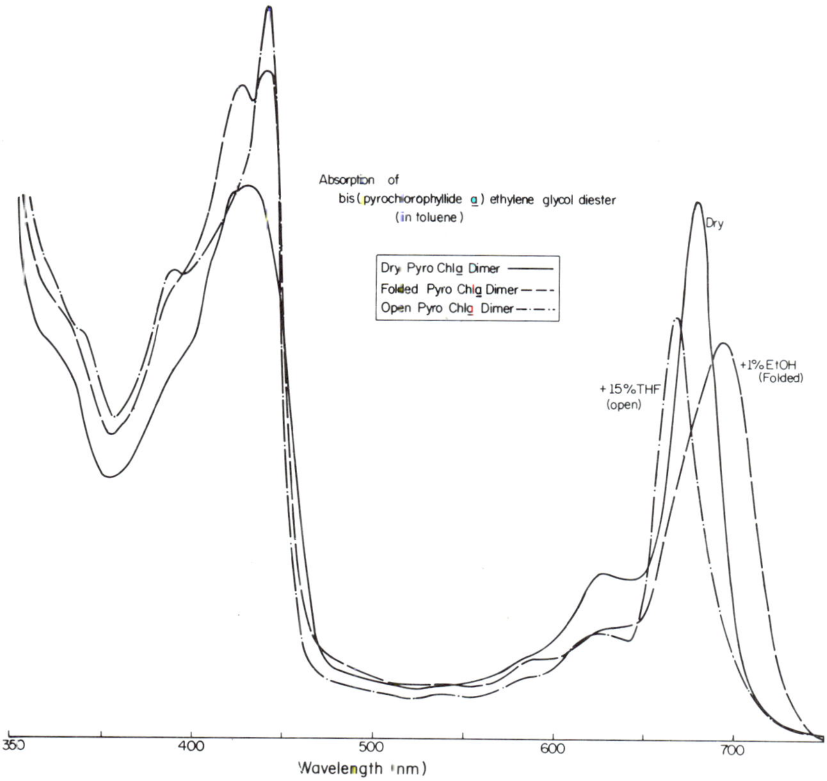

Fig. 15. Electronic absorption spectra of the PChl-*a* dimer in three different solvents

Both the [^{1}H] NMR and absorption spectra are consistent with the formation of aggregates in which the oxygen atom of the C-9 keto carbonyl of one chlorophyll coordinates to the magnesium atom of another chlorophyll in either an inter- or intramolecular fashion. Of course, the 2-acetyl group of the BChl-*a* macrocycle may also participate in these coordinative interactions. The broadened NMR spectra apparently result from relatively slow exchange between differing coordinated species.

On the other hand, treating these dry solutions with a large (10-100 x) molar excess of water, primary alcohols, or primary thiols, results in a dramatic sharpening of the proton resonances. This is shown for BChl-*a* dimer in Fig. 14. A comparison between the spectra in Fig. 14 reveals that several resonances of the dimer in the water-saturated solution are substantially shifted from their respective positions in the fully disaggregated species. These changes are presented in Table 2. A careful examination of the Table shows that these changes are paralleled in both the PChl-*a* and Chl-*a*

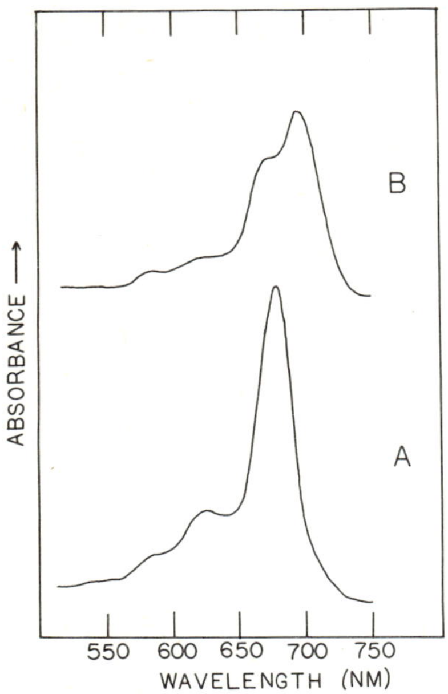

Fig. 16. Electronic absorption spectra of the Chl-*a* dimer in two different solvents. *A* CCl_4, *B* CCl_4 saturated with water

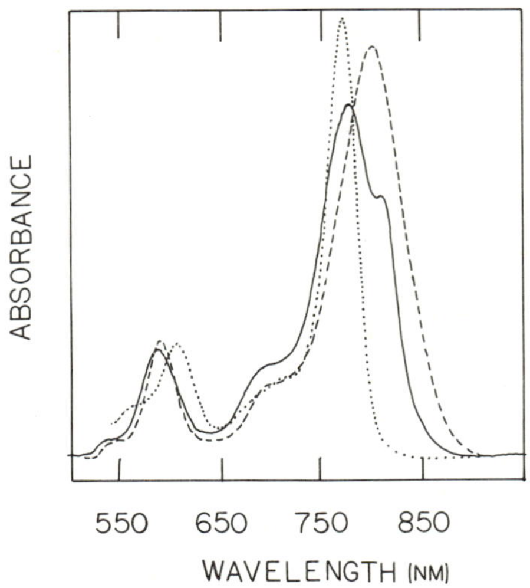

Fig. 17. Electronic absorption spectra of 10^{-5} M solutions of covalently linked BChl-*a* dimer in 10% pyridine in benzene (. . . . .), in dry benzene (——), and in water-saturated benzene (– – – –)

dimers. Since all three dimers display only one set of resonance in the dry benzene solution, the two macrocycles of each dimer must be equivalent on the [^{1}H] NMR time scale. In each case the 5-methyl group experiences a strong upfield shift. The protons associated with ring V also move upfield, while the 4b methyl resonance shifts downfield by about 0.5 ppm. Interestingly, the methyl resonance of the acetyl groups of the BChl-*a* dimer shifts very little from its position in the disaggregated dimer. As these chemical shift changes are due primarily to the influence of the diamagnetic ring current of one macrocycle on its dimeric partner, the [^{1}H] NMR data suggest that the macrocycles are on the average parallel to each other with the III and V rings of each macrocycle experiencing the greatest interannular overlap. The results further suggest that the acetyl groups of the BChl-*a* dimer do not participate appreciably in folding the dimer via hydrogen bonding interactions.

When methanol or ethanol is substituted for water giving an [^{1}H] NMR data suggest equivalent to that depicted in Fig. 14, protons α to the hydroxyl group experience large upfield shifts, while those β to the hydroxyl group are shifted to a somewhat lesser extent. At the point where the [^{1}H] NMR spectrum indicates that the folded structure dominates in solution, the integral of the methanol signal correspondends to two methanol molecules per molecule of dimer [36].

The collected [^{1}H] NMR data on these systems indicate that all three singly linked covalent dimers fold into the C_2 symmetric conformation depicted in Fig. 12 when dissolved in soft non-nucleophilic solvents containing an excess of a hydrogen bonding nucleophile.

Further evidence for the strong participation of the C-9 keto carbonyl group in the folded covalent dimer structure is obtained from infrared data [44]. When each covalently linked chlorophyll dimer is completely disaggregated, e.g., when it is dissolved in pyridine, the C-9 keto carbonyl stretching band appears at 1680-1690 cm^{-1}. Folding of each dimer into the conformation depicted in Fig. 12 results in the appearance of a new band at 1650-1660 cm^{-1} accompanied by the loss of the free keto carbonyl band. Under these same conditions the ester carbonyl band remains virtually unchanged. These spectral changes are illustrated for the Chl-*a* dimer in Fig. 18.

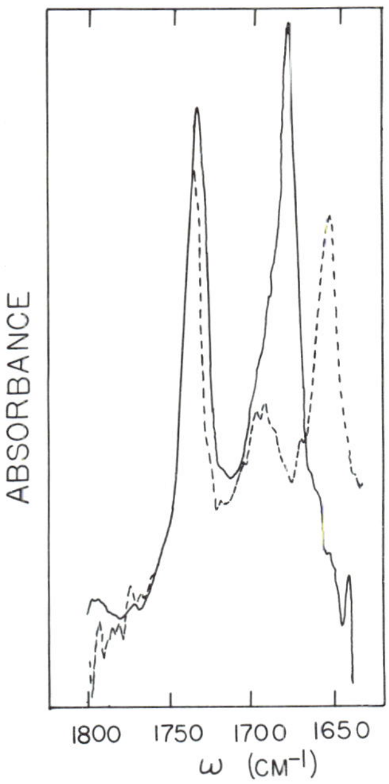

Fig. 18. Difference infrared absorption spectra of 10^{-3} M covalently linked Chl-*a* dimer in 10% pyridine in toluene against toluene (——), and in 0.1 M ethanol in toluene against toluene (– – – –)

The infrared data provide strong evidence for the involvement of the C-9 keto carbonyl in the hydrogen bonding interaction which folds the dimer into the structure shown in Fig. 12. The ester carbonyl groups do not participate in this interaction. It is also important to point out that these room-temperature infrared spectra parallel those obtained with free Chl *a* in toluene-ethanol media at cryogenic temperatures. Thus, not only do we have an idea of what the structure of the low-temperature aggregate might be, but we have a dimeric system with a well-defined average structure which can be examined as a model of P865 or P700 at physiological temperatures.

6 Photophysical Properties of Singly Linked Covalent Chlorophyll Dimers

The discussion in this section will examine work involving characterization of the photophysical processes which follow light excitation of the singly linked chlorophyll dimers. The folding of the PChl-*a* and the Chl-*a* dimers into the configuration depicted in Fig. 12 is accompanied by shifts of the longest wavelength absorption in their electronic spectra from 663 nm to 694 nm, Figs. 15 and 16, respectively. This wavelength shift closely mimics the 700 nm absorption of the phototrap of photosystem I in green plants.

The Chl-*a*-linked dimer however retains a shoulder at 667 nm of approximately one-third the oscillator strength of the 694 nm band. These relative oscillator strengths were determined by computer deconvolution of the spectrum. At equilibrium in vitro at room temperature the carbomethoxy group at C-10 is partially epimerized to form about 15% of the a′ isomer. Because this model is dimeric the stereoisomerism at C-10 results in a statistical distribution of isomers that consists of about 73% a-a isomer, 25% a-a′ isomer, and 2% a′-a′ isomer. The a-a and a′-a′ isomers are both able to fold into the closely spaced C_2 symmetric stacked structure proposed for this species (Fig. 12). In this structure both carbomethoxy groups point away from the interior of the structure. However, space-filling (CPK) models predict that the a′-a′ isomer which comprises about 25% of the equilibrium mixture should have difficulty folding into a similar structure because of steric compression of one carbomethoxy group between the macrocycles.

The 694 nm absorption of the PChl-*a* dimer does not exhibit a short wavelength shoulder. This result is consistent with the proposed greater degree of conformational freedom in the PChl-*a* dimer. Similarly, the BChl-*a* dimer does not exhibit a short wavelength shoulder due to the 99%+ stereochemical purity (natural isomer) at C-10 that is retained in solution.

Folding of the BChl-*a* dimer into the conformation depicted in Fig. 12 shifts the red-most absorption maximum from 775 nm to 803 nm. This shift is significantly different from the 865 nm absorption of the reaction center dimer in vivo. Nevertheless, the utility of this model lies in the fact that it elucidates the degree of electronic interaction possible for two BChl-*a* molecules under a given geometry. The presence of two additional BChl-*a* and two BPheo-*a* molecules in the bacterial reaction center may further shift the electronic transition of the reaction center dimer to longer wavelengths. Moreover, the influence of the local protein environment on the electronic transition is not accounted for in this model. Similar arguments may be advanced to account for the smaller discrepancy between the 694 nm absorption of the P700 models and P700 itself.

Both the PChl-*a* and Chl-*a* dimers display solvent-dependent fluorescence maxima and fluorescence lifetimes. One group of workers has determined several photophysical parameters of the singlet manifold of these compounds [45]. A selection of their results is presented in Table 3.

Folding of either the PChl-*a* or Chl-*a* dimer into the C_2 symmetric conformation depicted in Fig. 12 shifts the fluorescence maximum of each dimer from 680 nm to

Table 3. Absorption and fluorescence characteristics of Chl *a* and the dimers of Chl *a* and PChl *a* in various solvents

Compound	Conc., mM	Solvent	Maxima nM Absorption	Maxima nM Fluorescence	ϕf	τf ns
Chl *a*	1.0	Pyridine	671	684; 736	0.35	7.3 ± 0.1
Chl a_2 (open)	0.5	Pyridine	670	683; 734	0.21	5.8 ± 0.3
PChl a_2 (open)	0.5	Pyridine	669	685; 735	0.13	5.7 ± 0.2
Chl *a*	1.5	Benzene sat. with H_2O	667	667; 732		
Chl a_2 (folded)	0.5	Benzene sat. with H_2O	695	730		
Chl *a*	1.0	Toluene + 0.1 EtOH	665	876; 725	0.2	6.0 ± 0.3
Chl a_2 (folded)	0.5	Toluene + 0.1 EtOH	695	730		3.2 ± 0.1
PChl a_2 (folded)	0.5	Toluene + 0.1 EtOH	695	730	0.3	4.4 ± 0.3

730 nm. By comparison, the weak fluorescence usually attributed to P700 in photosystem I occurs near 720 nm. The fluorescence lifetimes of the folded dimers are somewhat shorter than those of the corresponding monomers, but certainly not dramatically different, while the dimers in their disaggregated states exhibit lifetimes close to those for the monomers.

The quantum yield of fluorescence for the folded PChl-*a* dimer is almost the same as that of the corresponding monomer. On the other hand, the quantum yields for the open Chl-*a* and PChl-*a* dimers are less than those of their respective monomers. The authors contend that the shorter lifetimes and reduced quantum yields for the dimers in their open configuration may be due to increased rates of intersystem crossing or enhanced internal conversion via coupling of rotational and vibrational modes associated with motion about the dimeric link. The relative rates for these processes could not be ontained from their data because of the sampling geometry employed in acquiring the data. The effects of concentration quenching and self-absorption could not be separated from the results.

These same authors also examined the Chl-*a* dimers and PChl-*a* under conditions of high intensity irradiation. Fluorescence lifetime shortening and in most cases stimulated emission were observed as the 337 nm nitrogen laser excitation intensity was increased. The authors conclude that care must be taken in photophysical measurements involving these pigments with regard to apparent fluorescence lifetime variations due to excessive excitation intensities.

One of the more interesting conclusions derived from studies of the fluorescence quantum yield of the Chl-*a* and PChl-*a* folded dimers is the fact that they fluoresce almost as efficiently as their respective monomers. This result indicates that two chlorophylls in this particular configuration do not exhibit significant concentration quenching of fluorescence. In fact, one could predict that inhibition of the charge transfer photochemistry of P700 at the primary acceptor level might lead to a highly fluorescent species. This speculation assumes of course that the in-vivo geometry is somewhat similar to that of the model. Work discussed in a later section will address this question more directly.

The excited singlet state properties of the open configuration of the dimers is interesting from the standpoint of chlorophyll-chlorophyll interactions in the antenna. For example, the low fluorescence quantum yields and short fluorescence lifetimes observed in vivo are attributed to the high concentration (about 0.1 M) of chlorophyll present in the antenna system. This high concentration leads to a decrease in fluorescence quantum efficiency and shortened fluorescence lifetimes via concentration quenching [46]. It has been suggested that the quenching mechanism involves Forster transfer followed by trapping at a pair of molecules closer than a critical distance [47, 48]. One view of the trap is any two chlorophyll molecules that are separated by less than 10 Å and that do not interact appreciably in the ground state, yet do interact strongly when one of the pair becomes excited so as to form an excimer or other complex that does not fluoresce in the monomer region of the spectrum [47, 48].

However, this cannot be the whole story. It is clear from the fluorescence quantum yields of the folded Chl-*a* and PChl-*a* dimers that simply having two chlorophyll molecules in close proximity does not necessarily result in a significant decrease in either fluorescence quantum yield or lifetime. Although it might be stating the obvious, the fact remains that the relative orientation of the chromophores with respect to one another has a profound effect of these parameters.

The effect of solvent and temperature on the singlet state photophysics of the PChl-*a* dimer has also been examined in a detailed fashion [49]. The PChl-*a* dimer was employed exclusively in this study to facilitate data analysis due to the absence of the isomerism problem associated with C-10 in the Chl-*a* dimer.

The two most striking aspects of the data involve large changes in the excited singlet state lifetime of the folded PChl-*a* dimer. First, this lifetime varies dramatically with a change in solvent, CH_2Cl_2 vs CCl_4, at a given temperature. In this instance the results obtained in CCl_4 are nearly identical to those obtained in other soft non-nucleophilic solvents, such as benzene or toluene. Second, similar dramatic lifetime variations are observed for folded PChl-*a* dimer in CH_2Cl_2 as a function of temperature. The fluorescence lifetime of the folded PChl-*a*dimer ($2 \cdot 10^{-4}$ M) in CH_2Cl_2 varies from 110 ps at room temperature to 4.6 ns at 200 K (Fig. 19). Below 200 K the lifetime lengthens slightly, reaching a limit of 5.8 ns. No change in lifetime can be detected down to 4 K. However, the quantum yield of fluorescence of these samples remains relatively constant in the 290 K to 200 K range (Table 4). The quantum yield data at room temperature is in close agreement with that measured at higher concentrations [45].

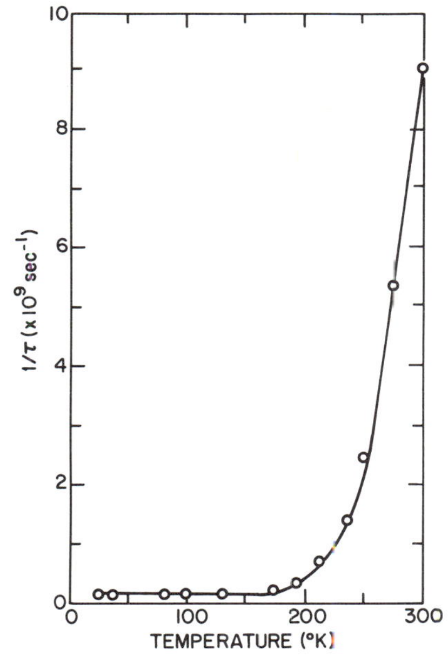

Fig. 19. The temperature dependence of the fluorescence lifetime (τ) of $2 \cdot 10^{-4}$ M bis(pyrochlorophyllide *a*) ethylene glycol diester folded with ethanol in methylene chloride

The excited state (difference) absorpltion spectrum taken 6 ps after excitation is shown in Fig. 20. It shows a narrow band of bleaching at 700 nm. An additional negative absorbance change is also seen at 440 nm. A small positive absorbance change was observed

Table 4. Quantum yield data for bis(pyrochlorophyllide *a*) ethylene glycol diester ($2 \cdot 10^{-4}$ M)

Solvent	Temperature (K)	Quantum yield
CCl_4	298	0.32
CH_2Cl_2	298	0.03
CH_2Cl_2	269	0.03
CH_2Cl_2	234	0.03
CH_2Cl_2	208	0.04
CH_2Cl_2	994	0.06
CH_2Cl_2	178	0.10

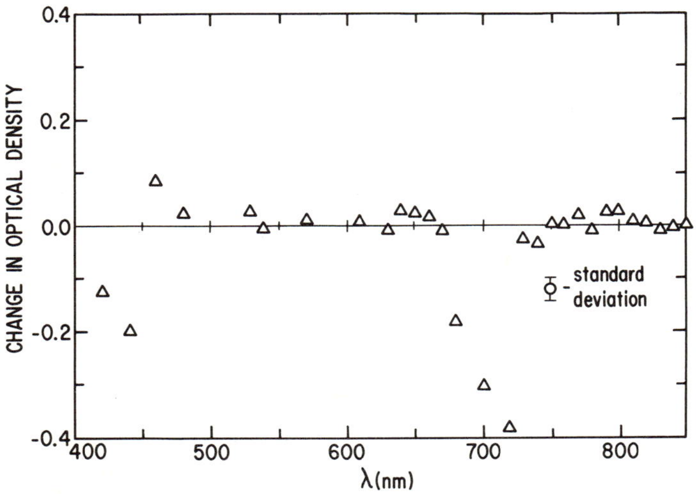

Fig. 20. Excited state difference spectrum of $2 \cdot 10^{-4}$ M bis(pyrochlorophyllide *a*) ethylene glycol diester folded with methylene chloride. Spectrum was recorded 6 ps following excitation

at 460 nm. As the sample was cooled from 290 K to 270 K, the bleaching centered at 700 nm became somewhat broader and a new positive optical density change appeared at 660 nm (Fig. 21).

All light-induced absorption changes for folded PChl-*a* dimer in CH_2Cl_2 decay with a lifetime of 110 ± 40 ps at room temperature. This decay rate is identical to the fluorescence lifetime of the folded PChl-*a* dimer under these solvation conditions. When the temperature is lowered, the lifetime of the excited state absorbance increased, while the magnitude of the 700 nm bleaching remained largely unchanged (Fig. 21). The 700 nm bleaching and the positive absorbance change at 660 nm each possess a lifetime that is identical once again to the fluorescence lifetime of the dimer.

An extensive series of control experiments showed that the short lifetimes were not due to stimulated processes as neither sample concentration, cell path length, view-

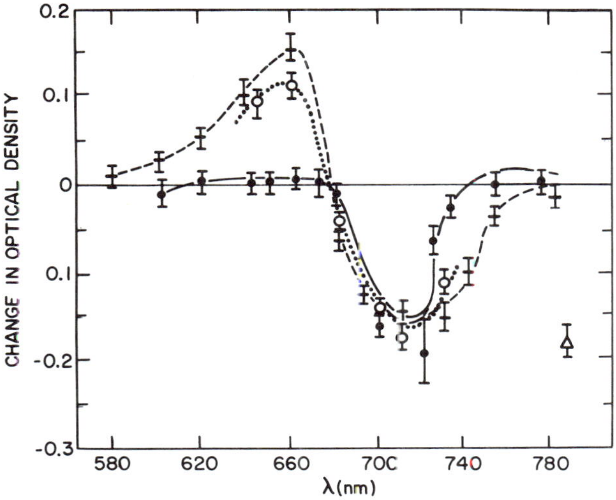

Fig. 21. Excited state difference spectrum recorded 6 ps following excitation of bis(pyrochlorophyllide *a*) ethylene glycol diester ($2 \cdot 10^{-4}$ M) folded with ethanol. At 298 K the solvents were methylene chloride (– – – –) and carbon tetrachloride (——). At 273 K the solvent was methylene chloride (. . . . .)

ing angle, or excitation intensity affected the excited state behavior of the folded PChl-*a* dimer in methylene chloride. Consequently, a systematic experimental characterization of the resultant excited states was undertaken to provide an explanation for these unusual properties.

Two problems were immediately eliminated from consideration. First, a change in the aggregation state of the PChl-*a* molecules could be the source of the lifetime variations. Yet, the excited singlet state lifetime of the folded PChl-*a* dimer did not depend on dimer concentration in the range 10^{-6} M to $5 \cdot 10^{-4}$ M. Second, chlorinated hydrocarbons often undergo photochemical reactions. Fast chemical processes could serve to shorten the lifetime of the dimer-excited singlet state. However, measurements performed in CCl_4, a chlorocarbon more reactive than CH_2Cl_2, were identical to those performed in toluene or hexane. Thus, these effects could not be attributed to photochemistry.

The remaining arguments were geared toward showing that the results can best be explained in terms of an excited state equilibrium involving two excited singlet states of PChl-*a* dimer, one state of which is fluorescent while the other is essentially nonfluorescent.

The PChl-*a* dimer fluorescence lifetime in CH_2Cl_2 increased by a factor of 30 over a temperature range that left the quantum yield for fluorescence relatively unchanged. A single excited level would require that both the radiative rate and the sum of the nonradiative rates each change by a factor of 30 as the temperature is varied from 298 to 200 K. While such a change in the nonradiative rate has been previously seen, there is no precedent to support a radiative rate change by that amount. Changes in radiative rate with temperature would be reflected in changes in Franck-Condon factors, which in turn would be revealed in changes in the fluorescence spectrum. No such changes are observed.

In order to explain the data one needs at least a two level scheme. These two states may reside on the same molecule or be representative of two distinct excited species in solution. For the case in which two distinct excited species exist in solution the rise of the 660 nm absorption change would indicate a shift in the relative equilibrium concentration of the two species.

If both species are fluorescent, the emission spectrum from each species would almost certainly be different. Thus, one would expect that the luminiscence spectrum would change with temperature. No temperature-dependent fluorescence spectra changes were observed and thus within this framework only one of the two species could be emissive. However, at every temperature examined both the fluorescence and absorption change decays were similar.

Absorption changes reflect changes in concentration of both species, while fluorescence is indicative of only the emissive species. Since both changes exhibit the same lifetimes, one would have to assume then that rapid energy transfer between the two species could account for the data. The concentration-independent fluorescence results make this possibility unlikely.

If one assumes that both excited singlet states reside on the same molecule, the effect of changing solvent or changing temperature is to shift the population of the excited PChl-*a* dimer molecules from one state of the other. A detailed analysis of the absorption change data suggests that these relative populations are under kinetic rather than thermodynamic control. This is indicated by the fact that a large change in relative populations occurs over a relatively small temperature range. The possibility exists that switching between two excited singlet states may be a feature which

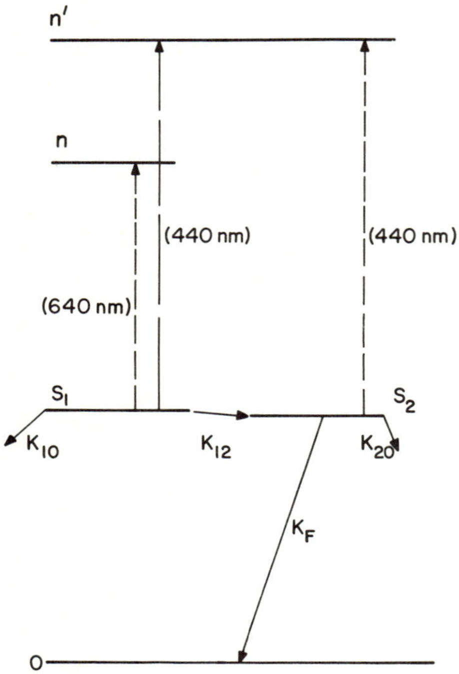

Fig. 22. A model for the excitation decay in bis(pyrochlorophyllide *a*) ethylene glycol diester. S_1 is the state originally populated. Transfer out of S_1 proceeds by nonradiactive paths. k_{12} represents transfer to the fluorescent state S_2. k_{10} represents all other decay rates. k_f is the radiative rate, and k_{20} represents all nonradiative paths from S_2. The *vertical arrows* represent the absorptions observed when the excited dimer is in S_1 or S_2

allows the reaction center to store excitation energy for a brief time, while waiting for an oxidized acceptor or reduced donor to be formed. This excited state scheme is depicted in Fig. 22.

Another photophysical study of the PChl-*a* dimer has examined the events that take place on a much longer time scale in a soft non-nucleophilic solvent. Linschitz et al. [50] photolyzed a benzene solution of PChl-*a* dimer folded with methanol, utilizing 30 ns pulses of light at 694 nm. The absorption change spectra they obtained are illustrated in Fig. 23. Strong bleaching at 695 nm and the Soret band are seen in addition to positive △ od changes at 500 nm and at 670 nm. The decay kinetics of these changes were independent of wavelength. The authors noted that the spectrum resembles a superposition of the triplet spectrum of Chl *a* or PChl *a* and the ground-state absorption spectrum of monomeric Chl *a* or PChl *a*. They concluded that the folded PChl-*a* dimer is converted to a half-excited species containing covalently linked triplet and ground-state PChl-*a* molecules. The close resemblance of the components of the transient spectrum to the separated species is interpreted as evidence that there is no strong interaction between the triplet and ground-state components of the dimer. The authors also find that photolysis of the PChl-*a* dimer initially in its open configuration also leads to a similar half-excited triplet species.

An analysis of the decay kinetics for these absorption changes yielded the mixed first and second order decay kinetics characteristic of triplet decay with $k_1 = 5 \cdot 10^3\ sec^{-1}$ and $k_g = 3 \cdot 10^9\ M^{-1}\ sec^{-1}$. The constant k_1 represents only an upper limit to the intrinsic rates of radiative and nonradiative decay. The authors note that the T-T annihilation rate constant k_g is close to that measured for monomeric Chl-*a*

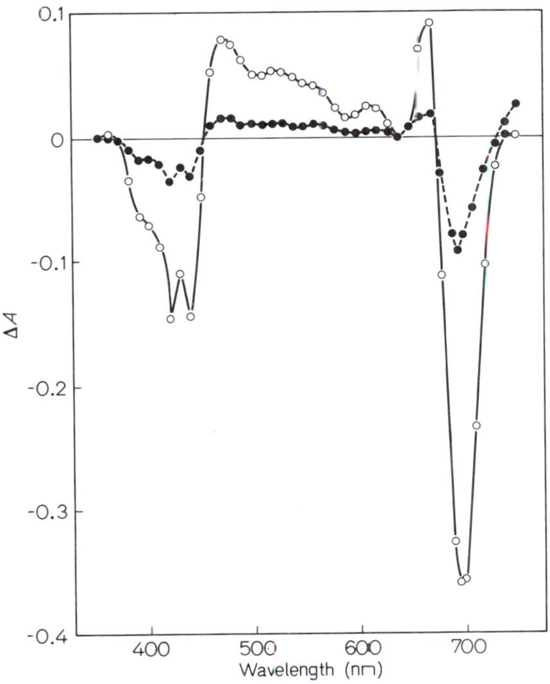

Fig. 23. Difference spectra of PChl-*a* dimer (6 μM) in benzene/0.5 M methanol. Immediately after the 30 ns flash (o) at 694 nm; 20 μs after the flash at 694 nm (•)

triplets and they conclude that these results together with the spectra essentially rule out interpretation of the observed transients as due to the formation of free radicals.

In another series of experiments these same authors showed that the decay of the PChl-*a* dimer half-excited triplet to the folded ground state is limited by the triplet decay rate with folding of PChl-*a* dimer ground state rapidly following.

7 Photochemical Properties of Singly Linked Covalent Chlorophyll Dimers

The most important property of any model of photoreaction center structure and function is its ability to mimic in an accurate fashion the photoinduced charge separation observed in vivo. An ultimate goal of these modeling studies is to devise systems in which the charge separation can be maintained over times long enough to carry out endoergic chemical and/or electrochemical processes. The singly linked chlorophyll dimers in their C_2 symmetric folded conformation have been subjected to a series of photoredox experiments in order to evaluate their ability to mimic photoreaction center charge separation photochemistry.

The PChl-*a*, Chl-*a*, and BChl-*a* folded dimers each undergo one electron oxidation with a variety of chemical oxidants [40, 41]. For cases in which the rate of chemical oxidation is very slow, illumination of solutions of the folded dimer and an oxidant with red light, 694 nm for PChl-*a* and Chl-*a* dimers, and 803 nm for BChl-*a* dimer, results in a rapid increase in the oxidation rate. For example, irradiation of a 10 μM solution of PChl-*a*, Chl-*a*, or BChl-*a* dimer in water-satured toluene containing an equivalent of iodine in a standard 1 cm path length optical cell results in complete bleaching of the 694 nm band for PChl-*a* and CHl-*a* dimer, and the 803 nm band for BChl-*a* dimer in a few seconds. This bleaching is accompanied by a positive absorption change near 800 nm for the PChl-*a* and Chl-*a* dimers, and at 1150 nm for the BChl-*a* dimer. These changes are illustrated for the PChl-*a* dimer in Fig. 24. The spectrum of each oxidized dimer is similar to that obtained via photooxidation of P700 and P865. A band near 800 nm has been assigned to $P700^+$, while a similar band for $P865^+$ occurs at 1250 nm. Other oxidizing agents, such as quinones, yield similar results.

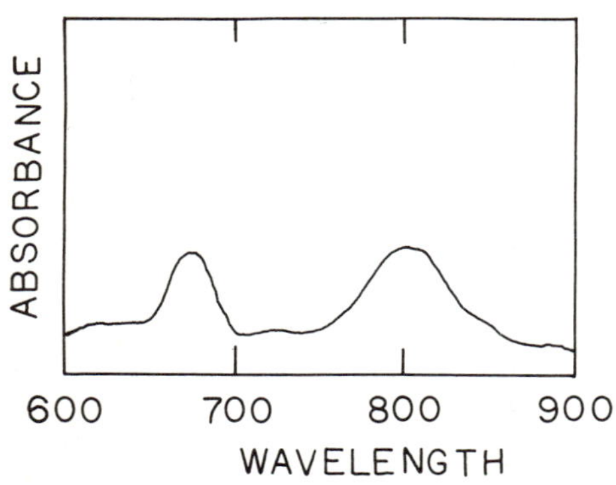

Fig. 24. Electronic absorption spectrum of the folded PChl-*a* dimer cation radical

If the singly linked folded dimers in millimolar concentrations are photooxidized with red light in the presence of an electron acceptor or treated with a fairly strong chemical oxidant such as $ZnTPP^{+} ClO_4^{-}$[52] while the sample is in the microwave cavity of an EPR spectrometer, the bleaching of the longest wavelength optical absorption is accompanied by the appearance of a Gaussian EPR line at g = 2.0026 of 7.5G line width for PChl-*a* and Chl-*a* dimers, and 10.6G line width for BChl-*a* dimer. These line widths are narrowed by about a factor of $1/\sqrt{2}$ from those exhibited by the well-known radical cations of the corresponding monomeric chlorophyll macrocycles [8]. The line width of the signal observed for the oxidized PChl-*a* and Chl-*a* dimers compares favorably with that observed for $P700^{+}$ in vivo [8]. Similarly, the line width of the oxidized BChl-*a* dimer signal is close to that observed for $P865^{+}$ in a number of photosynthetic bacteria [8].

Both the optical and EPR data for the radical cations of the folded singly linked chlorophyll dimers indicate that these compounds are reasonably good models for the photoreaction center chlorophyll dimers in vivo. However, some additional comments need to be directed to the difference between modeling ground-state properties versus properties of the dimer radical cations. The singly linked BChl-*a* dimer is a case in point.

The BChl-*a* dimer absorbs light at 803 nm in its folded configuration. Clearly, this value is nowhere near the 865 nm absorption of the phototrap in *R. sphaeroides,* yet the 803 nm signal photobleaches and the EPR line width of the cation radical produced upon one electron oxidation of the BChl-*a* dimer displays a strong absorption at 1150 nm, not very far removed energetically from the 1250 nm absorption observed in vivo [5].

Thus, the covalently linked BChl-*a* dimer mimics the properties of oxidized bacterial photoreaction center chlorophyll quite well, yet absorbs light in the ground state at 803 nm instead of 865 nm. At this point it is well to remember that we are probing the properties of an isolated pair of BChl-*a* molecules of a given average structure and that therein lies the strength of the model. Interaction of the in-vivo photoreaction center dimer with the additional two BChl-*a* and two BPh-*a* molecules present in the bacterial reaction center may shift the special pair optical transition to longer wavelengths. The important point to be made is that these additional chromophores need not be involved in the delocalization of the unpaired spin in the oxidized special pair.

The redox properties of the singly linked chlorphyll dimers have been examined in another independent fashion utilizing electrochemical techniques. The oxidation potentials of each dimer were determined by ac voltammetry at a platinum disc electrode (Table 5). The use of the ac technique permits a more accurate assessment of the half-wave potentials due to the employment of phase-selective double layer charging current rejection. The later effect is particularly severe in the low-polarity solvent used in the determination.

The oxidation of each singly linked dimer was examined in methylene chloride containing 0.1 M ethanol and 0.1 M tetrabutyl ammonium perchlorate as the supporting electrolyte. Methylene chloride is one of the few solvents in which the folded conformation of the dimers is maintained while providing a sufficiently conductive medium for electrochemical measurements. In each case the one electron oxidation potential of

Table 5. One-electron oxidation potentials of covalently linked chlorophyll dimers *vs* SCE

Compound	$E_{1/2}$ (Open conformation)	$E_{1/2}$ (Folded conformation)
PChl a_2	0.61	0.53
Chl a_2	0.61	0.55
BChl a_2	0.44	0.38

the dimers in their folded conformation was less positive by 50-75 mV than that of the corresponding monomeric chlorophyll. The geometrical configuration of the in-vivo phototrap is always such that the special pair is oxidized more readily than the surrounding antenna chlorophylls. Delocalization of the positive charge in the dimer radical cation over both macrocycles is reflected in the transition state for oxidation and thus in the half-wave potential.

The nature of the partially hydroxylic solvent required to fold the singly linked dimers precluded an estimation of their reduction potentials due to the known rapid protonation of the anion radicals of the chlorophylls.

In other experiments controlled potential pulse generation of the folded dimer radical cations at an optically transparent electrode yielded optical spectra for the cations which are virtually identical to those obtained via photooxidation of the folded dimers.

8 Biomimetic Charge Separation Photochemistry

The experiments described in the previous section all indicate that the singly linked chlorophyll dimers in their C_2 symmetric folded configuration are readily photooxidized. A much more demanding criterion for a model of photoreaction center chlorophyll dimers is the duplication of the kinetics of electron transfer known in detail for reaction centers from purple photosynthetic bacteria. Modeling of these kinetics is an important first step toward controlled charge separation in vitro.

The dimeric BChl-*a* species in reaction centers from *R. sphaeroides* absorbs light at 865 nm, yielding an excited state which rapidly transfers an electron to a molecule of bacteriopheophytin *a* (BPh *a*) in <10 ps [53]. In about 150 ps the electron is transferred to a quinone molecule which in turn transfers an electron to a secondary quinone in a few microseconds [54, 55]. In reaction centers, for which the quinones have been either extracted or chemically reduced prior to light excitation, BPh^- back-transfers an electron to (BChl $a)_2^+$ in 10-20 ns [56]. Thus, the in-vivo geometry of the reaction center is such that the reverse electron transfer is 5-10 thousand times slower than the forward reaction. The first in-vitro model reaction center that duplicates both the rapid light-induced charge transfer and the slow back-reaction of the bacterial photoreaction center has recently been reported [57].

In this effort to produce a functional model reaction center, the authors employed the PChl-*a* dimer because of its inherent stability toward photodegradation. Folding the PChl-*a* dimer into a conformation depicted in Fig. 12 results in a shift of its absorption maximum from 663 nm to 694 nm, while its fluorescence maximum shifts from 670 nm to 720 nm. In order to position pheophytin *a* which serves as the electron acceptor close to the dimer, the strategy employed was that of replacing the hydrogen-bonded short chain alcohol which secures the conformation of the dimer shown in Fig. 12 with a primary alcohol derivative of pheophytin *a*. The pyropheophorbide a ethylene glycol monoester, PPhide a-OH, a precursor of the PChl-*a* dimer, proved to be ideal for this purpose.

When dry 0.2 mM solutions of PChl-*a* dimer in toluene are treated with as little as a fourfold molar excess of PPhide a-OH at room temperature or a stoichiometric twofold molar excess at 0°, the PChl-*a* dimer readily folds into the structure depicted in Fig. 25. This folding is accompanied by a shift of the dimer electronic absorption maximum to 694 nm. More importantly, the 720 nm fluorescence of the dimer is almost completely quenched as shown in Fig. 26. If a similar experiment at the same concentrations is performed with PChl-*a* dimer folded with ethanol, while unbound pheophytin *a* is present in the solution, the fluorescence quantum yield of the dimer actually increases by about 10% over that in the absence of pheophytin *a*. This latter effect may be attributed to excitation energy transfer from excess pheophytin *a* in solution to the folded dimer which acts as an energy trap.

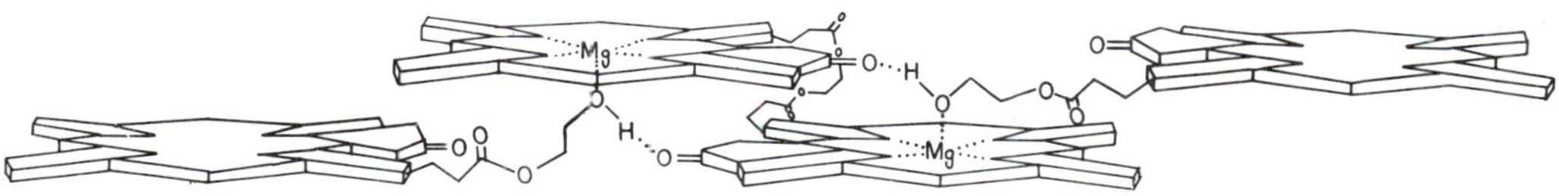

Fig. 25. Proposed structure of the $(PChl\,a)_2$ $(PPhide\,a\text{-}OH)_2$ reaction center model

In order to determine the mechanism and kinetics of the quenching of the PChl-*a* dimer fluorescence by PPhide a-OH in this model reaction center, the excited state difference spectrum of the model was determined as a function of time on the picosecond time scale. The spectrum illustrated in Fig. 27 shows the difference in optical density between the excited state and the ground state following a 6 ps flash at 530 nm.

The rise time for all Δod changes is <6 ps. These changes persist for between 10 and 20 ns with the subsequent return of the ground-state absorption spectrum. The spectrum in Fig. 27 displays a positive Δod change at 800 nm and strong bleaching at 660 nm. Positive Δod changes dominate the 440-630 nm region of the spectrum. The 800 nm feature in the spectrum is characteristic of a Chl a cation radical [52]. The one electron oxidation product of the PChl-*a* dimer folded with ethanol displays a broad electronic absorption near 800 nm. Similarly, P700 in green plants undergoes light-induced oxidation to yield a species which absorbs light near 800 nm [51]. The strong bleaching at 660 nm along with the smaller bleachings at 500 nm and at 530 nm can be ascribed to the disappearance of ground-state PPhide a-OH. One electron reduction of pheophytin a results in very similar spectral changes [58]. Thus, the overall spectrum in Fig. 27 may be adequately described as a superposition of the PChl-*a* dimer cation radical spectrum and the PPhide *a*-OH anion radical spectrum. Whether

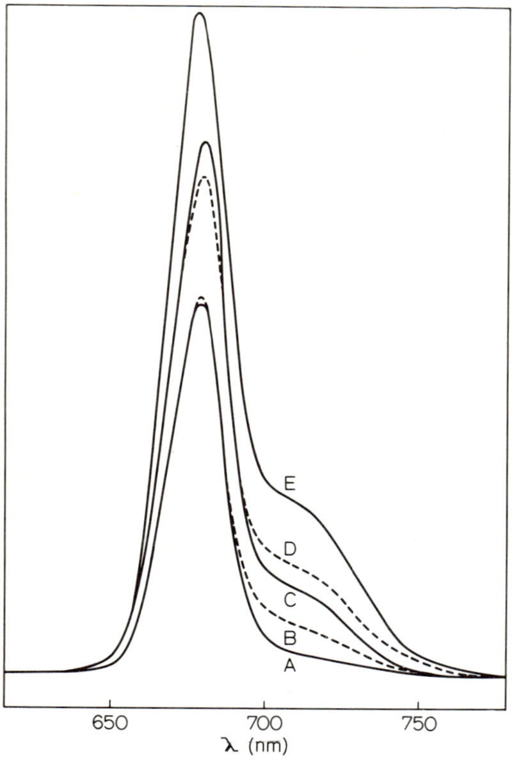

Fig. 26. Fluorescence spectrum of 6.9 · 10^{-4} M PChl-*a* dimer and 2.8 · 10^{-3} M PPhide *a* - OH in toluene with the indicated amount of ethanol

Front surface excitation at 420 nm
6.9 · 10^{-4} M Pyrochl *a* dimer
2.8 · 10^{-3} M Pyropheophorbide *a* glycol ester in toluene

A. 0 M EtOH
B. 6.8 · 10^{-3} M EtOH (10 fold excess)
C. 1.4 · 10^{-2} M EtOH (20 fold excess)
D. 3.1 · 10^{-2} M EtOH (50 fold excess)
E. 6.5 · 10^{-2} M EtOH (100 fold excess)

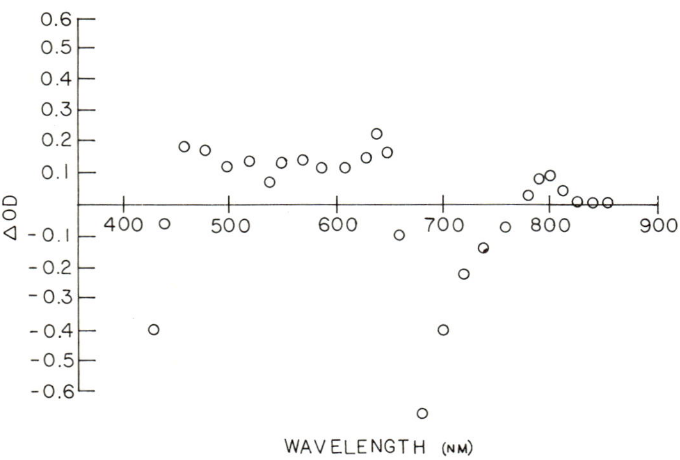

Fig. 27. Initial absorption changes following 6 ps excitation with 530 nm light of 0.2 mM PChl-*a* dimer and 0.8 mM PPhide *a*-OH in toluene at 20°

in fact the complex exists as a radical pair or as a charge transfer state has not yet been ascertained. The overall spectral features strongly parallel the type of spectral changes observed following flash excitation of bacterial reaction centers, even though the wavelengths of the Δod changes are different for the a-series plant chlorophylls. Thus, this model reaction center channels most of the light energy it absorbs into an electron transfer process which displays the same asymmetric kinetics that are observed in vivo.

It is clear from these experiments that positioning the PPhide *a*-OH electron acceptor close to the PChl-*a* dimer electron donor is important for efficient light-induced electron transfer. Yet, the kinetics of both the forward and reverse transfer depend critically on the energies of the states involved. Excitation of the PChl-*a* dimer into its first excited singlet state raises its energy to ∿1.8 eV above the ground state. Charge separation is then an energetically downhill process with the resultant ion pair state ∿1.5 eV above the ground state [59]. The ∿0.3 eV energy difference between the singlet and ion pair states is sufficient to provide for a rapid forward transfer rate, yet results in the reverse electron transfer being an activated process. Thus, the rate of an electron transfer leading back to the first excited singlet state should be substantially slower than the charge separation rate.

The only reasonable pathway remaining for the ion pair state is to cross over to a highly vibrationally excited state of the ground electronic state. Since $\Delta G << 0$, ∿-1.5 eV, the Franck-Condon factors for this process are small and thus the rate of return to the ground state is relatively slow (Fig. 28).

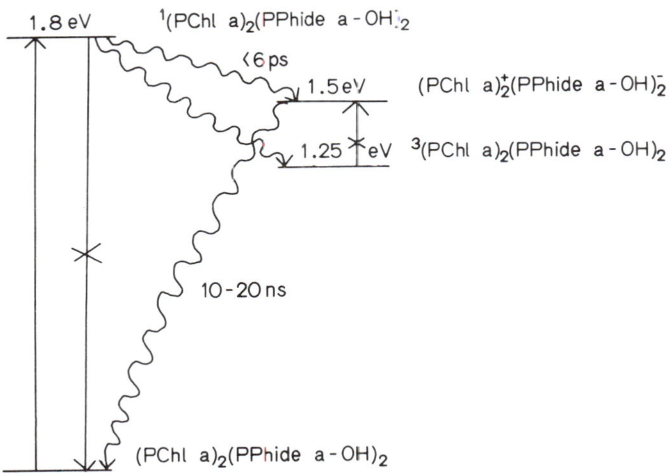

Fig. 28. Energy level scheme for light-induced charge separation in the $(PChl\ a)_2$ $(PPhide\ a\text{-}OH)_2$ system

Since the PPhide *a*-OH molecules may exhibit some conformational freedom due to rotation about the single bonds of their alkyl chains, the role of these changes following excitation and their effect on the reverse electron transfer rate needs to be further explored.

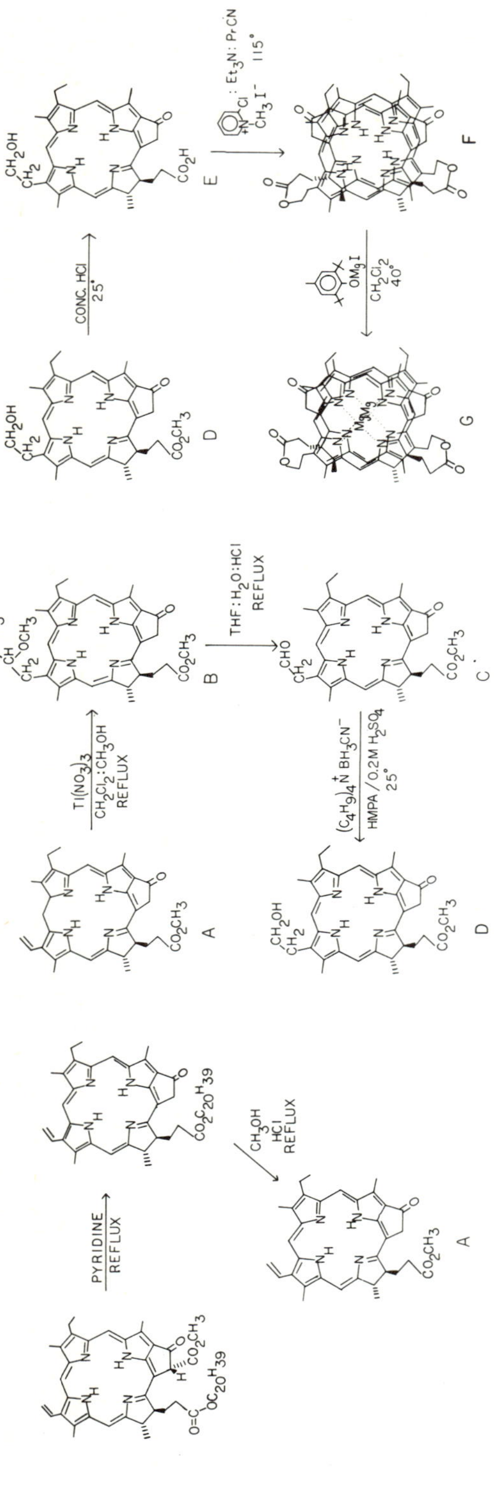
PYRIDINE
REFLUX
CH3OH
HCl
REFLUX
A
Tl(NO3)3
CH2Cl2:CH3OH
REFLUX
B
THF:H2O:HCl
REFLUX
C
HMPA/0.2M H2SO4
25°
D
CONC. HCl
25°
E
Et3N:PrCN
115°
F
CH2Cl2
40°
G

Scheme II

9 Doubly Linked Chlorophyll Cyclophane Models of Special Pair Structure

Since the confirmation of the singly linked chlorophyll dimers responsible for their biomimetic properties is both dynamic and highly solvent dependent, studies of their physical properties as a function of structure are necessarily restricted. As described in the beginning of this chapter, several workers have prepared bis(porphyrin) cyclophanes in order to solve similar difficulties with porphyrin dimer models [21-24]. Since porphyrins do not posses the optical, redox, or coordination properties unique to the chlorophylls, the synthesis of bis(chlorophyll) cyclophanes has been undertaken in order to study the photochemical and redox properties of special pair models of well-defined structure compatible with a broad range of solvation conditions.

The discussion in Sect. 7 pointed out that on the basis of data from the BChl-*a* singly linked dimer, it is quite possible that different numbers of chromophores and consequently different overall structures may be responsible for the optical properties of the special pair in contrast to its redox properties. The first bis(chlorophyll) cyclophane was prepared in order to test this hypothesis in addition to providing an entry point into a series of rigid chlorophyll special pair models [60]. This cyclophane was designed such that the Q_y transition dipole of one macrocycle is held fixed at a 90° orientation relative to that of the second macrocycle. The described structure has both macrocycles stacked directly on top of one another (Fig. 29).

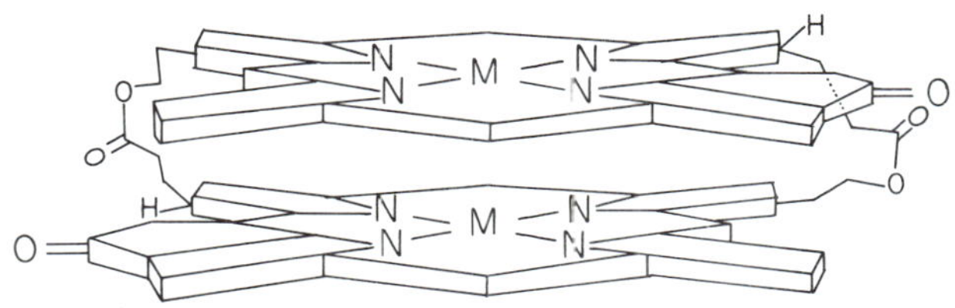

F M = H_2
G M = Mg

Fig. 29. Structure of the bis(chlorophyll) cyclophane

The preparation of this compound is illustrated in Scheme 2. Pheophytin *a* is converted in two well-known steps to methyl pyropheophorbide *a*, A. In some new chemistry the vinyl group of A was selectively oxidized with thallium nitrate [61]. Following hydrolysis of dimethyl acetal B the resulting aldehyde was selectively reduced to the terminal alcohol D [62]. Hydrolysis of the methyl ester yielded the hydroxy acid E. Double esterification with a coupling reagent developed by Mukaiyama [63] yielded the desired sandwich structure F. Magnesium insertion was readily carried out employing Eschenmoser's reagent [65] to yield G. Time-of-flight mass spectrometry on F yielded a strong parent ion at 1068 mass units as expected for the given structure.

The principal evidence for the relative orientation of the macrocycles with respect to one another comes from proton NMR spectroscopy. The proton NMR spectra of F and G indicate that both macrocycles in each cyclophane are equivalent and therefore support a symmetric structure. The protons at the periphery of the macrocycles experience small chemical shift changes relative to the corresponding monomeric methyl mesopyropheophorbide *a*, H and methyl mesopyrochlorophyllide *a*, I reference compounds (Table 6). Since these shifts result from the influence of the diamagnetic ring current of the adjacent macrocycle [64], their small magnitude suggests a structure for the cyclophane in which the macrocycles are stacked approximately center to center. Moreover, the pyrolic NH protons of F occur 0.75 ppm upfield of those in H, consistent with the positioning of these protons directly above the π system of the adjacent macrocycle. The data do not preclude the possibility that there may be a small angle of tilt between the planes which the macrocycles occupy. Space-filling models indicate that the angle of twist between the ring I - ring III axes may oscillate about 40° either side of the 90° average configuration. An x-ray structure of one of these molecules will yield a more accurate geometry.

Table 6. Proton magnetic resonance data

	Chemical shift, δ (ppm)			
Position	F[a]	H[a]	G[b]	I[b]
4b	1.27	1.64	1.47	1.59
8a	1.53	1.72	1.49	1.62
7a, 7b	1.6-2.3	1.6-2.3	1.8-2.4	1.8-2.4
1	3.05	3.18	2.87	3.05
3	3.10	3.22	3.23	3.07
4a	3.70	3.81	3.50	3.62
5	3.57	3.54	3.57	3.15
7d CO_3CH_3		3.59		3.67
-CH_2CH_2O-	3.75		3.76	
$CH_2\mathit{CH}_2O$-	3.90		3.86	
7	4.19	4.24	4.28	3.90
8	4.45	4.46	4.28	4.05
10	4.67	5.09	4.81	4.97
δ	7.97	8.38	8.38	8.27
α	8.79	9.13	8.90	9.29
β	9,22	9.40	9.62	9.73
NH	- 2.42	- 1.67		

[a]In chloroform-d. [b]In 10% pyridine-d_5 in benzene-d_6.

The optical absorption spectra of G and I are compared in Fig. 30. The Q_y transition in the spectrum of G occurs at 650 nm. Cooling the sample to 77K does not change the wavelength of the transition appreciably. The 658 nm Q_y transition of F behaves similarly. These transitions do not exhibit the solvent dependency characteristic of structural changes in the singly linked chlorophyll dimers. Moreover, the spectra of F and G are virtually superimposible on those of their monomeric counterparts H and I, respectively, both at 297 K and at 77 K. This result is consistent with a structure for F and G in which the Q_y transition dipoles are effectively uncoupled.

The fluorescence maxima of F and G are unchanged relative to monomeric H and I with maxima at 666 nm and 657 nm (420 nm excitation), respectively, in acetone. This is illustrated for G and I in Fig. 31. Despite the proximity of the macrocycles (∿4-6 Å) in the cyclophane structure, there is no evidence of exciton splitting. The cyclophane structure is sufficiently rigid so that optical spectroscopy of the singlet manifold over a wide range of temperatures reveals very little electronic interaction between the macrocycles. This is precisely the condition required to test the hypothesis concerning a lack of correlation between optical and redox properties of special pair chlorophyll in vivo.

The photoexcited triplet states of F and G have also been examined and present a rather surprising set of results. The triplets were generated by irradiation of 10^{-3} M compound in 10% pyridine-toluene glass with yellow light at the temperatures indicated in Table 7. The surprising feature of the data is the equivalence of the zero-field splitting parameters for each cyclophane and its corresponding monomer. If, as is initially expected, the triplet excitation is rapidly exchanging between both macrocycles, the C_2 symmetric structure of F and G with the Q_y transition dipoles at 90° to one another requires that E = O. However, this is not the case and it thus appears that even the triplet excitation is confined to one macrocycle. Varying the solvation conditions, e.g., 2-methyltetrahydrofuran, does not change the result. In addition, substitution of a bifunctional ligand, 1,4-diazabicyclo (2.2.2)octane, for the pyridine does not change the triplet EPR spectrum of the magnesium-containing cyclophane. Thus, the

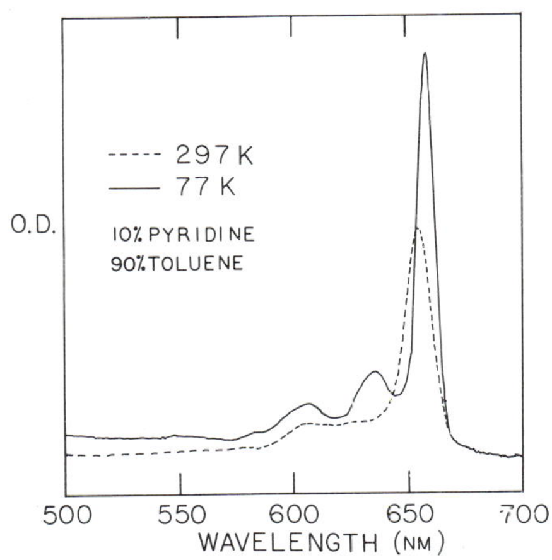

Fig. 30. Electronic absorption spectra for 10^{-5} M G and I. Note that the spectra for each compound at both 297 K and 77 K are completely superimpossible

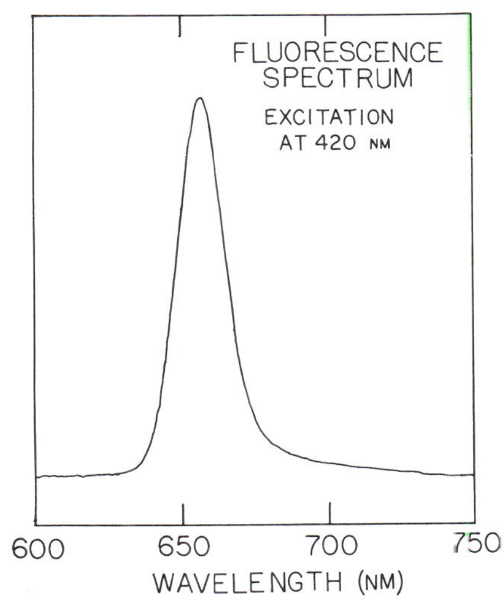

Fig. 31. Fluorescence emission spectra of 10^{-5} M G and I. Note that the emission spectra for each compound are superimpossible

essential asymmetry responsible for localization of the triplet on one macrocycle of the cyclophane must be more deep-seated.

Table 7. Zero-field splittings for selected compounds at 4.2 K in 10% pyridine-toluene glass

Compound	D	E
F	0.0345	0.0027
H	0.0345	0.0024
G	0.0300	0.0036
G[a]	0.0301	0.0038
G[b]	0.0302	0.0038
I	0.0301	0.0036

[a] 65 K. [b] 100 K

Several workers are currently engaged in attempts to determine the relative orientation of the macrocycles with respect to one another in the photoreaction center by comparing in vitro chlorophyll triplet data with that obtained by generating triplets of the special pair [65-70]. It is assumed in all cases that the excitation is exchanging between the macrocycles of the pair. Changes in the zero-field splitting parameter D in vivo relative to monomeric BChl *a* suggest that the triplet excitation in reaction centers from *R. sphaeroides,* in which the primary acceptor is reduced, is exchanging between two or more sites. Yet, there is no evidence in vivo based on triplet hyperfine interactions to confirm either the existence or extent of this exchange. In-vitro studies of chlorophyll dimer triplets face similar uncertainties.

The fact that the cyclophane special pair model yields a localized triplet should at the very least be regarded as a cautionary note in further conclusions concerning special pair structures derived via triplet state EPR data.

The cyclophane triplet EPR does not exhibit the unusual polarization of the in-vivo triplets [71]. This is not too surprising considering the fact that these polarizations are thought to arise via radical pair processes involving an electron acceptor [72].

The photoredox properties of the bis(chlorphyll) cyclophanes serve to test the original hypothesis. The cyclophane G undergoes photooxidation when irradiated with 650 nm light in the presence of electron acceptors, e.g., iodine, quinones, etc., and chemical oxidation by $ZnTPP^{+}\ ClO_4^{-}$[53] to yield a cation radical possessing a Gaussian EPR signal with a line width of 6.44G. Based on the treatment of Norris et al. [8], the narrowing of this line relative to the 9G line width of the corresponding monomer I indicates that once again spin is shared equally by the two macrocycles analogous to special pair chlorophyll in vivo.

The magnesium insertion reaction of the metal-free cyclophane may be interrupted to yield a cyclophane containing one chlorophyllide macrocycle and one pheophorbide macrocycle. The optical absorption spectrum of this species is essentially a superposition of the spectrum of H and I. The fluorescence spectrum is interesting from the standpoint that the 657 nm fluorescence of the magnesium-containing ligand is strongly quenched. The fluorescence quantum yield is only 10% that of the doubly metallated cyclophane. The intriguing possibility exists that the quenching may be due to electron transfer processes. The photochemistry of this species is currently under investigation.

Since the structure of the cyclophane is primarily solvent independent, a series of electrochemical determinations of the redox properties of this compound were carried out in polar media. The redox potentials for both the oxidation and reduction of G were determined accurately by ac voltammetry and are listed in Table 8. Cyclophane G undergoes reversible one electron oxidation 70 mV more easily than chlorophyll *a*. This result is consistent with the delocalization of unpaired electron density over both macrocycles in the cation radical of G. Characteristically, chlorophyll special pairs in vivo are more readily oxidized than bulk chlorophyll in the organism.

Table 8. Redox potentials for G and Chl a vs SCE

Compound	Oxidation[a]		Reduction[b]	
	$E^{1}_{1/2}$	$E^{11}_{1/2}$	$E^{1}_{1/2}$	$E^{11}_{1/2}$
G	0.54	0.77	- 1.29	- 1.77
Chlorophyll a	0.61	0.84	- 1.14	- 1.61

[a] In $CH_3CH_2CH_2CN$. [b] in DMF

Controlled potential oxidation of 8 with removal of one coulombic equivalent in a cell designed for optical observation of the electrolyzed species yielded the electronic spectrum of the cation radical. The spectrum is characterized by a broad maximum at 800 nm typical of chlorophyll cation radicals of the a series.

The reversible one electron reduction of G is more difficult than that of Chl a by 150 mV. This is indeed a surprising result. Cyclophanes of aromatic molecules characteristically are both oxidized and reduced more readily than their monomeric constituents [73]. This effect is due one again to charge delocalization over both macrocycles in the product being reflected in the transition state for interfacial electron transfer. The 10 G line width of the EPR signal from the radical anion of G generated via potassium reduction in HMPA is almost identical to that of the corresponding monomer radical anion. This result shows that in fact the anion is localized on one macrocycle.

The cyclophane models show that the optical and redox properties of special pairs in vivo may be indicative of very different features of the pair. The only evidence that photoreaction center chlorophyll in both purple photosynthetic bacteria and in green plants consists of a pair of chlorophyll molecules is derived from magnetic resonance experiments. On the other hand, all known reaction center chlorophylls possess a long wavelength optical transition which is red shifted relative both to that of bulk antenna chlorophyll in the organism and to that of the corresponding monomleric chlorophyll in vitro.

The results of these cyclophane studies to date support the proposal that special pair geometrics exist which adequately account for both the redox and spin delocalization properties of chlorophyll special pairs in vivo, yet do not give rise to unusually red-shifted optical spectra. Thus, the optical spectra exhibited by chlorophyll special pairs in vivo need not be exclusively an intrinsic property of the pair itself but may be influenced strongly both by the presence of additional chromophores and by important chlorophyll-protein interactions.

10 Concluding Remarks

The work described in this chapter serves to illustrate the point that useful information concerning the primary processes of photosynthesis may be obtained through a synthetic modeling approach. Not only is it possible to mimic key spectroscopic features of photoreaction centers, but it is also possible to duplicate the primary photochemical events of photosynthesis in models of defined structure. Future work in this area will explore additional steps of the photosynthetic pathway leading toward sustained charge separation in an in-vitro model photoreaction center.

Acknowledgment. The author wishes to thank the Division of Chemical Sciences, Office of Basic Energy Sciences of the Department of Energy, under whose auspices this chapter was written.

References

1. Govindjee, Govindjee, R.: In: Bioenergetics of Photosynthesis, Govindjee (ed). New York: Academic Press 1975
2. Borisov, A.Yu.: In: The Photosynthetic Bacteria, Clayton R.K., Sistrom W.R. (eds). New York: Plenum 1978
3. Clayton, R.K.: In: The Photosynthetic Bacteria, Clayton, R.K., Sistom, W.R. (eds.) New York: Plenum 1978
4. Loach, P.A., Androes, G.M., Maksin, F., Calvin, M.: Photochem. Photobiol. **2**, 443 (1963)
5. Parson, W.W., Cogdell, R.J.: Biochim. Biophys. Acta *416*, 105 (1975)
6. Bolton, J.R.: In: Primary Processes of Photosynthesis. Topics in Photosynthesis, Barber, J. (ed.), Vol. 2. Amsterdam: Elsevier 1977
7. McElroy, J.D., Feher, G., Mauzerall, D.C.: Biochim. Biophys. Acta **172**, 180-183 (1969)
8. Norris, J.R., Uphaus, R.A., Crespi, H.L., Katz, J.J.: Proc. Natl. Acad. Sci. USA **68**, 625-628 (1971)
9. Van Gockam, H.J., Pulles, M.P.J., Wessals, J.S.C.: Biochim. Biophys. Acta **408**, 331 (1975)
10. Feher, G., Hoff, A.J., Isaacson, R.A., Ackerson, L.C.: Ann. N. Y. Acad. Sci. **244**, 239-259 (1975)
11. Clayton, R.K.: Annu. Rev. Biophys. Bioeng. **2**, 131-156 (1973)
12. Hall, R.L., Kung, M.C., Fu, M., Hales, B.J., Loach, P.: Photochem. Photobiol. **18**, 505-520 (1973)
13. Okamura, M.Y., Steiner, L.A., Feher, G.: Biochemistry **13**, 1394-1403 (1974)
14. Thornber, J.P., Alberte, R.S.: In: The Enzymes of Biological Membranes, Martonosi, A. (ed.), pp. 163-190. New York: Plenum 1967

15. Schwartz, F.P., Goutermann, M., Muljiani, Z., Dolphin, D.: Bioinorg. Chem. **2**, 1-32 (1972)
16. Knox, R.: In: Bioenergetics of Photosynthesis, Govindjee (ed.). New York: Academic Press 1975
17. Norris, J.R., Levanon, H.: This volume
18. Anton, J.A., Kwong, J., Loach, P.: J. Heterocycl. Chem. **13**, 717-725 (1976)
19. Anton, J.A., Loach, P.A., Govindjee: Photochem. Photobiol. **28**, 235-242 (1978)
20. Little, R.G.: J. Heterocycl. Chem. **15**, 203-208 (1978)
21. Paine, J.B., III Dolphin, D.: Can. J. Chem. **56**, 1710-1712 (1978)
22. Paine, J.B., III Dolphin, D.: Can. J. Chem. **56**, 1712-1715 (1978)
23. Ogoshi, H. Sugimoto, H., Yoshida, Z.: Tetrahedron Lett. 169-172 (1977)
24. Collmann, J.P., Elliott, C.M., Halbert, T.R., Tovrog, B.S.: Proc. Natl. Acad. Sci. USA **74**, 18-22 (1977)
25. Kagan, N.E., Mauzerall, D., Merrifield, R.B.: J. Am. Chem. Soc. **99**, 5484-5486 (1977)
26. Chang, C.K.: J. Heterocycl. Chem. **14**, 1285-1288 (1977)
27. Petke, J.D., Maggiora, G.M., Shipman, L.L., Christoffersen, R.E.: J. Mol. Spectr. **73**, 311-331 (1978)
28. Petke, J.D., Maggiora, G.M., Shipman, L.L., Christoffersen, R.E.: J. Am. Chem. Soc. **99**, 7478-7489 (1977)
29. Katz, J.J., Norris, J.R., Shipman, L.L, Thurnauer, M.C., Wasielewski, M.R.: Annu. Rev. Biophys. Bioeng. 7, 393-434 (1978)
30. Shipman, L.L., Janson, T.R., Ray, J.G., Katz, J.J.: Proc. Natl. Acad. Sci. USA **72**, 2873-2876 (1975)
31. Ballschmiter, K., Katz, J.: J. Am. Chem. Soc. **91**, 2661-2677 (1969)
32. Closs, G.L., Katz, J.J., Pennington, F.C., Thomas, M.R., Strain, H.H.: J. Am. Chem. Soc. **85**, 3809-3821 (1963)
33. Wasielewski, M.R., Thompson, J.F., unpublished result
34. Cotton, T.M.: Thesis, Northwestern University 1976
35. Shipman, L.L., Cotton, T.M., Norris, J.R., Katz, J.J.: Proc. Natl. Acad. Sci. USA **73**, 1791-1794 (1976)
36. Boxer, S.G., Closs, G.L.: J. Am. Chem. Soc. **98**, 5406-5408 (1976)
37. Fong, F.K., Koester, V.J.: Biochim. Biophys. Acta **423**, 52-64 (1976)
38. Seely, G.R., in: The Chlorophylls, Vernon, L.P., Seely, G.R. (eds.), pp. 523-568. New York: Academic Press 1966
39. Baum, S.J., Burnham, B.F., Plane, R.A.: Proc. Natl. Acad. Sci. USA **62**, 1439-1442 (1964)
40. Wasielewski, M.R., Studier, M.H., Katz, J.J.: Proc. Natl. Acad. Sci. USA **73**, 4282-4286 (1976)
41. Wasielewski, M.R., Smith, U.H., Cope, B.T., Katz, J.J.: J. Am. Chem. Soc. **99**, 4172-4173 (1977)
42. Itoh, M., Hagiwara, D., Notani, J.: Synthesis 456-458 (1975)
43. Wasielewski, M.R.: Tetrahedron Lett. 1373-1376 (1977)
44. Wasielewski, M.R.: Unpublished result
45. Hindman, J.C., Kugel, R., Wasielewski, M.R., Katz, J.J.: Proc. Natl. Acad. Sci. USA **75**, 2076-2079 (1978)
46. Shapiro, S.L,, Kollman, V.H., Campillo, A.J.: FEBS Lett. **54**, 353-362 (1975)
47. Beddard, G.S., Carlin, J.E., Porter, G.: Chem. Phys. Lett. **43**, 27-32 (1976)
48. Beddard, G.S., Porter, G.: Nature **260**, 366-367 (1976)
49. Pellin, M.J., Wasielewski, M.R., Kaufmann, K.J.: J. Am. Chem. Soc. **102**, 1868-1873 (1980)
50. Periasamy, M., Linschitz, H., Closs, G.L., Boxer, S.G.: Proc. Natl. Acad. Sci. USA **75**, 2563-2566 (1978)
51. Hiyama T., Ke, B.: Biochem. Biophys. Acta **276**, 160-171 (1972)
52. Borg, D.C., Fajer, J., Felton, R.H., Dolphin, D.: Proc. Natl. Acad. Sci. USA **67**, 813-820 (1970)
53. Dutton, P.L., Prince, R.C., Tiede, D.M., Petty, K.M., Kaufmann, K.J., Netzel, T.L., Rentzepis, P.M.: Brookhaven Symp. **28**, 213-237 (1979)
54. Wraight, C.A.: Biochim. Biophys. Acta **459**, 525-531 (1977)
55. Vermeglio, A., Clayton, R.K.: Biochim. Biophys. Acta **461**, 159-165 (1977)

56. Parson, W.W., Clayton, R.K., Cogdell, R.J.: Biochim. Biophys. Acta **387**, 265-278 (1975)
57. Pellin, M.J., Kaufmann, K.J., Wasielewski, M.R.: Nature **278**, 54-55 (1979)
58. Fujita, I., Davis, M.S., Fajer, J.: J. Am. Chem. Soc. **100**, 6280-6282 (1978)
59. Eestimated from the difference between the one electron oxidation potential of the folded PChl-*a* dimer (0.50V *vs* SCE) and the one electron reduction potential of PPhide *a*-OH (-1.02V *vs* SCE) in methylene chloride.
60. Wasielewski, M.R., Svec, W.A., Cope, B.T.: J. Am. Chem. Soc. **100**, 1961-1962 (1978)
61. Kenner, G.W., McCombie, S.W., Smith, K.M.: Justus Liebigs Ann. Chem. 1329-1338 (1973)
62. Hutchins, R.O., Kandasamy, D.: J. Am. Chem. Soc. **95**, 6131-6133 (1973)
63. Mukaiyama, T., Usui, M., Saigo, K.: Chem. Lett. 49-50 (1976).
64. Abraham, R.J., S.C., Fell, C.M., Smith, K.M.: Org. Magn. Reson. **9**, 376-373 (1977).
65. Clarke, R.H., Connors, R.E., Schaafsma, T.J., Kleibeuker, J.F., Platenkamp, R.J.: J. Am. Chem. Soc. **98**, 3674-3677 (1976)
66. Clarke, R.H., Hofeldt, R.J.: J. Chem. Phys. **61**, 4582-4587 (1975)
67. Hagele, W., Drissler, F., Schmid, D., Wolf, H.C.: ESR studies of the photoexcited triplet states of chlorophyll *a* and chlorophyll *b* in poly (methyl methacrylate) and methyltetrahydrofuran at 4.2K. Presented at Third Int. Seminar on Excitation Energy Transfer in Condensed Matter, Prague, Czechoslovakia 1976.
68. Kleibeuker, J.F., Schaafsma, T.J.: Chem. Phys. Lett. **29**, 116-122 (1974)
69. Levanon, H., Scherz, A.: Chem. Phys. Lett. **31**, 119-124 (1975)
70. Thurnauer, M.C., Norris, J.R.: Chem. Phys. Lett. **47**, 100-105 (1977)
71. Dutton, P.L., Leigh, J.S., Reed, D.W.: Biochim. Biophys. Acta **292**, 654-664 (1973)
72. Thurnauer, M.C., Katz, J.J., Norris, J.R.: Proc. Natl. Acad. Sci. USA **72**, 3270-3274 (1975)
73. Gerson, F., Kaupp, G., Ohya-Nishiguchi, H.: Angew. Chem. Int. Ed. **16**, 657-658 (1977)

Chapter 8
Light Path of Carbon Reduction in Photosynthesis

Francis K. Fong[1]

1 Introduction

The Chl-*a* light reaction leading to photosynthesis

$$CO_2 + 2H_2O \xrightarrow[\text{Chl } a]{\text{visible light}} \frac{1}{n}(CH_2O)_n + H_2O + O_2 + \text{stored energy} \qquad (1)$$

involves an interplay of primary and secondary processes, whereby water is oxidized to evolve molecular oxygen, and the hydrogen from water ends up reducing CO_2 to yield carbohydrates. In this chapter we review the current progress in work on Chl-*a* photocatalyzed synthesis of organic compounds from H_2O and CO_2. It is shown that quantum storage in photosynthesis is reasonably described as a reversal of dark reactions in involving the 3-phosphoglyceric acid molecule. The Chl-*a* photoactivated reequilibration of physiological products in a direction opposite to that of an otherwise irreversible (dark) pathway provides a useful definition of carbon reduction in photosynthesis.

It is well established that the photosynthetic path of carbon, as given by the reductive phosphate (RPP) cycle, terminates in the dark with the ribulose diphosphate (RuDP) carboxylase reaction. In this reaction one CO_2 molecule is assimilated by one molecule of RuDP followed by hydrolytic splitting of the C_6 intermediate to yield two molecules of phosphoglyceric acid (PGA). The evidence for this reaction is overwhelming and, for more than 20 years, biochemical studies of the RPP cycle have added strength to the assumption that the dark carboxylation reaction stoichiometry is also obtained under normal conditions of photosynthesis.

In spite of the above assumption, it might be suggested that the RPP cycle is initiated by a photoactivated reduction of PGA to produce RuDP in the light and then sustained by photoreductive assimilation of CO_2 by RuDP. Fortunately, controlled experiments to be discussed later permit unambiguous sequencing of the photosynthetic events. Among these experiments are the [^{14}C]-labeling results obtained by Bassham and Calvin [1], who found that the stoichiometry of CO_2 assimilation by RuDP in the light differed markedly from that in the dark. The relative decay rates of RuDP under dark and light conditions are indicative of the photochemical origin of the RuDP carboxylation reaction during steady operation of the RPP cycle. More important, characterization of the photochemical properties of the Chl-*a* water photolysis and carbon assimilation reactions in in-vitro experiments, together with biochemical

1 Department of Chemistry, Purdue University, West Lafayette, IN 47907, USA

evidence, provides the basis for postulating the light path of carbon reduction in terms of the reversal of the mechanisms for carbohydrate degradation in the dark.

A good deal of interest in the past two decades has been focused on the presence in plant photosynthesis of two different Chl-*a* reaction centers known as P700 and P680 [2-5]. Current interpretations of the P700 and P680 photoreactions are incomplete and often contradictory. Among these interpretations is the series scheme, in which the P700 and P680 light reactions, respectively referred to as the PS1 and PS2 reactions, are one quantum photoevents connected in series by a chain of dark redox couples [6, 7]. The PS1 reaction is said to be responsible for the reduction of ferredoxin and NADP (nicotinamide adenine dinucleotide phosphate), leading to the thermochemical reduction of PGA in the RPP cycle. The O_2 evolution from water splitting occurs as a consequence of the PS2 reaction. A second interpretation [8] of plant photosynthesis was formulated based on the energy upconversion postulate [9, 10], in which the P680 light reaction is connected to ferredoxin reduction in a single two-photon process. The molecular details are given in terms of the structural and photochemical properties of hydrated Chl *a* [11, 12]. The overall stoichiometry of this reaction is given by the expression [8].

$$2H_2O + 4Fd_{ox} + 2ADP + 2P_i \xrightarrow[(\mathrm{Chl}\,a \cdot 2H_2O)_2]{h\nu} 4Fd_{red} + 2ATP + O_2 \qquad (2)$$

where the transfer of four electrons from water to ferredoxin is supposedly attended by the formation of ATP (adenosine triphosphate) in photophosphorylation [13]. The P700 reaction center is modeled after the monohydrate dimer, $(\mathrm{Chl}\,a{\cdot}H_2O)_2$ [12], and is presumably responsible for cyclic photophosphorylation according to the proposals by Arnon [14].

It is generally recognized that Mn is engaged in the P680 O_2 evolution process [15-17]. Proton release according to the half reaction

$$2H_2O \xrightarrow{\mathrm{Mn}} 4H^+ + O_2 + 4e \qquad (3a)$$

is believed to occur asynchronously with O_2 evolution [16-21]. The reducing side of the P680 reaction, however, remains an open debate. Reaction (2), proposed on the basis of the in vitro demonstration of water photolysis by $(\mathrm{Chl}\,a{\cdot}2H_2O)_n$, [8, 22, 23], appears to be supported by recent observations [24] on the photoreaction properties of ferredoxin. In the series scheme plastoquinone is assumed to be the P680 primary electron acceptor, in contrast to in-vitro experiments that demonstrated the preferential photoreduction of H_2O by hydrated Chl *a*in the presence of benzoquinone [25]. Other difficulties concerning the series scheme have also been noted. For example, the assignment of PS1 as the sole agent for ferredoxin reduction is not easily reconciled with the observation that electrons ending up in NADP do not appear to involve P700 directly [26].

In the discussion to follows, summarized in Fig. 1, it is shown that the apparent inconsistencies may be eliminated by minor rearrangements of the existing schemes. Features of the above proposals for Chl-*a* light reaction, combined with physiological effects of primary electron acceptors obtained since these proposals were made, provide a plausible photochemical mechanism for the reduction of carbon in photosynthesis.

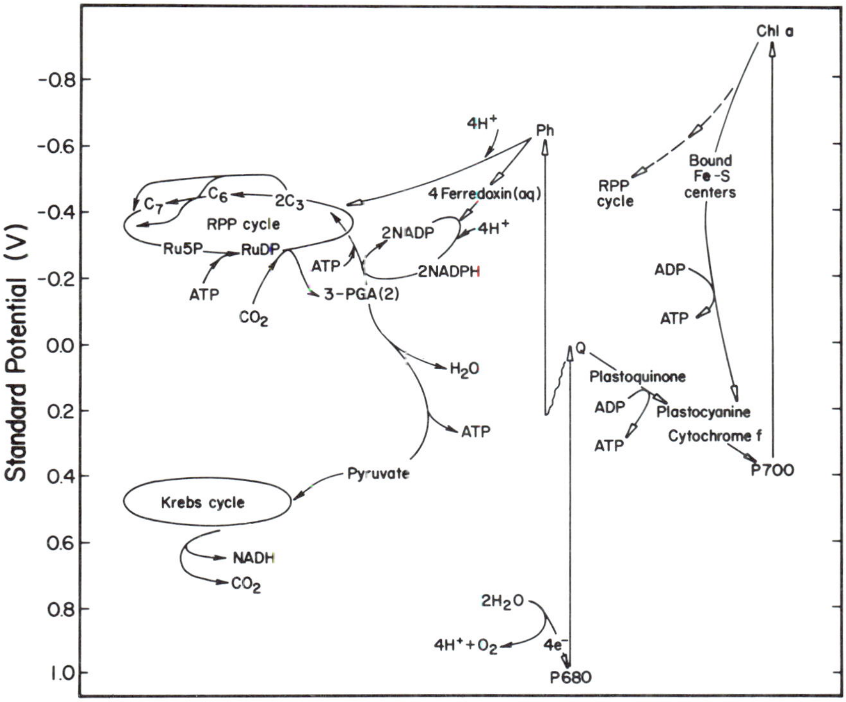

Fig. 1. Schematic representation of plant photosynthesis. The *open arrows* indicate the path of electron and/or [H] flow in the P680 and P700 light reactions

2 Scope

Areas of active photosynthesis research have developed largely from enzymological observations of reaction products and/or reaction intermediates leading to these products. By contrast, as we noted in the above introduction, physiological functions of the Chl-*a* light reactions are less well understood. Relationships between biochemical observations and the underlying Chl-*a* photon mechanisms are desirable. In order to arrive at some focused definition of quantum storage in photosynthesis, traditional questions in plant physiology are examined in terms of in-vitro Chl-*a* photochemical and catalytic interactions specialized to the water-splitting and carbon assimilation reactions.

2.1 Origin of O_2 Evolution

An early statement defining the scope of photosynthesis research was made in 1796 by Ingenhousz, who concluded [27] that the green parts of plants absorb from "carbonic acid in the sunshine, the carbon, throwing out at that time the oxygen alone, and keeping the carbon itself as nourishment." As recently as 1956 Warburg suggested that a bound CO_2 molecule transferred an oxygen atom to the carbonyl oxygen of the chlorophyll to form a peroxide and then a molecular oxygen [28]. That water may be the source of molecular oxygen was suggested by several authors around 1930 [29-32]. Indirect experimental support for this suggestion was provided by workers [33, 34] who showed that no correlation could be established between the uptake of CO_2 and the appearance of O_2. A direct approach to the question was made by Ruben et al. [35a] in a mass spectrometric determination of photosynthetic O_2 evolution from [^{18}O]-enriched H_2O. However, in their study there was a question of the exchange of oxygen from CO_2 with that from H_2O [35b]. In this respect it seems desirable to examine the intrinsic ability of Chl *a* to photocatalyze water splitting in a CO_2-depleted environment.

2.2 Light and Dark Paths of Carbon

The realization of H_2O, instead of CO_2, as the origin of O_2 evolution in photosynthesis led Ruben to the suggestion [36] that CO_2 fixation in plants may occur as a dark reaction. The dark path of carbon was subsequently proposed based on the detailed analyses of [^{14}C]-labeled photosynthetic products [37], and became known as the RPP cycle or the "Calvin-Benson cycle". The identification [38] of the C_5 ketose, ribulose diphosphate, as the primary CO_2 acceptor for the carboxylation reaction leading to PGA, phosphoglyceric acid, was supported by the discovery [39] in photosynthetic tissues of the enzyme, ribulose diphosphate carboxylase, required for the production of PGA from dark CO_2 assimilation by RuDP. In the past this observation has led to the widespread belief that PGA is the first stable product of photosynthesis. However, the dark RuDP carboxylase reaction stoichiometry is not obtained under normal conditions of photosynthesis [1]. It has been established that the RuDP carboxylase reaction terminates the RPP cycle, as the concentration of RuDP decays to vanishingly low levels within 30 s after the light is turned off [1]. Consequently, the requirement of RuDP as the primary CO_2 acceptor in photosynthesis is tantamount to the conclusion that RuDP is the first stable product of photosynthesis prior to establishment of the RPP cycle. It is desirable to delineate the three stages of the carbon reduction path, i.e., initiation, steady-state operation, and termination of the RPP cycle in dark $\leftrightarrow$ light transitions.

The RPP cycle may be described as a sequence of photoactivated events in a direction opposite to the dark metabolic paths of hexoses and pentoses, as shown in Fig. 1. In all living systems the 3-PGA molecule is a precursor common in the biosynthesis of carbohydrates, fats, amino acids, and proteins. It is the intermediate through which the oxidation of C_6 carbohydrates is brought to the level of CO_2 in the dark. In plants, according to Fig. 1, PGA also serves as a precursor to photosynthetic CO_2 assimilation, providing a switching mechanism for reversing the oxidative paths of carbon to sustain the cycle of energy consumption and storage in the living world. At the onset of dark-light transition PGA is reduced by NADPH, assisted by ATP, to yield trioses and then C_6, C_5, and C_7 carbohydrates. The production of RuDP follows, making it possible for the photoreductive assimilation of CO_2 in steady-state operation of the RPP cycle.

The reductive carbon cycle outlines a specific path as indicated by the time sequence for the appearance of [^{14}C]-labeled phosphorelated hydroxyacids and sugars and by the distributions of radioactivity in the [^{14}C]-labeled product molecules [37]. The specificity of the dark carbon path rules out an earlier proposal [40, 41] that photosynthesis occurs through the production of formaldehyde from H_2O and CO_2 in the light, followed by autocondensation of CH_2O to yield carbohydrates [42]. However, the ability of Chl *a* to photocatalyze water splitting in vitro [22, 23] suggests the possibility of in-vitro photosynthesis of organic matter from the Chl-a water-splitting reaction using metal surfaces as catalyst in place of biological enzymes. This possibility is investigated in connection with the postulated photoreduction of CO_2 by P680 to the level of carbohydrates in plant photosynthesis.

2.3 Submolecular Interactions of Chl-*a* Light Reactions

In addition to the O_2 evolution and the carbon path, the submolecular mechanisms of Chl-*a* light reactions in photosynthesis have been the subject of extensive discussion Early workers speculated on the possibility of atomic rearrangement following photoexcitation of the chlorophyll. In 1953 Levitt proposed [43] that the primary process of photosynthesis may be viewed as a photoactivated electron transfer from the chlorophyll to an acceptor [44, 45]. Levitt's suggestion was later established on experimental grounds [46-54] although in-vitro chlorophyll research in the fifties through the early seventies was, for the most part, stimulated by solid-state concepts that are related to the Levitt proposal but somewhat less relevant to the primary photosynthetic mechanism. In 1949 Katz [55] suggested that photoconduction and energy migration in the two-dimensional crystals of chlorophyll may play a central role in the primary photoact of photosynthesis. According to this suggestion a mobile electron is excited in an ordered array of chlorophyll molecules and then migrates to some center where it becomes trapped. The trapped electron would become effectively a chemical reducing agent. The hole, likewise trapped elsewhere as an oxidant, then sets the stage for further reactions in photosynthesis. The proposed generation of trapped electrons and holes recalls a similar theory by Mott and Gurney [56] to account for the primary photographic process. In 1965 Bardley and Calvin [57] suggested that oxidizing and reducing centers in photosynthesis are formed through the creation of electron-hole pairs and their subsequent spatial separation to different points within the chloroplast lamellar structure. These various proposals were followed by experimental studies in which both chlorophyll *a* and methyl chlorophyllide *a* were found to be photoconductive in films deposited from solvents. Nelson reported [58] the observation of a current of 10^{-12} A under strong illumination with 5 V applied across the sample film. Nelson's results were confirmed and extended by Arnold and Maclay [59], Terenin et al. [60]. Rosenberg and Camiscoli [61], McCree [62], and Albrecht and co-workers [63-67]. It was found that the photocurrent increases with the number of monolayers and applied voltage, and that about 10^9 quanta were needed to produce one electron [62]. This low value suggests that photoconduction as a primary mechanism for energy transduction in photosynthesis is rather unlikely [68]. In current work the emphasis has shifted from photoconduction and has focused on the electron donor-acceptor interactions of photoexcited Chl *a*.

The specificity of Chl-*a* pathways in photosynthesis is underscored by the large number of observations not directly relevant to the photocatalytic mechanisms of Chl *a* in reaction (1). For example, the spectroscopic identification of the Chl-*a* triplet state is possible only when the chlorophyll is rendered photosynthetically nonfunctional [69-77]. The oxidation of chlorins to porphines by quinones, virtually a quantitative reaction in vitro, apparently does not occur in vivo [78]. Monomeric Chl *a* in homogeneous polar solutions readily undergoes photodegradation in the presence of O_2 [79], in contrast to the relative photochemical stability of in-vivo chlorophylls. Among the best characterized in-vitro photochemical redox reactions of the chlorophyll are those collectively known as the Krasnovski reactions [80], which are apparently of little consequence in plant photosynthesis [78]. The design of in-vitro experiments in addressing the photosynthesis problem is thus narrowed to the investigation of in-vitro Chl-*a* properties suitable for modeling the in-vivo behavior. The goal is to reproduce in vitro the observables of photosynthetic reaction centers, including the photolytic splitting of water and carbon dioxide reduction as direct consequences of the Chl-*a* light reaction.

3 Model for Chlorophyll Light Reactions in Photosynthesis

The photophysical as well as photochemical properties of reaction center chlorophyll complexes in vivo differ markedly from those of the light-harvesting chlorophylls. The nomenclature, P680 and P700, for the two photosystems in green plants reflects the long-wavelength shifts of the photoreactive chlorophylls. The location of P700 on the standard potential scale in Fig. 1 indicates its low oxidation potential relative to the P680. The latter chlorophyll complex is positioned in Fig. 1 so as to denote its ability to oxidize water in the oxygen evolution process (E_{red} = 0.81 V at pH 7).

3.1 Long-Wavelength Shifts of Chlorophyll Aggregates

It was once thought that the long-wavelength shifts and photoredox properties of P680 and P700 were exclusively due to protein-chlorophyll interactions. Recent work on photoreactive aggregates of hydrated chlorophyll suggests the possibility that the physical and photochemical properties of the chlorophyll are significantly influenced by the stereospecific interactions in coordination and hydrogen bonds involving the Mg atom and cyclopentanone carbonyl groups at C9 and C10 of the Chl-*a* molecules. The in-vivo properties attributed to the P680 and P700 reaction centers are respectively reflected in the corresponding properties of dihydrate and monohydrate

Chl-*a* adducts. Relevant features of the in-vitro findings are detailed in Sects. 4-7, providing guidelines for modeling the P680 and P700 light reactions in photosynthesis.

Chlorophyll owes its optical activity to the asymmetric carbons at C7, C8, and C10 [81]. Steric repulsion froces the C10 carbomethoxy group to the side of the chlorin ring opposite the C7 propionic acid phytol ester [81b]. The pigments Chl *a'* and Chl *b'* have been shown to be interconvertible with Chl *a* and Chl *b*, respectively, by heating [82]. They are probably C10 epimers of the natural pigments [83]. It turns out that Chl *a* derives its unique photophysical and photochemical properties from its various aggregation states [84-87]. Unlike the typically weak interactions in porphyrin free bases and metalloporphyrins [88-95], strong and highly stereospecific [96] metal to side-chain carbonyl (at C7, C9, and C10) interactions play a dominant role in chlorophyll aggregation.

The interest in Chl-*a* aggregation has been sustained through decades of investigation by the belief that some aggregated forms of Chl *a* may play a role in photosynthetic organisms [53, 54, 97-101]. The possible existence of dimeric chlorophyll in photosynthesis was suggested by S.S. Brody [99] who found a 720 nm fluorescence band in Chlorella and Porphyridium at liquid air temperature (80 K) and proposed that this in-vivo fluorescence may be attributed to a dimeric form of Chl *a* similar to the aggregate that gives rise to the 715 nm fluorescence band observed in ethanolic solutions of Chl *a* at low temperatures. The observation of in-vivo 720 nm fluorescence in Porphyridium was confirmed by other workers [102, 103]. However, the role of Chl-*a* dimer participation in photosynthesis was regarded unlikely by some workers who favored a hypothetical Chl-*a*-cytochrome complex as the origin of the long-wavelength fluorescence [104].

The intimate molecular association of Chl *a* with water was established long ago by Willstätter und Stoll [105], who gave the semihydrate Chl $a \cdot 0.5\ H_2O$ for the empirical formula of chlorophyll. In the ensuing decades it became apparent that the water content of nonpolar Chl-*a* solutions cannot be reduced to a level lower than that of the order of the chlorophyll concentration [106-108]. The early findings were recently confirmed by the conclusion that the "driest" chlorophyll in nonpolar solutions exists as the monohydrate Chl $a \cdot H_2O$ [12, 109]. The problem of Chl-*a* aggregation in nonpolar solutions is thus defined by the range of Chl a-H_2O equilibria [110]

$$\begin{array}{ccccccc}
\overset{3}{(\text{Chl}\ a_2)_n} + 2\ \text{mH}_2\text{O} & \rightleftharpoons & \overset{1}{2\ \text{mChl}\ a \cdot \text{H}_2\text{O}} & \underset{+2\ \text{mH}_2\text{O}}{\overset{-2\ \text{mH}_2\text{O}}{\rightleftharpoons}} & \overset{2}{2\ \text{mChl}\ a \cdot 2\ \text{H}_2\text{O}} & \rightleftharpoons & \overset{4}{(\text{Chl}\ a \cdot 2\ \text{H}_2\text{O})_{n=2m}} \\
 & & \updownarrow & & {\scriptstyle +2\ \text{mH}_2\text{O}} \updownarrow {\scriptstyle -2\ \text{mH}_2\text{O}} & & \\
 & & \underset{5}{\text{m}(\text{Chl}\ a \cdot \text{H}_2\text{O})_2} & & \underset{6}{\text{m}(\text{Chl}\ a \cdot 2\ \text{H}_2\text{O})_2 \cdot 2\ \text{H}_2\text{O}} & \rightleftharpoons & \underset{6'}{2\ \text{m}(\text{Chl}\ a \cdot 2\ \text{H}_2\text{O}) \cdot \text{H}_2\text{O}}
\end{array} \quad (4)$$

The characteristic red absorption maxima and the most probable bonding interactions in the molecular complexes *1-6* are listed in Table 1.

Table 1. C9 and C10 carbonyl bonding interactions in Chl *a*-H_2O aggregates. [a]Exo interactions are given as *dashed lines.* A detailed discussion of exo and endo Chl a-H_2O interactions is given in Section 4.3. Repetitive interactions are denoted by "etc."

Index number	Empirical formula	Probable interactions	Red absorption maxima in nonpolar solvent (nm)
1	Chl *a* • H_2O	Mg···· O(H)H	665(300 K)
2	Chl *a* • 2 H_2O	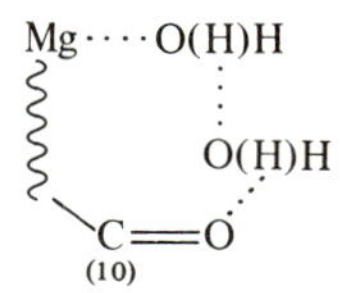	~663(300 K)
3	$(Chl\ a_2)_n$	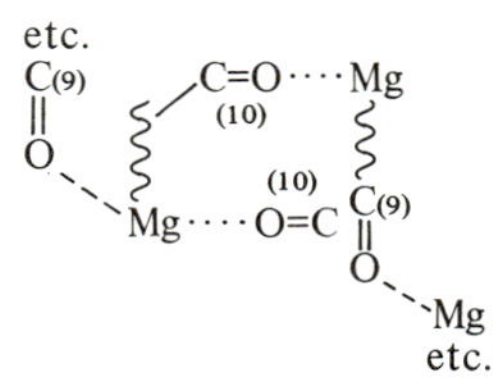	678(300 K)
4	$(Chl\ a \cdot 2\ H_2O)_{2n}$	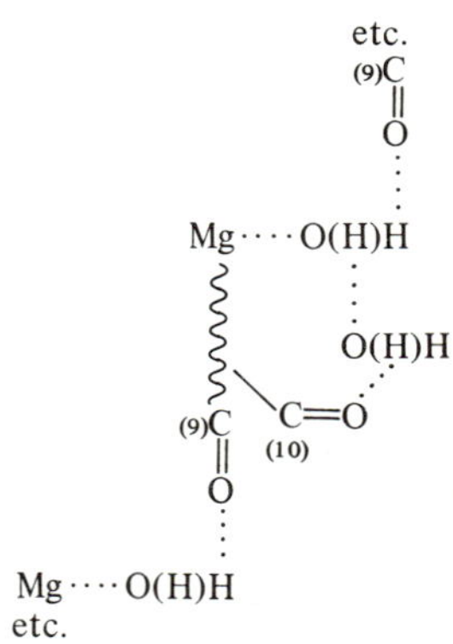	743(300 K)
	$(Chl\ a \cdot 2\ H_2O)_2$	Dimeric segment of above	695(77 K)

Index number	Empirical formula	Probable interactions	Red absorption maxima in nonpolar solvent (nm)
5	$(Chl\ a \cdot H_2O)_2$	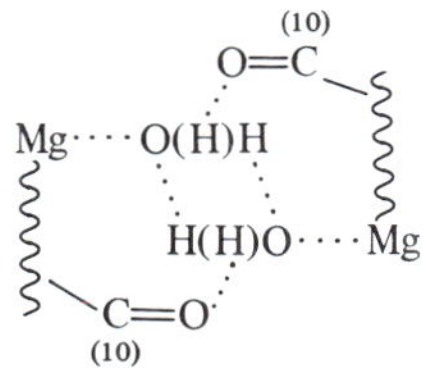	702.5(121 K)
6[a]	$(Chl\ a \cdot 2\ H_2O)_2 \cdot 2\ H_2O$	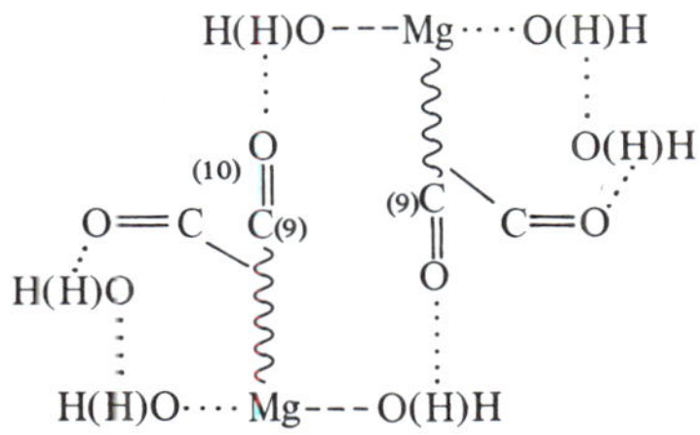	695(77 K)
6'	$(Chl\ a \cdot 2\ H_2O) \cdot H_2O$	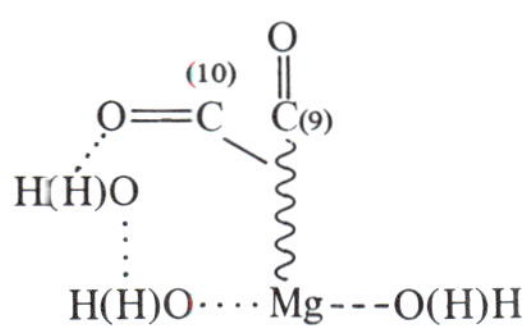	660(300 K)

[a] Similar exo interactions involving the C7 propionic ester carbonyl are equally probable. The C7 exo dimer has a 77 K red absorption maximum at 716 nm

3.2 Postulates

In in-vitro work the in-vivo photosynthetic reactions are modeled in terms of the structural and photochemical interactions between Chl *a* and H_2O: (1) The P680 reaction center is modeled [8] after the structural interactions in Chl *a* dihydrate (see Fig. 2), which are derived from the x-ray diffraction results [111] on single crystalline ethyl chlorophyllide *a* dihydrate. (2) The P700 reaction center and the corresponding bacterial reaction center, P870, are given [9] by the model structure of $(Chl\ a \cdot H_2O)_2$ shown in Fig. 3. In this structure the two parallel chlorin rings of the Chl-*a* molecules are complementarily held in position by two H_2O molecules, each being involved in a hydrogen bond to the carbonyl of the ring V carbomethoxy group of one Chl-*a* molecule, and an oxygen coordination bond to the magnesium atom of the second Chl-*a* molecule. (3) The primary photoexcitation process is modeled after

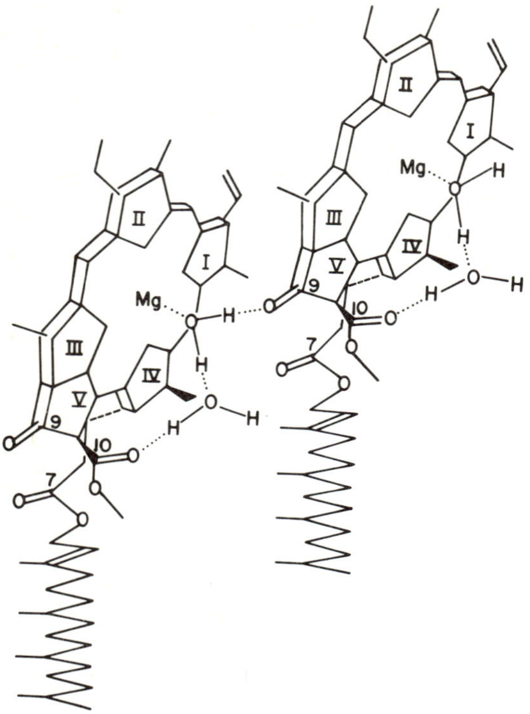

Fig. 2. Model structure of P680 reaction center chlorophyll

the energy fusion concept [9]. Through annihilation interactions between the reaction center and antenna Chl *a*, the lowest excited singlet and triplet states of the chlorophyll are envisaged to be upconverted to a two-quantum equivalent. Photo-

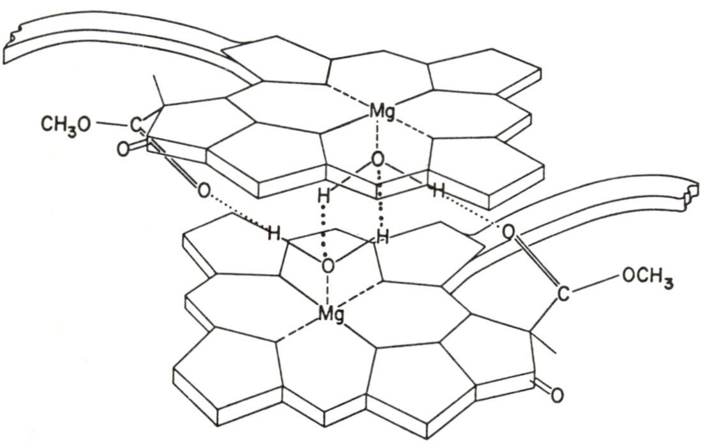

Fig. 3. Model structure of P700 reaction center chlorophyll

chemistry can occur either off a singly excited manifold or the upconverted charge-transfer state of the chlorophyll [10]. (4) The reduction of CO_2 as well as the oxidation of H_2O are modeled as consequences of the primary light reaction between Chl *a* and H_2O.

The P680 model originated from the observation [8] that, in the presence of a suitable catalyst such as Pt, the 743 nm-absorbing dihydrate polymer, $(\text{Chl}\ a \cdot 2H_2O)_n$, is capable of photocatalyzing the water-splitting reaction using visible light [8, 22]

$$2H_2O \xrightarrow[(\text{Chl}\ a \cdot 2H_2O)_{n \geq 2}]{\text{visible light}} 2H_2 + O_2 \tag{5}$$

It appears reasonable to suppose that $(\text{Chl}\ a \cdot 2H_2O)_2$, the smallest molecular assembly characterized by the bonding interactions found in $(\text{Chl}\ a \cdot 2H_2O)_n$, may be representative of the P680 reaction center adduct. However, the corresponding monomeric complex, Chl $a \cdot 2H_2O$, if it exists in a stabilized form under photosynthetic conditions, cannot be ruled out as a viable model for P680 (see Sect. 8). The absorption maximum of $(\text{Chl}\ a \cdot 2H_2O)_n$ occurs at 743 nm, whereas the dihydrate dimer $(\text{Chl}\ a \cdot 2H_2O)_2$ has an absorption maximum at 695 nm [96, 112]. The absorption maximum of monomeric Chl $a \cdot 2H_2O$ is expected to occur to the blue of 695 nm. From difference spectral determinations, the absorption maximum of P680 was found to occur within the 682-695 nm wavelength region [4, 5].

In both models for P680 and P700 the Chl-*a* C10 ester carbonyl

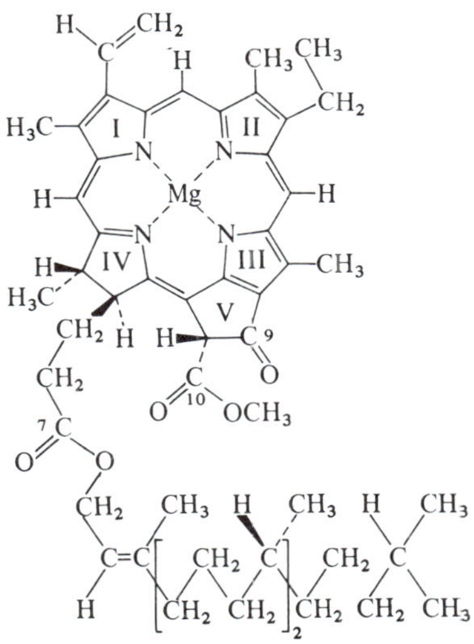

group is the site for stereospecific [96] interactions that define the symmetry types of $(\text{Chl}\ a \cdot H_2O)_2$ structures. Interestingly, *Chlorobium* chlorophylls, in which the C-10 ester carbonyl is replaced by hydrogen atom [113-115] are found only as photochemically inactive antenna pigments in green photosynthetic bacteria.

The upconversion postulate [9,10] is based on the fact that one-quantum events involving red light are energetically insufficient to sustain the chemical transformations in reaction (2). The proposal appears to be compatible [8] with the generally accepted two quanta per electron requirement in plant photosynthesis, as well as the observation of singlet-triplet annihilation as a dominant photophysical pathway in Chl *a* [116].

3.3 Path of Electrons from Water

The model for P680 light reaction given by reaction (2) appears to be consistent with Arnon and co-workers' finding of an unexpectedly high affinity of ferredoxin for "electrons from water" [24]. In electrochemical terms, the electrons from reaction (3a) are transferred, through the chlorophyll, to the electron acceptor of P680. The cathodic complement to reaction (3a) is the half reaction

$$e + H_2O \rightarrow [H] + OH^- \qquad (3b)$$

Invoking the presence of an electron acceptor, A, we write

$$e + H^+ + A \rightarrow HA \qquad (3c)$$

or simply,

$$e + H^+ \xrightarrow{A} [H]. \qquad (3d)$$

In reactions (3b, d) [H] denotes the hydrogen from water splitting acquired by reduction products of the chlorophyll light reaction (see Sect. 8.2). In reaction (3c) the transfer of electron to A is followed by an uptake of H^+ to yield HA. In the àbsence of CO_2 the recombination of [H] produced by Chl-*a* water photolysis on a metal surface results in H_2 evolution [22]. In this case the metal provides surface active sites for mediating the path of [H]. The formation of atomic hydrogen intermediates has been similarly postulated for metal catalyzed reactions such as conventional water electrolysis and alkene hydrogenation reactions. In the presence of CO_2 the production of [H] in water photolysis by photoexcited Chl *a* results in the in-vitro reduction of CO_2 to yield formaldehyde and formic acid [117]. In the discussions to follow, it is shown that the [H] intermediate concept of CO_2 photoreduction also appears to be applicable to observations of carbon reduction in plant photosynthesis.

Through manipulation of equilibria (4), it is possible to obtain preparations of $(\text{Chl}\ a \cdot 2H_2O)_{n \geqslant 2}$ and $(\text{Chl}\ a \cdot H_2O)_2$. Comparison can then be made of the properties of the in-vitro preparations with those of photosynthetically active Chl *a* in vivo, such as the quantum requirement, absorption spectra, redox potentials, and characteristic photochemical behvaior of P700 and P680. The goal is to demonstrate the simultaneous evolution of O_2 from H_2O cleavage and CO_2 reduction to yield C_1 products [117]:

$$CO_2 + H_2O \xrightarrow[\text{Chl}\ a\cdot 2H_2O,\ Pt]{\text{visible light}} H_2CO + O_2 \qquad (6)$$

$$2CO_2 + 2H_2O \xrightarrow[\text{Chl}\ a\cdot 2H_2O,\ Pt]{\text{visible light}} 2H_2CO_2 + O_2. \qquad (7)$$

The results obtained in the study of reactions (5-7) suggest the ability of electrons from the Chl-*a* water photolysis reaction to reduce CO_2 to carbohydrates and/or plant acids, consistent with the photoreduction of CO_2 in the postulated light path of carbon shown in Fig. 1.

4 Dimer Model of P700

The delineation of in-vitro Chl *a* bonding interactions and photochemical properties analogous to those of P680 and P700 is made possible by specialized preparatory and instrumental procedures. The necessary stages leading to molecular modeling of P700 are described in the following.

4.1 Chlorophyll Purification

High purity chlorophyll can be readily prepared from fresh spinach. Methods for monitoring the water contents of Chl-*a* preparations have been developed [12] in order to minimize possible ambiguities due to the lack of sample quality control. The procedure [11] for Chl *a* extraction and purification employed in the experimental studies reviewed in this chapter differs slightly from conventional methods [118]. The details of this procedure are given as follows.

About 100 g of pulverized spinach leavers are dissolved in 250 ml of acetone. The acetone extract is filtered through four layers of cheese cloth in a conventional Buchner. The filtrate is diluted with about 62.5 ml of water and passed through a 5 cm polyethylene powder column covered by 1 cm of glass beads. A pale yellow filtrate that separated from the bulk of deep-green fraction at the top of the column is discarded. The column is dried by aspiration for about 15 s and subsequently covered with ethyl ether. Gentle suction is applied with a suction flask along with periodic addition of small amounts of ether until the bulk of the chlorophyll is washed down into the flask. The ether solution is evaporated to near dryness, and the residue is redissolved in pure n-pentane. The resulting solution is swirled with water in a separatory funnel to remove remaining traces of acetone. The water is then drawn off and the n-pentane solution is evaporated to dryness.

The residue thus obtained is redissolved in a minimum volume of n-pentane. The solution is poured onto a powdered sugar column. The column is then eluted with 0.5% n-propanol in n-pentane. The first fraction of yellow pigments is discarded. The green band that followed, containing a mixture of Chl *a*, Chl *a'*, and pheophytin *a*, is collected. The chlorophyll eluent is shaken with water several times to remove the n-propanol, and then evaporated to dryness.

The residue from the preceding procedure is dissolved in a small quantity of 4 : 1 n-pentane: ether. The solution is poured onto a second sugar column. The narrow band, about 2-4 cm length in, is washed with small portions of n-pentane and eluted with 0.5% n-propanol in n-pentane. The chlorophyll-containing sugar column is transferred into a beaker containing 600 ml of 4% ethanol in n-pentane. The sugar is filtered from the solution and washed with several portions of the ethanolic pentane.

The resulting solution, about 1 liter in volume, is evaporated down to about 100 ml. The solution is shaken with water in a separatory funnel to remove the ethanol. On cooling to 0°C the chlorophyll precipitated out as the crystalline dihydrate (Chl *a* · $2H_2O)_n$. Further purification is accomplished by reprecipitation. Due to the stereospecific nature of Chl *a*-H_2O interactions, the preparation of the chlorophyll in the crystalline dihydrate form appears to be a convenient method to minimize any contamination of Chl *a* by Chl *a'*. The Chl *a* and Chl *a'* are epimeric about the asymmetric center at C10.

The purity and water content of the Chl *a* samples are monitored by optical spectroscopy. Absorption spectra of the purified chlorophyll *a* in anhydrous ether (AR grade) solution have absorption maxima at 4286 ± 1 Å and 6604 ± 1 Å with blue/red absorbancy ratios of 1.29 ± 0.01. These data have been consistently reproduced using the above recrystallization technique, and they agree with established criteria for chlorophyll *a* purity [118]. Any contamination by Chl *a'* is detected by liquid chromatography, and such contamination is not observable in samples prepared by the procedure outlined. The dihydrate crystals are readily converted to the monohydrate by heating under vacuum conditions at 80°C for several hours. Convenient nonpolar solvents include mixtures of n-pentane and methylcyclohexane. In order to control the water content the hydrocarbon solvents are routinely dried by distillation over lithium aluminium hydride [12].

4.2 Mg. . .O(H)H Interactions

It was suggested [12] that, when rigorous dehydration procedures are not applied, Chl *a* in nonpolar solutions commonly exists as the dihydrate, Chl *a* · $2H_2O$. One of the two water molecules of hydration in Chl *a* · $2H_2O$ can be removed on heating the chlorophyll sample at 80°C in vacuo, leaving the remaining tightly bound water molecule, presumably bonded to the Chl *a* Mg atom through the Mg. . .O(H)H interaction. It is recognized [96] that this interaction provides the key to Chl *a* hydration behavior and to intermolecular aggregation of hydrated Chl *a*. In this connection, efforts were made [11] to characterize the proposed interaction by X-ray photoelectron spectroscopy [119] (XPS), which has proved to be a powerful tool in the study of porphyrins and related molecules because of the important structural details inherent in the N 1 s, C 1 s, and metal spectra [120]. An extension of XPS determinations to a characterization of Chl *a*-H_2O interactions was made with the threefold objective [121, 122]: First, by XPS determination of the O 1 s spectrum, a direct experimental observation may be made of the chlorophyll monohydrate. Second, using XPS as monitor the chlorophyll dehydration process may be followed as the temperature is incrementally raised above 80°C, the temperature commonly employed as the upper limit in most drying procedures [12]. Third, by a comparison between the hydration properties of Chl *a* and pheophytin *a*, it may be possible to determine the binding site of H_2O in Chl *a* · H_2O.

Film preparation was accomplished using the sample preparation equipment available on a modified Hewlett-Packard 5950A spectrometer [123]. An atomically clean gold surface was transferred from the sample preparation chamber with a residual pressure $5 \cdot 10^{-9}$ Torr into the attached N_2 atmosphere box. About 10^{15} Chl *a* molecules cm^{-2} were deposited on this gold surface by allowing 2 μl of a $3 \cdot 10^{-4}$ solution of Chl *a* in highly purified byturonitrile to evaporate. The "dry" film was then inserted into the analyzer section of the ESCA instrument which has a residual pressure of 2 10^{-9} Torr. The film was thick enough so that no gold peaks were visible in the XPS spectrum. Slight charging effects (∿1-2 eV) were partially compensated by using an electron gun that floods the sample with low-energy (<1 eV) electrons. Measurement of accurate peak positions is not possible using this approach, but the spectral distributions can be obtained with high accuracy. The gold blank was sepctroscopically examined (Fig. 4h) in order to exclude possible contamination of the sample spectrum from an adventitious source.

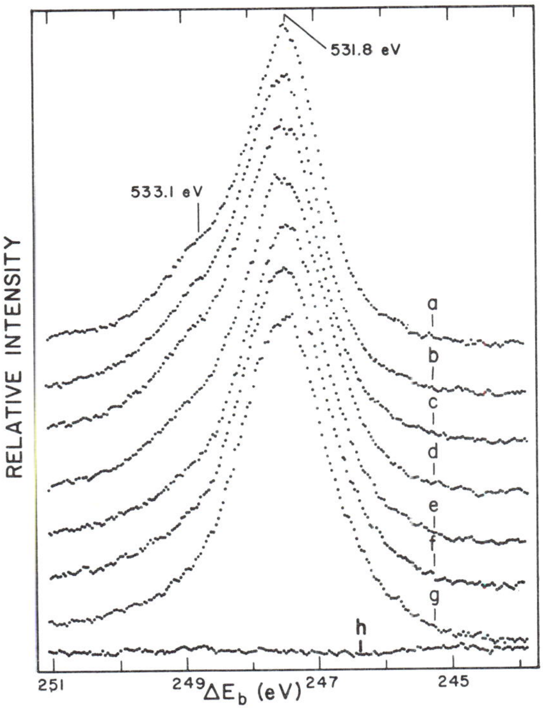

Fig. 4. Oxygen 1 s spectrum of Chl *a*. The ΔE_b is defined as the difference between the binding energy of the O 1 s peak and the C 1 s peak: *a* 30°C, *b* 90°C, *c* 120°C, *d* 150°C, *e* 180°C, *f* 210°C, *g* 250°C, *h* gold blank

The O 1 s spectrum (measured at 30°C) of Chl *a*, given in Fig. 4 a, is referenced to the corresponding C 1 s binding energy (284.3 eV at 30°C) to offset the charging effects that are, within experimental errors, the same for both the C 1 s and O 1 s spectra in the temperature range 30-125°C. The high binding energy shoulder (533.1 eV at 30°C) is attributable to the oxygen of the Chl-*a* water of hydration [121]. The value 533.1 eV for the O 1 s binding energy appears to be indistinghishable from that found for condensed H_2O [119], or H_2O present in various types of hydrated samples [124]. The main band centered at 531.8 eV appears to be an overlap of spectral contributions attributable to the five oxygen atoms of the Chl *a* C-7 propionic ester, C10 carbomethoxy, and C9 keto groups. This value agrees well with the O 1 s value of 531.4 eV reported for oxygens in sodium benzoate [125].

The attribution of the high binding energy shoulder at 533.1 eV to the presence of water of hydration is supported by the sequence of XPS Chl *a* O 1 s spectra measured at different temperatures in the 30-250°C range. No discernible changes are observed as the sample temperature is varied from 30° to 120°C (Fig. 4a-c). At temperatures exceeding 120°C the high binding energy shoulder begins to diminish and, at 250°C, this shoulder appears to have vanished quantitatively (Fig. 4d-g).

A computer deconvolution of the O 1 s spectrum (30-120°C) reveals an approximate 5 : 1 area ratio for the main band and the high binding energy shoulder, consistent with the monohydrate stoichiometry of Chl *a* · H_2O [12, 84]. The sample chlorophyll after heat treatment at 250°C was redissolved in diethyl ether. The absorption spectrum of this solution had maxima at 428.2 and 660.2 nm, reproducing those expected of pure Chl *a* solutions in ether [12]. However, a blue-red peak absorbancy ratio of 3.2 (instead of the corresponding value 1.29 ± 0.01 observed for the sample prior to the heat treatment) was obtained. In addition, an onset at 380 nm of a pronounced absorption edge (absent in the case of pure Chl *a*) into the near-ultraviolet region was also found. These optical measurements are indicative of probable Chl *a* degradation as a result of the dehydration of the tightly bound water of hydration [121]. In this sense, the "tightly bound" water refers to the remaining water of hydration as the dihydrate Chl *a* · $2H_2O$ loses one H_2O molecule to yield the corresponding monohydrate. It appears reasonable to suppose that the tightly bound water molecule plays a role in the stabilization of the five coordinated Mg atoms in Chl *a* · H_2O, and that the removal of this water may be concomitant to degradation of the chlorophyll molecule.

The water molecule in Chl *a* · H_2O is thus envisaged to be associated with the Mg atoms, as given in *A*, and is displaceable at room temperature only in the presence of excess polar groups or molecules such as acetone

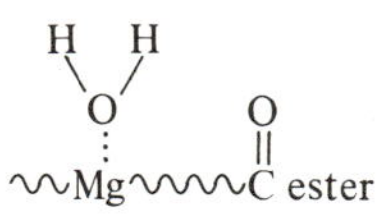

(*A*)

or the carbonyl groups of other Chl *a* molecules which compete for the Mg binding site. It would seem possible to illustrate this point by a comparative study of Chl *a* and its Mg-free derivative pheophytin *a* (Ph *a*), in which the Mg atom is replaced by two protons. The absence of the bound water molecule in Ph *a* observed under conditions identical to those under which Chl *a* · H_2O is stable should establish the Mg atoms as the binding site for the water molecule in Chl *a* · H_2O.

In Fig. 5 and in Table 2 the data are summarized for the O 1 s spectra of Ph-*a* and two different hydrated Chl-*a* aggregates at the ambient XPS probe temperature (50 ± 5°C). Figure 5 shows the O 1 s peak from a Ph-*a* film which was prepared from a rigorously dried solution of n-pentane. The peak has a full width at half maximum (FWHM) of 1.4 eV and is quite symmetrical. The O 1 s spectrum from Chl *a*, which was

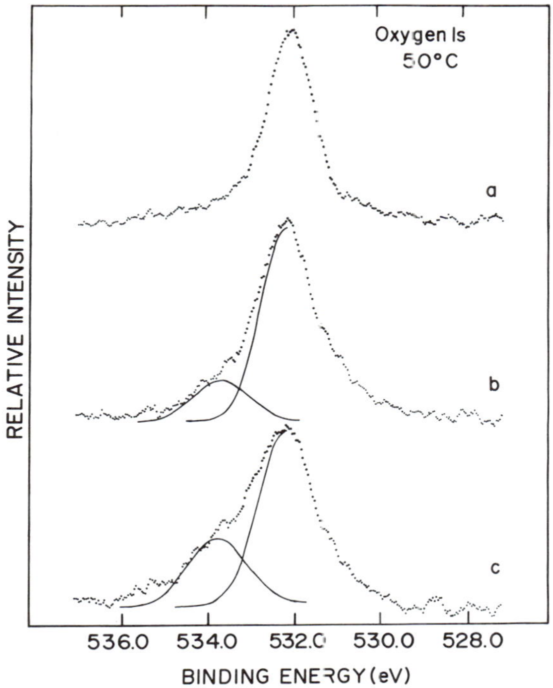

Fig. 5. Oxygen 1 s spectra of *a*, pheophytin *a*, *b* dried chlorophyll *a*, and *c*, hydrated chlorophyll *a*. These results establish the Mg atom as the binding site of the water hydration in Chl *a* · H_2O, and the dihydrate origin of the 743 nm absorbing polycrystalline chlorophyll. The lack of any shake-up transitions that accompany the O 1 s excitation of the five oxygen atoms in the chlorophyll is manifested by the nearly Gaussian shape of the Ph-*a* spectrum in *a*, making possible a meaningful determination of the water content of the samples in *b* and *c*. The present determination of P743 as the dihydrate polycrystal, (Chl *a* · $2H_2O)_n$, is in agreement with the x-ray diffraction work by Strouse et al. [111] on the 743 nm absorbing polymer of ethyl chlorophyllide dihydrate. The assignment of the monohydrate stoichiometry in *b* is consistent with the observation [109], in Fig. 4, of the tightly bound water in Chl-*a* monohydrate at temperatures less than 120°C

Table 2. The oxygen 1 s and nitrogen 1 s XPS binding energies for pheophytin *a* and chlorophyll *a*[a]

Sample	Main O 1 s peak[b]	Shoulder O 1 s peak[b]	N 1 s peak	Observed[c] shoulder area	Calculated shoulder area
Ph *a*	532.1	- - -	398.2, 400.0	- - -	- - -
Chl *a*	532.1	533.7	398.0	17.8 ± 3.2%	16.6% (Chl a · H_2O)
Chl *a* 743	532.1	533.7	398.0	29.5 ± 1.3%	28.6% (Chl a · $2H_2O$)

[a] Values reported in eV

[b] Error limits are estimated at ± 0.2 eV by averaging the values obtained from different films prepared under identical conditions

[c] Error limits computed from the deconvolution of 3 spectra taken from films prepared under identical conditions

prepared under indentical conditions, is shown in Fig. 5b. The sample-handling procedures were identical for these two samples. The peak is somewhat broader (FWHM = 1.5 eV) and slightly unsymmetrical, evincing a small shoulder on the higher binding energy side assigned to a single bound water of hydration as in Fig. 4. In Fig. 5c, we show the O 1 s peak from a 743 nm-absorbing species of Chl *a* which exists as the insoluble, crystalline aggregate (Chl $a \cdot 2H_2O)_n$ at room temperature. The XPS peak shows considerable broadening (FWHM – 1.8 eV) and is decidedly unsymmetrical, with a broad shoulder clearly apparent on the higher binding energy side.

Computer deconvolution of the spectra in Fig. 5b and c reveals that the shoulder arises from a fairly broad peak centered at 533.7 eV in both cases, with respective contributions to the total area under the curve of 18% and 29%. It should also be noted that the major peak occurs, within experimental error, at the same binding energy (532.1 eV) in all three spectra. From the experimental results listed in Table 2, it may be concluded that the assignment of the high-binding energy shoulder to the waters of hydration in Chl $a \cdot 2H_2O$ and Chl $a \cdot H_2O$ is correct, and that Ph *a* exists as an anhydrous species under experimental conditions identical to those under which the chlorophyll exists as the monohydrate.

The N 1 s spectra for these compounds are shown in Fig. 6. For Ph *a*, the expected N 1 s doublet is observed, characteristic of the porphyrin ring structure without a complexed metal ion [126]. The peaks apparently arise from the protonated and unprotonated nitrogens, demonstrating the localized, rather than bridged, nature of the N-H bond. The dviation from the expected 1 : 1 area ratio for the two peaks has been observed previously on other free-base porphyrins, and is thought to arise from a $\pi \rightarrow \pi^*$ shake-up transition which accompanies the lower binding energy N 1 s excitation and which overlaps with the higher binding energy N 1 s peak [126].

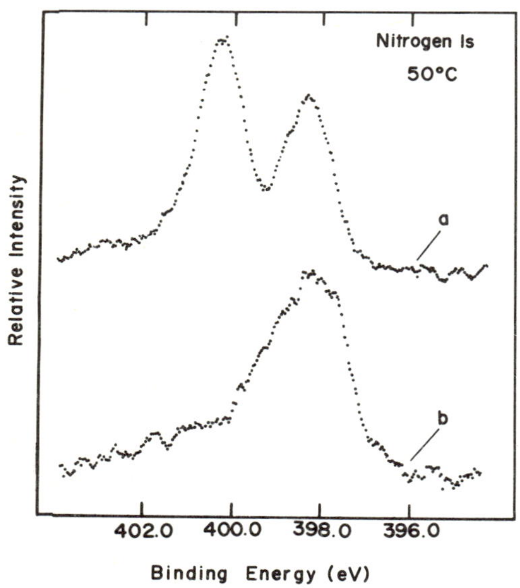

Fig. 6. Nitrogen 1 s spectra of *a*, pheophytin *a* and *b*, chlorophyll *a*. The deviation from the expected 1 : 1 area ratio for the two peaks is attributable to a $\pi \rightarrow \pi^*$ shake-up transition accompanying the lower binding energy N 1 s excitation

Incorporation of the Mg atom into the nitrogen core induces dramatic alterations in the N 1 s spectrum as illustrated for the Chl-*a* aggregates in Fig. 6b. The Ph-*a* doublet collapses into a skewed singlet as the electron density on all four nitrogens becomes nearly indistinguishable, indicative of more or less degenerate canonical structures for the π-electronic resonance in the macrocycle of Chl *a*. The electron density on the four nitrogens is not expected to be exactly indentical, however, which may be the reason for the observed skew. Such differences have been observed qualitatively using [^{15}N] NMR techniques [127]. Both the monohydrate and dihydrate of Chl *a* exhibit identical N 1 s spectra, suggesting that the hydration processes does not significantly alter the nitrogen electron density. The obvious differences between the spectra in Fig. 6a and b have also proven useful in providing an analytical check for the decomposition of Chl-*a* samples to Ph *a*.

4.3 Model P700 Structure and Properties

This XPS measurements thus yielded direct spectroscopic information concerning the state of hydration in Chl-*a* aggregates. The lack of any observable water signal from Ph-*a* films prepared in a manner identical to that of the Chl-*a* films reflects the remarkable affinity of Chl *a* for water resulting from the presence of the Mg atom in Chl *a*. Based on structure *A* for the monohydrate, Chl *a* · H_2O, a logical structure for the dihydrate, Chl *a* · $2H_2O$, would be

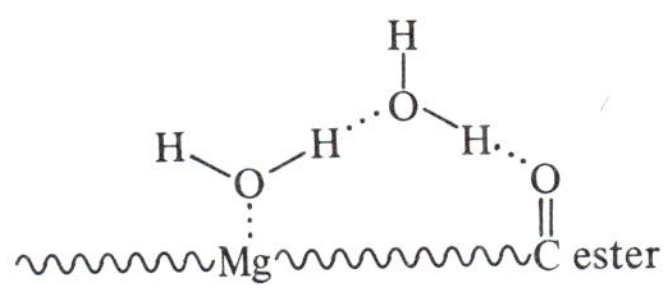

(*B*)

in which the second water molecule of hydration, through hydrogen bonds, bridges the C10 carbomethoxy carbonyl group and the tightly bound water molecule. Structure *B* agrees with the unit interactions in signle crystalline ethyl chlorophyllide *a* · $2H_2O$ obtained from x-ray diffraction determination [111]. The assignment of *A* and *B* to the monohydrate and dihydrate, respectively, thus appears to be completely consistent with the XPS observation of O 1 s intensities from bound water in Chl *a* and with other independent experimental observables [12, 111].

Upon minor rearrangement of the H_2O molecules in *B* one obtaines the structure

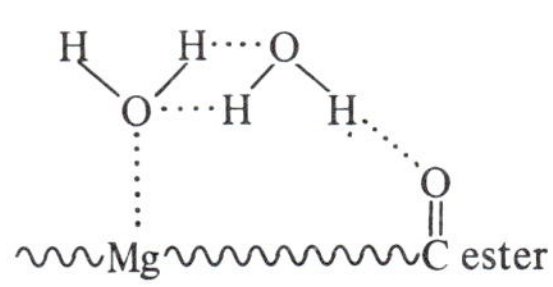

(*C*)

It was noted [85] that a symmetrical addition of two Chl *a* · H_2O monomeric units of *A* is equivalent to the addition of a second Chl-*a* molecule to *C*:

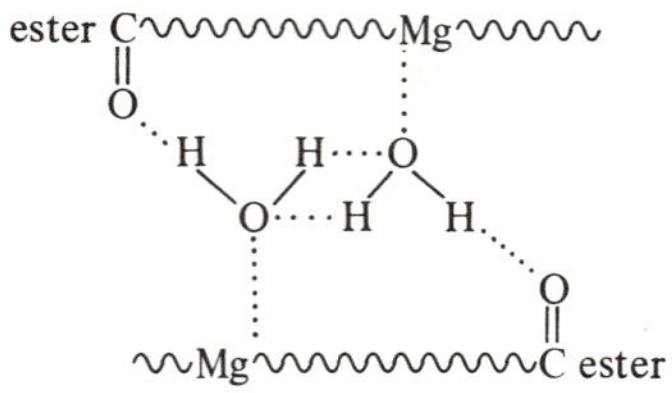

(*D*)

In both *C* and *D*, the coordination interactions occur on the same side of the chlorin plane as the C10 ester group. They are known as endo interactions and are most likely enhanced by the observed [111] 0.4 Å displacement of the Mg atom in the same direction as the C10 ester group. Aggregating interactions occur on the opposite side of the carbomethoxy group (exo interactions) when the C10 ester group is either sterically blocked [96, 128] and/or removed [129], resulting in the formation of the dimeric adduct.

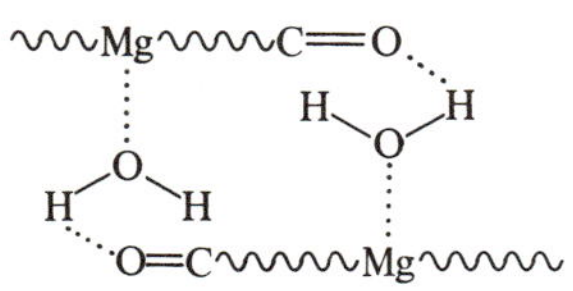

(*E*)

Due to the deprivation of the two additional hydrogen bonds in *D* and the opposite out-of-plane displacement of the Mg atom, the exo interactions in *E* readily undergoes photodissociation upon illumination [130].

In *D* and *E* the angles θ subtended by the $S_0 \rightarrow S_1$ Q_y transition moments (along the N1-N3 axes) in the two Chl-*a* molecules are 60° [86] and 180° [129], respectively. On the basis of standard exciton theory, it was shown [86, 87] that the $\theta = 60°$ value for *D* accounts for the observed 3 : 1 intensity ratio for the exciton components of the 700 nm band of *D*. On the other hand, the $\theta = 180°$ value for *E* appears to be responsible for the single Gaussian component observed for the Q_y transition in *E* [86, 98]. Structure *D* is readily obtained [12] on cooling of a nonpolar solution of Chl *a* · H_2O. Structure *E* may be prepared synthetically according to the method by Boxer and Closs [129]. The optical absorption spectra of *D* and *E* are compared in Fig. 7. It is noted [131, 132] that the two principal red absorption bands of *D* at 654 and 701 nm are in good agreement with the corresponding bands of P700. In contrast, it is observed in Fig. 5 that the optical properties of *E* are distinctly different from those of *D*.

Further comparison of the photochemical properties of *D*, *E* and P700 was made [131]. It was noted that on photoactivation *D* is oxidized, whereas *E* is photochemically incative in the absence of an electron acceptor such as I_2 [133] or tetranitromethane [128]. On photooxidation *D* gives rise to an electron spin resonance signal [131] having a g-value of 2.003 and a line width of 7.5 G, which are indistin-

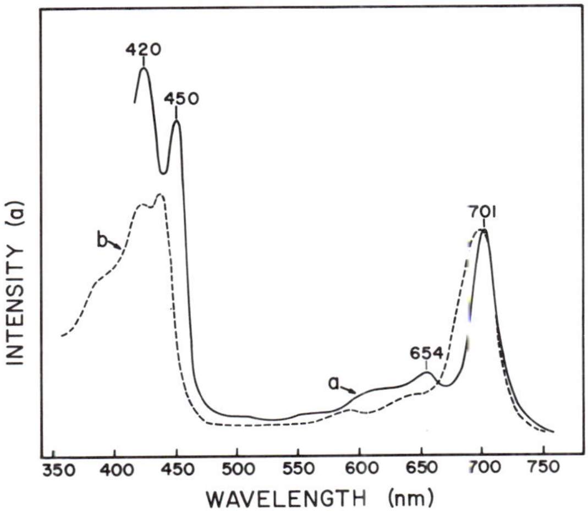

Fig. 7. Comparison of optical spectra of $(Chl\ a \cdot H_2O)_2$ given by endo C10 ester carbonyl bonding interactions *(solid line)*, and the analogous exo dimer involving C9 keto carbonyl bonding interactions

guishable from the corresponding ESR parameters of P700. The reduction potential of D is $\lesssim 0.5$ V [12], comparable to the corresponding value, 0.43 V [2], of P700. Based on the various comparisons it might be suggested that D provides a reasonable model for P700, whereas E may be a suitable model for a long-wavelength form of light-harvesting Chl a, such as the 695 nm-absorbing antenna Chl-a complex [134].

5 P680 Model and Water Splitting

The $(Chl\ a \cdot 2H_2O)_2$ model for P680 is supported by the in-vitro observations [131] in nonpolar solutions of reversible ESR signals attributable to Chl-a photooxidation and reduction according to the following cycle:

light $\quad (Chl\ a \cdot 2H_2O)_{n \geqslant 2} + H_2O \rightarrow (Chl\ a \cdot 2H_2O)_2^{+} \cdot + OH^- + [H]$ (8)

dark $\quad (Chl\ a \cdot 2H_2O)_{n \geqslant 2}^{+} + 1/2\ H_2O \rightarrow H^+ + 1/2\ [O]$ (9)

where [H] and [O] denote intermediate precursors to H_2 and O_2 evolution. These observations appear to be corroborated by the detection, using Pt electrodes, of photogalvanic currents arising from the half reactions [8, 110]

Chl a-Pt photocathode: $\quad 2H_2O + 2e \rightarrow H_2 + 2OH^-$ (8a)

Chl a-free Pt anode: $\quad 2H_2O \rightarrow 4H^+ + O_2 + 4e^-$. (9a)

The presence of Pt results in the liberation of [H] and [O] in (8) and (9) as molecular H_2 and O_2. The detection of the evolved gases by mass spectrometry was made on illumination of platinized $(Chl\ a \cdot 2H_2O)_{n \geqslant 2}$ films immersed in [^{2}H] and [^{18}O]-enriched water [22]. The rate of water photolysis varies over a broad range as physical parameters for the sample cell are varied. Notably, the presence of O_2 inhibits reactions (8) and (9) [8]. The mass spectrometric observation of the in-vitro Chl-*a* water-splitting reaction is possible only under oxygen-depleted conditions, as described below.

A Pt foil was platinized by passing a 30-mA current for 10 min through a $7 \cdot 10^{-2}$ M chloroplatinic acid solution containing $6 \cdot 10^{-4}$ M lead acetate. A layer of polycrystalline chlorophyll, containing $1.5 \cdot 10^{17}$ Chl-*a* molecules, was deposited on the platinized electrode surface by electroplating [8]. The Chl-*a*-plated electrode was then platinized again in the same chloroplatinic acid solution by passing the 30 mA current for 15 s. The resulting sample electrode was baked at 60°C under atmospheric pressure for several hours in order to rid of any adventitious gaseous occlusion during the platinization procedure. The photolytic experiments were performed by illuminating the Pt-Chl-*a* electrodes thus prepared in rigorously deaerated aqueous electrolytes.

The action spectrum of the photogalvanic response of the platinized Chl-*a* electrode at pH7, measured in a cell [8] employing as the second half cell a platinized electrode not covered with Chl *a*, shows a 740 nm maximum indicative of the fact that $(\text{Chl } a \cdot 2H_2O)_n$ is primarily responsible for the observed photogalvanic effects. Prolonged illumination using the entire output from a 1000 W tungsten halogen lamp led to the observation of gaseous evolution from the Pt-Chl-a electrode. In mass spectrometric analyses (Figs. 8 and 9) it was ascertained that both H_2 and O_2 were evolved at the Pt-Chl-*a* electrode. "Blank" control experiments were carried out without the platinized electrodes under conditions identical to those of the photolytic runs.

After the light-induced reaction of a platinized Chl-*a* sample (immersed in a pH 3 citrate-phosphate, KC1 containing buffer solution) was allowed to proceed for a given time period, the gaseous content over the electrolyte solution, on cooling to room temperature, was evacuated directly into the sample chamber of a mass spectrometer. Mass spectrometric analyses of the light-induced products from various mixtures of $H_2[^{16}O]$, $D_2[^{16}O]$, and $H_2[^{18}O]$ are compared (Fig. 8) with those of electrolyzed samples (30 mA passed for 10 min) containing the corresponding isotopic water mixtures. The results of a light experiment (90 min illumination) using 1 : 1 D_2O-H_2O are shown in Fig. 8 a. Assuming that the mass 4 (D_2^+) line resulted from water splitting and using the mass 20 (D_2O^+) line as an internal reference for calibration, it was estimated, from a comparison of the relative intensity ratios of lines at masses 20 (D_2O^+) and 4 (D_2^+) observed in the photolytic and electrolytic runs, that water photolysis occurs at a rate of $9 \cdot 10^{-6}$ mol/h, corresponding to a gaseous (H_2 and O_2) evolution rate of 0.3 ml/h. At sufficiently high instrumental resolution, it was possible to differentiate the H_2^+ line from the interference at mass 2 due to D^+ obtained from fragmentation of deuterated compounds. The determination of the molecular species H_2^+, HD^+, and D_2^+ thus made for the photolysis and electrolysis of 3 : 1 D_2O-H_2O is shown in Fig. 8b. The presence of molecular oxygen in the light-induced splitting of water is ascertained by using 5 : 1 $H_2[^{16}O]$-$H_2[^{18}O]$ (see Fig. 8c). The lines at masses 34 ($[^{16}O][^{18}O]^+$) observed for the photolytic sample are in agreement with the corresponding lines observed from electrolysis. From statistical considerations the observed 10 : 1 intensity ratios for the mass 34 ($[^{16}O][^{18}O]^+$) and 36 ($[^{18}O]_2^+$) lines are compatible with a scrambilng of the 5 : 1 $[^{16}O]$-$[^{18}O]$ atoms in the oxygen evolution process.

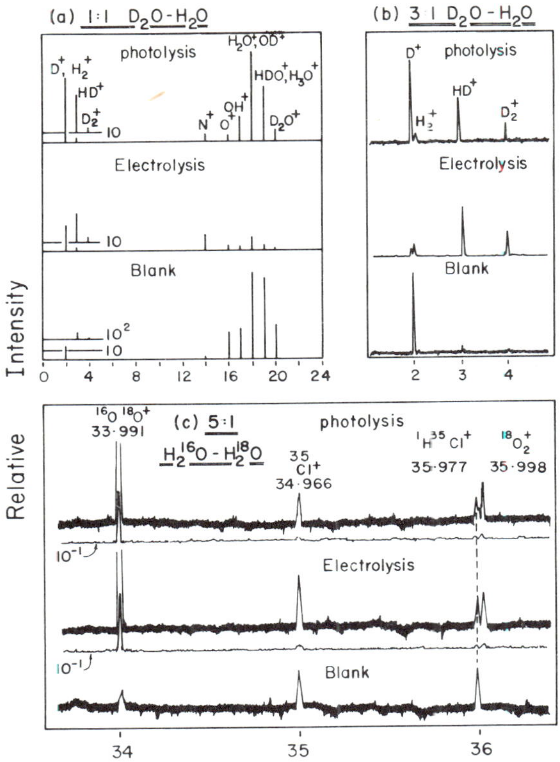

Fig. 8. Mass spectrometric analyses of the photolytic and electrolytic products of *a* 1 : 1 D_2O-H_2O, *b* 3 : 1 D_2O-H_2O, and *c* 5 : 1 $H_2[^{16}O]$-$H_2[^{18}O]$. The mass 34 line in the blank run in *c* is due to the 0.204% natural abundance of $[^{18}O]$. The observed 34 : 36 mass intensity ratios of ∼10 for the photolytic and electrolytic runs in *c* are in agreement with the expected value for 5 : 1 $H_2[^{16}O]$-$H_2[^{18}O]$. The lines at mass numbers 35 and 36 in *c* corresponding to Cl^+ and HCl^+, respectively, are attributable to the KCl (0.45 M) in the buffered (pH 3) electrolyte solution. The mass spectra in this figure were obtained with the ion source focused at mass 2. The scale about the indicated mass numbers in *b* has been expanded to allow representation of the spectral resolution at mass 2. The low noise level in the electrolytic run in *b* is due to the relatively strong signals from passing a 30 mA current for 10 min. The H^+ line at mass 1 is not detected owing to instrumental limitation

A mass spectrometric study of the simulteneous generation of H_2 and O_2 from 63.6 : 36.4 $H_2[^{18}O]$-$H_2[^{16}O]$ in a pH 11 sodium hydroxide-phosphate buffer, in which no KCl was added, is reproduced in Fig. 9. The relative line intensities at masses 2 (H_2^+), 34 ($[^{16}O][^{18}O]^+$), and 36 ($[^{18}O]_2^+$), absent in the blank, are in excellent agreement between the photolytic and electrolytic runs. The experiment was performed under a positive pressure of Ar in order to generate the $[^{36}Ar]^+$ (0.336% natural abundance) line, which serves as a useful internal standard in the mass 36 region. The intensity ratio of the lines at masses 28 ($[^{14}N]_2^+$) and 32 ($[^{16}O]_2^+$) in the blank agrees with that of an air sample. Based on this ratio, a correction was made for the air contamination effect on the line intensity at mass 32 ($[^{16}O]_2$) of the photolysis sample spectrum. The resulting pattern for the observed isotopic distribution of molecular oxygen is in excellent agree-

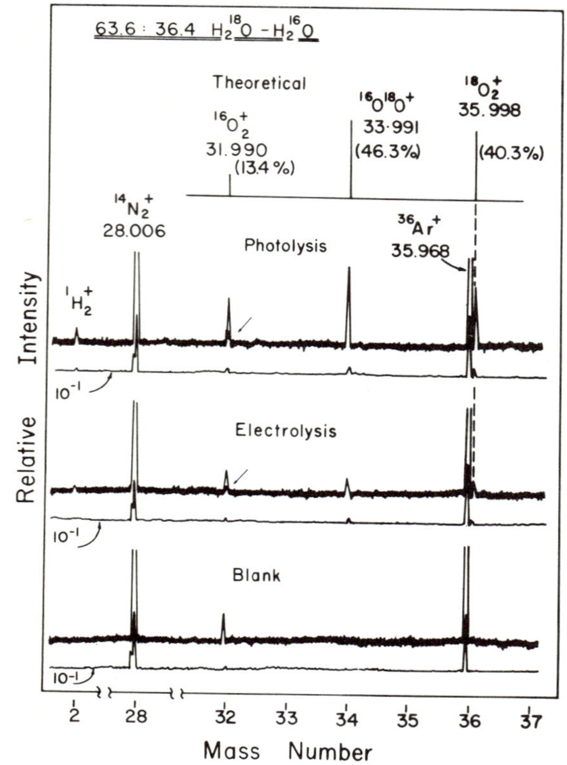

Fig. 9. Mass spectrometric study demonstrating the simultaneous generation of molecular hydrogen and oxygen in water photolysis. The light reaction was carried out for 20 min. The control experiment corresponds to the electrolysis of $5 \cdot 10^{-7}$ mol of H_2O. The *arrows* at mass 32 in the photolytic spectrum represent the $[^{16}O]_2^+$ signal due to water photolysis after correction for air contamination in the sample. After a similar correction in the electrolytic spectrum, the isotopic distributions of O_2 obtained in photolysis and electrolysis are seen to be in agreement with the theoretical pattern. The scale about the indicated mass numbers has been expanded by a factor of 3 to facilitate the representation of the line intensities ans spectral resolution. The mass spectra in this figure were obtained with the ion source focused at the $[^{40}Ar]^+$line

ment with the expected pattern (see Fig. 9). The photolytic sample was measured after 20 min of illumination. The electrolysis was carried out by passing a 20 mA current for 5 s. Calibrating the line intensities at masses 34 ($[^{16}O][^{18}O]^+$) and 36 ($[^{18}O]_2^+$) of the photolytic sample against the corresponding intensities of the electrolytic sample, one arrives at a water-splitting rate of 10^{-5} mol/h. The agreement of this photolysis rate determination with that given in the preceding paragraph may be fortuitous. The quality of the Chl-*a*-Pt interface is expected to play a role. Also the accumulation of O_2 in reaction 10 is expected to inhibit the photolytic process over prolonged periods in excess of several minutes of continuous operation. From the same calibration, using the line intensity at mass 28 (N_2^+) and the known mass intensity ratio or 28(N_2^+) : 32($[^{16}O]_2^+$) for air, it was estimated that atmospheric contamination in the photolytic sample amounted to $3 \cdot 10^{-7}$ mol of $[^{16}O]_2$, approximately an order of magnitude lower than the $2.5 \cdot 10^{-6}$ mol of molecular oxygen, occurring variously in the forms of $[^{16}O]_2$, $[^{16}O][^{18}O]$, and $[^{18}O]_2$, produced in 20 min of the light-induced water-splitting reaction.

6 Carbon Reduction by Water

The observation in Fig. 9 of O_2 evolution suggests rather unambiguously, the Chl-*a*-water-splitting reaction as the source of the oxygen evolved. The question of possible CO_2 reduction by the hydrogen from in-vitro water photolysis to yield C_1 reduction products was examined experimentally [117].

At Pt/Chl-*a* electrode prepared in the manner described in Sect. 5, was placed in a cell containing a mixture of $H_2[^{16}O]$, $D_2[^{16}O]$, and/or $H_2[^{18}O]$. The sample cell containing doubly distilled $[^{18}O]$- and/or D-enriched water was deaerated by purging with Ar gas for 20 min. Following CO_2 saturation of the water, the cell was closed and the sample was illuminated with the visible output from a 1000 W xenon arc lamp. After 30 min of illumination, the gaseous mixture above the aqueous solution was analyzed mass-spectrometrically.

The effect of CO_2 on the product distribution of the Pt/Chl-*a* light reaction is illustrated by the mass spectral results in Fig. 10. It is evident that the hydrogen evolution obtained in the absence of CO_2 (Fig. 10A) is suppressed in the presence of CO_2 (Figs. 10B and C). Appearance of $[^{16}O][^{18}O]^+$ (mass 34) and $[^{18}O]_2^+$ (mass 36) lines in the $H_2[^{18}O]$-labeled solution ($[^{16}O]/[^{18}O] = 1.82$), as before, establishes that O_2 is liberated from water as the oxidation product. The mass 34/36 ratio is 3.2, in agreement with the random scrambling $[^{16}O][^{18}O]/[^{18}O]_2$ ratio of 3.6 based on the known $[^{18}O]/[^{16}O]$ content in the water. There is no evidence for the formation of H_2O_2 on account of the lack of intensity increases at masses 35 ($DH[^{16}O]_2^+$), 37 ($DH[^{16}O][^{18}O]_2^-$), 38 ($D_2[^{16}O][^{18}O]^+$, $H_2[^{18}O]_2^+$), 39 ($HD[^{18}O]_2^+$), or 40 ($D_2[^{18}O]_2^+$).

From a comparison of Fig. 10B and C it appears that thermal radiation from the xenon arc lamp produces little effect on the product distribution. It was shown [135] that Pt in water produces H_2 but no O_2 upon heating, and that the concurrent production of H_2 and O_2 occurs only when both light and Chl *a* are present. The observation of O_2 in the presence of CO_2 upon illumination in Fig. 10B and C thus suggests a photochemical pathway involving the chlorophyll. The effects in Fig. 10B and C are not observed using visible light and a Chl-*a*-free Pt sample under otherwise identical conditions.

Figures 10B and C also show pronounced increases in signal intensities at masses 30 and 31. The correlative assignment of mass 31 to their $[^{13}C]$ analogs is consistent with the observed mass 31/30 intensity ratio of 0.009 ± 0.002. The natural abundance ratio of $[^{13}C]/[^{12}C]$ is 0.011. The line at mass 30 is assigned to $[^{12}C][^{18}O]^+$ and/or $D[^{12}C][^{16}O]^+$, $H_2[^{12}C][^{16}O]^+$. The $[^{12}C][^{18}O]^+$ fragment arises from the exchange of $[^{12}C][^{16}O]_2$ with $H_2[^{18}O]$ in the $[^{18}O]$-labeled water mixture.

The possible generation of CO_2 reduction products is manifested by the variation of the product distribution with χ_D, the mol fraction of $D_2[^{16}O]$ in H_2O that is not isotopically enriched in $[^{18}O]$ (see Fig. 11). Both D^+ and mass 30 intensities evidently

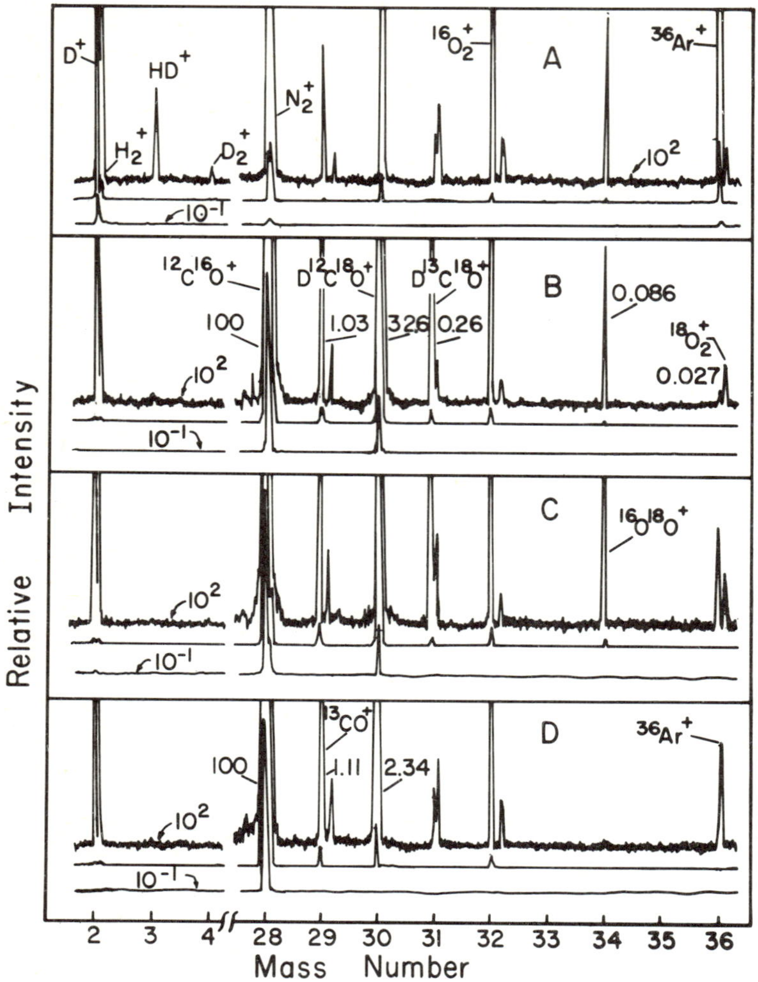

Fig. 10. Comparison of the mass spectrometric determination of products from Chl-*a* light reaction in Ar-purged H_2O with product analyses using isotope-enriched CO_2-saturated water having 1 : 1 H/D and 1.82 : 1 $[^{16}O]/[^{18}O]$: *A* platinized (Chl *a* · $2H_2O)_n$ in Ar-purged water illuminated for 30 min using red light; *B* platinized (Chl *a* · $2H_2O)_n$ in CO_2-saturated water illumination by light from a 1000 W xenon arc lamp passed through a 2 in water filter; *C* same conditions as in *B* except for the elimination of IR radiation from the xenon arc lamp using a 10 in water filter; *D* CO_2-saturated water illuminated by the white light for 30 min. The intensity ratios of the mass 34 ($[^{16}O][^{18}O]^+$) and 36 ($[^{18}O]_2{}^+$) lines in *B* and *C* are 3.2 and 3.8, respectively, in fair agreement with the corresponding ratio of 4.2 obtained in *A*. The predicted ratio from a random scrambling of $[^{16}O]$ and $[^{18}O]$ derived from water is 3.64. The mass 29 ($[^{13}C][^{16}O]^+$) intensity is found to be $\sim$1% of the mass 28 ($[^{12}C][^{16}O]^+$) intensity consistent with the natural abundance of $[^{13}C]$. The indicated intensity values are relative to a mass 29 signal of unity in *B-D*. The generation of the oxygen lines at masses 34 and 36 is accompanied by growth at masses 30 ($D[^{12}O][^{16}O]^+$, $[^{12}O][^{18}O]^+$) and 31 ($D[^{13}C][^{16}O]^+$, $[^{13}C][^{18}O]^+$, indicative of reduction product fragmentation. A simultaneous loss of the mass 3 (HD^+) and mass 4 ($D_2{}^+$) as well as the mass 2 ($H_2{}^+$) signal is evident from comparison of *A* with *B* and *C*. The appearance of the $[^{36}Ar]^+$ line in *B-D* is the result of the Ar-degassing procedure described in the text

increase linearly with χ_D (see Fig. 11a-d). The observed mass 30/28 intensity ratio of 0.006 in pure D_2O (Fig. 9d) corresponds to a total mass 30 line intensity about three times the expected contribution from $[^{12}C][^{18}O]^+$ owing to the natural abundance of $[^{18}O]$ (∼0.2%). The rise in the mass 30 line with χ_D indicates that $D[^{12}C][^{16}O]^+$ is the primary source of contribution to the observed intensity enhancement. Standardized mixtures of DCOOD in 3 ml of CO_2-saturated D_2O were examined by mass spectrometry. Under 1 atm of CO_2, a sample/blank intensity ratio of 5 ± 1 is obtained from the D^+ line from solutions 0.001 and 0.01 M in DCOOD. No significant intensity changes at mass 30 were observed. However, on reduction of the CO_2 pressure by a factor of ∼2 in the 0.001 M DCOOD sample and increasing the DCOOD concentration to 0.01 M, a 50-fold increase in the D^+/H_2^+ intensity ratio was obtained, as was a 2-fold increase in the mass 30/29 ratio relative to a blank prepared at 1 atm in CO_2. The simultaneous increases of the D^+ and mass 30 lines thus observed are comparable in magnitude to the corresponding increases observed in Fig. 11. This observation suggests that the CO_2 pressure is reduced during the Pt/Chl *a* light reaction, resulting in the $D[^{12}C][^{16}O]^+$ fragment.

The mass spectral analysis in Fig. 12 obtained using CO_2-saturated $D_2[^{16}O]$ provides evidence for the photosynthesis of formic acid. Line intensities at masses 2, 30, 46, 47, and 48 increase significantly relative to the blank. The sample signals suggest the presence of deuterated formic acid fragments: $D[^{12}C][^{16}O]_2^+$ (mass 46), $D[^{13}C][^{16}O]_2^+$ (mass 48), D^+ (mass 2), and $D[^{12}C][^{16}O]^+$ (mass 30). Significantly, the line intensity at mass 30 exceeds that expected from formic acid alone. The experimental mass spectrum obtained from CO_2-saturated solution of 0.1 M $H_2[^{12}O][^{16}O]_2$ in $H_2[^{16}O]$ gives an $H[^{12}C][^{16}O]^+/H_2[^{12}C][^{16}O]_2^+$ ratio of 0.05. The observed mass 30/48 ratio in Fig. 12 corresponding to the analogous $D[^{12}C][^{16}O]^+/D_2[^{12}C][^{16}O]_2^+$ ratio is 4.8. Previous studies [136] rule out this large change being due solely to isotope effects in the cracking patterns of HCOOH and DCOOD. This observation and that of the enhanced rise in D^+ compared with $D[^{12}C][^{16}O]^+$ in Fig. 11e thus suggest the presence of more than one reduction product: possibly deuterated formaldehyde, formic acid, and methylene glycol [137]. Using conventional electrolysis results as a calibration, an approximate photochemical yield of $0.8 \cdot 10^{-6}$ mol of O_2 was obtained for a 30 min sample exposure to visible light [117] This is on the order of, but lower than, the expected O_2 yield based on the reaction stoichiometry of (3) and the corresponding yield of $2.7 \cdot 10^{-6}$ mol of reduction products, estimated [117] from calibrated mass spectral line intensities. Spectrophotometric analyses on sample purity indicated no significant Chl-*a* degradation after light reaction.

7 Two-Photon Activation of Water Splitting

The minimum quantum requirement in plant photosynthesis is 8 quanta/O_2 molecule evolved or 2 quanta/electron transferred in the primary light reaction [138]. According to the P680 model given by reaction 2, this requirement may be attributable to the quantum requirement for the photoreaction of (Chl *a* · $2H_2O)_2$ with water. A study [23] was made of the flux dependence of the photogalvanic current, i_p, obtained from

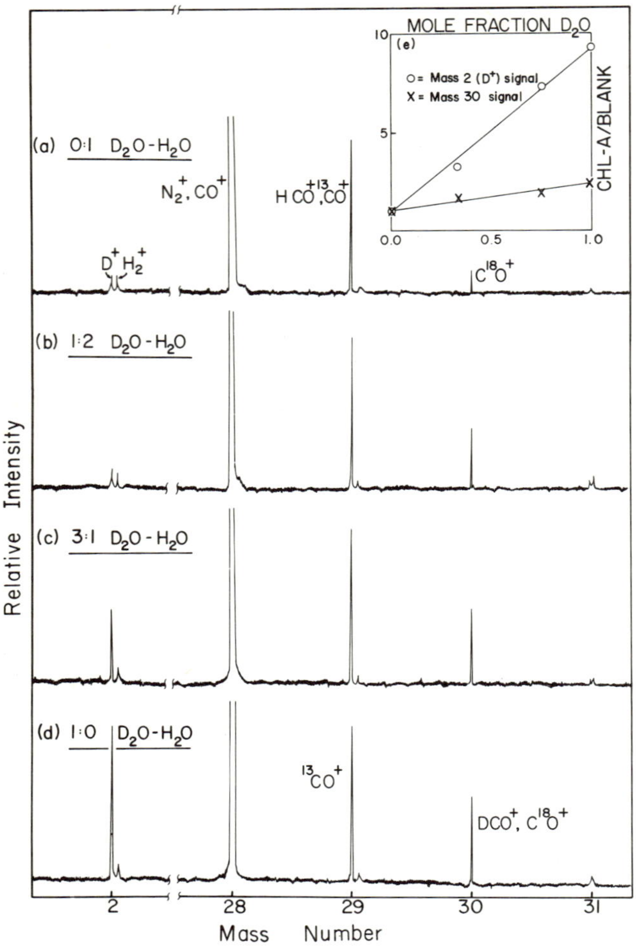

Fig. 11. Mass spectral determination of the dependence of product distribution on the mol fraction of D_2O in H_2O. The Pt/Chl-*a* sample was subjected to 30 min of illumination with IR and short-wavelength filtered light with CO_2-saturated solutions containing *a* pure $H_2[^{16}O]$; *b* 1 : 2 $D_2[^{16}O]$-$H_2[^{16}O]$; *c* 3 : 1 $D_2[^{16}O]$-$H_2[^{16}O]$; and *d* pure $D_2[^{16}O]$. The spectra in the mass 28-31 and mass 2 regions were obtained with the ion source focused at mass 44 and 2, respectively. The mass 29 ($[^{13}C][^{16}O]^+$) signal was observed to be ∿1% of the mass 28 ($[^{12}C][^{16}O]^+$) signal consistent with the natural abundance ratio of $[^{13}C]/[^{12}C]$. Mass intensities in the 28-31 region were scaled to a constant mass 29 signal. In the mass 2 region the H_2^+ signal was arbitrarily chosen as the reference. The monotomic trend of increasing D^+ and mass 30 ($D[^{12}C][^{16}O]^+$) line intensities with increasing χ_D, the mole fraction of D_2O, is evident in *a-d*. In *e* the ratio, Chl *a*/blank, is that of the scaled signal intensity at masses 30 and 2 obtained from the illuminated sample and its corresponding unilluminated balnk. The variations of Chl *a*/blank intensity ratios at masses 30 and 2 are linear in χ_D

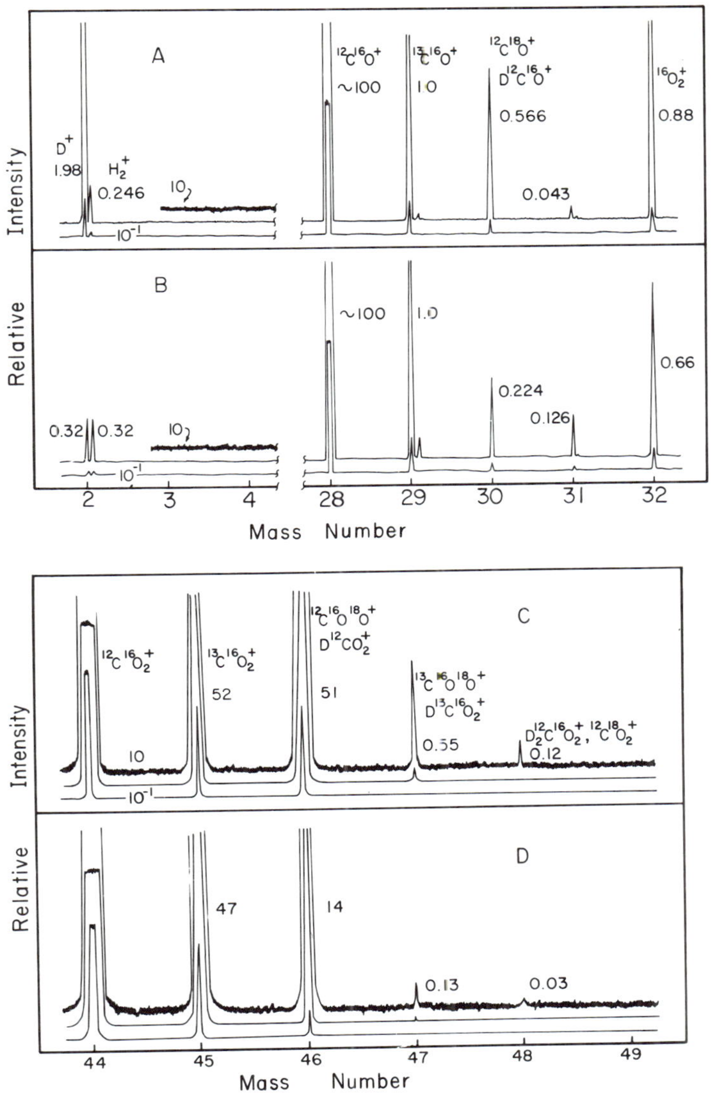

Fig. 12. Mass spectrometric analysis of product distribution from an illuminated Pt/Chl-*a* sample in CO_2-saturated $D_2[^{16}O]$ using IR and short-wavelength filtered light from a 1000 W xenon arc lamp. In *B* and *D* are shown the blank (unilluminated) signals in the mass regions 2-4, 28-32, and 44-49, respectively. Figures *A* and *C* are the corresponding signals obtained on sample illumination. The indicated line intensities are scaled relative to a mass 29 ($[^{13}C][^{16}O]^+$) line intensity of unity in the mass regions 28-31 and 44-50. The enhancement of the mass 46 line in *C* is attributed to $D[^{12}C][^{16}O]_2^+$

reactions 8 and 9. In this study the Pt/Chl-*a* electrode was prepared in the manner described in Sect. 5, except that the exposed surface of the Chl-*a* deposit on Pt was not platinized to prevent short circuiting of the photogalvanic action. The Pt/Chl-*a* electrode thus prepared was used in a photogalvanic assembly made up of two half cells separated by a glass frit, one containing the Chl-*a*/Pt electrode and the other a Chl-*a*-free Pt electrode. Both electrodes were immersed in a 0.1 M aqueous solution of KC1.

The photogalvanic response of $(\text{Chl}\,a \cdot 2H_2O)_n$ is greatly enhanced when the Chl-*a*/Pt and Chl-*a*-free half cells are, respectively, kept under acidic and basic conditions. In the present experiment, pH values of 2.2 and 12.0 were maintained for the Chl-*a*/Pt and Chl-*a*-free half cells using citrate-phosphate buffers. The cell assemblies were deoxygenated by bubbling argon through the individual half cells for 1 h.

The illumination of the Chl-*a*/Pt electrode resulted in a flow of electrons from the Chl-*a*-free electrode (anode) to the Chl-*a*/Pt electrode (cathode). The initial (t = 0) readings of the photogalvanic response were recorded. The photogalvanic response indicates a linear flux dependence at higher light intensities. To establish the power dependence of the $(\text{Chl}\,a \cdot 2\,H_2O)_n$ light reaction in the strong flux limit, we employed the total output from a 1000 W tungsten-halogen lamp. The incident flux was varied by placing a series of five neutral density filters in the path of the exciting light at several lamp power supply voltage settings. The flux at each voltage setting was calibrated. The resulting experimentel points, represented as $\log i_p$ vs $\log I_0$, where i_p and I_0 are the photogalvanic current and incident flux, respectively, and fitted to the empirical equation

$$i_p = [(1.58 \cdot 10^{-29})I_0/(1 + 10^{15} I_0^{-1})]^{1/2} \qquad (10)$$

are shown in Fig. 13. It is apparent that the $(\text{Chl}\,a \cdot 2H_2O)$-water photoreaction rate varies as the square root of I_0 in the strong light limit.

With proper consideration of the two-photon activation mechanism, the present observations may be interpreted in terms of a steady-state scheme in which a generalized chlorophyll light reaction is given in a sequence of six consecutive steps with corresponding rate. W, given as follows:

initial excitation of photoactive aggregate (Chl *a*)*

$$(\text{Chl}\,a) + h\nu \rightarrow (\text{Chl}\,a)^* \qquad W_a = k_a I_0 \qquad (11)$$

radiative or radiationless decay

$$(\text{Chl}\,a)^* \rightarrow (\text{Chl}\,a)\ (+\,h\nu') \qquad W_b = k_b[(\text{Chl}\,a)^*] \qquad (12)$$

upconversion to a tautomeric charge transfer (CT) state

$$(\text{Chl}\,a)^* + h\nu \rightarrow \text{CT} \qquad W_c = k_c I_0 [(\text{Chl}\,a)^*] \qquad (13)$$

nonradiative decay of the tautomeric state

$$\text{CT} \rightarrow (\text{Chl}\,a) \qquad W_d = k_d[\text{CT}] \qquad (14)$$

reaction with the primary electron acceptor A

$$\text{CT} + \text{A} \rightarrow (\text{Chl}\,a)^+ + \text{A}^- \qquad W_e = k_e[\text{CT}][\text{A}] \qquad (15)$$

back-reaction or regeneration

$$(\text{Chl}\,a)^+ + \text{A}^- \rightarrow (\text{Chl}\,a) + \text{A} \qquad W_f = k_f[(\text{Chl}\,a)^+]^2 \qquad (16)$$

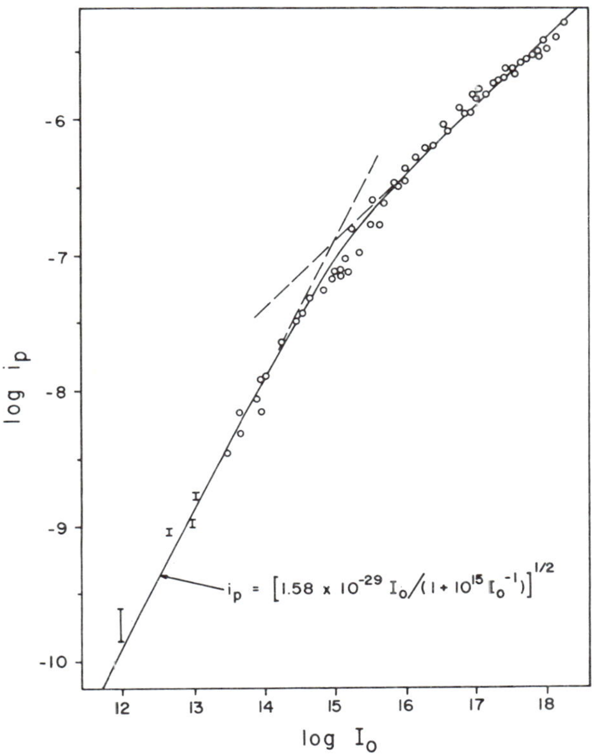

Fig. 13. The flux dependence of the $(Chl\ a \cdot 2H_2O)_n$ -H_2O photogalvanic response. The low-flux data points are given in error bars (99% confidence intervals) corresponding to $\sigma = \pm 2.5$ and noise excursions about the mean. The doubling of the exponential flux dependence from semilinearity in the strong light limit to linearity in the weak light limit is attributed to the two-photon upconversion activation in step 13. The fluxes were estimated by a power integration of the lamp function

The steady-state solution of steps 11-16 results in the expression for the photogalvanic current

$$i_p \propto [(\mathrm{Chl}\ a)^+] = \left[\frac{k_c k_e[A]}{K_f(k_d + k_e[A])} \times \frac{k_a I_0}{(k_c + k_b I_0^{-1})}\right]^{1/2} \qquad (17)$$

which reproduces the empirical fit [Eq. (10)] for the photoresponse of $(\mathrm{Chl}\ a \cdot 2H_2O)_n$. We observe that the presence of the term $k_b I_0^{-1}$ in Eq. (17), resulting from the photochemical upconversion (step 13), is responsible for the doubling in the exponential flux dependence from semilinearity in the weak light limit. In the absence of step 13 the corresponding steady-state solution yields a semilinear dependence throughout the entire range of fluxes, a condition that has long been established for one-photon light reactions involving monomeric Chl *a* [139].

8 Primary and Secondary Processes of Photosynthesis

The results described above suggest the capability of a single chlorophyll photosystem to catalyze water splitting and CO_2 reduction. The proposed P680 light reaction in Fig. 1 is accordingly shown to be responsible for the photoreduction of CO_2 to carbohydrates as well as oxygen evolution process. The suggested cyclic path for P700 and electron flow between P680 and P700 provide the energy currency for photophosphorylation. The postulated mechanisms for the P680 and P700 reactions are described in this section.

8.1 Comparison of Models for P680 and P700

Delocalization of the positive charge in (Chl $a \cdot 2H_2O)_n$, *E* is possible

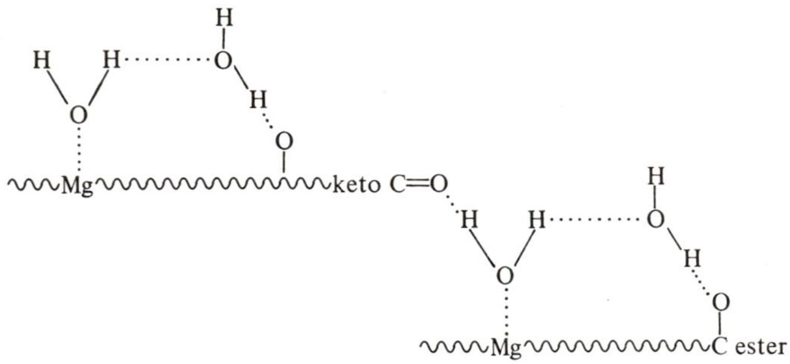

(*F*)

through π-electron conjugation of the CO keto group with the chlorin in macrocycles. The Mg. . .O(H)H. . .O(H)H. . .O = C10 linkage in *F* hinders epimerization of Chl *a* at C10 and also inhibits photoenolization of the ring V β-ketoester group. The reduction potential of (Chl $a \cdot 2H_2O)_{n \geqslant 2}$ is found [23] to be $\geqslant$0.91 V in an acetone-H_2O mixture, greater than that, $\lesssim$0.5 V, of (Chl $a \cdot H_2O)_2{}^+$ [12]. This findings underscores the Chl-*a* monohydrate and dihydrate models for P700 and P680, which are respectively noted for their low ($\sim$0.4 V)^{2c} and high (>0.82 V) reduction potentials as shown in Fig. 1. The lower bound of the latter potential is given by the reduction potential of the water oxidation half reaction at pH7.

Aggregates of Chl $a \cdot 2H_2O$ are the only known complexes of Chl *a* capable of photocatalyzing the water-splitting and CO_2 fixation reactions. That the Mg-H_2O-ester carbonyl interactions in *F* may be responsible for this capability is suggested by the fact that hydrated pheophytin *a*, in which the Chl-*a* Mg atom is replaced by H atoms, is photoreductively degraded by hydrogen from the water-splitting reaction [140].

The dimeric origin of P700 is strengthened by the agreement of the properties of P700 with those of (Chl $a \cdot H_2O)_2$, as described in Sect. 4.3. The available information on the P680 reaction center is less unambiguous. The P680 complex is tentatively given [8] by the dimer, (Chl $a \cdot 2H_2O)_2$. On the other hand, in view of the variability of the difference absorption maxima attributed [4, 5] to P680, the possible involvement of monomeric Chl *a* in P680 cannot be ruled out. One primary function of the chlorophyll in photosynthesis is the conversion of light into an electrochemical potential given by the redox potential of Chl $a^{\dagger}$. The latter is measured by the difference between the ground-state energy of Chl $a^{\dagger}$ and that of Chl *a*. In (Chl $a \cdot 2H_2O)_n^{+}\cdot$, where n is large, the positive sign is delocalized over many monomeric units of the aggregate, resulting in a variation of the ground-state energy of (Chl $a \cdot 2H_2O)_n^+$ and hence a corresponding dependence of the reduction potential on the Chl a aggregate size. In this respect, the quantum efficiencies of the in-vitro water-splitting and CO_2 fixation reactions are expected to be enhanced if the chlorophyll can be stabilized as (Chl $a \cdot 2H_2O)_2$ or monomeric Chl $a \cdot 2H_2O$, instead of polymeric (Chl $a \cdot 2H_2O)_n$, on the metal substrate.

8.2 Light Reaction Sequence

The light-dark cycles of photosynthesis, in vivo and in vitro, consist of the reaction sequence: primary processes that generate the reducting and oxidizing equivalents, [H] and Chl $a^{+\cdot}$, followed by secondary processes leading to CO_2 reduction and O_2 evolution, respectively. The primary processes presumably initiate with charge separation within the reaction center complex and then electron transfer to the acceptor, A. The charge separation possibly occurs on a sub-picosecond time scale. The electron transfer is a much slower multistep event, culminating in the appearance of Chl $a^{+}\cdot$, extending into the millisecond time domain in plant photosynthesis [141] as well as in Chl-a water photolysis in vitro [142].

In view of (3c), the electron transfer is written

$$\text{Chl } a + \text{A} + \text{H}^+ \xrightarrow{h\nu} \text{Chl } a^{+}\cdot + \text{HA} \tag{18}$$

or, equivalently,

$$\text{Chl } a + \text{H}^+ \xrightarrow[\text{A}]{h\nu} \text{Chl } a^{+}\cdot + [\text{H}]. \tag{19}$$

We observe that the proton obtained as a product in (3a) is consumed as a result of Chl-a photooxidation in (18), that the production of Chl $a^{+}\cdot$ in (19) results in the O_2 evolution in (3a), and that the electron from water splitting in (3a) returns Chl $a^{+}\cdot$ to the reduced state. A combination of the events attending reactions (3a) and (19) may be given by the overall stoichiometry

$$2\text{H}_2\text{O} \xrightarrow[\text{Chl } a,\ \text{A, Mn}]{4h\nu} 4[\text{H}] + \text{O}_2. \tag{20}$$

The production of [H] results in carbon reduction in the RPP cycle (see Fig. 1). The secondary reactions following the primary light reaction in plant photosynthesis are commonly discussed in the literature in terms of "electron flow" initiating from the P680 water-splitting reaction. Electron flow involves no mass transfer, whereas atomic hydrogen transfer is manifested by isotope effects. Hydrogen transfer according to (3a) and (20) accompanies the flow of electrons from the water-splitting reaction. That the rate of water splitting in vivo may be limited by hydrogen transfer according to reaction (19) is reflected by the diminution of the oxygen evolution rate observed [143] when D_2O instead of H_2O is employed.

By analogy to the above description for plant photosynthesis, the corresponding secondary processes in vitro are catalyzed by a suitable metal surface which, in place of Mn and A in reaction (20),provides active sites for O_2 evolution as well as paths for [H] transfer and/or H_2 evolution. The demonstration of O_2 evolution from CO_2-depleted water (Fig. 8) supports the intrinsic ability of the chlorophyll to photodecompose water. The finding of simultaneous evolution of O_2 from water splitting and CO_2 reduction (Figs. 9-11) suggests the formation of metal-bound atomic hydrogen, [H], as an intermediate pathway in (3b). Molecular hydrogen and CO_2 do not react to an appreciable extent under ordinary conditions. The water-splitting and CO_2 reduction effects shown in Figs. 8-11 are not observed in the absence of the platinum. There thus exist a correspondence between the postulated reaction sequences for photosynthesis in vitro and in vivo.

8.3 Photochemical Reduction of CO_2

Despite their obvious differences thermochemical dark reactions are not easily distinguishable from secondary processes of photosynthetic light reactions. For example, molecular hydrogen in the presence of ferredoxin, a flavoprotein enzyme and added hydrogenase, is capable of substituting for light in the reduction of NADP by chloroplasts [144]. The mechanisms of CO_2 fixation in plant photosynthesis are derived from relating dark and light reactions based on the kinetic behavior of transients observed during transitions between conditions of light and darkness. That the path of carbon in the light quantitatively differs from the dark reaction was clear in the kinetic work published by Calvin and his associates [1], although the significance of the original data was not fully discussed by these workers. The proposal by Calvin that CO_2 reduction in photosynthesis occurs after dark production of PGA in RuDP carboxylation is evidently contrary to established experimental observations.

The increase in PGA concentration observed [145] on transition from conditions of light to darkness (see Fig. 14) is attributed to the dark RuDP carboxylase reaction for addition of CO_2 to the RuDP followed by hydrolytic splitting to give two molecules of PGA [39]:

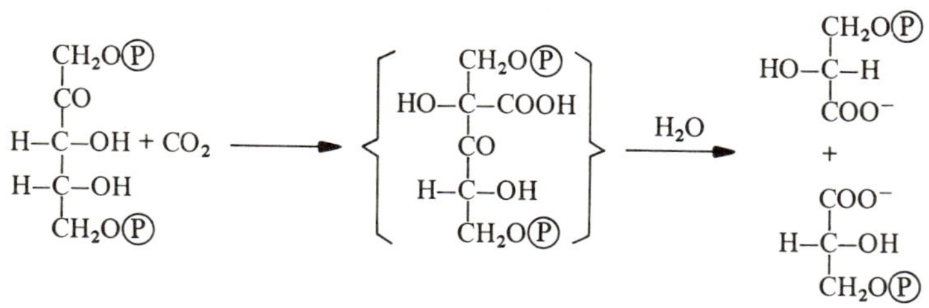

(21)

Ribulose–1,5–diphosphate | Unstable C_6 intermediate | Two 3–phosphoglyceric acid

The formation of RuDP from ribulose 5-phosphate with ATP and phosphoribulokinase, as well as other biochemical steps in the RPP cycle, are considered dark reactions. However, many of these reactions are known to be regulated by light. The RuDP formation, for example, is blieved to be limited by the reactivation of biphosphotases in the light after a dark period [146].

The physiological free energy of a biochemical reaction is written

$$\Delta G^S = -RT \ln (k_b/k_f) \tag{22}$$

where k_f and k_b are rates for the forward and backward reactions given by Arrhenius temperature behavior for activated processes. Secondary processes of the light reaction and dark reactions are both temperature dependent according to Eq. (22). Secondary light processes result from a physiological response to the photoproduction of Chl $a^{+\cdot}$ and [H]. Thus O_2 evolution in photosynthesis is a manifestation of water oxidation by Chl $a^{+\cdot}$, whereas carbon reduction is a consequence of [H] accumulation from the light reaction.

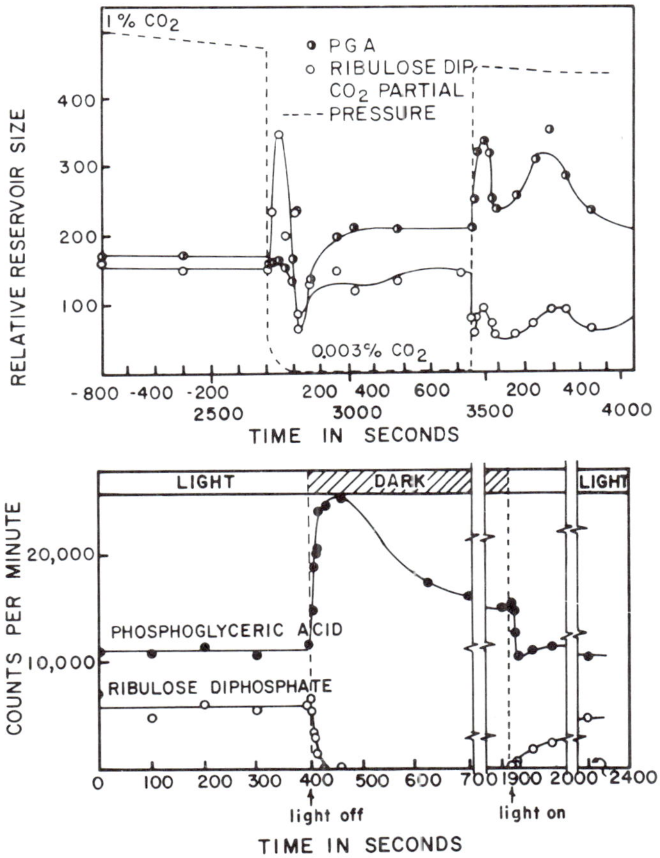

Fig. 14. Kinetic changes in the reservoir sizes of 3PGA and RuDP as a function of CO_2 partial pressure and conditions of light and dark. The 30 s time observed for RuDP carboxylation in the dark *(lower figure)* compared with the $\lesssim 1$ s RuDP decay time obtained under conditions of light *(upper figure)* rules out Calvin's proposal that dark carboxylation reaction is a necessary precursor to carbon reduction. The results shown in this figure indicate that PGA is an obligatory reactant in plant photosynthesis (see Sect. 9.3). The results presented are taken from ref. [1]

The earliest recorded conception of carbon reduction in photosynthesis was proposed in 1843 by Liebig [147], who suggested that plant acids are intermediate between CO_2 and sugars. This theory seems reasonable on account of the fact that plant acids are intermediate in redox level between CO_2 and carbohydrates. Liebig's theory was later disputed by Baeyer [148], who advanced the theory of CO_2 photoreduction to formaldehyde followed by autocondensation of the formaldehyde to carbohydrate.

That photosynthesis is a complex reaction involving multistep "dark" processes following the primary light reaction was first recognized in the work of Blackman, who demonstrated in 1905 [149] that the photochemical steps in photosynthesis resulted in one or more temperature-dependent reactions. Subsequent developments gravitated toward the theory that water is first decomposed by photolysis, and that the reducing equivalent, [H], from the water photolysis is then used to reduce CO_2.

The concept of carbon reduction in photosynthesis as a temperature-dependent process closely related to the light reaction was formulated in the 1950s by Calvin and co-workers, who suggested that the "dark" reactions and the photochemical steps are connected by cofactors derived from the photochemical reactions. These workers observed that, under conditions of light, the RPP cycle evidently follows a pathway other than the dark reaction given by (21). During photosynthesis, a sudden increase in CO_2 pressures from low levels results in an increase in the relative reservoir size of PGA and a corresponding decrease in that of RuDP (see Fig. 14). The increase in carbon in PGA under these conditions is approximately three-fifths the decrease in total carbon of RDP [1]. Contrary to the stoichiometry of PGA formation in (21), there is apparently an increase of only one mole of PGA for a decrease of one mol in RuDP. These observations suggest that about half of the expected PGA is either immediately reduced or not formed in the light, leading Bassham and Calvin to propose in 1957 the existence of a secondary light reaction (L) in addition to reaction (21) (D) [1]:

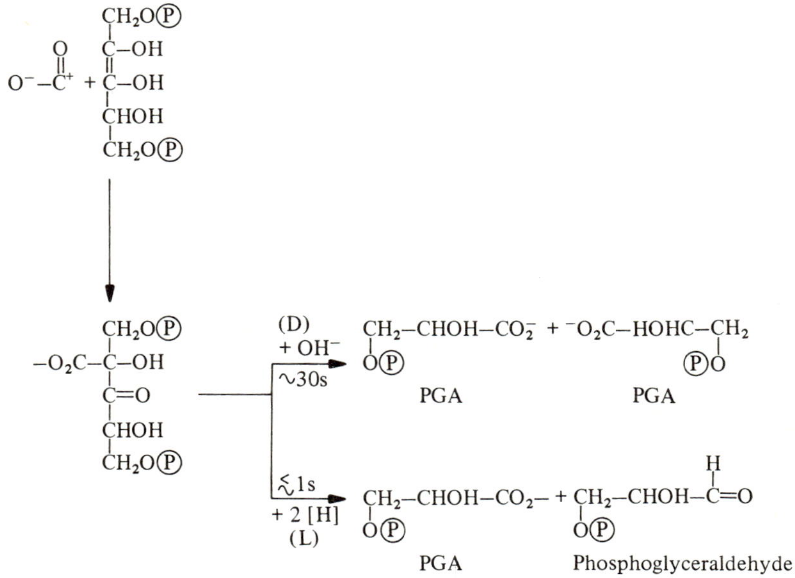

(23)

Unfortunately, in spite of the evidence, reaction (23) (L) was omitted in further discussion of the RPP cycle by these authors. The proposal of a light path of carbon reduction other than reaction (21) is supported by kinetic data. The decay of RuDP in the dark reaction, (23) (D), occurs in about 30 s, whereas the corresponding decay time for RuDP in the light reaction, (23) (L), is about 1 s (see Fig. 14). This comparison rules out Calvin's proposal of dark RuDP carboxylation as a precursor to CO_2 reduction under conditions of light. The former reaction would otherwise have been the rate limiting step, in which case RuDP decay as a result of CO_2 assimilation in the light would have been given by a lifetime in excess of 30 s, contrary to the observed behavior.

8.4 Time Sequence and Branching of Electron Flow from Water

The P680 light reaction is schematically shown in Fig. 1. The water-splitting reaction is the source of electrons. Pheophytin, Ph, is reportedly an intermediary electron carrier in the P680 light reaction. The reduction potential for the Ph/Ph^- couple associated with P680 has been determined to be −0.61 V [150]. It follows that two red quanta are required to raise the electron from the redox level of P680, positioned at about 1.1 V in Fig. 1, to the level of Ph/Ph^-.

The establishment of Ph and Q as electron acceptors for P680, at reduction potentials respectively corresponding to the two-quantum and one-quantum photoexcited states of P680, supports the proposal that photochemistry may occur off both the do doubly and singly excited states of the chlorophyll [10]. Given the observed time sequence of P680 photochemistry described in Sects. 82 and 8.3, and elsewhere [151-154], it is possible to suggest a plausible mechanism for electron flow originating from the water-splitting reaction:

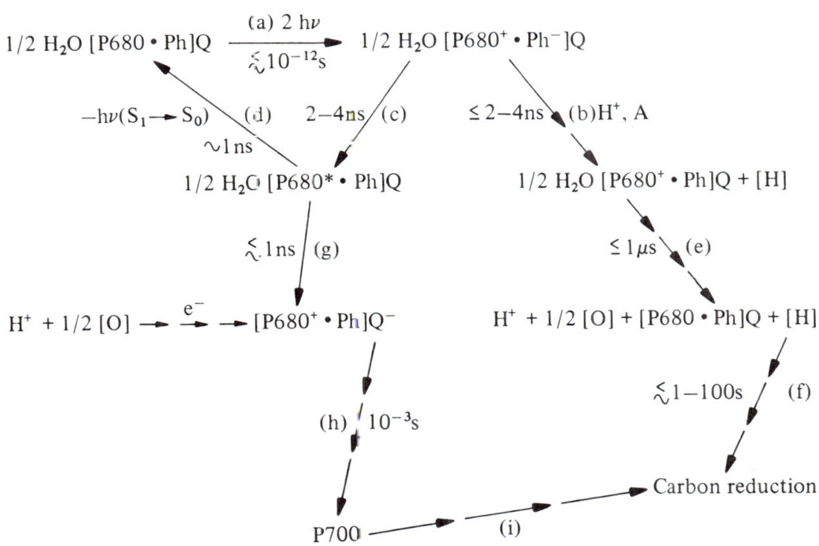

(24)

The two-photon activation of [P680$^+$ · Ph$^-$] according to (24a) results in a branching of electron flow from water, leading to carbon reduction on the one hand, and, on the other, to communication through appropriate redox couples between P680 and P700. The time sequence of reaction events in (24) indicates that electron flow between P680 and P700 arises from the back decay of [P680$^+$ · Ph$^-$] to the first excited state of the chlorophyll, denoted in (24) by P680*. The reduction of Q would thus be reflected by a concomitant increase in P680 luminescence via (24d).

The above description appears to be fully corroborated by experimental observations. The variable fluorescence of P680, attributable [155-157] to [P680$^+$ · Ph$^-$] recombination through (24c, d), has been found to accompany [158] the reduction of Q in (24g), and disappear [155-157, 159, 160] on photoreduction of Ph in (24a). The lifetime of [P680$^+$ · Ph$^-$] has been determined [157] to be 2-4 ns at 295 K. The quantum yield for [P680$^+$ · Ph$^-$]Q$^-$ accumulation at 295 K on prior reduction of Q has been estimated to be 0.002-0.005 [155, 156, 160] consistent with the ratio of the [P680$^+$ · Ph$^-$] recombination time and the time, $\leqslant 1$ μs, for reduction of P680$^+$in reaction (24e) by electrons from water splitting [152-154].

The upper-bound value of the time for initiating reaction sequence (24b) is given by the measured lifetime, 2-4 ns, of [P680$^+$ · Ph$^-$]. The proposed photoreduction of water-soluble ferredoxin, and then NADP, by P680 as a result of (24b) is supported by the recent work of Arnon et al. [24]. The electron flow to P700, in (24h), following the reduction of Q by P680 is the same as that given in the traditional series scheme. The time range, $\lesssim 1$-100 s, for carbon reduction in (24f) is given by the kinetic data reproduced in Fig. 14. In transition from darkness to light 3-PGA reduced is by NADPH, in the presence of ATP, to produce RuDP (see Figs. 1 and 14). The accumulation of RuDP toward to a steady-state concentration under conditions of light takes about 100 s. The time for photoreductive carboxylation involving RuDP under normal photosynthetic conditions is $\lesssim 1$ s. Reaction sequence (24i) is discussed in Sect. 9).

9 Further Conclusions

Observations on the light reaction path of photosynthesis are thus summarized in (24) and Fig. 1. This summary leads to several conclusions that appear to be worthy of further consideration.

9.1 Spatial Relationships of P680 and P700

In addition to the description of P680 and P700 in terms of primary and secondary events spanning several decades in time domain, the P680 and P700 photosystems may be characterized by properties that underscore the differences between the physical environments of the two photosynthetic reaction centers. Ferredoxin, NADP, as well as constituents of the RPP cycle, are mostly water soluble. It is accordingly suggested that the P680 carbon reduction and water oxidation reactions take place in an aqueous or water-rich environment. By contrast, it was reported [161, 162] that the P700 light reaction engages bound Chl-*a* and Fe-S centers as electron acceptors [161,

162] that do not require proton uptake in photoreduction, indicative of the hydrophobic milieu of the P700 reaction center. We may thus conclude that P680 and P700 are spatially separated, even while their respective photoreactions are evidently connected by secondary electron flow in (24g, h). It is known that "PS1 particles" enriched in P700 are readily isolated from the remainder of the photosynthetic apparatus. The chlorophyll monohydrate and dihydrate models for P700 and P680, as we have shown above, are respectively obtained in vitro under anhydrous and excess-water conditions, and thus appear to be entirely compatible with the above-described hydrophobic and hydrophyllic origins of the two reaction center complexes.

In the series scheme NADP reduction has been traditionally associated with P700, in contrast to the biphotonic model for P680 light reaction described in this work. In this connection Q, instead of Ph, was thought to be primary electron acceptor, so the engagement of P700 seemed obligatory in order to raise the [H] from H_2O to the reduction potential of NADP/NADPH. Frequent mention has been made of the poisoning of NADP reduction and/or water oxidation by inhibitors such as 3-(3', 4'-dichlorophenyl)-1, 1-dimethylurea (DCMU) and dibromothymoquinone (DBMIB) as evidence in support of the series scheme, although it seems likely that the presence of extraneous inhibitors may introduce effects other than those intended. The use of DBMIB, for example, is thought to interrupt the electron flow between PS2 and PS1 beyond plastoquinone. However, it is known that DBMIB also quenches PS2 fluorescence, indicative of irreversible P680 oxidation by DBMIB.

9.2 Quantum Requirement of Oxygen Evolution

The photosynthesis conversion of CO_2 to carbohydrate reverses the oxidative path of C_6 degradation, requiring a free energy input of 686 kcal per mol of C_6 formed. The minimum quantum requirement, obtained in weak light, is 8 quanta for the evolution of one O_2 molecule. Experimental measurements fall in the range 8-12 quanta/O_2 molecule evolved. The biphotonic model for P680, given in Fig. 1 and reaction (24) in terms of the branched electron or [H] flow from water to CO_2 reduction and P700, provides a reasonable account for the observations. At low light levels it is possible for the reaction sequence, (24a, b, e, f), to proceed without appreciable back decay of [$P680^+ \cdot Ph^-$], in which case the 8-quanta/O_2 requirement is obtained. Higher values determined for the quantum requirement possibly reflect the effects arising from saturation of the forward channel to carbon reduction under strong illumination, leading to [$P680^+ \cdot Ph^-$] recombination and then electron flow to P700. A maximum requirement of 12 quanta/O_2 is expected in this case, provided that P700 reaction, unlike the biphotonic P680 reaction, is given by a one-photon mechanism. Observables comparable to P680 recombination liminescence and separate stages of P680 ionization involving Ph and Q, manifesting two-photon effects of the P680 secondary effects, are absent in the photochemistry of P700. While biphotonic origin of P680 seems reasonably established under some conditions, the photon mechanism responsible for the P700 reaction remains somewhat tenuous. The electron flow from the water, through (24h) to the P700 light reaction necessitates a noncyclic flow of electrons from the P700 to acceptors other than these within the cyclic path of photophosphorylation indicated in Fig. 1. The highly hegative reduction potential of P700 suggests that P700 as well as P680, is capable of activating the carbon reduction cycle as indicated in (24i).

The description of structural and photoreactive properties of P680 and P700 given above is supported by the in-vitro reproduction, detailed in Sects. 4-7, of phenomenological effects attributed to green plant photosynthesis. The characterization and preparation of (Chl $a \cdot H_2O)_2$ and (Chl $a \cdot H_2O)_n$, respectively as models for P700 and P680, make available a useful handle in probing detailed mechanisms for the photophysical pathways. Using time-resolved methods, we [163] recently demonstrated that the biphotonic excitation of (Chl $a \cdot 2H_2O)_n$ arises from annihilation interactions between the first excited singlet and the lowest triplet states of the chlorophyll. Similar studies on in-vitro preparations of (Chl $a \cdot H_2O)_2$ will hopefully provide information relevant to the photon mechanism of P700, and hence the quantum requirement for O_2 evolution in plant photosynthesis as discussed in the preceding paragraph.

9.3 PGA Reduction as Mechanism for Photoregulation

Based on the kinetic effects observed on dark ↔ light transitions (Fig. 14), three stages of the RPP cycle are defined:

(1) cycle initiation on transition from dark to light

$$\begin{array}{l} 5\text{PGA} \xrightarrow[5\text{ ATP}]{10[\text{H}]} 5\text{C}_3 \\ 2\text{C}_3 \longrightarrow \text{C}_6 \\ \text{C}_6 + 2\text{C}_3 \longrightarrow \text{C}_5 + \text{C}_7 \\ \text{C}_7 + \text{C}_3 \longrightarrow 2\text{C}_5 \\ 2\text{C}_5 \xrightarrow{2\text{ATP}} 2\text{RuDP} \\ \hline 5\text{ PGA} + 10[\text{H}] \xrightarrow{7\text{ATP}} \text{C}_5 + 2\text{RuDP} \end{array} \qquad (25)$$

(2) cycle sustenance

$$\begin{array}{l} 3\text{ CO}_2 + 3\text{ RuDP} \xrightarrow{6[\text{H}]} 3\text{C}_3 + 3\text{PGA} \\ 3\text{PGA} \xrightarrow[3\text{ATP}]{6[\text{H}]} 3\text{C}_3 \\ 2\text{C}_3 \longrightarrow \text{C}_6 \\ \text{C}_6 + 2\text{C}_3 \longrightarrow \text{C}_5 + \text{C}_7 \\ \text{C}_7 + \text{C}_3 \longrightarrow 2\text{C}_5 \\ 3\text{C}_5 \xrightarrow{3\text{ATP}} 3\text{ RuDP} \\ \hline 3\text{ CO}_2 + 12[\text{H}] \xrightarrow{6\text{ATP}} \text{C}_3 + 3\text{H}_2\text{O} \end{array} \qquad (26)$$

and (3) cycle termination on transition from light to dark

$$2RuDP + 2CO_2 \rightarrow 4PGA. \quad (27)$$

The carbon balance in (26) is that obtained under stable conditions of photosynthesis. Combining (25) and (27) the carbon balance from cycle initiation and termination on completion of the dark → light → dark sequence is obtained

$$PGA + 2CO_2 + 10[H] \xrightarrow{7ATP} C_5 + 2H_2O \quad (28)$$

from which it is clear that PGA is consumed, along with CO_2, as a primary reactant upon photoinitiation and completion of the RPP cycle. It is further observed that the mechanisms underlying reaction (28) may also provide an effective means for regulating the photosynthetic rate in response to variations in intensity of the incident light.

In the above discussions we have placed the series scheme of photosynthesis within the frame of observations made since that scheme was formulated about 20 years ago. The cycles of P680 and P700 light reactions are connected with the path of carbon reduction. An important question is the extension of the in-vivo investigations to controlled experiments in vitro, so that elements of the elegant apparatus may eventually be reassembled outside of the living plant. There is no in-vitro analog of the photoregulatory mechanism involving PGA. Metal surfaces afford only crude approximations of the enzymatic carriers of [H] in vivo. Also, inhibitive effects of molecular oxygen on the Chl-*a* water-splitting and carbon reduction reactions in vitro are not easily prevented. The difficulties reflect, in part at least, our incomplete understanding of the in-vivo process. At present better experimental and theoretical methods relevant to the study of photobiological and photochemical effects are being developed. In addition, photosynthetic effects are being considered in a more rigorously defined framework of photoelectrochemistry, so one may reasonably look forward to connecting the remaining conceptual gaps in this area of research.

References

1. Bassham, J.A., Calvin, M.: The Path of Carbon in Photosynthesis, pp 51-55. Englewood Cliffs, N.J.: Prentice-Hall 1957

2a. Hill, R., Bendall, F.: Nature **186**, 136 (1960)

2b. Duysens, L.N.M., Amesz, J., Kamp, B.M.: Nature **190**, 510 (1961)

2c. Kok, B.: Biochim. Biophys. Acta **48**, 526 (1961)

3. Boardman, N.K.B., Anderson, J.M.: Nature **203**, 166 (1954)

4. Döring, G., Stiehl, H.H., Witte, H.T.: Z. Naturforsch. **22**, 639 (1967)

5. Van Gorkum, H.J., Tamminga, J.J., Haveman, J.: Biochim. Biophys. Acta **347**, 417 (1974)

6. Photosynthesis I. In: Encyclopedia of Plant Physiology, N.S., Vol. 5, Trebst, A., Avron, M. (eds.). Berlin-Heidelberg-New York: Springer 1975

7. Govindjee: Bioenergetics of Photosynthesis, New York: Academic Press 1975

8. Fong, F.K., Polles, J.S., Galloway, L., Fruge, D.R.: J. Am. Chem. Soc. **99**, 5802 (1977)

9. Fong, F.K.: Proc. Natl. Acad. Sci. USA, **71**, 3692 (1974)

10. Fong, F.K.: J. Am. Chem. Soc. **98**, 7840 (1976)

11. Brace, J., Fong, F.K., Karweik, D.H., Koester, V.J., Shephard, A., Winograd, N.: J. Am. Chem. Soc. **100**, 5203 (1978)
12. Fong, F.K., Koester, V.J.: Biochem. Biophys. Acta **423**, 52 (1976)
13. Arnon, D.I., Tsujimoto, H.Y., McSwain,,B.D.: Proc. Natl. Acad. Sci. USA **51**, 1274 (1964)
14. Arnon, D.I., Tsujimoto, H.Y., McSwain, B.D.: Nature **214**, 562 (1967)
15. Cheniae, G.M., Martin, I.M.: Biochim. Biophys. Acta **197**, 219 (1970)
16. Wydrzynski, T., Zumbulyadis, N., Schmidt, P.G., Gutowsky, H.S., Govindjee: Proc. Natl. Acad. Sci USA **73**, 1196 (1976)
17. Kusonoki, M., Kitaura, K., Morokuma, K., Nagata, C.: FEBS Lett. **117**, 179 (1890)
18. Juliot, P., Barbieri, G., Chabeud, R.: Photochem. Photobiol. **10**, 309 (1969)
19. Kok, B., Forbush, B., McGloin, M.: Photochem. Photobiol. **11**, 457 (1970)
20. Fowler, C.F.: Biochim. Biophys. Acta **462**, 414 (1977)
21. Saphon, S., Crofts, A.R.: Z. Naturforsch. **32C**, 617 (1977)
22. Fong, F.K., Galloway, L.: J. Am. Chem. Soc. **100**, 3594 (1978)
23. Galloway, L., Roettger, J., Fruge, D.R., Fong, F.K.: J. Am. Chem. Soc. **100**, 4635 (1978)
24a. Tsujimoto, H.Y., Hiyama, T., Arnon, D.I.: Biochem. Biophys. Res. Commun. **93**, 215 (1980)
24b. Arnon, D.I., Tsujimoto, H.Y., Tang, G.M.-S.: Proc. Natl. Acad. Sci. USA **77**, 2676 (1980)
25. Fong, F.K., Winograd, N.: J. Am. Chem. Soc. **98**, 2287 (1976)
26a. Rurainski, H.J., Randles, J., Hoch, G.E.: Biochim. Biophys. Acta **205**, 254 (1970)
26b. Rurainski, H.J., Randles, J., Hoch, G.E.: FEBS Lett. **13**, 98 (1971)
27. Ingenhousz, J.: Essay on the Food of Plants and Renovation of Soils. London: 1976
28. Warburg, O., Krippahl, G.: Z. Naturforsch. **11B**, 179 (1956)
29. Wurmser, R.: Oxidations et Reductions. Paris: Presses University of France 1930
30. Shibata, K.: Carbon and Nitrogen Assimilation, Gest, H., Togasaki, R.K. (translators). Tokyo: Japan Science Press 1975 (1931)
31. Van Niel, C.B.: Arch. Mikrobiol. **3**, 1 (1931)
32. Van Niel, C.B.: Cold Spring Harbor Symp. Quant. Biol. **3**, 138 (1935)
33. Hill, R., Scarisbrick, R.: Nature **146**, 61 (1940)
34. Emerson, R., Lewis, C.M.: Am. J. Bot. **28**, 789 (1941)
35a. Ruben, S., Randall, M., Kamen, M., Hyde, J.: J. Am. Chem. Soc. **63**, 877 (1941)
35b. Metzner, H.: J. Theor. Biol. **51**, 301 (1975)
36. Ruben, S.: J. Am. Chem. Soc. **65**, 279 (1943)
37. Bassham, J.A., Benson, A.A., Kay, L.D., Harris, A.Z., Wilson, A.T., Calvin, M.: J. Am. Chem. Soc. **76**, 1760 (1954)
38. Benson, A.A.: J. Am. Chem. Soc. **73**, 2971 (1951)
39. Weisbach, A., Horecker, B.L., Hurwitz, J.: J. Biol. Chem. **218**, 795 (1956)
40. Warburg, O. Krippahl, G., Lehman, A.: Am. J. Bot. **56**, 961 (1969)
41. Walker, J.F.: Formaldehyde. New York: Reinhold 1964
42. Weiss, A.H., Socha, R.F., Likholobov, V.A., Sakharov, M.M.: Chem. Technol. Fuels Oils USSR 643 (1980)
43. Levitt, L.S.: Science **118**, 696 (1953)
44. Levitt, L.S.: Appl. Spectrosc. **14**, 161 (1960)
45. Van Niel, C.B.: Annu. Rev. Plant Physiol. **12**, 1 (1962)
46. Arnon, D.I.: Nature **184**, 10 (1959)
47. Commoner, B., Townsed, J., Pake, G.: Nature **174**, 689 (1954)
48. Commoner, B., Heise, J., Townsend, J.: Proc. Natl. Acad. Sci. USA **42**, 710 (1956)
49. Commoner, B., Heise, J., Lippincott, B., Norberg, R., Passinneau, J., Townsend, J.: Science **126**, 57 (1957)
50. Sogo, P., Pon, N., Calvin, M.: Proc. Natl. Acad. Sci. USA **43**, 387 (1957)
51. Warden, J.T., Bolton, J.R.: Acc. Chem. Res. **7**, 189 (1974)
52. Clayton, R.K.: Proc. Nat. Acad. Sci. USA **69**, 44 (1966)
53. Katz, J.J., Norris, J.R.: Curr. Top. Bioenerg. **5**, 41 (1973)
54. Feher, G., Hoff, A.J., Isaacson, R.A., Ackerson, J.C.: Ann. New York Acad. Sci. **244**, 239 (1975)
55. Katz, E., in: Photosynthesis in Plants, Loomis, W.E., Frank, J. (eds.), p. 291. Ames: Iowa State College Press 1949

56. Gurney, R.W., Mott, N.F.: Proc. R. Soc. **A164**, 151 (1938)
57. Bradley, D.F., Calvin, M.: Proc. Natl. Acad. Sci. USA **41**, 563 (1955)
58. Nelson, R.C.: J. Chem. Phys. **27**, 864 (1957)
59. Arnold, W., Maclay, H.K.: Brookhaven Symp. Biol. **11**, 1 (1958)
60. Terenin, A.N., Putzeiko, A.K., Akimov, I.: Discussions Faraday Soc. **27**, 83 (1959)
61. Rosenberg, B., Camiscoli, J.F.: J. Chem. Phys. **35**, 982 (1961)
62. McCree, K.J.: Biochim. Biophys. Acta **102**, 90, 96 (1965)
63. Tang, C.W., Albrecht, A.C.: J. Chem. Phys. **62**, 2139 (1975)
64. Tang, C.W., Albrecht, A.C.: J. Chem. Phys. **63**, 953 (1975)
65. Tang, C.W., Douglas, F., Albrecht, A.C.: J. Phys. Chem. **79**, 2923 (1975)
66. Tang, C.W., Albrecht, A.C.: Mol. Cryst. Liq. Cryst. **25**, 53 (1974)
67. Tang, C.W., Albrecht, A.C.: Nature **254**, 507 (1975)
68. Ke, B., in: The Chlorophylls, Vernon, L.P., Seeley, G.R. (eds.). New York: Academic Press 1966
69. Müller, A., Fork, D.C., Witt, H.T.: Z. Naturforsch. **13b**, 142 (1963)
70. Norris, J.: Photochem. Photobiol. **23**, 449 (1976)
71. Parson, W.W., Codgell, R.J.: Biochim. Biophys. Acta **416**, 105 (1975)
72. Codgell, R.J., Menger, T.G., Parson, W.W.: Biochim. Biophys. Acta **408**, 189 (1975)
73. Okamura, M.Y., Isaacson, R.A., Feher, G.: Proc. Natl. Acad. Sci. USA **72**, 3491 (1975)
74. Wraight, C.A., Leigh, J.S., Dutton, P.L., Clayton, R.K.: Biochim. Biophys. Acta **333**, 401 (1974)
75. Rockley, M.G., Windsor, M.W., Cogdell, R.J., Parson, W.W.: Proc. Natl. Acad. Sci. USA **72**, 2251 (1975)
76. Dutton, P.L., Leigh, J.S., Seibert, M.: Biochem. Biophys. Res. Commun. **46**, 406 (1972)
77. Leigh, J.S., Dutton, P.L.: Biochim. Biophys. Acta **357**, 67 (1974)
78. Aronoff, S., in: The Chlorophylls, Vernon, L.P., Seeley, G.R. (eds.). New York: Academic Press 1966
79. Harbour, J.R., Bolton, J.R.: Photochem. Photobiol. **28**, 231 (1978)
80. Seeley, G., in: The Chlorophylls, Vernon, L.P., Seeley, G.R. (eds.). New York: Academic Press 1966
81a. Stoll, A., Wiedemann, E.: Helv. Chim. Acta **16**, 307 (1973)
81b. Fleming, I.: Nature **216**, 151 (1967)
82. Strain, H.H., Manning, W.M.: J. Biol. Chem. **146**, 275 (1942)
83. Strain, H.H.: J. Agr. Food Chem. **2**, 1222 (1954)
84. Fong, F.K.; Koester, V.J.: J. Am. Chem. Soc. **97**, 6888 (1975)
85. Fong, F.K.: J. Am. Chem. Soc. **97**, 6890 (1975)
86. Fong, F.K., Koester, V.J., Polles, J.S.: J. Am. Chem. Soc. **98**, 6406 (1976)
87. Koester, V.J., Fong, F.K.: J. Phys. Chem. **80**, 2310 (1976)
88. Abraham, R.J., Burbridge, P.A., Jackson, A.H., MacDonald, D.B.: J. Chem. Soc. **B**, 620 (1966)
89. Abraham, R.J., Barnett, G.H., Bretschneider, E.S., Smith, K.M.: Tetrahedron **29**, 553 (1973)
90. Doughty, D.A., Dwiggins, C.W., Jr.: J. Phxs. Chem. **73**, 423 (1969)
91. LaMar, G.N., Viscio, D.B.: J. Am. Chem. Soc. **96**, 7354 (1974)
92. Abraham, R.J., Eivazi, F., Pearson, H., Smith, K.M.: J. Chem. Soc. Chem. Commun. **697** (1977)
93. Abraham, R.J., Eivazi, F., Pearson, H., Smith, K.M.: J. Chem. Soc. Chem. Commun. **699** (1977)
94. Janson, T.R., Katz, J.J.: Magn. Reson. Rev. **6**, 209 (1972)
95. Boucher, L.J., Katz, J.J.: J. Am. Chem. Soc. **89**, 4703 (1967)
96. Fong, F.K., Koester, V.J., Galloway, L.: J. Am. Chem. Soc. **99**, 2372 (1977)
97. Krasnovsky, A.A., Voinovskaya, K.K., Kosobutskaya, L.M.: Dokl. Akad. Nauk SSSR **85**, 389 (1952)
98. Krasnovsky, A.A., Kosobutskaya, L.M.: Dakl. Akad. Nauk SSSR **91**, 343 (1952)
99. Brody, S.S.: Science **128**, 838 (1958)
100. Brody, S.S., Brody, M.: Biochim. Biophys. Acta **54**, 495 (1961)
101. Lavorel, J.: J. Phys. Chem. **61**, 1864 (1957)

102. Duysens, L.N.M.: Nature **168**, 548 (1951)
103. French, S., Young, V.K., in: Research in Photosynthesis. New York: Wiley-Interscience 1957
104. Kamen, M.D.: Primary Processes in Photosynthesis, pp. 120, 142. New York: Academic Press 1963
105. Willstätter, R., Stoll, A.: Investigations on Chlorophyll, Schertz, F.M., Merg, A.R. (translators), p. 31. Lancaster, Pennsylvania: Science Press 1928
106. Livingston, R., in: Photosynthesis in Plants. Ames, Iowa: Iowa State College Press 1949
107. Livingston, R., Watson, W.F., McArdle, J.: J. Am. Chem. Soc. **71**, 1542 (1949)
108. Rabinowitch, E., in: Photosynthesis I, p. 450. New York: Interscience 1945
109. Winogard, N., Shepard, A., Karweik, D.H., Koester, V.J., Fong, F.K.: J. Am. Chem. Soc. **98**, 2369 (1976)
110. Fetterman, L.M., Galloway, L., Winograd, N., Fong, F.K.: J. Am. Chem. Soc. **99**, 653 (1977)
111. Chow, H.-C., Serlin, R., Strouse, C.E.: J. Am. Chem. Soc. **97**, 7230 (1975)
112. Fong, F.K., Wassam, W.A.: J. Am. Chem. Soc. **99**, 2375 (1977)
113. Hughes, D.W., Holt, A.S.: Can. J. Chem. **40**, 171 (1962)
114. Holt, A.S., Hughes, D.W., Kende, H.J., Purdie, J.W.: Plant Cell Phsyiol. Tokyo **4**, 49 (1963)
115. Stanier, R.Y., Smith, J.H.C.: Biochim. Biophys. Acta **41**, 478 (1960)
116. Knox, R.S., Ghosh, V.J.: Photochem. Photobiol. **22**, 149 (1975)
117. Fruge, D.R., Fong, G.D., Fong, F.K.: J. Am. Chem. Soc. **101**, 3694 (1979)
118. Strain, H.H., Svec, W.A.: In: The Chlorophylls, Vernon, L.P., Seeley, G.R. (eds.). New York: Academic Press 1966
119. Siegbahn, K. Nordling, C., Johannsson, G., Hedman, J., Heden, P.F., Hamrin, U., Geluis. Bergmark, T., Werme, L.O. Manner, R., Baer, Y.: ESCA Applied to Free Molecules, p. 85. Amsterdam: North-Holland 1969
120. Zeller, M.V., Hayes, R.G.: J. Am. Chem. Soc. **95**, 3855 (1973)
121. Winograd, N., Shepard, A., Karweik, D.H., Koester, V.J., Fong, F.K.: J. Am. Chem. Soc. **98**, 2369 (1976)
122. Brace, J.G., Shepard, A., Karweik, D.H., Koester, V.J., Fong, F.K., Winograd, N.: J. Am. Chem. Soc. **100**, 5203 (1978)
123. Kim, K.S., Baitinger, W.E., Amy, J.W., Winograd, N.: J. Electron Spectrosc. Relat. Phenom. **5**, 351 (1974)
124. Brundle, C.R., Carley, A.F.: Faraday Disc. Chem. Soc. **60**, 51 (1975)
125. Siegbahn, K., Nordling, C., Fahlman, A., Nordberg, R., Harmin, K., Hedman, G., Johansson, G., Bergmark, T., Karlsson, S., Lindgren, I., Lindberg, B.: ESCA: Atomic, Molecular and Solid State STructure Studied by Means of Electron Spectroscopy. Uppsala: Almquist and Wiksells 1967
126. Karweik, D., Winograd, N.: Inorg. Chem. **15**, 2336 (1976)
127. Boxer, J.J., Closs, G.L., Katz, J.J.: J. Am. Chem. Soc. **96**, 7058 (1974)
128. Shipman, L.L., Cotton, T.M., Norris, J.R., Katz, J.J.: Proc. Natl. Acad. Sci. USA **73**, 1791 (1976)
129. Boxer, S.G., Closs, G.L.: J. Chem. Soc. **98**, 5406 (1976)
130a. Perlasamy, N., Linschitz, H., Closs, G.L., Boxer, S.G.: Proc. Natl. Acad. Sci. USA **75**, 2563 (1978)
130b. Perlasamy, N., Linschitz, H.: J. Am. Chem. Soc. **101**, 1056 (1979)
131. Fong, F.K.: Hoff, A.J., Brinkman, F.A.: J. Am. Chem. Soc. **100**, 619 (1978)
132. Breton, J.: Biochim. Biophys. Acta **459**, 66 (1977)
133. Wasielewski, M.R., Studier, M.H., Katz, J.J.: Proc. Natl. Acad. Sci. USA **73**, 4282 (1976)
134. Thornber, J.P., Alberte, R.S., Hunter, R.A., Schiozawa, J.A., Kan, K.-S.: Brookhaven Symp. Biology **28**, 132 (1976)
135. Galloway, L., Fruge, D.R., Haley, G.M., Coddington, A.B., Fong, F.K.: J. Am. Chem. Soc. **101**, 229 (1979)
136. Ropp, G.A., Melton, C.E.: J. Am. Chem. Soc. **80**, 3509 (1958)
137. Walker, J.F.: Formaldehyde, p. 49, 104. New York: Reinhold 1953
138a. Bioenergetics of Photosynthesis, Govindjee (ed.). New York: Academic Press 1975

138b. Photosynthesis I. Encyclopedia of Plant Physiology, N.S. Vol. 5, Trebst, A., Avron, M. (eds.). Berlin-Heidelberg-New York: Springer 1977
139. McBrady, J.J., Livingston, R.: J. Phys. Colloid Chem. **51**, 775 (1947)
140. Galloway, L., Fruge, D.R., Fetterman, L.M., Fong, F.K., in: Solar Energy Chemical Conversion and Storage, Hautala, R.R., King, R.B., Kutal, G.R. (eds.), p. 51. New Jersey: Humana Press 1979
141. Jung, W., in: see Ref. 138b
142. Showell, M.S., Fong, F.K.: J. Am. Chem. Soc., **104**, 2773 (1982)
143a. Craig, F.N., Trelease, S.F.: Am. J. Bot. **24**, 232 (1937)
143b. Curry, J., Trelease, S.F.: Science **82**, 18 (1935)
144. Shin, M., Tagawa, K., Arnon, D.I.: Biochem. Z. **338**, 85 (1963)
145. Calvin, M., Massini, P.: Experientia **8**, 445 (1952)
146. Bassham, J.A., in: see Ref. 138b
147. Liebig, J.: Ann. Chemie **46**, 58-97 (1843). In: Rabinowitch, E.I.: Review in Photosynthesis and Related Processes, I, 246. New York: Interscience 1945
148. Baeyer, A.: Ber. Dtsch. Chem. Ges. **3**, 63-75 (1870). In: see Ref. 147
149. Blackman, F.F., Mattaei, G.L.G.: Proc. R. Soc. (London) B **76**, 402-445 (1905)
150. Klimov, V.V., Allakhverdiev, S.I., Demeter, S., Krasnovsky, A.A.: Dokl. Akad. Nauk **249**, 227-230 (1980)
151. Klimov, V.V., Dolan, E., Ke, B.: FEBS Lett. **112**, 97 (1980)
152. Gläser, M., Wolf, Ch., Renger, G.: Z. Naturforsch. **31C**, 712 (1976)
153. Van Best, J.A., Mathis, P.: Biochim. Biophys. Acta **503**, 178 (1978)
154. Sonneveld, A., Rademaker, H., Duysens, L.N.M.: Biochim. Biophys. Acta **548**, 536 (1979)
155. Klimov, V.V., Klevanik, A.V., Shuvalov, V.A., Kranovsky, A.A.: FEBS Lett. **82**, 183 (1977)
156. Klevanik, A.V., Klimov, V.V., Shuvalov, V.A., Kranovsky, A.A.: Dokl. Acad. Nauk **236**, 241 (1977)
157. Klimov, V.V., Allakhverdiev, S.I., Pashchenco, V.Z.: Dokl. Akad. Nauk **242**, 1204 (1978)
158. Duysens, L.M.N., Sweers, H.E., in: Studies on Microalgae and Photosynthetic Bacteria, Miyachi, S. (ed.), p. 353. Tokyo: University Press 1963
159. Klimov, V.V., Allakhverdiev, S.I., Kranovsky, A.A.: Dokl. Akad. Nauk **249**, 485 (1980)
160. Klimov, V.V., Allakhverdiever, S.I., Shutilova, N.I., Krasnovsky, A.A.: Sov. Plant Physiol. 27/2 (1980)
161. Ke, B., Shuvalov, V.A., Dolan, E.: Frontiers of Biological Energetics: From Electrons to Tissues, Dutton, E.L., et al (ed.). New York: Academic Press 1978
162. Shuvalov, V.A., Dolan, Ed., Ke, B.: Proc. Natl. Acad. Sci. USA Biophys. **76**, 770 (1979)
163. Alfano, A.J., Fong, F.K.: J. Am. Chem. Soc. **104**, 2767 (1982)

Notes Added in Proof for Chapter 4 (pp. 80–151)

p. 84. At the time of writing of this chapter, knowledge of the primary reactions in green photosynthetic bacteria was fragmentary. In the meantime, in a series of papers Swarthoff et al. [342–346] have charted electron transport components in the green bacterial photosystem to a point that our understanding of their electron transport chain is now comparable to that of electron transport in the purple bacteria. The primary donor, P840, is a Bchl dimer. Another Bchl *a* molecule serves as intermediary acceptor, whereas two bound iron-sulfur proteins function as primary and secondary acceptors. Between P840 and the intermediary Bchl *a* (which is the analog of Bph *a* in the purple bacteria) another porphyrin-like acceptor is located, the identity of which is as yet uncertain (it is probably not a Bph *c* or Bchl *c* molecule). It appears that the green bacterial photosystem has features of the purple bacterial photosystem and of PS I of plant photosynthesis, a finding that has obvious evolutionary ramifications.

p. 100. We have found recently (A.J. Hoff and L. Nan, unpublished experiments) that gelatine gels may conveniently substitute for polyacrylamide gels for orienting particles by pressure. An important advantage of gelatine is the absence of oxidizing agents in the preparation procedure, permitting the study of LD under reducing conditions.

p. 105. It has recently been possible to discriminate between the two models discussed in Section 2.1.5 by triplet-singlet ground state optical difference spectroscopy, employing the novel technique of absorption-detected magnetic resonance (ADMR) of the triplet state [347]. It was found [348] that the triplet state of the bacterial donor is localized on one of the Bchl molecules on an optical timescale. The other Bchl of the dimer absorbs at 810 nm in *Rps. sphaeroides* and at 838 nm in *Rps. viridis* when the triplet state is present. It was concluded that the two Bchl molecules of the primary donor are in the ground state strongly excitonically coupled. No bleachings other than at the longest wavelength absorption band (890 and 990 nm for the two organisms, respectively) were found. This means that the model proposed by Shuvalov et al. is essentially correct, i.e., the Bchl are coupled such that one of the exciton bands is forbidden, the allowed one constituting the longest wavelength absorption band.

p. 107. Recent sub-picosecond experiments, employing excitation of Bchl at 610 nm, suggest that one of the accessory B800 pigment is intermediate between P860 and Bph [64, 349, 350]. It was found that a bleaching at 795 nm disappears in about 5 ps, concomitant with the appearance of bleachings at 545, 778 and 790 nm. It should be kept in mind, however, that for 610 nm pulses also B800 is excited. If the energy transfer rate to P860 is 5 ps, then this could equally well explain the observations in [64, 349, 350]. Support for one of the B800 taking part in electron transport came from a study of the temperature dependence of the triplet-singlet absorption difference spectrum of *Rps. sphaeroides* R-26 [351, 352] and of *Rps. viridis* [352] between 293 and 77 K. From the disappearance of a bleaching at 797 nm with lower temperature it was concluded that at 293 K, the triplet excitation is shared between P860 and one B800. In view of the data presented in [348], however, it is not clear whether this temperature effect might not be equally well explained by temperature dependent features (band-shifts, monomeric absorption in $P860^T$) of the ΔA spectrum of $P860^T$.

p. 108. See notes p. 107.

p. 116. Recently, we have found (H.J. den Blanken and A.J. Hoff, unpublished experiment) that the decayrates of monomeric Bchl *a* in vitro are considerably faster than previously reported [374] (see also [373]). With the new in-vitro rates, the exciton model applied to $P860^T$ does give consistent results.

p. 121. Setif and Mathis [353] have reexamined that E_m of P700 in chloroplast lamellae, and in subchloroplast particles of different preparation method at 294 and 77 K. In "intact" chloroplasts, and in D-144 and TSF-1 particles (see table 3.1) the E_m (P700) = 490 mV at pH 7. In particles obtained with harsher treatments, the E_m was lower by up to 60 mV. The E_m of particles proved to be sensitive to the measuring temperature, being almost 100 mV lower at 77 K than at 294 K. According to the authors, bulk chlorophyll oxidation did not occur under their experimental conditions, so that the value of 490 mV seems at present to be preferable over the published values.

p. 122. See note p. 128.

p. 128. Recently, doubts have been raised concerning the dimeric nature of P700 [354, 355]. The second moment of the ESR line of fully deuterated ^{13}C-enriched (89%) Chl a^+ and $P700^+$ was found to be almost identical. Similar analysis for $^2H^{13}C$ – Bchl a^+ and $^2H^{13}C$ – $P860^+$ yielded a difference of a factor of two, as expected for the odd electron of $P860^+$ being shared by two Bchl *a* molecules [354]. The contrasting result for P700 strongly suggests that the primary donor of photosystem 1 is a *monomer.* The reduction by a factor of 1.4 of the ESR linewidth of $P700^+$ compared to Chl a^+ was explained by a shift in spin density distribution over the carbon skeleton of the Chl *a*, possibly because P700 is in the *enol* form [355]. The linewidth in $^1H^{12}C$ compounds is primarily due to hyperfine interaction with protons, which sense only a limited part of the total spin density. If one accepts the above interpretation, then the ENDOR data [112–115] are coincidental, i.e., the hf splittings of the CH_3 groups do not reflect the overall spin distribution. As mentioned in Section 3.2.1, a similar suggestion was made earlier [87].

p. 131

1. A study by Setif et al. [356] in which flash-induced optical difference spectra of photosystem 1 particles were measured at 21°C and 10 K has recently confirmed the suggestion in Section 3.3.1 that the ΔA spectrum of the 1.3 ms decay component is due to the triplet state $P700^T$.

2. An ESR spectrum attributed to Q^- has recently been observed in subchloroplast particles enriched in photosystem 2 [357]. It resembles the ESR signal of the quinone-iron complex of the photosystem of purple bacteria, having an apparent g-value of 1.82. Upon treatment with chaotropic and complexing agents resulting in a low Fe/

P680 ratio, a narrow (9 G) ESR signal at g = 2.0044(3) becomes visible [358, 359], similar to the signal of the reduced bacterial acceptor from which the iron is dissociated. Apparently, in the native system Q is complexed to a high-spin iron ion.

p. 132. A triplet possessing the characteristic polarization pattern resulting from radical pair recombination has now also been observed in photosystem II particles [360, 361]. Its zero field splitting parameters are identical to those of monomeric ^{3}Chl *a*, in support of the suggestion that P680 is a monomeric Chl *a* molecule [317].

p. 136. See note p. 139.

p. 139. The temperature dependence of the reaction $P^+Bph^-UQ_1 \rightarrow P^+Bph.UQ_1^-$ has been investigated more extensively in the range 300 – 4.2 K, with the surprising result that in the domain 5 – 50 K the rate first steeply increases and then decreases with increasing temperature [362]. This temperature dependence could be rather well explained using the multimode version of the theory of vibrationally-assisted tunneling [270, 300, 363–365]. Here, vibrations and phonon-interactions with the environment are grouped in low-frequency (soft)modes, 10 – 200 cm^{-1}, and higher frequency (hard)modes, $\nu > 200$ cm^{-1}. By judicially choosing δE, Δ, and the active vibrational modes as a function of temperature, the temperature-activated part of the rate vs. temperature curve and the part with negative temperature coefficient could both be fitted. Quite clearly, detailed study of the temperature dependence of the primary reactants will yield important data on the mechanistic details of the charge-separation process and subsequent dark reactions. The refinements resulting from such work will lead to better estimates of tunneling matrix elements and distances between reaction sites.

p. 141

1. A similar conclusion has been reached by Tiede and Dutton [366], who also studied the UQ_1^- line in Fe-depleted reactions centers in reconstituted multilayer membranes. Their work shows a somewhat better resolution of the orientation dependent g-value of the ESR line than [334].

2. It has recently proven possible to isolate the LM unit while retaining the $UQ_1.Fe^{2+}$ complex [367]. The electron transfer from UQ_1 to UQ_2, however, was impaired. Upon reconstitution with the H subunit this transfer was restored, suggesting that the H subunit is required for proper binding of UQ_2 to the active site. The removal of H rather than that of iron may have caused the loss of $UQ_1 \rightarrow UQ_2$ electron transfer observed in [336].

3. The topology of the reaction center components in the membrane has been further studied by crossed-immunoelectrophoresis [368], tritium labeling [369], ^{125}I-labeling and protease treatment [370], and lipophilic fluorescent [371] and ^{125}I-markers [372]. The results are not unambiguous. With immunoelectrophoresis, it was found that whole RC were exposed to the outer surface of chromatophores (i.e., the cytoplasmic aspect) [368]. ^{3}H-labeling of chromatophores resulted in heavily labeled H subunit and less heavily labeled L subunit, of membrane vesicles (right-side-out) in less heavily labeled H [369]. This would indicate that H is exposed to the cytoplasmic side (weak peri-

plasmic side labeling in membrane vesicles might have resulted from a fraction of vesicles having "wrong-side"-out); whereas L is located at the periplasmic side. The results for M were inconclusive. Radio-iodination of reaction center components was done with the oxidizing agent chloroglycoluril (coated on glass tube walls) and $Na^{125}I$ [370]. In this way only solvent-exposed polypeptides are labeled with ^{125}I. Using chromatophores and membrane vesicles, it was found that H and M were transmembrane subunits, whereas L was much less labeled in either preparation than the H and M subunits. All subunits had more sites exposed on the vesicle outer surface (periplasmic aspect) than on the chromatophore surfaces. Protease treatment, however, revealed cleavage of H subunits in chromatophores but not in vesicles, whereas L and M were not cleaved in either preparation [370]. The lipophilic fluorescent marker fluorescamine mainly labels M, and to a lower extent H, but not L [371]. However, with the lipophylic radioactive marker 5–^{125}Iodonaphtyl–1–azide, the specific labeling of L and M was about equal and much higher than that of H [372]. From these results [371, 372] one surmises that M is located predominantly and H only partially inside the membrane.

The conflicting results reviewed above point to the major difficulty in the labeling experiments: it is difficult to quantitate exposure when only certain groups (amino acids or polypeptide regions) can be labeled or cleaved. Different labels have different affinity for different parts of the protein surface. Absence of labeling or cleavage is no proof that the subunit is not exposed; conversely, heavy labeling or cleavage does indicate exposure but tells us little on the extent. All in all, one might conclude that both L and M are transmembranal, with H to a considerable extent exposed to the cytoplasmic waterphase. The L subunit might lie at the perplasmic side of the membrane and be exposed to the water phase to only a small extent.

Added Bibliography

Recent Book

B9. Clayton, R.K. Photosynthesis: Physical mechanisms and chemical patterns, IUPAB Biophysics Series, Vol. 4, Cambridge: Cambridge University Press, 1980.

Review Articles

R19. DeVault, D.: Quantum mechanical tunneling in biological systems, Quart. Rev. Biophys. **13**: 387–564 (1980)

R20. Jortner, J.: Dynamics of electron transfer in bacterial photosynthesis. Biochim. Biophys. Acta **594**: 193–230 (1980)

R21. Sauer, K.: Photosynthesis – the light reactions, Annu. Rev. Phys. Chem. **30**: 155–178 (1979)

R22. Maggiora, G.M., Assessment of reaction center special-pair chlorophyll models. Int. J. Quant. Chem. **16**: 331–352 (1979)

R23. Hoff, A.J.: ODMR Spectroscopy in Photosynthesis II. The reaction center triplet in bacterial photosynthesis. In: Triplet State ODMR Spectroscopy: Techniques and applications to biological systems (Clarke, R.H., ed.), Chap. 9, New York: John Wiley & Sons, 1982

R24. Hoff, A.J.: Magnetic field effects on photosynthetic reactions. Quart. Rev. Bioph. **14**: 599–665 (1981)

Added References

342. Swarthoff, T., Amesz, J.: Biochim. Biophys. Acta **548**: 427–432 (1979)
343. Swarthoff, T., Van der Veek-Horsley, K.M., Amesz, J.: Biochim. Biophys. Acta **635**: 1–12 (1981)
344. Swarthoff, T., Gast, P., Hoff, A.J.: FEBS Lett. **127**: 83–86 (1981)
345. Swarthoff, T., Gast, P., Hoff, A.J., Amesz, J.: FEBS Lett. **130**: 93–98 (1981)
346. Swarthoff, T., Gast, P., Van der Veek-Horsley, K.M., Hoff, A.J., Amesz, J.: FEBS Lett. **131**: 331–334 (1981)
347. Den Blanken, H.G., Van det Zwet, G., Hoff, A.J.: Chem. Phys. Lett. **85**: 335–338 (1982)
348. Den Blanken, H.G., Hoff, A.J.: Biochim. Biophys. Acta (in press)
349. Holten, D., Hoganson, C., Windsor, M.W., Schenck, C.C., Parson, W.W., Migus, A., Fork, R.L., Shank, C.V., Biochim. Biophys. Acta **592**: 461–477 (1980)
350. Pulles, M.P.J., Van Gorkom, H.J., Verschoor, G.A.M.: Biochim. Biophys. Acta **440**: 98–106 (1976)
351. Shuvalov, V.A., Parson, W.W.: Proc. Natl. Acad. Sci. USA **78**: 957–961 (1981)
352. Shuvalov, V.A., Parson, W.W.: Biochim. Biophys. Acta **638**: 50–59 (1981)
353. Setif, P., Mathis, P.: Arch. Biochem. Biophys. **204**: 477–485 (1980)
354. Wasielewski, M.R., Norris, J.R., Crespi, H.L., Harper, J.: J. Am. Chem. Soc. **103**: 7664–7665 (1981)
355. Wasielewski, M.R., Norris, J.R., Shipman, L.L., Lin, C.P., Svec, W.A.: Proc. Natl. Acad. Sci. USA **78**: 2957–2961 (1981)
356. Setif, P. Hervo, G., Mathis, P.: Biochim. Biophys. Acta **638**: 257–267 (1981)
357. Nugent, J.H.A., Diner, B.A., Evans, M.C.W.: FEBS Lett. **124**: 241–244 (1981)
358. Klimov, V.V., Allakhverdiev, S.I., Krasnovskii, A.A.: Dokl. Akad. Nauk. **249**: 485–488 (1979)
359. Klimov, V.V., Dolan, E., Shaw, E.R., Ke, B.: Proc. Natl. Acad. Sci. USA **77**: 7227–7231 (1980)
360. Rutherford, A.W., Mullet, J.E.: Biochim. Biophys. Acta **635**: 225–235 (1981)
361. Rutherford, A.W., Paterson, D.R., Mullet, J.E.: Biochim. Biophys. Acta **635**: 204–214 (1981)
362. Schenck, C.C., Parson, W.W., Holten, D.W., Windsor, M.W.: Biophys. J. **36**: 479–489 (1981)
363. Sarai, A.: Biochim. Biophys. Acta **589**; 71–83 (1980)
364. Kakitani, T., Kakitani H.: Biochim. Biophys. Acta **635**: 498–514 (1981)
365. Kuznetsov, A.M., Ulstrup, J.: Biochim. Biophys. Acta **636**: 50–57 (1981)
366. Tiede, D.M., Dutton, P.L.: Biochim. Biophys. Acta **637**: 278–290 (1981)
367. Debus, R.J., Okamura, M.Y., Feher, G.: Biophys. J. **33**: 19a (1981)
368. Collins, M.L.P., Mallon, D.E., Niederman, R.A.: J. Bacteriol. **143**: 221–230 (1980)
369. Francis, G.A., Richards, W.R.: Biochemistry **19**: 5104–5111 (1980)
370. Bachmann, R.C., Gillies, K., Takemoto, J.Y.: Biochemistry **20**: 4590–4596 (1981)
371. Bachofen, R.: FEBS Lett. **107**: 409–412 (1979)
372. Odermatt, E., Snozzi, M., Bachofen, R.: Biochim. Biophys. Acta **591**: 372–380 (1980)
373. Hoff, A.J., Cornelissen, B.: Mol. Phys. **45**: 413–425 (1982)
374. Clarke, R.H., Connors, R.E., Frank, H.A.: Biochem. Biophys. Res. Commun. **71**: 671–675 (1976)

Author Index

Subject Index

Volume 19
M. B. Mathews

Connective Tissue

Macromolecular Structure and Evolution
1975. 31 figures. XII, 318 pages
ISBN 3-540-07068-0

Volume 20
M. A. Lauffer

Entropy-Driven Processes in Biology

Polymerization of Tobacco Mosaic Virus Protein and Similar Reactions
1975. 90 figures. X, 264 pages
ISBN 3-540-06933-X

Volume 21
R. C. Burns, R. W. F. Hardy

Nitrogen Fixation in Bacteria and Higher Plants

1975. 27 figures. X, 189 pages
ISBN 3-540-07192-X

Volume 22
H. J. Fromm

Initial Rate Enzyme Kinetics

1975. 88 figures, 19 tables. X, 321 pages
ISBN 3-540-07375-2

Volume 23
M. Luckner, L. Nover, H. Böhm

Secondary Metabolism and Cell Differentiation

1977. 52 figures, 7 tables. VI, 130 pages
ISBN 3-540-08081-3

Volume 24

Chemical Relaxation in Molecular Biology

Editors: J. Pecht, R. Rigler
With contributions by numerous experts
1977. 141 figures, 50 tables. XVI, 418 pages
ISBN 3-540-08173-9

Volume 25

Advanced Methods in Protein Sequence Determination

Editor: S. B. Neddleman
With contributions by numerous experts
1977. 97 figures, 25 tables. XII, 189 pages
ISBN 3-540-08368-5

Volume 26
A. S. Brill

Transition Metals in Biochemistry

1977. 49 figures, 18 tables. VIII, 186 pages
ISBN 3-540-08291-3

Volume 27

Effects of Ionizing Radiation on DNA

Physical, Chemical and Biological Aspects
Editors: A. J. Bertinchamps, J. Hüttermann, W. Köhnlein, R. Téoule
With contributions by numerous experts
1978. 74 figures, 48 tables. XXII, 383 pages
ISBN 3-540-08542-4

Volume 28
A. Levitzki

Quantitative Aspects of Allosteric Mechanisms

1978. 13 figures, 2 tables. VIII, 106 pages
ISBN 3-540-08696-X

Volume 29
E. Heinz

Mechanics and Energetics of Biological Transport

1978. 35 figures, 3 tables. XV, 159 pages
ISBN 3-540-08905-5

Volume 31

Membrane Spectroscopy

Editor: E. Grell
With contributions by numerous experts
1981. 146 figures. XI, 498 pages
ISBN 3-540-10332-5

Volume 32

Chemical Recognition in Biology

Editors: F. Chapeville, A.-L. Haenni
With contributions by numerous experts
1980. 210 figures, 39 tables. XX, 430 pages
ISBN 3-540-10205-1

Volume 33
E. Heinz

Electrical Potentials in Biological Membrane Transport

1981. 15 figures. X, 85 pages
ISBN 3-540-10928-5

Volume 34
Y. A. Vinnikov

Evolution of Receptor Cells

Cytological, Membranous and Molecular Levels
Translated from the Russian by N. Bobrov
With a Foreword by S. Price
1982. 36 figures. XII, 142 pages
ISBN 3-540-11083-6

Springer-Verlag
Berlin
Heidelberg
New York